The Early Universe

NATO ASI Series

Advanced Science Institutes Series

A Series presenting the results of activities sponsored by the NATO Science Committee, which aims at the dissemination of advanced scientific and technological knowledge, with a view to strengthening links between scientific communities.

The series is published by an international board of publishers in conjunction with the NATO Scientific Affairs Division

A	Life Sciences	Plenum Publishing Corporation
B	Physics	London and New York
C	Mathematical and Physical Sciences	D. Reidel Publishing Company Dordrecht, Boston, Lancaster and Tokyo
D	Behavioural and Social Sciences	Martinus Nijhoff Publishers
E	Applied Sciences	Dordrecht, Boston and Lancaster
F	Computer and Systems Sciences	Springer-Verlag
G	Ecological Sciences	Berlin, Heidelberg, New York, London,
H	Cell Biology	Paris, and Tokyo

Series C: Mathematical and Physical Sciences Vol. 219

The Early Universe

edited by

W. G. Unruh

and

G. W. Semenoff

Department of Physics,
The University of British Columbia,
Vancouver, Canada

Springer-Science+Business Media, B.V.

Proceedings of the NATO Advanced Study Institute on
The Early Universe
Victoria, B.C., Canada
August 17-30, 1986

Library of Congress Cataloging in Publication Data

NATO Advanced Study Institute on The Early Universe (1986: Victoria, B.C., Canada)
   The early universe / edited by W. G. Unruh.
   p. cm. — (NATO ASI series. Series C, Mathematical and physical sciences; vol. 219)
   "Proceedings of the NATO Advanced Study Institute on The Early Universe, Victoria,
B.C., Canada, August 17–30, 1986"—T.p. verso.
   Includes index.
   ISBN 978-94-010-8281-5   ISBN 978-94-009-4015-4 (eBook)
   DOI 10.1007/978-94-009-4015-4

   1. Cosmology—Congresses. 2. Nuclear astrophysics—Congresses. I. Unruh,
W. G. II. North Atlantic Treaty Organization. Scientific Affairs Division. III. Title.
IV. Series: NATO ASI series. Series C, Mathematical and physical sciences; no. 219.
QB980.N38   1985
523.1—dc 19   87–23715
                                                                              CIP

CONTENTS

PREFACE

In this century and especially in the past ten years, the study of the early universe has become not only respectable, but one of the most exciting fields of physics. The change in attitude is best epitomised by words attributed to Rutherford, "If someone in my lab begins to talk of the universe, I tell him it is time to leave".

With the publication of the Principia 300 years ago, Newton introduced the possibility that various phenomena in the physical world could be explained, could be shown to obey simple, comprehensible laws. Could one, however, say the same about the universe as a whole? The question was asked by many people, but attempts to answer always floundered on a paraphase of Augustine's question, "What was God doing before he created the world?" Because both time and space as a-priori concepts existed independently of the universe and of the matter therein, one was left with one of two uncomfortable choices. One could believe that the universe suddenly came into existence in an act outside the laws of physics that now govern the world. This of course removed any discussion of that creation from the realm of science. The other alternative was to believe in an eternal universe, in which case one was faced with severe difficulties in applying known physical laws to the universe as a whole. Some examples of such problems were the finite and short lifetime of our sun ($\sim 10,000$ years – much less than the already known lifetime of the earth) if it was fueled by known processes like carbon combustion, the instability of Newtonian gravity which should have caused the collapse of all matter into tiny dense cold lumps, and Obler's paradox, if one believed that the stars etc. really were eternal and spatially uniform. Rutherford's attitude is thus understandable in the face of these two equally untenable positions.

It was the advent of Einstein's theory of General Relativity, in which both space and time lost their a-priori status and became participants in change and development with the rest of physics, that gave the possibility of a resolution. The origin of the universe lost its extra scientific status, and the possibility was opened that even that event could be explained. Furthermore, because the state of the universe was different in its youth than now, one could hope to see in its present state remnants of that youth – remnants which could give us a clue as to what that youth was like; remnants which could change the study of the early universe from mere speculation to verifiable science.

As Toulmin emphasizes in his lectures in this volume, the study of cosmology now is in a very similar state to the study of geology at the beginning of the 19th century. That again was a field which had had a somewhat unsavoury reputation in science, and which at that time underwent an explosion of knowledge. S.J. Gould in his recent book, <u>Time's Arrow, Time's Cycle</u>, (Harvard U. Press, 1987) argues that, contrary to popular myth, that explosion was driven by a fruitful combination of theoretical speculation followed by the search for confirmation of those speculations.

Cosmology is in a similar state. Again, the discovery of the 3° microwave background radiation would probably have been treated as an obscure puzzle had not Gamow, Dicke and others predicted just such a radiation as a relic of the youth of the universe. Such obscurity was after all the fate of the measured anomolously high temperature of interstellar CN clouds, which is now realized to be caused by that relic microwave radiation (see P. Thaddeus, Ann. Rev. Astr., Vol. 10, p.310 (1971).

It was to examine the present state of our speculations and theories about the early universe that the NATO Advanced Study Institute (ASI) was held at Lester B. Pearson College on Vancouver Island, B.C., Canada, in August of 1986. The papers in this volume which arose out of that ASI give a cross section of the whole gamut of work being done in the area as of early 1987. It is a field in its infancy with many exciting vistas having been discovered, but much work still needing to be done.

I. Moss, in his article on the Quantum origin of the Universe, tells about some of the most speculative areas of research. That anyone would even contemplate a unique origin for the universe would have seemed foolish and pointless even a few years ago, yet now various authors have offered glimpses of such a possibility. The area is exciting not just from a cosmological perspective, but also because it forces us to grapple with and confront issues in quantum measurement theory long relegated, by many physicists, to philosophy.

One of the most exciting developments of recent years has been the fertile interchange between particle physics and the study of the early universe. From the first successes with the explanation of the He/H ratio in primeval matter by nucleosynthesis in the early universe (reviewed in M. Turner's paper), the interchange has produced tentative explanations for many of the old cosmological puzzles. In return, cosmology has become one of the very few testing grounds for many of the most exotic of modern particle physics theories (see for example G. Gelmini's paper on supersymmetry and the early universe). One of the the prime examples has been the idea of inflation (see again M. Turner's paper). That the very large scale structure, from the homogeneity and isotropy of the universe to the clustering and formation of galaxies, could have come about from the behaviour of matter fields on scales of 10^{-28}cm is inconceivable in any theory but gravity. The scale changing properties of gravity, the fact that by the creation of space produced by General Relativity, phenomenon on some tiny scale could ultimately be responsible for the largest scale phenomena, brings about a unity of physics and of the physical world only dimly perceived by Laplace in his idea of "monads".

This scale changing property of gravity also contains the seeds of serious questions which must be answered before one can finally accept inflation as a valid answer to the cosmological questions. The present homogeneity of space come about in inflation because of the assumed homogeneity of the initial state of matter on scales much less than 10^{-33}cm. (That is the scale on which one quanta of any field would have enough energy to produce a black hole with size of order the Compton wavelength of the quantum.) This assumption neglects any effects quantum gravity would have, and assumes that any spacetime foamlike structure, which many people expect to occur on those short scales, would, when expanded up by the growth of the universe, become a smooth vacuum-like field. Any reply to such questions must probably await a more complete theory of quantum gravity.

The realm of particle physics has also produced a raft of exotic particles as relics of the youth of the universe whose presence could have a major influence on the behaviour of the present universe. In this context see the articles by G. Gelmini on cosmions and the solar neutrino problem, and the paper by A. Drukier on the possible strategies for detecting such particles.

If study of the early universe were limited to the speculative issues discussed so far, the subject would not have had the scientific impact it has, and would probably still merit Rutherford opprobrium. It is now however an experimental science as well, with predictions of measurable quantities and tests of these predictions.

The earliest of these tests was the cosmic background radiation, and this holds promise of being also one of the newest and most sensitive tests as well. With observations of the spatial fluctuations in the temperature on the order of one part in 10^5 becoming possible, and observations of departures from the high frequency thermal spectrum being made, the technique holds promise of becoming one of the most sensitive and unambiguous tests of the various models of the very early universe. It is the very largest scale structures probed by this test which would be expected to have evolved least since the beginning, and thus be the most unambiguous relic. These tests and their implications are covered in the articles by R. Bond and J. Wall.

Over the past 15 years, P.J.E. Peebles has advocated the power of galaxy-galaxy and other correlation functions as another of the classic tests of cosmological models. One of the key difficulties has always been that on the currently observable scales, non linear gravitational evolution and clustering will have made a significant impact. The use of present data to infer the initial conditions is thus uncertain unless one learns how to handle this highly complex evolution. M. Wise in his article presents an initial attempt at handling such evolution analytically. The use of diagramatic techniques from particle physics greatly simplifies the approach, although the development of the technique to make quantitive estimates is still needed. This gap is being filled by the very powerful numerical approach outlined by S. White in his paper. There seems to be at least the promise that many of the unusual and novel structures seen in the heavens may just be the result of the non linear gravitational dynamics acting on simple initial fluctuations left over from say the inflationary era.

The necessity of the computer in modern cosmology is becoming more and more evident. Although many of the concepts are simple (such as Einstein's theory of General Relativity), their use to make physical predictions seems very complex. Furthermore, the universe allows no chance for manipulative experiments to isolate key concepts for study. It is here that the powerful numerical techniques described by S. White and by T. Piran are coming into their own.

The model of inflation has certainly exerted a strong pull on the theoretical approach to cosmology. P.J.E. Peebles reminds us in his paper that one should not forget other phenomenological models, in his case for the formation of galaxies.

Finally, J. Wall reminded us that theorists must be both tied to, and suspicious of, astronomical observations. The observations are all extremely difficult, while the desire for some results to use to test, to narrow down the vast range of current models, is great. We must never lose sight of the difficulties inherent in the attempt to make such measurements, nor of the need to discover and develop new and imaginative tests and observations such as are suggested by J. Barrow in his paper.

There is one major topic of growing interest, that of cosmic strings, which was covered in the school but is not represented here. There already exist a number of review articles to which we refer the reader, for example, A. Vilenkin's paper in Physics Reports, 121, 263 (1985).

This is the place to thank all of those individuals and organizations who made this book and the ASI possible. I would firstly like to thank the NATO Scientific Affairs Division for its generous support both of the conference and of this book. Without their support the conference would have been much poorer, and would not have had the outreach that it had.

Secondly I thank the Canadian Institute for Advanced Research for its support not only of the ASI but also for its ongoing support of the study of the early universe via the institution in its program on Cosmology.

In addition, the Canadian Institute for Theoretical Astrophysics, the University of British Columbia, TRIUMF and NSERC also contributed funds, which were greatly appreciated.

I would also like to thank the field workers who helped in organizing and running the ASI. C. Zator was secretary and on her shoulders fell most of the daily organization and problem solving necessary in any Conference. J. Ng, P. Hickson and N. Weiss were very helpful in the organization of the school while A. Roberge, T. Zannias and K. Heiderich bore the brunt of implementing the organizational decisions during the ASI. Finally I would like to thank N. Squire and all the staff at Lester B. Pearson College of the Pacific for their professional hosting of the Conference. Without their expert help and work the ASI would have been much more difficult for organizers and participants alike.

W.G. Unruh
January 1987

LIST OF PARTICIPANTS

H. Atmanspacher
West Germany

Xavier Barcons
Spain

James Bardeen
U.S.A.

J.D. Barrow
United Kingdom

Alan Bentley
U.S.A.

Arvind Borde
U.S.A.

Erez Braun
Israel

G. Bublik
U.S.A.

Jane Charlton
U.S.A.

M. Clutton-Brock
Canada

Ruth Daly
U.S.A.

Nathalie Deruelle
France

Margaret Douglass
U.S.A.

A. Drukier
U.S.A.

Shlomit Finkelstein
U.S.A.

David Finkelstein
U.S.A.

Mike Fitchett
Canada

Graciela Gelmini
Italy

L. Gonzales-Mestres
France

J. Halliwell
United Kingdom

David Hansel
France

D. Harari
U.S.A.

David Hartwick
Canada

Karen Heiderich
Canada

Conrad Hewitt
Canada

Paul Hickson
Canada

Christian Holm
U.S.A.

Lucas Hsu
Canada

Werner Israel
Canada

Nick Kaiser
United Kingdom

R. Laflamme
United Kingdom

P. Laguna-Castillo
U.S.A.

Ofer Lahav
United Kingdom

J. Lattanzio
Canada

Eric Linder
U.S.A.

Steve Lonsdale
United Kingdom

Sabino Matarrese
Italy

Paulo Moniz
Portugal

Ian Moss
United Kingdom

John Ng
Canada

Ornella Pantano
Italy

P.J.E. Peebles
U.S.A.

Qiu-he Peng
Holland

T. Piran
Israel

Simon Radford
U.S.A.

Soo-Jong Rey
U.S.A.

André Roberge
Canada

Sirpa Saarinen
United Kingdom

Maria Sakellariadou
U.S.A.

Robert Schaefer
U.S.A.

Albert Stebbins
U.S.A.

J.F. Sygnet
France

S. Toulmin
U.S.A.

M. Turner
U.S.A.

W.G. Unruh
Canada

T. Vachaspati
U.S.A.

Alex Vilenkin
U.S.A.

Robert Wald
U.S.A.

J. Wall
United Kingdom

Rachel Webster
Canada

Nathan Weiss
Canada

R. van de Weygaert
The Netherlands

Simon White
U.S.A.

Larry Widrow
U.S.A.

M. Wise
U.S.A.

Basilis Xanthopoulos
Greece

Thomas Zannias
Greece

THE QUANTUM ORIGIN OF THE UNIVERSE

Ian Moss
Department of Theoretical Physics
University of Newcastle upon Tyne
Newcastle upon Tyne NE1 7RU
U.K.

ABSTRACT. Here is a description of some of the modern day beliefs about the origin of the universe and quantum gravity. There are occasional glimpses of the wide areas of ignorance which should provide fruitful ground for the growth of new ideas.

1. INTRODUCTION

In this article I shall try to explain some recent ideas about the origin of the universe. Detractors[1] have described this subject as a fringe area of physics. My own point of view is that this is a very active frontier and we only use models which are reasonable extrapolations of the high energy physics used successfully in particle accelerators or in astrophysics.

Because of the success of the inflationary scenario[2] as an explanation of the large scale structure of the universe, it will form an important part of the model. We shall distinguish between the following:

An inflationary model is a physical model (e.g. a choice of Lagrangian) which can lead to inflationary solutions.

An inflationary scenario is a descripton of the universe which involves a period of inflation.

Whether or not an inflationary model leads to an inflationary scenario depends upon the initial conditions. By taking an arbitrary universe and evolving it both backwards and forwards in time we can show that even in inflationary models there are solutions in which the large scale structure is never consistent with the observed universe. Therefore, to explain the present day state of the universe we must make some assumptions about the initial conditions.

One attempt at a description of the origin of the universe in an inflationary model would be to suppose that the universe was initially chaotic[3]. We find ourselves today inside a region which expanded expontentially, in the process smoothing out any irregularties. This description explains the large scale structure of the universe, though it remains incomplete without some description of the initial chaos.

1

W. G. Unruh and G. W. Semenoff (eds.), The Early Universe, 1–18.

In another attempt, the universe is supposed to come into existence by quantum tunnelling from nothing[4]. This is suggested by the existence of a solution of Einstein's equations[5] which could play the same role as a liquid bubble in a supercooled phase transition. Though this argument is really no more than an analogy, there are some points in common with the quantum cosmological approach described below.

The approach we shall use is based upon Hawking's suggestion that "spacetime is finite but unbounded"[6]. This is realised by the Hartle-Hawking prescription for the wave function of the universe[7]. This state is a function of the geometry of 3-dimensional hypersurfaces Σ described by a metric tensor g_{ij} and matter fields ϕ. The state is defined by

$$\psi(g,\phi) = \Sigma \, e^{iS[^4g, \, ^4\phi]} \tag{1}$$

where the sum extends over all spacetimes and matter configurations such that the spacetime is compact with no boundary other than Σ. This is shown schematically in fig. 1.

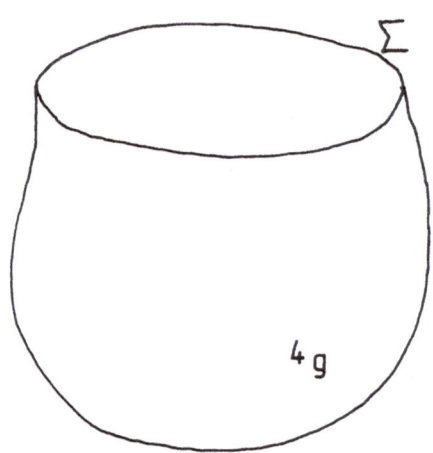

Figure 1. The compact spacetime ^4g with boundary Σ.

The consequences of using this state in an inflationary model are shown in fig. 2. The wave function is dominated by Elucidean geometries (with metric signature ++++) when the universe is small. Since Euclidean geometries have no causal structure, the origin of the universe cannot be assigned to a particular point and there is no physical singularity. This picture is similar to the creation of the universe from nothing.

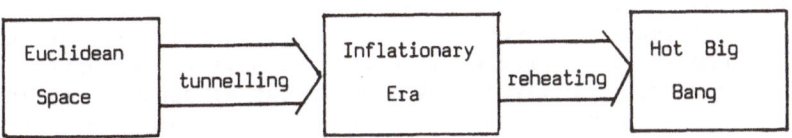

Figure 2. The history of the universe as presented here.

For larger spatial volumes the wave function becomes dominated by
semi-classical Lorentzian spacetimes and it makes sense to talk about
the evolution of the universe. Because the matter field lives on a
closed manifold it is constrained in a way which naturally leads to
inflation. It is during this inflationary era that the universe
becomes spatially flat and homogeneous. Eventually, when the inflation
ends the universe reheats to recover the conventional hot big bang
picture.

A major consequence, then, of the Hartle-Hawking state is that the
inflationary scenario is produced for inflationary models. The use of
an inflationary model is itself essential here, otherwise the state
would lead to a dimunitive universe whose overall size was comparable
only to the planck length, 10^{-33} cm.

In investigating the origin of the universe we shed light on some
old fundamental questions but we also raise some new ones. Take the
choice of Lagrangian, for example. There is, as yet, no compelling
model for producing inflation. The quantum cosmological viewpoint
helps by providing suitable initial conditions for certain inflationary
models. Thermal effects, for example, do not appear before the
inflationary era. Nevertheless, the models still have undesirable
'fix-ups' to obtain reasonably small quantum fluctuations or
satisfactory reheating.

The gravitational part of the action presents particular
difficulties. The Einstein-Hilbert action probably leads to an
inconsistent quantum theory. Certainly, modifications of Einstein's
theory such as superstring theory should be considered in this context.
For the present, we will have to stick with the Einstein-Hilbert action
in the hope that it is a good approximation whenever the radius of
curvature of spacetime is larger than the Planck length. This is
analogous to using the Coulomb potential to describe a Hydrogen atom
where we fix a boundary condition on the wave function at the centre
despite the fact that we know that the Coulomb potential is invalid
inside of the nucleus.

Besides the problems in choosing a Lagrangian, we are confronted with considerable problems of interpretation. The observer is necessarily part of the system, as in the "Many Worlds" interpretation of quantum mechanics. In this picture, the collapse of the wave function associated with a measurement becomes a splitting of the wave function into non-interacting branches. With quantum cosmology it is particularly difficult to see what causes the splitting and which universe from the superposition we live in. We have to guarantee that different observers, who may not even be in causal contact initially, should ultimately agree on all the values of the cosmological parameters.

As an example, consider two causally separate observers S and V, who each make a measurement of the expansion rate of the universe. They then climb into their rocket ships and race towards the latest NATO workshop on cosmology. When they meet over a beer they agree to play a simple game. Each holds his hands behind his back, with a number of fingers corresponding to his value of the expansion rate, say one finger for each 10 of his units. On the count of three they each reveal their hands. We would be surprised to see S holding up five fingers and V holding up ten, since we expect them both to hold up an equal number of fingers. They are, after all, living in the same universe. The problem is how do we prove by a physical argument that they each display the same number of fingers?

2. CANONICAL QUANTUM GRAVITY

The Einstein-Hilbert action for gravity,

$$S_g = \int R \, (^4g) \, (-\det {}^4g)^{\frac{1}{2}} \, d^4x \tag{2}$$

is a function of the spacetime metric ^4g. In the canonical approach to quantum gravity we decompose spacetime into 3-dimensional spatial hypersurfaces Σ_t. The phase space[8,9] consists of 3-metrics g_{ij} which describe the geometry of the hypersurfaces, together with their conjugate momenta p^{ij}. Under this canonical decomposition, the action takes the form

$$S_g = \int (\dot{g}p - NH_g - N_i H^i_g) \, d^4x \tag{3}$$

where the indices on g and p are implicit. The functions H_g and H^i_g are given by

$$H_g(g,p) = G_{ijkl} \, p^{ij} \, p^{kl} - g^{\frac{1}{2}} R(g) \tag{4}$$

$$H^i_g(g,p) = -2p^{ij}|_j \tag{5}$$

where bar denotes the g-covariant derivative with the Ricci scalar R, and G_{ijkl} is the DeWitt mettric,

$$G_{ijkl} = \tfrac{1}{2} g^{-\frac{1}{2}} (g_{ik} \, g_{jl} + g_{il} \, g_{jk} - g_{ij} \, g_{kl}) \tag{6}$$

Exercise 2.1 Perform the canonical decomposition of gravity

(a) read up about curvature 2-forms[10]

(b) Define an orthonormal tetrad w^i,

$$w^o = -Ndt$$

$$w^i = w^i_{\ l}(N^l\ dt + dx^l)$$

where $g_{lK} = w^i_{\ l}\ w^k_{\ K}\ \delta_{ik}$.

(c) Decompose the connection 1-form $w^o_{\ i}$ into $K_i + A_i$ where

$K_i = K_{ij}\ w^j$ is the extrinsic curvature and $A_i = a_i w^o$.

(d) Use the vanishing torsion to obtain K and A in terms of N and N^i.

(e) Decompose the curvature 2-form into a 3-dimensional curvature 2-form and the extrinsic curvature.

(f) Use these results in the action. The momentum p^{ij} is related to K by

$$p^{ij} = g^{\frac{1}{2}}(-K^{ij} + g^{ij}K^k_{\ k}).$$

Variation of the action with respect to the lapse and shift functions N and N_i gives the constraint equations $H=0$ and $H^i = 0$. These equations can be viewed as a subset of Einstein's gravitational field equations. The remaining Einstein equations follow from the variation of the action with respect to g and p. These equations can be expressed in terms of a covariant derivative $\mathcal{G}$ formed from the metric $\mathcal{G} = N\ G$ on configuration space. The solutions to Einstein's equations describe trajectories $g_{ij}(t)$ in configuration space which are geodesies of this derivative.

We can easily extend the action to include any kind of matter fields ϕ, with conjugate momenta π. Then we replace H_g by

$$H(g,p,\phi,\pi) = H_g(g,p) + H_m(g,\phi,\pi) \tag{7}$$

as in exercise 2.2.

Exercise 2.2. Show that the canonical decomposition of the action for a massive scalar field coupled to gravity takes the form

$$S = \int (\dot{g}p + \dot{\phi}\pi - NH - N_i H^i)d^4x$$

where

$$H_m = \tfrac{1}{2}g^{-\frac{1}{2}}\pi^2 + \tfrac{1}{2}g^{\frac{1}{2}}g^{ij}\phi_{|j}\phi_{|j} + \tfrac{1}{2}g^{\frac{1}{2}}m^2\phi^2$$

$$H_m^i = g^{ij}\phi_{|j}\pi$$

as in equation (7).

We can construct the quantum theory by introducing wave functions $\Psi(g,\phi)$ and replacing p by the operator $i\delta/\delta g$. The constraints must then be realised by[8,9]

$$H\psi = 0 \tag{8}$$

$$H^i\psi = 0 \tag{9}$$

Since the operator H^i is associated with infinitesimal coordinate redefinitions on the hypersurfaces Σ, equation 9 represents the independence of ψ from the choice of coordinates. Equation 8 is known as the Wheeler-DeWitt equation. There is a non-trivial factor ordering problem associated with this equation because the DeWitt metric depends upon g. A particularly elegant solution of the factor ordering problem is to use the $\mathcal{G}$ covariant derivative, where $\mathcal{G} = (N\,G_{ijkl},\ NG_{\phi\phi})$. This gives

$$NH_g = -\nabla_g^2 + U \tag{10}$$

where $U = Ng^{\frac{1}{2}}R(g)$ and

$$\nabla_g^2 = NG^{-\frac{1}{2}}\frac{\delta}{\delta g_{ij}}\,G_{ijkl}\,G^{\frac{1}{2}}\frac{\delta}{\delta g_{kl}} \tag{11}$$

With this factor ordering the Wheeler-DeWitt equation is invariant under coordinate redefinitions.

If we choose instead to quantise the theory by path integrals, then the transition amplitude is given by

$$<g',\phi'\,|\,g,\phi> = \int d[g,\phi]\ e^{iS[{}^4g,\,{}^4\phi]} \tag{12}$$

where the 4-geometry ${}^4g$ interpolates between the 3-metrics g and g'. This amplitude satisfies the Wheeler-DeWitt equation provided that we use the configuration space metric $\mathcal{G}$ to define the path integral measure,

$$d[\,g,\phi] = \pi\,(\det \mathcal{G})^{\frac{1}{2}}\,dg_{ij}(x)\,d\phi(x)\,dN(x)\,dN_i(x) \tag{13}$$

which is invariant under coordinate redefinitions.

Because of the vanishing of the superhamiltonian H it is impossible to introduce a $\partial\psi/\partial t$ term on the right hand side of the Wheeler-DeWitt equation, unlike the normal Schrodinger equation. Instead, the Wheeler-DeWitt equation is a dynamical equation because H forms a hyperbolic operator. This is only possible because H is not a positive definite Hamiltonian.

The fact that ψ does not depend upon time is simply an expression of general coordinate invariance, because time is a coordinate label. Physical questions about time development have to be addressed by choosing some degrees of freedom to form clock subsystems against which the time development of the remaining system can be measured. As we shall see later, when the gravitational field behaves semi-classically, then it is possible to measure the passing of time by the evolving geometry.

Ideally, we should also address questions of interpretation by introducing model observers into the wave function. I would recommend this as an interesting problem to work on. For the present we shall look in on the system as outside observers, which I like to call the Olympian viewpoint. We take $|\psi(g,\phi)|^2$ to be a joint probability distribution. Physical information is obtained from the correlation of measurable quantities.. To calculate these correlations we need to know both the wave function and the effect of measurement on the wave function described by projection operators P. The probability of measuring A given B is

$$P(A|B) = P(A \wedge B)/P(B) = \int \psi^* P_A P_B \psi / \int \psi^* P_B \psi \tag{14}$$

As an example, consider the measurement of a scalar field at a given value of the spatial volume $2\pi^2 a^3$ in a homogeneous cosmological model. The projection operators are simply δ functions, so that we have

$$P(\phi|a) = |\psi(a,\phi)|^2 / \sum_\phi |\psi(a,\phi)|^2 \tag{15}$$

and the expected value of ϕ would be

$$<\phi>_a = \sum_\phi \phi P(\phi|a) \tag{16}$$

This quantity can be thought of as the time development of the average value of ϕ.

We can sometimes interpret the wave function in terms of semi-classical trajectories. In one such limit we recover the familiar Schrodinger equation. Consider wave functions of the form

$$\psi(g, \phi) = \psi_g(g)\psi_m(\phi,\tau) \tag{17}$$

Gravitational wave functions ψ_g can be constructed which are sharply peaked around a solution $g(\tau)$ of[8] Einstein's equations with a back reaction term. The wave function can then be made into an approximate solution of the Wheeler-DeWitt equation. The function ψ_m must satisfy.

$$H_m \psi_m = i \, \delta\psi_m/\delta\tau \tag{18}$$

in the background gravitational field.

Exercise 2.3 Derive Schrodinger's equation

    (a)   Revise Hamilton–Jacobi theory.

    (b)   Define the semi-classical field by
$$H_g(g, \delta \bar{S}/\delta g) + <H_m> = 0$$

where τ is given by $\delta g_{ij}/\delta \tau = G_{ijkl} \delta \bar{S}/\delta g_{ij}$ and

$$<H_m> = \int d[\phi] \psi_m(\phi, \tau) \, H_m(g, \phi, i\delta/\delta\phi) \psi_m(\phi, \tau)$$

    (c)   Try a solution to the Wheeler–DeWitt equation of the form
$$\psi_g(g) = A \exp i\bar{S}(g, g_0).$$

    (d)   Neglect the $\delta^2 A/\delta g^2$ term and show that

$$A = (\det \delta^2 \bar{S}/\delta g \delta g_0)^{\frac{1}{2}} \; e^{i\int <H_m> d^3 x d\tau}$$

A general solution of the Wheeler–DeWitt equation will develop into a superposition of WKB components with coefficients which give an idea of the relative importance of each semi-classical metric g, but the problem of how we come to see just one component remains unresolved.

3.   MINISUPERSPACE

The canonical approach to quantum gravity being developed here gives an alternative perturbative technique to those which are available in an explicitly covariant approach. We are able to expand the metric g_{ij} about a homogeneous metric $\bar{g}_{ij}$, which for example may be the metric on a 3-sphere. We write[11]

$$g_{ij} = a^2(\bar{g}_{ij} + \epsilon\Sigma\gamma_n Q^n \bar{g}_{ij} + \epsilon\Sigma\gamma_n Q^n{}_{Sij} + \epsilon\Sigma\gamma_n Q^n{}_{Vij} + \epsilon\Sigma\gamma_n Q^n{}_{Tij}) \quad (19)$$

where QS, QV and QT are formed from scalar, vector and tensor harmonics on the 3-sphere. In this approach the a degree of freedom is treated non-perterbatively. The lapse and shift functions are also integrated out leaving just six perterbed fields, compared with the ten covariant fields.

The zero'th order approximation in ϵ of this type is known as a minisuperspace approximation[12]. Unfortunately there are problems with renormalisability at quadratic order. In an analytic regularisation scheme, such as ζ-function regularisation, where no subtraction of divergences is required the result may just be consistent, though it would be far better to approach this problem with a renormalisable theory. I recommend this problem to the reader.

As a simple example, consider gravity with a cosmological constant Λ in the minisuperspace limit of equation 19. The action is

$$S = \frac{1}{16\pi G} \int_m (R-2\Lambda) dv + \frac{1}{16\pi G} \int_{\partial m} 2K dv = \int L \, dt \quad (20)$$

The presence of a boundary requires the additional term in the action which depends on the extrinsic curvature K of the boundary. Substituting the metric equation 19 gives L = T−U, where

$$T = -\tfrac{1}{2}N^{-1}\kappa^{-2}a\dot{a}^2 \tag{21}$$

$$U = \tfrac{1}{2}\kappa^{-2}\ aN(H^2a^2-1) \tag{22}$$

The constants $\kappa^2 = 2G/3\pi$ and $h^2 = \tfrac{1}{3}\Lambda$. From equation 21 we can read off the configuration space metric $G^{aa} = -\tfrac{1}{2}N^{-1}\kappa^{-2}a$. Therefore the Wheeler-Dewitt equation 10 is

$$(\tfrac{1}{2}N\kappa^2a^{-\frac{1}{2}}\ \frac{d}{da}\ a^{-\frac{1}{2}}\ \frac{d}{da}\ +\ U)\psi\ =\ 0 \tag{23}$$

The potential U and a solution ψ are sketched in fig. 3. Since the expansion rate is related to the momentum $iN^{-1}\delta/\delta a$ we can identify the oscillatory part of the wave function with real values of N and it matches a classical deSitter spacetime of radius h^{-1}. The exponential part of the wave function matches a 4-sphere of radius h^{-1}

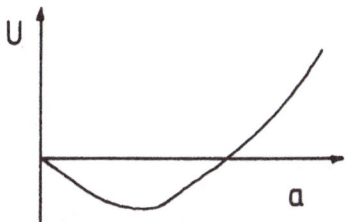

Figure 3a.

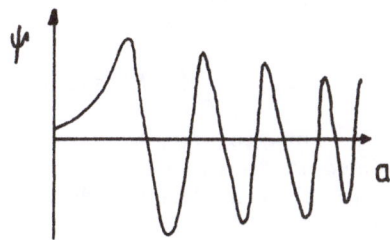

Figure 3b.

Another way of quantising this theory is to use the path integral. Then

$$<a'|a>\ =\ \int d[a,N]e^{iS} \tag{24}$$

where

$$d[a,N]\ =\ \Pi(a(x)/N(x))^{\frac{1}{2}}\ da(x)\ dN(x) \tag{25}$$

This amplitude satisfies the Wheeler-DeWitt equation.

Exercise 3.1 Quantise L from first principles and rederive the Wheeler-DeWitt equation 23.

Exercise 3.2 Check that satisfies the Wheeler-DeWitt equation. The method can be adapted from ref.13. When the paths are replaced by discrete variables a_n, you should note that replacing $a(t)$ by $a_{n-1} + f(a_n - a_{n-1})$ for any function f leads to an arbitrary factor ordering.

4. THE HARTLE-HAWKING WAVE FUNCTION

The unique choice of quantum state given by the Hartle-Hawking[7] prescription is

$$\psi(g,\phi) = \int d[g,\phi] \; e^{iS[^4g, ^4\phi]} \tag{26}$$

where ^4g and $^4\phi$ are defined on a manifold M which is compact with a boundary Σ. The induced metric on Σ must be g_{ij}. This prescription is minimal in the sense that there are no other boundaries on which we need extra boundary conditions.

In the minisuperspace approximation the Hartle-Hawking state has many desirable properties, such as regularity at spacetime singularities. It induces vacuum fluctuations and has a tendency to produce inflation, as was mentioned alongside fig. 2.

For the simple model of gravity with a cosmological constant introduced in section 3 the wave function can be evaluated exactly. The path integral is

$$\psi(a) = \int d[a,N] \; \exp \; i \int (-\tfrac{1}{2}N^{-1}\kappa^{-2}a \; \dot{a}^2 \; -\tfrac{1}{2}\kappa^{-2}aN(h^2a^2-1)dt \tag{27}$$

the a integral can be eliminated by gauge fixing, but for the present purposes it is instructive to examine the integral over the lapse function N in a steepest descents approximation. The saddle points are shown in fig. 4. When $ah < 1$, then N is imaginary at the saddle point and they represent Euclidean spacetimes. On the other hand, when $ah > 1$, the saddle points lie on the real axis and Lorentzian spacetimes dominate the path integral. The wave function has a form similar to the one shown in fig. 3b.

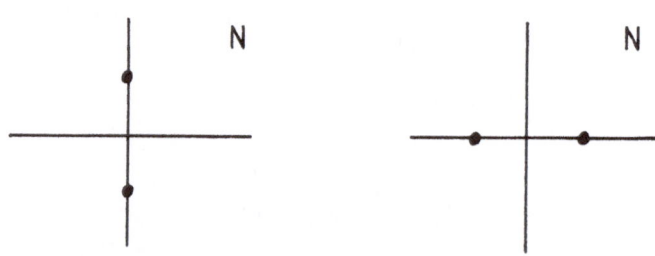

Figure 4. The saddle points used to approximate the path integral are shown in the complex N plane.

In the original Hartle-Hawking prescription for the wave function, the path integral is taken over Euclidean metrics with an analytic continuation of the conformal degrees of freedom to make the integral convergent. This can be shown to lead to a real rather than a complex wave function[14]. In practice, an examination of the lapse function integral indicates which contour to choose for a real and finite result.

Because the wave function is real it represents an equal superposition of expanding and contracting spacetimes. This is consistent with the time symmetry of de Sitter space, but in more general examples we have to conclude that for each evolving universe represented by the wave function, there is another universe with the direction of time reversed. These two universes are indistinguishable to their inhabitants because the sign of the time coordinate is merely a mathematical convention.

We can extend this model to manufacture simple inflationary models by including a scalar field $\phi(t)$. Then the Lagrangian becomes

$$L = -\tfrac{1}{2}N^{-1}\kappa^{-2}a\,\dot{a}^2 + \tfrac{1}{4}N^{-1}a^3\,\dot{\phi}^2 - \tfrac{1}{2}\kappa^{-2}a\,N(a^2V(\phi)-1) \qquad (28)$$

where $V(\phi)$ is the scalar field potential. The classical equations of motion are

$$3a^{-2}(\dot{a}^2 + 1) = 3\kappa^2\rho \qquad (29)$$

$$\ddot{\phi} + 3\frac{\dot{a}}{a}\dot{\phi} + V'(\phi) = 0 \qquad (30)$$

where $\rho = \tfrac{1}{2}\dot{\phi}^2 + V$. The condition for inflation is that the timescale of evolution of the scalar field is larger than the expansion timescale. This happens for $V''(\phi) \ll h^2$, where h is the expansion rate, $h = \dot{a}/a$.

To quantise this model we introduce wave functions $\psi(a,\phi)$. From equation 28 we can read off the configuration space metric components.
$G^{aa} = -\tfrac{1}{2}N^{-1}k^{-2}a$ and $G^{\phi\phi} = \tfrac{1}{4}N^{-1}a^3$ to construct the Wheeler-DeWitt equation,

$$\left(\tfrac{1}{2}N\kappa^2 a^{-\frac{1}{2}}\frac{\partial}{\partial a}\,a^{-\frac{1}{2}}\frac{\partial}{\partial a} - Na^{-3}\frac{\partial^2}{\partial\phi^2} + U\right) = 0 \qquad (31)$$

where U is the potential term in equation 28. A more convenient choice of coordinates to take is $x = a\,\sinh\phi$, $y = a\,\cosh\phi$, in which the characteristics of the wave equation are straight lines[15]. The (x,y) plane is plotted in fig. 5.

We wish to calculate the Hartle-Hawking state, but rather than perform the path integral in general, it is much simpler to find the wave function analytically near to a=0 and then extend the solution to the whole x,y plane using the Wheeler-DeWitt equation. Once the wave function has been found we can examine which cosmologies dominate in the semi-classical limit, where $\psi \sim A\exp iS$. An important question that we wish to address is whether or not inflation takes place. Now, since the evolution timescale is related to the momentum p_ϕ, the condition for inflation is that $p_\phi \ll P_a$. In the semi-classical limit $p = \nabla S$ and we have simply to see whether ∇S lies in the direction of increasing a.

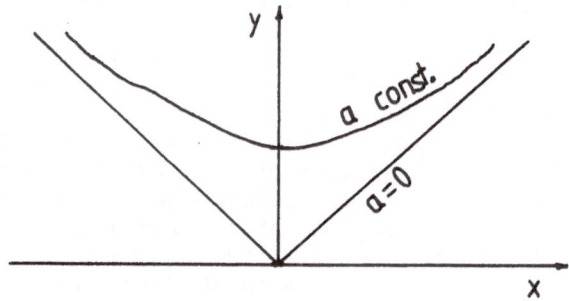

Figure 5. The xy plane with lines of constant a.

The wave function has been calculated in various inflationary models[16,17]. In fig. 6 we see the result for a massive scalar field, $V(\phi) = \frac{1}{2}m^2\phi^2$. All of the features which discussed so far are present in this wave function, including regularity at a=0 and inflationary behaviour for large ϕ.

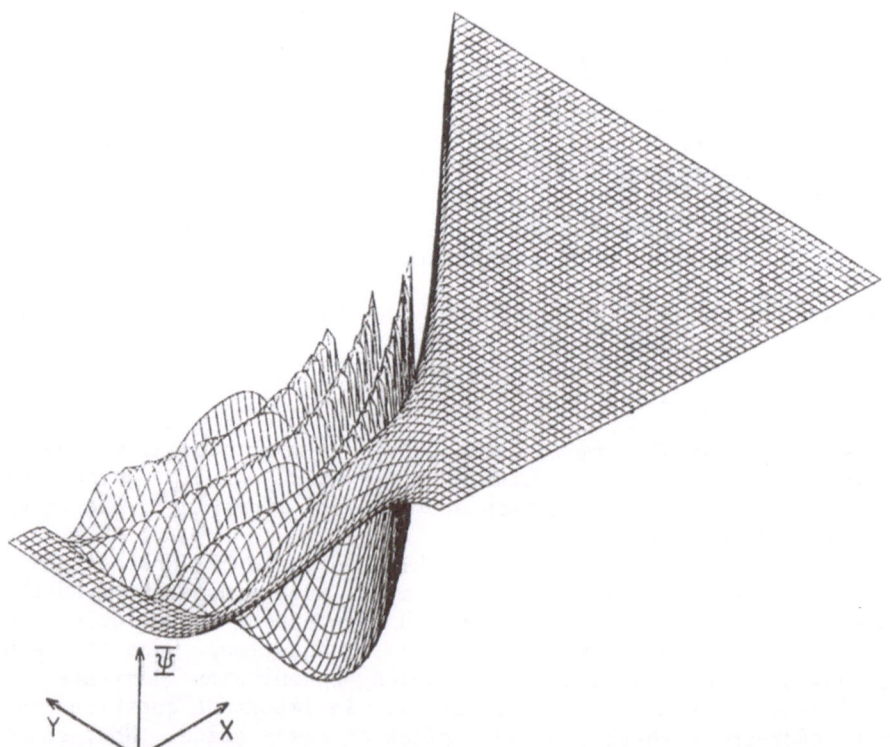

Figure 6. The wave function of the universe for the model described above.

5. ANISOTROPIC UNIVERSES

The limits on the isotropy of the universe today are very restrictive. It is an important test of our theories of the origin of the universe that they are consistent with these limits.

We shall continue to use homogeneous cosmological models. The most general closed anisotropy model has the Bianchi IX metric[10],

$$ds^2 = -dt^2 + a^2(e^{2\beta})_{ij} \, w^i \, w^j \tag{32}$$

where w^i are spatial 1-forms, and $\beta = \text{diag} \, (\beta_+ + \sqrt{3}\beta_-, \, \beta_+ - \sqrt{3}\beta_-, \, -2\beta_+)$. If the matter is again a scalar field, then the wave function for this model has the form $\Psi(a, \beta_+, \beta_-, \phi)$.

The strategy for analising this wave function is firstly to take an inflationary model. Then the path integral is calculated to give the wave function when a is small. Finally the Wheeler-DeWitt equation is integrated or, if it is more convenient, a semi-classical approximation is used to find the behaviour of the anisotropy of subsequent times.

Such a wave function is shown in fig. 7[23]. The wave function is shown as a function of $\beta_\pm$ just prior to inflation. The anisotropy is peaked around zero with a spread comparable to one in Planck units. There is no sign of the chaotic behaviour which classical Bianchi IX models display[24]. This is because the wave function is dominated by Euclidean spacetimes at early times.

The initial anisotropy represented by this wave function can be evolved forwards in time in the semi-classical approximation. The resulting anisotropy would be difficult to measure using present observational techniques. One can show that $\nabla T/T \sim 10^{-56}/I$, where I is the expansion factor of the inflationary stage.

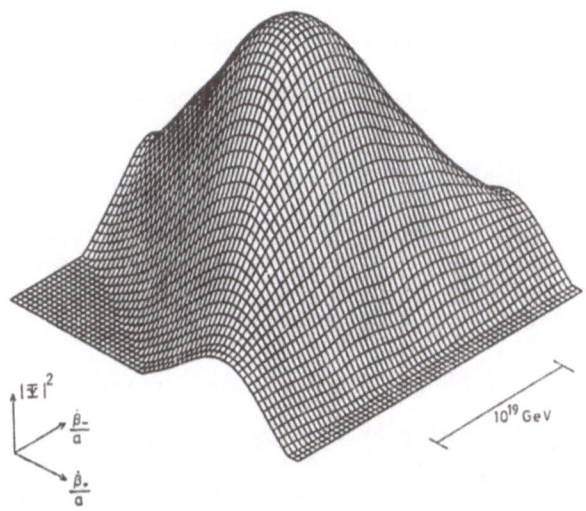

Figure 7. The probability for various values of the anisotropy is shown prior to the inflationary era.

14

6. CHANGES IN TOPOLOGY

We could imagine transition amplitudes between disconnected metrics, such as the one shown in fig. 8 which represents $<g_1,g_2|g>$. In particle theory, the analogous diagram would be an interaction vertex which indicates the necessity for second quantisation. In the case of the amplitude $<g_1,g_2|g>$, the Wheeler-DeWitt equation breaks down because the transition between the initial and final geometries is discontinuous. We can use the path integral instead, but there are theorems in differential topology which imply that no Lorentzian 4-geometry exists which can match on to topologically distinct 3-geometries. This is a fundamental different beween Euclidean and Lorentzian formulations of the path integral.

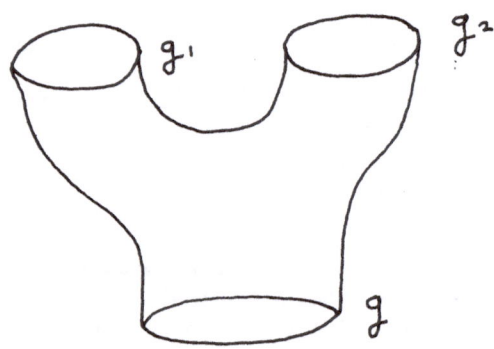

Figure 8. A change in topology.

There are two distinct ways in which we can proceed:

(i)   Sum over all of the unobservable components of the 3-geometry.

(ii) Extend the Hilbert space to include $|g> + |g_1g_2> + ..$

Case (i) is based upon an idea of Hawking [25] and Page [26], and leads to transitions from pure states to density matrices. Consider the joint transition amplitude $P(|g_1> \to |g_1'>, <g_2| \to <g_2'|)$. Allowing for changes in topology with a sum over the unobserved components,

$$\underline{p} = <g_1'\ g_1><g_2'|\ g_2> + \Sigma\ <g'_1,g|\ g_1><g_2|\ g,g_2'> + \ ....$$ (33)

We can represent this diagramatically by fig. 9, where g is the metric on an internal surface shown by the dotted line. Each diagram corresponds to a Euclidean path integral with boundary metrics $g_1,\ g_2\ g_1',\ g_2'$.

A pure state has the form $\rho = |\psi><\psi|$ where $|\psi> = \Sigma \psi(g)|g>$. The development of this state is described by a superscattering operator $, where

$$\$\rho = \sum_{g_1 g_2} \sum_{g_1' g_2'} \psi(g_1)\psi(g_2) |g_1><g_2| \ P(|g_1><g_2|+|g_1'><g_2'|) \tag{34}$$

From the expansion of P given in equation 33 we can see that this expression does not factorise in general and therefore $\$\rho$ does not represent a pure state.

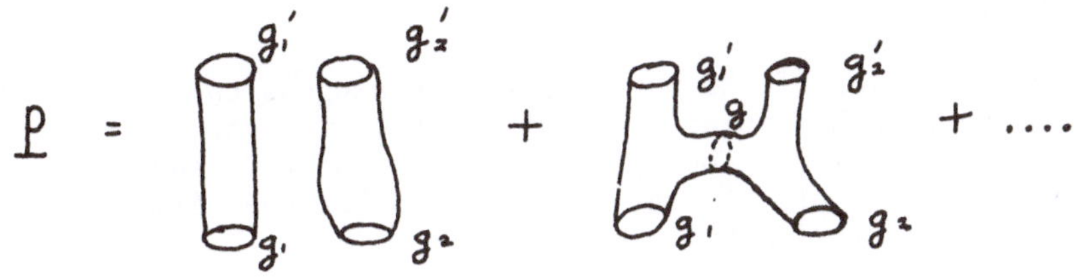

Figure 9. The transition probability P split up into various topological components. The metric g can be placed on internal slices, as shown by the dotted line.

In quantum cosmology, the development of pure to mixed states can be intepreted as branching of the wave function. For example, a pure state $|RW1> + |RW2>$ representing a superposition of Robertson–Walker universes 1 and 2 can evolve to the density matrix $|RW1><RW1| + |RW2><RW2|$ in which all of the quantum interference terms between the universes vanish. This may free us from having to involve a complicated observer in order to split the wave function.

Case (ii) involves a second quantisation of the Wheeler-DeWitt equation. We proceed by constructing an action whose classical equations of motion are the constraint equations. Next, we introduce interaction terms corresponding to fig. 8 and finally, the theory is quantised by path integration over a suitable class of wave functions ψ.

Some of the gauge freedom can be fixed at this stage, introducing ghost fields θ and θ^i into the wave function $\psi(g,\theta,\theta^i)$. The remaining BRS symmetry imposes a constraining $Q\psi = 0$, which replaces the constraints 8 and 9. Using the methods of Fradkin and Vilkowiski 27 we can construct Q in terms of H, H^i and their commutators, with the result that

$$Q = \int d^3x (H + H_i\theta^i + \hat{\theta}\nabla_i\theta\theta^i + \tfrac{1}{2}\hat{\theta}_i(\nabla_k\theta^i)\theta^k - \tfrac{1}{2}\hat{\theta}_i(\nabla_k\theta^k)\theta^i) \tag{35}$$

The anticommuting fields $\hat{\theta}$ are antighost fields which are canonically conjugate to θ.

A suitable action is given by

$$S = \int d[g,\theta,\theta^i](\psi Q\psi + \frac{2}{3}\psi\psi\psi) \tag{36}$$

Variation of the first term in the action leads to the "free" constraint equation $Q\psi = 0$ which is equivalent ot the Hamiltonian and momentum constraints. The second term contains cubic interaction terms as shown in fig. 2, but quartic and higher terms could also be included.

The action (36) gives a theory which is no longer equivalent to the Wheeler-DeWitt formulation of quantum gravity. The semi-classical limit is modified also. From the variation of equation 36 we get

$$Q\psi + \psi\psi = 0 \tag{37}$$

From a decomposition of the wave function analogous to equation 17, we no longer obtain Schrodinger's equation but we get a non linear generalisation of it. The particular form of this equation depends upon how the product of ψ fields is defined. For the simplest choice $(\psi(g,\theta))^3$, the result is that

$$H_m \psi_m = i \delta\psi_m/\delta\tau - \psi_g \psi_m^2 \tag{38}$$

It is well known that such non-linear terms can induce effects which resemble the collapse of the wave function[28]. This can happen when there exist solitonic solutions of equation 38 which typical solutions to the linearised equation can evolve into. In the case of quantum cosmology, it may be possible for the selection of a universe to occur without the intervention of an observer.

Non-linear terms can also arise in an analagous situation where a many electron wave function is approximated by the wave function of a single electron. An example of this would be the Hartree-Fock approximation for a Helium atom, in which a single electron is viewed as if it were moving in its own charge distribution Such non-linear effects are also important in solid state physics, where the soliton solutions representing photons have been observed.

REFERENCES

1. W.G. Unruh, public communication.

2. A.H. Guth, Phys. Rev. $\underline{D23}$ 347 (1981).

3. A.D. Linde, Phys. Lett. $\underline{129B}$ 177 (1983).

4. A. Vilenkin, Phys. Lett. $\underline{117B}$, 25 (1982).

5. S.W. Hawking and I.G. Moss, Phys. Lett. $\underline{110B}$ 35 (1982).

6. S.W. Hawking, Pontif Accad. Sci. Varia. $\underline{48}$ 463 (1982).

7. J.B. Hartle and S.W. Hawking, Phys. Rev. $\underline{D23}$ 2960 (1983).

8. B.S. DeWitt, Phys. Rev. $\underline{169}$ 1113 (1967).

9. J.A. Wheeler, in Battelle Recontres, eds C. DeWitt and J.A. Wheeler (Benjamin, New York, 1968).

10. C.W. Misner, K.S. Thorne and J.A. Wheeler, "Gravitation" (W.H. Freeman, San Fransisco, 1970).

11. J. Halliwell and S.W. Hawking, Phys. Rev. $\underline{D31}$ 1777 (1985).

12. C.W. Misner, in "Majic Without Majic", ed J. Klauder (Freeman, 1972).

13. R.P. Feynman, Rev. Mod. Phys. $\underline{20}$ 267 (1948).

14. S.W. Hawking, Nuc. Phys. $\underline{B239}$ 257 (1984).

15. S.W. Hawking and J. Luttrell, Nuc. Phys. $\underline{B247}$ 250 (1984).

16. I.G. Moss and W.A. Wright, Phys. Rev. $\underline{D29}$ 1067 (1984).

17. U. Carow and S. Watamura, Phys. Rev. $\underline{D32}$ 1290 (1985).

18. S.W. Hawking and J.C. luttrell, Phys. Lett. $\underline{B143}$ 83 (1984).

19. A. Vilenkin, Phys. Rev. $\underline{D33}$ 3560 (1986.

20. S. Lonsdale, Phys. Lett. $\underline{B175}$ 312 (1986).

21. J. Halliwell, "Classical and quantum cosmology of the Salam-Sezgin model" (Cambridge preprint (1986).

22. W.A. Wright, Ph.D. thesis (Newcastle, 1985).

23. I.G. Moss and W.A. Wright, Phys. Let. $\underline{154B}$ 115 (1985).

24. E.M. Lifshitz and I.M. Khalatnikov, Sov. Phys. Usp. <u>6</u> 495 (1964).

25. S.W. Hawking, "The density matrix of the universe" (Cambridge preprint 1986).

26. D.N. Page, "Density matrix of the universe" (Pennysylvania preprint 1986).

27. E.S. Fradkin and G.A. Vilkoviski, Phys. Lett. <u>55B</u> 224 (1975).

28. D. Bohm and J. Bub, Rev. Mod. Phys. <u>38</u> 453 (1966).

COSMOLOGY AND PARTICLE PHYSICS

Michael S. Turner
NASA/Fermilab Astrophysics Center
Fermi National Accelerator Laboratory
Batavia, Illinois 60510
and
Departments of Physics and Astronomy and Astrophysics
Enrico Fermi Institute
The University of Chicago
Chicago, Illinois 60637

INTRODUCTION

In the past five years or so progress in both elementary particle physics and in cosmology has become increasingly dependent upon the interplay between the two disciplines. On the particle physics side, the $SU(3)_C \times SU(2)_L \times U(1)_Y$ model seems to very accurately describe the interactions of quarks and leptons at energies below, say, 10^3 GeV. At the very least, the so-called standard model is a satisfactory, effective low energy theory. The frontiers of particle physics now involve energies of much greater than 10^3 GeV—energies which are not now available in terrestrial accelerators, nor are ever likely to be available in terrestrial accelerators. For this reason particle physicists have turned both to the early Universe with its essentially unlimited energy budget (up to 10^{19} GeV) and high particle fluxes (up to $10^{107} \mathrm{cm}^{-2}\mathrm{s}^{-1}$), and to various unique, contemporary astrophysical environments (centers of main sequence stars where temperatures reach 10^8 K, neutron stars where densities reach $10^{14} - 10^{15} \mathrm{gcm}^{-3}$, our galaxy whose magnetic field can impart 10^{11} GeV to a Dirac magnetic charge, etc.) as non-traditional laboratories for studying physics at very high energies and very short distances.

On the cosmological side, the hot big bang model, the so called standard model of cosmology, seems to provide an accurate accounting of the history of the Universe from about 10^{-2} s after 'the bang' when the temperature was about 10 MeV, until today, some 10-20 billion years after 'the bang' and a temperature of about 3 K ($\simeq 3 \times 10^{-13}$GeV). Extending our understanding to earlier times and higher temperatures, requires knowledge about the fundamental particles (presumably quarks and leptons) and their interactions at very high energies. For this reason, progress in cosmology has become linked to progress in elementary particle physics.

In these lectures I will try to illustrate the two-way nature of the interplay between these fields by focusing on a few selected topics. In Lecture 1 I will review the standard cosmology, especially concentrating on primordial nucleosynthesis, and discuss how the standard cosmology has been used to place constraints on the properties of various particles. Grand Unification makes two striking predictions: (1)

19

W. G. Unruh and G. W. Semenoff (eds.), The Early Universe, 19–113.
© 1988 by D. Reidel Publishing Company.

baryon number nonconservation; (2) the existence of stable, superheavy magnetic monopoles. Both have had great cosmological impact. In Lecture 2 I will discuss baryogenesis, the very attractive scenario in which the B, C, CP violating interactions in GUTs provide a dynamical explanation for the predominance of matter over antimatter, and the present baryon-to-photon ratio. Baryogenesis is so cosmologically attractive, that in the absence of observed proton decay it has been called 'the best evidence for some kind of unification'. Monopoles on the other hand started out as a cosmological disaster; however, efforts to solve the problem of monopole overproduction in the standard cosmology led to one of the most exciting payoffs of the Inner Space/Outer Space connection: the inflationary Universe scenario. In the third lecture I will discuss how the very early ($t \lesssim 10^{-34}$sec) dynamical evolution of a very weakly coupled scalar field which is initially displaced from the minimum of its potential has the potential to explain a handful of very fundamental cosmological facts—facts which can be accommodated by the standard cosmology, but which are not 'explained' by it.

By selecting just a few topics I have left out some other very important and interesting topics—monopoles and astrophysics, axions, galaxy formation, the deconfinement/chiral transition of QCD, supersymmetry/supergravity and cosmology, superstrings, and cosmology in extra dimensions—to mention just a few. I refer the interested reader to other reviews of the early Universe (refs. 1) and papers on some of these topics (refs. 2 and 3).

LECTURE 1:
THE STANDARD COSMOLOGY AND ITS SUCCESSES

The hot, big bang cosmology—the so-called standard cosmology, neatly accounts for the (Hubble) expansion of the Universe, the 2.7 K microwave background radiation (see Figs. 1.1, 1.2), and, through primordial nucleosynthesis, the cosmic abundances of the light elements D and ^4He (and in all likelihood ^3He and ^7Li as well; see Fig. 1.3). The most distant galaxies and QSO's observed to date have redshifts in excess of 3—the current record holders are: for galaxies $z = 3.2$ (ref. 4) and for QSO's $z = 4.0$ (ref. 5). The light we observe from an object with redshift $z = 3$ left that object only 1–2 Byr after the bang. Observations of even the most distant galaxies and QSO's are consistent with the standard cosmology, thereby testing it back to times as early as 1 Byr (see, e.g., ref. 6). The surface of last scattering for the microwave background is the Universe at an age of a few $\times 10^5$ yrs and temperature of about 3000 K. Measurements at wavelengths from 0.05 cm to 80 cm indicate that it is consistent with being radiation from a blackbody of temperature 2.75 K $\pm$ 0.05 K (see Fig. 1.1 and ref. 7). Measurements of the isotropy indicate that the temperature is uniform to a part in 1000 on angular scales ranging from $10''$ to $180°$—to a part in 10^4 after the dipole component is removed (see Fig. 1.2 and ref. 8). The observations of the microwave background test the standard cosmology back to times as early as 100,000 yrs. According to the standard cosmology, when the Universe was 0.01 sec–300 sec old, corresponding to temperatures of $\simeq$ 10 MeV–0.1 MeV, conditions were right for the synthesis of a number of light nuclei. The predicted abundances of D, ^3He, ^4He, and ^7Li are consistent with their observed abundances provided that the baryon-to-photon ratio is

$$\eta \equiv n_b/n_\gamma \simeq (4-7) \times 10^{-10} \qquad (1.1)$$

The baryon-to-photon ratio and the fraction of critical density contributed by baryons are related by: $\Omega_b h^2/T_{2.7}^3 \simeq 3.53 \times 10^7 \eta$ where $T_{2.7}$ is the microwave temperature in units of 2.7K and h is the present value of the Hubble constant in units of 100 km s^{-1} Mpc^{-1}. The allowed range for η corresponds to: $0.014 \lesssim \Omega_b h^2/T_{2.7}^3 \lesssim 0.025$, implying that baryons alone cannot provide the closure density. The concordance of theory and observation for D and ^4He is particularly compelling evidence in support of the standard cosmology as there are no known contemporary astrophysical sites which can simultaneously account for the primordial abundances of both these isotopes (see Fig. 1.3; see ref. 9 for further discussion of primordial nucleosynthesis). In sum, all the available evidence indicates that the standard cosmology provides an accurate accounting of the evolution of the Universe from 0.01 sec after the bang until today, some 15 or so Byr later—quite a remarkable achievement!

I will now briefly review the standard cosmology (more complete discussions of the standard cosmology are given in ref. 6). Throughout I will use high energy physics units, where $\hbar = k = c = 1$. The following conversion factors may be useful.

$$1\text{GeV}^{-1} = 0.197 \times 10^{-13}\text{cm}$$
$$1\text{GeV}^{-1} = 0.658 \times 10^{-24}\text{sec}$$
$$1\text{GeV} = 1.160 \times 10^{13} K$$
$$1\text{GeV}^4 = 2.32 \times 10^{17}\text{gcm}^{-3}$$
$$1M_\odot = 1.99 \times 10^{33} g \simeq 1.2 \times 10^{57}\text{baryons}$$
$$1pc = 3.26 \text{ light} - \text{year} \simeq 3.09 \times 10^{18}\text{cm}$$
$$1\text{Mpc} = 3.09 \times 10^{24}\text{cm}$$
$$G_N = 6.673 \times 10^{-8}\text{cm}^3\text{g}^{-1}\text{sec}^{-2} \equiv m_{pl}^{-2}$$
$$(m_{pl} = 1.22 \times 10^{19}\text{GeV})$$

On large scales ($\gg$ 100Mpc) the Universe is isotropic and homogeneous, as evidenced by the uniformity of the 2.7 K background radiation, the x-ray background, and counts of galaxies and radio sources, and so the standard cosmology is based on the maximally-symmetric Robertson–Walker line element

$$ds^2 = -dt^2 + R^2(t)[dr^2/(1-kr^2) + r^2 d\theta^2 + r^2 \sin^2\theta d\phi^2] \qquad (1.2)$$

where ds^2 is the square of the proper separation between two space-time events, k is the curvature signature (and can, by a suitable rescaling of R, be set equal to -1, 0, or

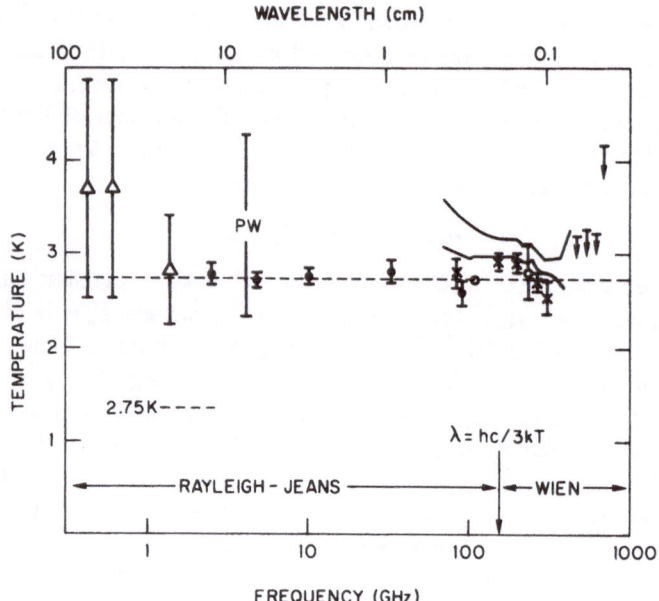

Figure 1.1: Summary of microwave background temperature measurements from $\lambda \simeq 0.05$ to 80 cm (see refs. 7). Measurements indicate that the background radiation is well described as a 2.75 ± 0.05K blackbody. PW denotes the discovery measurement of Penzias and Wilson.

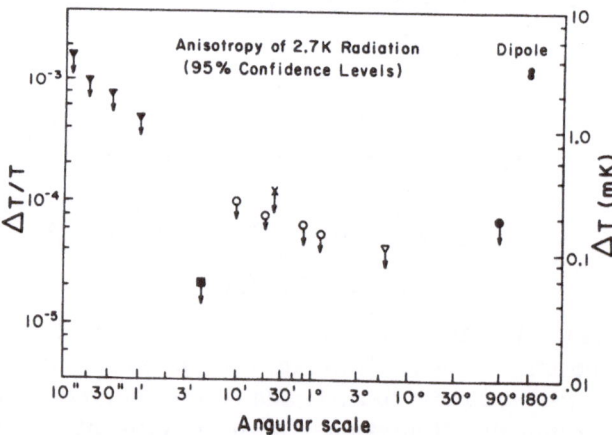

Figure 1.2: Summary of microwave background anisotropy measurements on angular scales from 10″ to 180° (see ref. 8). With the exception of the dipole measurements, the rest are 95% confidence upper limits to the anisotropy.

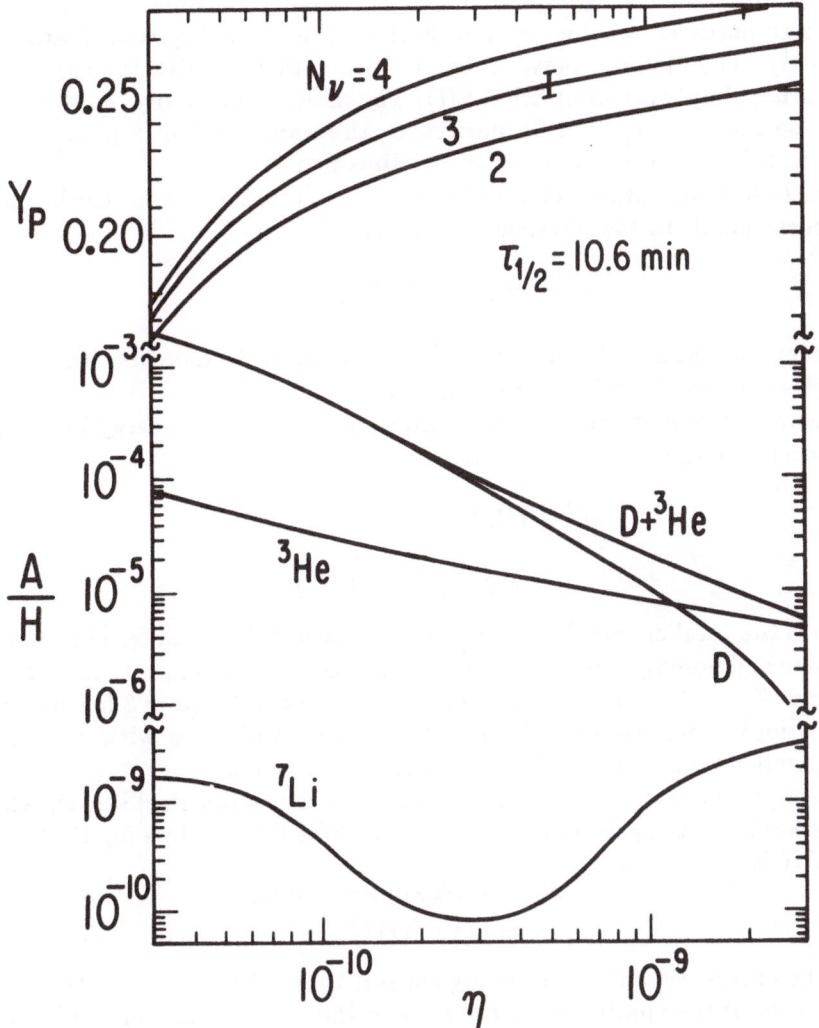

Figure 1.3: Big bang nucleosynthesis predictions for the primordial abundances of D, ^{3}He, ^{4}He, and ^{7}Li. Y_p = mass fraction of ^{4}He, shown for $N_\nu = 2$, 3, 4 light neutrino species. Present observational data suggest: $0.23 \leq Y_p \leq 0.25$, $(D/H)_p \geq 1 \times 10^{-5}$, $[(D+^3 He)/H]_p \leq 10^{-4}$, and $(^7 Li/H)_p \simeq (1.1 \pm 0.4) \times 10^{-10}$. Concordance requires $\eta \simeq (4-7) \times 10^{-10}$. For further discussion see ref. 9.

+1), and $R(t)$ is the cosmic scale factor. The expansion of the Universe is embodied in $R(t)$—as $R(t)$ increases all proper (i.e., physical—as measured by meter sticks) distances scale with $R(t)$. The coordinates r, θ, and φ are comoving coordinates: test particles initially at rest will have constant comoving coordinates, and the velocity of NR test particles moving with respect to the comoving coordinates decrease ($\propto R(t)^{-1}$). The distance between two objects comoving with the expansion, e.g., two galaxies, simply scales up with $R(t)$. The momentum of any freely-propagating particle decreases as $1/R(t)$. In particular, the wavelength of a photon $\lambda \propto R(t)$, i.e., is redshifted by the expansion of the Universe

The coordinate distance at which curvature effects become noticeable is $|k|^{-1/2}$, which corresponds to the physical (or proper) distance

$$R_{curv} \simeq R(t)|k|^{-1/2} \tag{1.3}$$

—which one might call the curvature radius of the Universe. Note that R_{curv} also just scales with the cosmic scale factor $R(t)$.

The evolution of the cosmic scale factor and of the stress energy in the Universe are governed by the Friedmann equations:

$$H^2 \equiv (\dot{R}/R)^2 = 8\pi G\rho/3 - k/R^2 \tag{1.4}$$

$$d(\rho R^3) = -p\,d(R^3) \tag{1.5}$$

where ρ is the total energy density and p is the isotropic pressure. [The assumption of isotropy and homogeneity require that the stress-energy tensor take on the perfect fluid form: $T^\mu_\nu = diagonal(-\rho, p, p, p)$.] Because $\rho \propto R^{-n}$ ($n = 3$ for matter, $n = 4$ for radiation) it follows from Eqn(4) that model Universes with $k < 0$ expand forever, while those with $k > 0$ must necessarily recollapse.

The expansion rate H (also known as the Hubble parameter) sets the characteristic timescale for the growth of $R(t)$: H^{-1} is the e-folding time for R. The present value of H is

$$H_0 = 100h \text{ km sec}^{-1}\text{Mpc};$$
$$\simeq h(10^{10}\text{yr})^{-1}$$

where the observational data strongly suggest that $0.4 \leq h \leq 1$ (ref. 10).

The sign of the spatial curvature k—and the ultimate fate of the Universe can be determined from measurements of ρ and H:

$$k/H^2R^2 = \rho/(3H^2/8\pi G) - 1$$
$$\equiv \Omega - 1 \tag{1.6}$$

where $\Omega = \rho/\rho_{crit}$ and $\rho_{crit} = 1.88h^2 \times 10^{-29}\text{gcm}^{-3} = 1.05 \times 10^4 h^2\text{eVcm}^{-3}$. The curvature radius, R_{curv}, is related to Ω by

$$(R_{curv}/H^{-1})^2 = 1/|\Omega - 1| \tag{1.7}$$

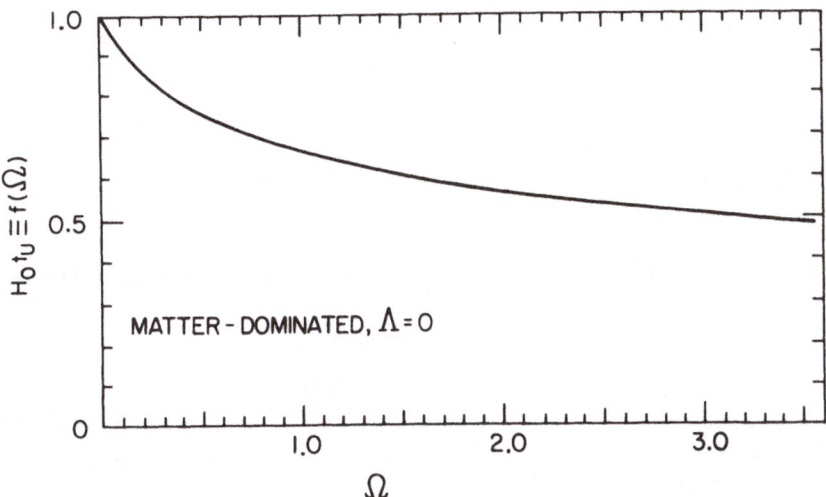

Figure 1.4a: The age of a matter-dominated, $\Lambda = 0$ model universe in Hubble units, $f(\Omega) = H_0 t_u$, as a function of Ω.

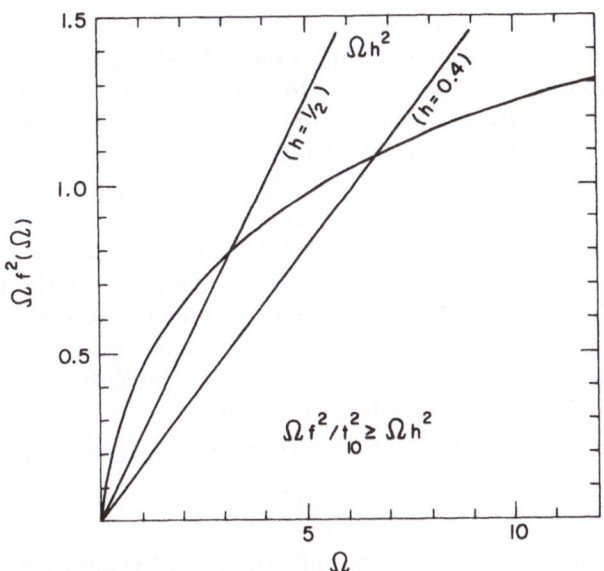

Figure 1.4b: The functions Ωf^2 and Ωh^2 ($h = 0.4$ and 0.5). The function $\Omega f^2 / t_{10}^2$ bounds Ωh^2 from above. For $t_{10} \geq 1$ and $h \geq 0.4(0.5)$, this implies $\Omega h^2 \leq 1.1(0.8)$. The age of the Universe $t_u = t_{10} 10$ Gyr.

A reliable and definitive determination of Ω has thus far eluded cosmologists. Based upon the luminous matter in the Universe (which is relatively easy to keep track of) we can set a lower bound to Ω

$$\Omega \geq \Omega_{LUM} \simeq 0.01$$

Based on dynamical techniques—which all basically involve Kepler's third law in one guise or another, the observational data seem to indicate that the material that clusters with visible galaxies on scales $\leq$ 10-30 Mpc accounts for

$$\Omega_{GAL} \simeq 0.1 - 0.3$$

Although Ω can, in principle, be determined by measurements of the deceleration parameter q_0

$$
\begin{aligned}
q_0 &\equiv -(\ddot{R}/R)/H^2, \\
&= \Omega(1 + 3p/\rho)/2,
\end{aligned}
\tag{1.8}
$$

because of the difficulty of reliably determining q_0, the observations probably only restrict Ω to be less than a few[10]. [For a more thorough discussion of the amount of matter in the Universe see ref. 11.]

The best upper limit to Ω comes from the age of the Universe. The age of the Universe is related to the Hubble time H^{-1} by

$$t_u = f(\Omega)H_0^{-1} \tag{1.9}$$

where $f(\Omega)$ is a monotonically decreasing function of Ω; $f(0) = 1$ and $f(1) = 2/3$ for a matter-dominated Universe and $1/2$ for a radiation-dominated Universe. The dating of the oldest stars and the elements strongly suggest that the Universe is at least 10 Byr old—the best estimate being around 15 Byr old[12]. From Eqn(9) and $t_u \geq t_{10}10$ Byr it follows that $\Omega f^2/t_{10}^2 \geq \Omega h^2$. The function Ωf^2 is monotonically increasing and bounded above by $\pi^2/4$, implying that *independent of h*, $\Omega h^2 \leq 2.5/t_{10}^2$. Requiring $h \geq 0.4$ and $t_{10} \geq 1$, it follows that $\Omega h^2 \leq 1.1$ (see Fig. 1.4).

The energy density of the Universe quite naturally splits up into that contributed by relativistic particles—today the microwave photons and cosmic neutrino backgrounds, and that contributed by non-relativistic particles—baryons and whatever else! The energy density contributed by non-relativistic particles decreases as $R(t)^{-3}$—just due to the increase in the proper volume of the Universe, while that of relativistic particles varies as $R(t)^{-4}$—the additional factor of R being due to the fact that the momenta of relativistic particles are redshifted by the expansion. [Both of these results follow directly from Eqn(1.5).]

The energy density contributed by relativistic particles at temperature T is

$$\rho_R = g_*(T)\frac{\pi^2}{30}T^4 \tag{1.10}$$

where $g_*(T)$ counts the effective number of degrees of freedom (weighted by their temperature) of all the relativistic particle species (those with $m \ll T$):

$$g_*(T) = \sum_{Bose} g_B(T_i/T)^4 + 7/8 \sum_{Fermi} g_F(T_i/T)^4, \tag{1.11}$$

here T_i is the temperature of the species i, and T is the photon temperature.

Today the energy density contributed by relativistic particles (photons and three neutrino species) is very small ($g_* = 3.36$)

$$\Omega_{\gamma\ 3\nu}h^2 \simeq 4 \times 10^{-5}T_{2.7}^4$$

However, because $\rho_R \propto R^{-4}$, while $\rho_{NR} \propto R^{-3}$, at early times the energy density contributed by relativistic particles dominated that of non-relativistic particles. To be specific, the Universe was radiation-dominated for

$$t \leq t_{EQ} \simeq 4 \times 10^{10}\text{sec}(\Omega h^2)^{-2}T_{2.7}^6,$$
$$R \leq R_{EQ} \simeq 4 \times 10^{-5}R_{today}(\Omega h^2)^{-1}T_{2.7}^4,$$
$$T \geq T_{EQ} \simeq 5.8\text{eV}(\Omega h^2)T_{2.7}^3.$$

Therefore, at very early times Eqn(1.4) simplifies to

$$\begin{aligned} H = (\dot{R}/R) &= (4\pi^3 g_*/45)^{1/2}T^2/m_{pl}, \\ &= 1.66g_*^{1/2}T^2/m_{pl} \end{aligned} \tag{1.12}$$

[Note since the curvature term varies as $R(t)^{-2}$ it too is negligible compared to the energy density in relativistic particles.] For reference, $g_*(\text{few MeV}) = 10.75$ (γ, $e^{\pm}$, $3\nu\bar{\nu}$); $g_*(100\text{GeV}) \simeq 110$ (γ, $8G$, $W^{\pm}Z$, 3 families of quarks and leptons, and 1 Higgs doublet).

So long as thermal equilibrium is maintained, the second Friedmann equation, Eqn(1.5), implies that the entropy per comoving volume, $S \propto sR^3$, remains constant. Here s is the entropy density which is dominated by the contribution from relativistic particles, and is

$$s = (\rho + p)/T \simeq (2\pi^2/45)g_*T^3. \tag{1.13}$$

The entropy density is just proportional to the number density of relativistic particles. Today the entropy density is just 7.04 times the number density of photons. The constancy of S means that $s \propto R^{-3}$, or that the ratio of any number density to s is just proportional to the number of that species per comoving volume ($N \propto nR^3 \propto n/s$). The baryon number-to-entropy ratio is

$$n_B/s \simeq (1/7)\eta \simeq (6 - 10) \times 10^{-11}$$

and since today the number density of baryons is much greater than that of antibaryons, this ratio is also the net baryon number per comoving volume—which is conserved so long as the rate of baryon-number non-conserving reactions is small.

The constancy of S implies that

$$T \propto g_*(T)^{-1/3}R(t)^{-1}. \tag{1.14}$$

Whenever g_* is constant, this means that $T \propto R(t)^{-1}$. Together with Eqn(1.12) this gives

$$R(t) = R(t_0)(t/t_0)^{1/2},$$

$$t \simeq 1/2H^{-1} \simeq 0.3g_*^{-1/2}m_{pl}/T^2, \qquad (1.15)$$

$$\simeq 2.4 \times 10^{-6}\text{sec } g_*^{-1/2}(T/\text{GeV})^{-2}.$$

Finally, let me mention one more important feature of the standard cosmology, the existence of particle horizons. In the standard cosmology the distance a photon could have traveled since the bang is finite, meaning that at a given epoch the Universe is comprised of many causally-distinct domains. Photons travel on paths characterized by $ds^2 = 0$; for simplicity and without loss of generality consider a trajectory with $d\varphi = d\theta = 0$. The coordinate distance traversed by a photon since 'the bang' is

$$\int_0^t dt'/R(t')$$

which corresponds to the physical distance (measured at time t)

$$d_H(t) = R(t) \int_0^t dt'/R(t'). \qquad (1.16)$$

If $R(t) \propto t^n$ and $n < 1$, then the horizon distance $d_H(t)$ is finite and $d_H(t) = t/(1-n) = nH^{-1}/(1-n) \simeq t$.

Note that even if $d_H(t)$ diverges (e.g., if $R(t) \propto t^n$ with $n > 1$), the Hubble radius H^{-1} still sets the scale of the 'Physics Horizon'. All physical distances scale with $R(t)$. Thus microphysical processes operating on a timescale $\gtrsim H^{-1}$ will have their effects distorted by the expansion, strongly suggesting that a coherent microphysical process can only operate over a time interval of order H^{-1}. Then, causally-coherent microphysical processes can only operate on distances $\leq$ the Hubble radius, H^{-1}. The intuitive notion that the Hubble radius acts as the 'Physics Horizon' is borne out quantitatively time and time again, and so it is useful to think of H^{-1} as the maximum scale for microphysical processes.

During the radiation-dominated era $n = 1/2$ and $d_H(t) = 2t$; the entropy and baryon number within the horizon at a given time are easily computed:

$$S_{HOR} = (4\pi/3)t^3 s,$$

$$\simeq 0.05g_*^{-1/2}(m_{pl}/T)^3,$$

$$N_{B-HOR} = (n_B/s)S_{HOR},$$

$$\simeq 10^{-12}(m_{pl}/T)^3,$$

$$\simeq 10^{-2}M_\odot(T/\text{MeV})^{-3}.$$

We can compare these numbers to the entropy and baryon number contained within the present horizon volume:

$$S_U \simeq 10^{88},$$

$$N_{BU} \simeq 10^{78}.$$

Evidently, in the standard cosmology the comoving volume which corresponds to the part of the Universe which is presently observable contained many, many horizon volumes at early times. This is an important point to which we shall return shortly.

Although our verifiable knowledge of the early history of the Universe only takes us back to $t \simeq 10^{-2}$ s and $T \simeq 10$ MeV (the epoch of primordial nucleosynthesis), nothing in our present understanding of the laws of physics suggests that it is unreasonable to extrapolate back to times as early as $\simeq 10^{-43}$ s and temperatures as high as $\simeq 10^{19}$ GeV. At high energies the interactions of quarks and leptons are asymptotically free (and/or weak), justifying the dilute gas approximation made in Eqn(1.10). At energies below 10^{19} GeV, quantum corrections to general relativity are expected to be small. I hardly need to remind the reader that 'reasonable' does not necessarily mean 'correct'. Making this extrapolation, I have summarized 'The Complete History of the Universe' in Fig. 1.5.

1.1. PRIMORDIAL NUCLEOSYNTHESIS

At present the most stringent test of the standard cosmology is big bang nucleosynthesis. Here I will briefly review primordial nucleosynthesis, discuss the concordance of the predictions with the observations, and mention one example of how primordial nucleosynthesis has been used as a probe of particle physics—constraining the number of light neutrino species.

Two fundamental assumptions underlie big bang nucleosynthesis: General Relativity is valid and the Universe was once hotter than a few MeV. An additional assumption (which, however, is not necessary) is that the lepton number of the Universe, $n_L/n_\gamma = (n_{e^-} - n_{e^+})/n_\gamma + (n_\nu - n_{\bar\nu})/n_\gamma \simeq \eta + (n_\nu - n_{\bar\nu})/n_\gamma$, like the baryon number ($\eta \simeq 4 - 7 \times 10^{-10}$), is small ($\lesssim 1$). Having swallowed these assumptions, the rest follows like 1-2-3.

Frame 1: $t \simeq 10^{-2}$sec, $T \simeq 10$MeV. The energy density of the Universe is dominated by relativistic species: $\gamma, e^+e^-, \nu_i\bar\nu_i$ ($i = e, \mu, \tau, \ldots$); $g_* \simeq 10.75$ (assuming 3 neutrino species). Thermal equilibrium is maintained by weak interactions ($e^+ + e^- \leftrightarrow \nu_i + \bar\nu_i, e^+ + n \leftrightarrow p + \bar\nu_e, e^- + p \leftrightarrow n + \nu_e$) as well as electromagnetic interactions ($e^+ + e^- \leftrightarrow \gamma + \gamma, \gamma + p \leftrightarrow \gamma + p$, etc.), both of which are occurring rapidly compared to the expansion rate $H = \dot{R}/R$. Thermal equilibrium implies that $T_\nu = T_\gamma$ and that $n/p = \exp(-\Delta m/T)$; here n/p is the neutron to proton ratio and $\Delta m = m_n - m_p$. At high temperatures $\gtrsim 0.3$ MeV) all the light isotopes are in nuclear statistical equilibrium with very small abundances, due to the very high photon to nucleon ratio, $\eta^{-1} \simeq 10^{10}$:

$$n_D/n_B \simeq \eta(T/m_N)^{3/2} \exp(2.2\,\text{MeV}/T),$$
$$\simeq 10^{-13}$$
$$n_{He}/n_B \simeq 0.2\eta^3(T/m_N)^{9/2} \exp(28\,\text{MeV}/T),$$
$$\simeq 4 \times 10^{-39},$$

where the abundances were evaluated for $T \simeq 10$ MeV, and 2.2 MeV and 28 MeV

30

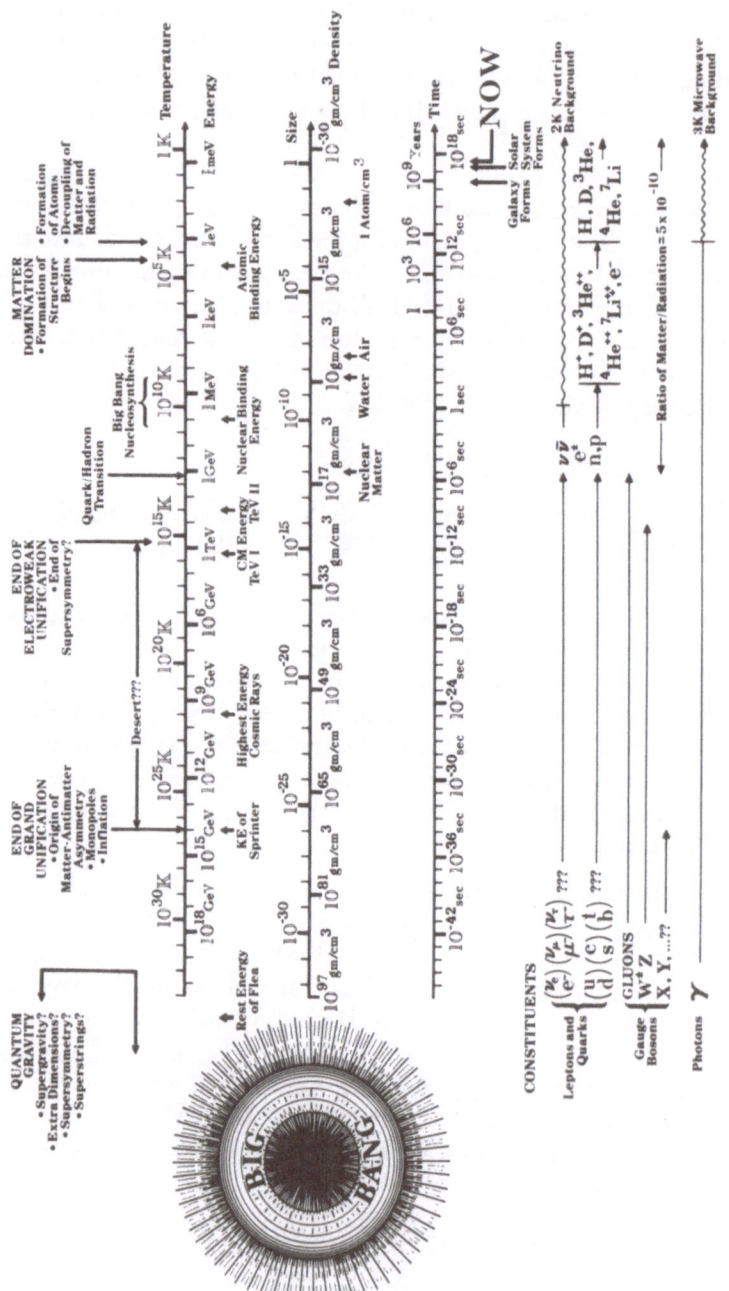

Figure 1.5: 'The Complete History of the Universe' according to the standard cosmology and currently fashionable ideas in elementary particle physics.

are the binding energies of D and ^4He respectively. The fact that very little nucleosynthesis has taken place when $T \simeq 1$ MeV is clearly traceable to the large value of η^{-1}. Of course, had the binding energies of the light nuclei been more like 30-100 MeV rather than 3-30 MeV nucleosynthesis would already be occurring when $T \simeq 10$ MeV. Since D is the first stepping stone in the path of nucleosynthesis, the absence of substantial nucleosynthesis until much lower temperatures is usually blamed on its small equilibrium abundance and binding energy, and this phenomenon is often referred to as 'the deuterium bottleneck'.

Frame 2: $t \simeq 1$ sec, $T \simeq 1$ MeV. At about this temperature the weak interaction rates become slower than the expansion rate and thus weak interactions effectively cease occurring. The neutrinos decouple and thereafter expand adiabatically ($T_\nu \propto R^{-1}$). This epoch is the surface of last scattering for the neutrinos; detection of the cosmic neutrino seas would allow us to directly view the Universe as it was 1 sec after 'the bang'. From this time forward the neutron to proton ratio no longer 'tracks' its equilibrium value, but instead 'freezes out' a value $\simeq 1/6$, decreasing very slowly due to occasional free neutron decays and other β-reactions. A little bit later ($T \simeq m_e/3$), the $e^\pm$ pairs annihilate and transfer their entropy to the photons, heating the photons relative to the neutrinos, so that from this point on $T_\nu \simeq (4/11)^{1/3}T_\gamma$. The so-called 'deuterium bottleneck' continues to operate, preventing nucleosynthesis.

Frame 3: $t \simeq 200$ sec, $T \simeq 0.1$ MeV. If ^4He were to track its equilibrium abundance, then n_{He}/n_B would reach order unity at a temperature of about 0.3 MeV. However, the equilibrium abundances of ^3H, ^3He, and D are too small at this temperature to allow ^4He to be produced rapidly enough to achieve its equilibrium value. At a temperature of about 0.1 MeV, there is sufficient ^3H, ^3He, and D to produce ^4He at a rate comparable to the expansion rate, and nucleosynthesis begins in earnest. Essentially all the neutrons present are quickly incorporated into ^4He nuclei. As the D and ^3He are depleted, the rates at which they are burned into ^4He fall ($\Gamma \propto n(^3He)$ or $n(D)$); eventually they drop below the expansion rate and so trace amounts of D and ^3He remain unburned. Substantial nucleosynthesis beyond ^4He is prevented by the lack of stable isotopes with A = 5 and 8, and by coulomb barriers. A small amount of ^7Li is synthesized by $^4He(t,\gamma)^7Li$ (for $\eta \lesssim 3 \times 10^{-10}$) and by $^4He(^3He,\gamma)^7Be$ followed by the eventual β-decay of ^7Be (via electron capture) to ^7Li (for $\eta \gtrsim 3 \times 10^{-10}$).

The nucleosynthetic yields depend upon η, N_ν (which I will use to parameterize the number of light ($\lesssim 1$ MeV) species present, other than γ and $e^\pm$), and in principle all the nuclear reaction rates which go into the reaction network. In practice, most of the rates are known to sufficient precision that the yields only depend upon a few rates. ^4He production depends only upon η, N_ν, and $\tau_{1/2}$, the neutron half-life, which determines the rates for all the weak processes which interconvert neutrons and protons. The mass fraction Y_p of ^4He produced increases monotonically with increasing values of η, N_ν, and $\tau_{1/2}$—a fact which is simple to understand. Larger η means that the 'deuterium bottleneck' breaks earlier, when the value of

n/p is larger. More light species (i.e., larger value of N_ν) increases the expansion rate (since $H \propto (G\rho)^{1/2}$), while a larger value of $\tau_{1/2}$ means slower weak interaction rates ($\propto \tau_{1/2}^{-1}$)—both effects cause the weak interactions to freeze out earlier, when n/p is larger. The yield of ^4He is determined by the n/p ratio when nucleosynthesis commences, $Y_p \simeq 2(n/p)/(1 + n/p)$, so that a higher n/p ratio means more ^4He is synthesized. At present the value of the neutron half-life is only known to an accuracy of about 2%: $\tau_{1/2} = 10.6 \, \text{min} \pm 0.2 \, \text{min}$. Since ν_e and ν_μ are known (from laboratory measurements) to be light, $N_\nu \geq 2$. Based upon the luminous matter in galaxies, η is known to be $\gtrsim 0.3 \times 10^{-10}$. If all the mass in binary galaxies and small groups of galaxies (as inferred by dynamical measurements) is baryonic, then η must be $\gtrsim 2 \times 10^{-10}$.

To an accuracy of about 10%, the yields of D and ^3He only depend upon η, and decrease rapidly with increasing η. Larger η corresponds to a higher nucleon density and earlier nucleosynthesis, which in turn results in less D and ^3He remaining unprocessed. Because of large uncertainties in the rates of some of the reactions which create and destroy ^7Li, the predicted primordial abundance of ^7Li is only accurate to within about a factor of 2.

In 1946 Gamow[13] suggested the idea of primordial nucleosynthesis. In 1953, Alpher, Follin, and Herman[14] all but wrote a code to determine the primordial production of ^4He. Peebles (in 1966) and Wagoner, Fowler, and Hoyle (in 1967) wrote codes to calculate the primordial abundances[15]. Yahil and Beaudet[16] (in 1976) independently developed a nucleosynthesis code and also extensively explored the effect of large lepton number ($n_\nu - n_{\bar\nu} \simeq 0(n_\gamma)$) on primordial nucleosynthesis. Wagoner's 1973 code[17] has become the 'standard code' for the standard model. In 1981 the reaction rates were updated by Olive et al.[18], the only significant change which resulted was an increase in the predicted ^7Li abundance by a factor of $0(3)$. In 1982 Dicus et al.[19] corrected the weak rates in Wagoner's 1973 code for finite temperature effects and radiative/coulomb corrections, which led to a systematic decrease in Y_p of about 0.003. Figs. 1.3, 1.6 show the predicted abundances of D, ^3He, ^4He, and ^7Li, as calculated by the most up to date version of Wagoner's 1973 code[20]. The numerical accuracy of the predicted abundances is about 1%. Now let me discuss how the predicted abundances compare with the observational data. [This discussion is a summary of the collaborative work in ref. 20.]

The abundance of D has been determined in solar system studies and in UV absorption studies of the local ($\lesssim 200$ pc) interstellar medium (ISM). The solar system determinations are based upon measuring the abundances of deuterated molecules in the atmosphere of Jupiter and inferring the pre-solar (i.e., at the time of the formation of the solar system) D/H ratio from meteoritic and solar data on the abundance of ^3He. These determinations are consistent with a pre-solar value of $(D/H) \simeq (2 \pm 0.5) \times 10^{-5}$. An average ISM value for $(D/H) \simeq 2 \times 10^{-5}$ has been derived from UV absorption studies of the local ISM with individual measurements spanning the range $(1 - 4) \times 10^{-5}$. Note that these measurements are consistent with the solar system determinations of D/H.

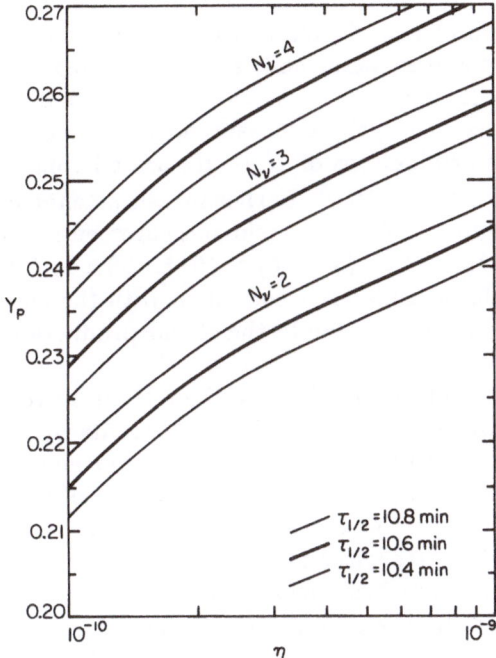

Figure 1.6: The predicted primordial abundance of ^4He. Note that Y_p increases with increasing values of $\tau_{1/2}, \eta$, and N_ν. Hence lower bounds to η and $\tau_{1/2}$ and an upper bound to Y_p imply an upper bound to N_ν. Taking $\tau_{1/2} \geq 10.4$ min, $\eta \geq 4 \times 10^{-10}$ (based on $D + {}^3He$ production), and $Y_p \lesssim 0.25$, it follows that N_ν must be ≤ 4.

The deuteron being very weakly-bound is easily destroyed and hard to produce, and to date, it has been difficult to find an astrophysical site where D can be produced in its observed abundance[21]. Thus, it is generally accepted that the presently-observed deuterium abundance provides a *lower* bound to the primordial abundance. Using $(D/H)_p \gtrsim 1 \times 10^{-5}$ it follows that η must be less than about 10^{-9} in order for the predictions of primordial nucleosynthesis to be concordant with the observed abundance of D. [Note: because of the rapid variation of $(D/H)_p$ with η, this upper bound to η is rather insensitive to the precise lower bound to $(D/H)_p$ used.] This implies an upper bound to $\Omega_b : \Omega_b \lesssim 0.035 h^{-2} T_{2.7}^3 K \leq 0.19$—baryons alone cannot close the Universe. One would like to also exploit the sensitive dependence of $(D/H)_p$ upon η to derive a *lower* bound to η for concordance; this is not possible because D is so easily destroyed. However, as we shall soon see, this end can be accomplished instead by using the abundances of both D and ^3He.

The abundance of ^3He has been measured in solar system studies and by observations of the $^3He^+$ hyperfine line in galactic HII regions (the analog of the 21 cm line of H). The abundance of ^3He in the solar wind has been determined by analyzing gas-rich meteorites, lunar soil, and the foil placed upon the surface of the moon by the Apollo astronauts. Since D is burned to ^3He during the sun's approach to the main sequence, these measurements represent the pre-solar sum of D and ^3He. These determinations of $D + {}^3He$ are all consistent with a pre-solar $[(D + {}^3He)/H] \simeq (4.0 \pm 0.3) \times 10^{-5}$. Earlier measurements of the $^3He^+$ hyperfine line in galactic HII regions and very recent measurements lead to derived present abundances of ^3He : $^3He/H \simeq (3 - 20) \times 10^{-5}$. The fact that these values are higher than the pre-solar abundance is consistent with the idea that the abundance of ^3He should increase with time due to the stellar production of ^3He by low mass stars.

^3He is much more difficult to destroy than D. It is very hard to efficiently dispose of ^3He without also producing heavy elements or large amounts of ^4He (environments hot enough to burn ^3He are usually hot enough to burn protons to ^4He). In ref. 20 we have argued that in the absence of a Pop III generation of very exotic stars which process essentially all the material in the Universe and in so doing destroy most of the ^3He without overproducing ^4He or heavy elements, ^3He can have been astrated (i.e. reduced by stellar burning) by a factor of no more than $f_a \simeq 2$. [The youngest stars, e.g., our sun, are called Pop I; the oldest observed stars are called Pop II. Pop III refers to a yet to be discovered, hypothetical first generation of stars.] Using this argument and the inequality

$$[(D + {}^3He)/H]_p \leq (D/H)_{ps} + f_a({}^3He/H)_{ps}$$
$$\leq (1 - f_a)(D/H)_{ps} + f_a(D + {}^3He/H)_{ps};$$

the pre-solar abundances (noted by 'ps') of D and $D + {}^3He$ can be used to derive an upper bound to the primordial abundance of $D + {}^3He$: $[(D + {}^3He)/H]_p \lesssim 8 \times 10^{-5}$. [For a very conservative astration factor, $f_a \simeq 4$, the upper limit becomes 13×10^{-5}.] Using 8×10^{-5} as an upper bound on the primordial $D + {}^3He$ production implies that for concordance, η must be greater than 4×10^{-10} (for the upper bound of 13×10^{-5}, η must be greater than 3×10^{-10}). To summarize, consistency between the predicted big bang abundances of D and ^3He, and the derived abundances observed today requires η to lie in the range $\simeq (4 - 10) \times 10^{-10}$.

Until very recently, our knowledge of the ^7Li abundance was limited to observations of meteorites, the local ISM, and Pop I stars, with a derived present abundance of $^7Li/H \simeq 10^{-9}$ (to within a factor of 2). Given that ^7Li is produced by cosmic ray spallation and some stellar processes, and is easily destroyed (in environments where $T \gtrsim 2 \times 10^6 K$), there is not the slightest reason to suspect (or even hope!) that this value accurately reflects the primordial abundance. Recently, Spite and Spite[22] have observed ^7Li lines in the atmospheres of 13 unevolved halo and old disk stars (Pop II) with very low metal abundances ($Z \simeq Z_\odot/12 - Z_\odot/250$), whose masses span the range of $\simeq (0.6-1.1)M_\odot$. [Note that $Z \equiv$ mass fraction of metals, a metal being any isotope with $A \geq 4$. $Z_\odot \simeq 0.02$.] Stars less massive than about $0.7M_\odot$ are expected to astrate (by factors $\geq 0(10)$) their ^7Li abundance during their approach to the MS, while stars more massive than about $1M_\odot$ do not seem to significantly astrate ^7Li in their outer layers. Indeed, they see this trend in their data, and deduce a primordial ^7Li abundance of: $^7Li/H \simeq (1.12 \pm 0.38) \times 10^{-10}$. Remarkably, this is the predicted big bang production for η in the range $(2-5) \times 10^{-10}$. If we take this to be the primordial ^7Li abundance, and allow for a possible factor of 2 uncertainty in the predicted abundance of Li (due to estimated uncertainties in the reaction rates which affect ^7Li), then concordance for ^7Li restricts η to the range $(1-7) \times 10^{-10}$. Note, of course, that their derived ^7Li abundance is the pre-Pop II abundance, and may not necessarily reflect the true primordial abundance (e.g., if a Pop III generation of stars processed significant amounts of material).

In sum, the concordance of big bang nucleosynthesis predictions with the derived abundances of D and ^3He requires $\eta \simeq (4-10) \times 10^{-10}$; moreover, concordance for D, ^3He, and ^7Li further restricts η: $\eta \simeq (4-7) \times 10^{-10}$.

In the past few years the quality and quantity of ^4He observations has increased markedly. In Fig. 1.7 all the ^4He abundance determinations derived from observations of recombination lines in HII regions (galactic and extragalactic) are shown as a function of metalicity Z (more precisely, 2.2 times the mass fraction of ^{16}O). [Astronomers refer to ionized hydrogen as 'HII'; an HII region then is a region of ionized hydrogen; typical sizes are 10's of pc and typical temperatures $\gg 10,000$ K.]

Since ^4He is also synthesized in stars, some of the observed ^4He is *not* primordial. Since stars also produce metals, one would expect some correlation between Y and Z, or at least a trend: lower Y where Z is lower. Such a trend is apparent in Fig. 1.7. From Fig. 1.7 it is also clear that there is a large primordial component to ^4He : $Y_p \simeq 0.22-0.26$. Is it possible to pin down the value of Y_p more precisely?

There are many steps in going from the line strengths (what the observer actually measures), to a mass fraction of ^4He (e.g., corrections for neutral ^4He, reddening, etc.). In galactic HII regions, where abundances can be determined for various positions within a given HII region, variations are seen within a given HII region. Observations of extragalactic HII regions are actually observations of a superposition of *several* HII regions. Although observers have quoted statistical uncertainties of $\Delta Y \simeq \pm 0.01$ (or lower), from the scatter in Fig. 1.7 it is clear that the systematic uncertainties must be larger. For example, different observers have derived ^4He abundances of between 0.22 and 0.25 for I Zw18, an extremely metal-poor dwarf emission line galaxy.

Perhaps the safest way to estimate Y_p is to concentrate on the ^4He determinations for metal-poor objects. From Fig. 1.7 $Y_p \simeq 0.23-0.25$ appears to be consistent with all the data (although Y_p as low as 0.22 or high as 0.26 could not be ruled out). Recently Kunth and Sargent[23] have studied 13 metal-poor ($Z \lesssim Z_\odot/5$) Blue Compact galaxies. From a weighted average for their sample they derive a *primordial abundance* $Y_p \simeq 0.245 \pm 0.003$; allowing for a 3σ variation this suggests

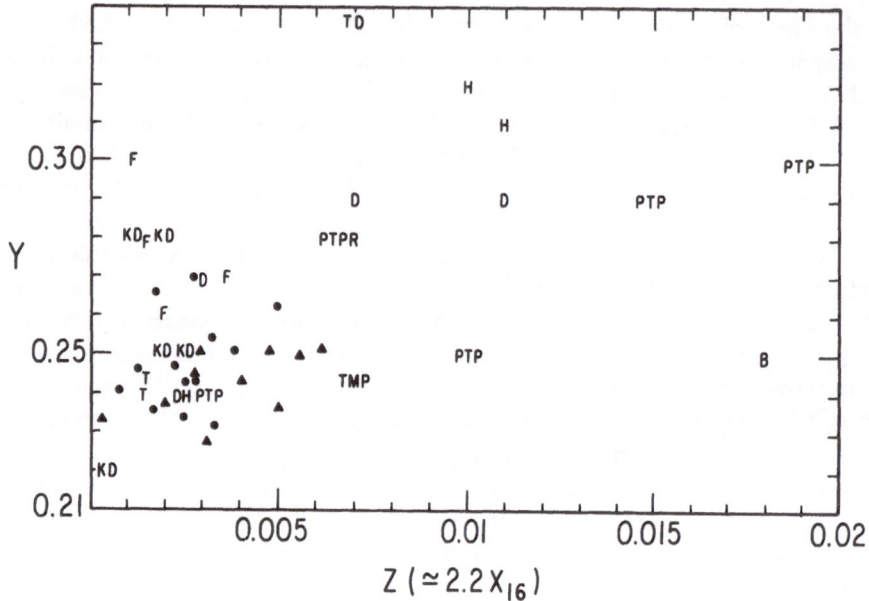

Figure 1.7: Summary of ^4He abundance determinations (galactic and extragalactic) from recombination lines in HII regions vs. mass fraction of heavy elements Z ($\simeq$ 2.2 mass fraction of ^{16}O). Note, observers do not usually quote errors for individual objects—scatter is probably indicative of the uncertainties. The triangles and filled circles represent two data sets of note: circles $\simeq$ 13 very metal poor emission line galaxies (Knuth and Sargent[23]); triangles - 9 metal poor, compact galaxies (Lequeux et al.[23]).

$0.236 \leq Y_p \leq 0.254$.

For the concordance range deduced from D, ^3He, and ^7Li ($\eta \geq 4 \times 10^{-10}$) and $\tau_{1/2} \geq 10.4$ min, the predicted ^4He abundance is

$$Y_p \geq \begin{cases} 0.230 & N_\nu = 2. \\ 0.244 & N_\nu = 3. \\ 0.256 & N_\nu = 4. \end{cases}$$

[Note, that $N_\nu = 2$ is permitted only if the τ-neutrino is heavy ($\gtrsim$ few MeV) and unstable; the present experimental upper limit to its mass is 160 MeV.] Thus, since $Y_p \simeq 0.23$ - 0.25 (0.22 - 0.26?) there are values of η, N_ν, and $\tau_{1/2}$ for which there is agreement between the abundances predicted by big bang nucleosynthesis and the primordial abundances of D, ^3He, ^4He, and ^7Li derived from observational data.

To summarize, the only isotopes which are predicted to be produced in significant amounts during the epoch of primordial nucleosynthesis are: $D, ^3He, ^4He$, and ^7Li. At present there is concordance between the predicted primordial abundances of all four of these elements and their observed abundances for values of $N_\nu, \tau_{1/2}$, and η in the following intervals: $2 \leq N_\nu \leq 4$; $10.4 \text{ min} \leq \tau_{1/2} \leq 10.8 \text{ min}$; and $4 \times 10^{-10} \leq \eta \leq 7 \times 10^{-10}$ (or 10×10^{-10} if the ^7Li abundance is not used). This is a truly remarkable achievement—note that the predicted abundances span some nine orders of magnitude—and strong evidence that the standard model is valid back as early as 10^{-2} sec after 'the bang'.

The standard model will be in serious straits if the primordial mass fraction of ^4He is unambiguously determined to be less than 0.22. What alternatives exist if $Y_p \leq 0.22$? If a generation of Pop III stars that efficiently destroyed ^3He and ^7Li existed, then the lower bound to η based upon D, ^3He, (and ^7Li) no longer exists. The only solid lower bound to η would then be that based upon the amount of luminous matter in galaxies (i.e., the matter inside the Holmberg radius): $\eta \geq 0.3 \times 10^{-10}$. In this case the predicted Y_p could be as low as 0.15 or 0.16. Although small amounts of anisotropy increase[24] the primordial production of ^4He, recent work[25] suggests that larger amounts could decrease the primordial production of ^4He. Another possibility is neutrino degeneracy; a large lepton number ($n_\nu - n_{\bar\nu} \simeq 0(n_\gamma)$) drastically modifies the predictions of big bang nucleosynthesis[26]. Finally, one might be forced to discard the standard cosmology altogether. [For the most up-to-date review of primordial nucleosynthesis see ref. 27.]

1.2. PRIMORDIAL NUCLEOSYNTHESIS AS A PROBE

If, based upon its apparent success, we accept the validity of the standard model, we can use primordial nucleosynthesis as a very powerful probe of cosmology and particle physics. For example, concordance requires: $4 \times 10^{-10} \lesssim \eta \lesssim 7 \times 10^{-10}$ and $N_\nu \leq 4$. This is the most precise determination we have of η and implies that

$$0.014h^{-2}T_{2.7}^3 \lesssim \Omega_b \lesssim 0.024h^{-2}T_{2.7}^3 \qquad (1.17)$$

$$0.014 \leq \Omega_b \lesssim 0.14 \qquad (1.18)$$

$$n_B/s \simeq n/7 \simeq (6-10) \times 10^{-11}.$$

If, as some dynamical studies suggest, $\Omega > 0.14$, then some other non-baryonic form of matter must account for the difference between Ω and Ω_b. [For a recent review of the dynamical measurements of Ω, see refs. 28, 29.] Numerous candidates have been proposed for the dark matter, including primordial black holes, axions, quark nuggets, photinos, gravitinos, relativistic debris, massive neutrinos, sneutrinos, monopoles, pyrgons, maximons, etc. [A discussion of some of these candidates is given in ref. 30.]

With regard to the limit on N_ν, Schvartsman[31] first emphasized the dependence of the yield of ^4He on the expansion rate of the Universe during nucleosynthesis, which in turn is determined by g_*, the effective number of massless degrees of freedom. As mentioned above the crucial temperature for ^4He synthesis is $\simeq 1\,\text{MeV}$— the freeze out temperature for the n/p ratio. At this epoch the massless degrees of freedom include: $\gamma, \nu\bar\nu, e^\pm$ pairs, and any other light particles present, and so

$$g_* = g_\gamma + 7/8(g_{e^\pm} + N_\nu g_{\nu\bar\nu}) + \sum_{Bose} g_i(T_i/T)^4 + 7/8 \sum_{Fermi} g_i(T_i/T)^4$$

$$= 5.5 + 1.75 N_\nu + \sum_{Bose} g_i (T_i/T)^4 + 7/8 \sum_{Fermi} g_i (T_i/T)^4. \tag{1.19}$$

Here T_i is the temperature of species i, T is the photon temperature, and the total energy density of relativistic species is: $\rho = g_* \pi^2 T^4/30$. The limit $N_\nu \leq 4$ is obtained by assuming that the only species present are: γ, $e^\pm$, and N_ν neutrinos species, and follows because for $\eta \geq 4 \times 10^{-10}$, $\tau_{1/2} \geq 10.4 min$, and $N_\nu \geq 4$, the mass fraction of ^4He produced is ≥ 0.254 (which is greater than the observed abundance). More precisely, $N_\nu \leq 4$ implies

$$g_* \lesssim 12.5 \tag{1.20}$$

or

$$1.75 \gtrsim 1.75(N_\nu - 3) + \sum_{Bose} g_i (T_i/T)^4 + 7/8 \sum_{Fermi} g_i (T_i/T)^4. \tag{1.21}$$

At most one additional light ($\lesssim$ MeV) neutrino species can be tolerated; many more additional species can be tolerated if their temperatures T_i are $< T$. [Big bang nucleosynthesis limits on the number of light ($\lesssim$ MeV) species have been derived and/or discussed in refs. 32.]

The number of neutrino species can also be determined by measuring the width of the Z° boson: each neutrino flavor less massive than $0(M_Z/2)$ contributes $\simeq$ 190 MeV to the width of the Z°. Preliminary results on the width of the Z° imply that $N_\nu \lesssim 0(10)$[33] and $e^+ e^-$ collider experiments looking for $e^+ e^- \rightarrow X + \gamma$ ($X =$ undetected) also imply $N_\nu \lesssim 0(10)$. Note that while big bang nucleosynthesis and the width of the Z° both provide information about the number of neutrino flavors, they 'measure' slightly different quantities. Big bang nucleosynthesis is sensitive to the number of light ($\lesssim$ MeV) neutrino species, and all other light degrees of freedom, while the width of the Z° is determined by the number of particles less massive than about 50 GeV which couple to the Z° (neutrinos among them). This issue has been recently reviewed in ref. 34.

Given the important role occupied by big bang nucleosynthesis, it is clear that continued scrutiny is in order. The importance of new observational data cannot be overemphasized: extragalactic D abundance determinations (Is the D abundance universal? What is its value?); more measurements of the ^3He abundance (What is its primordial value? Is it true that ^3He has not been significantly astrated?); continued improvement in the accuracy of ^4He abundances in very metal poor objects (Recall, the difference between $Y_p = 0.22$ and $Y_p = 0.23$ is crucial); and further study of the ^7Li abundance in very old stellar populations (Has the primordial abundance of ^7Li already been measured?). Data from particle physics will prove useful too: a high precision determination of $\tau_{1/2}$ (i.e., $\Delta \tau_{1/2} \leq \pm 0.05$ min) will all but eliminate the uncertainty in the predicted ^4He primordial abundance (in the standard cosmology); an accurate measurement of the width of the recently-found Z° vector boson will determine the total number of neutrino species (less massive than about 50 GeV) and thereby bound the total number of light neutrino species.

All these data will not only make primordial nucleosynthesis a more stringent test of the standard cosmology, but they will also make primordial nucleosynthesis a more powerful probe of the early Universe.

1.3. DEPARTURE FROM THERMAL EQUILIBRIUM, 'FREEZE-OUT' AND THE MAKING OF A RELIC SPECIES

Thus far I have tacitly assumed that the Universe is always in thermal equilibrium. Of course if that were always the case, the Universe would be a very boring place (Fe nuclei at 3K today!). In the strictest sense the Universe can never be in thermal equilibrium because of its expansion. However, if particle reaction rates (Γ) are rapid compared to the expansion rate ($\Gamma \gg H$), then the Universe will pass through a succession of nearly equilibrium states characterized by $T \propto g_*^{-1/3} R^{-1}$. Interestingly, a massless (or very relativistic, $T \gg m$), non-interacting species which initially has an equilibrium phase space distribution will continue to do so with a temperature $T \propto R^{-1}$. The same is true for a very massive (i.e., $m \gg T$), non-interacting species, except that $T \propto R^{-2}$. [Both of these facts are straightforward to verify and follow from $p \propto R^{-1}$ and $V \propto R^{-3}$.]

Photons and ionized matter remain in thermal equilibrium until $T \simeq \frac{1}{3}$eV, $R \simeq 10^{-3} R_{today}$, and $t \simeq 6.5 \times 10^{12} (\Omega h^2)^{-1/2}$ sec, when it becomes energetically favorable for the ions and electrons present to form neutral atoms (at which time the scattering cross section for photons drops precipitously from the Thomson cross section $\sigma_T \simeq \alpha^2/m_e^2 \simeq 0.66 \times 10^{-24}$cm^2). This is the so-called decoupling or recombination epoch. After decoupling, the photons freely expand, and the expansion preserves their thermal distribution with a temperature $T \propto R^{-1}$.

A given particle species can only remain in 'good thermal contact' with the photons if the reactions that are important for keeping it in thermal equilibrium are occurring rapidly compared to the rate at which T is decreasing (which is set by the expansion rate $-\dot{T}/T \simeq \dot{R}/R = H$). Roughly-speaking the criterion is

$$\Gamma \gtrsim H, \tag{1.22}$$

where $\Gamma = n\langle\sigma v\rangle$ is the interaction rate per particle, n is the number density of target particles and $\langle\sigma v\rangle$ is the thermally-averaged interaction cross section. When Γ drops below H, that reaction is said to 'freeze-out' or 'decouple'. The temperature T_f (or T_d) at which $H = \Gamma$ is called the freeze-out or decoupling temperature. [Note that if $\Gamma = aT^n (n > 2)$ and the Universe is radiation-dominated so that $H = (2t)^{-1} \simeq 1.67 g_*^{1/2} T^2/m_{pl}$, then the total number of interactions which occur for $T \lesssim T_f$ is just: $\int_{T_f}^{o} \Gamma dt \simeq (\Gamma/H)|_{T_f}/(n-2) \simeq (n-2)^{-1}$]. If the species in question is very relativistic ($T_f \gg m_i$) (or very non-relativistic ($T_f \ll m$)) when it decouples, then its phase space distribution (in momentum space) remains thermal (i.e., Bose-Einstein or Fermi-Dirac) with a temperature $T_i \propto R^{-1}$ (or $T_i \propto R^{-2}$).

[It is interesting to note that based upon just the known interactions, one would not expect the Universe to be in thermal equilibrium during its earliest epochs. At high temperatures, the cross section for renormalizable interactions which transfer significant momentum ($\simeq$ few T) and which are mediated by light gauge bosons scale as: $\sigma \simeq \alpha^2/T^2$ ($\alpha =$ appropriate gauge coupling constant), and the number density of particles $n \simeq g_* T^3$. Taking the expansion rate to be: $H \simeq g_*^{1/2} T^2/m_{pl}$, it follows that $(\Gamma \simeq n\langle\sigma v\rangle) \geq H$, only for $T \lesssim g_*^{1/2}\alpha^2 m_{pl} \simeq 10^{16}\text{GeV}$ (for $g_*^{1/2}\alpha^2 \simeq 10^{-3}$).]

Now consider the evolution of the temperature of a decoupled, relativistic species relative to that of the photons. For the decoupled species $T_i \propto R^{-1}$. However, due to the entropy release when various massive species annihilate (e.g., $e^\pm$ pairs when $T \simeq 0.1\,\text{MeV}$), the photon temperature does not always decrease as R^{-1}. [More precisely, when various massive species transfer their entropy to the EM plasma.] Entropy conservation ($S \propto g_* T^3 =$ constant) of course, can be used to calculate its evolution; if g_* is decreasing, then T will decrease less rapidly than R^{-1}. As an example consider neutrino freeze-out. The cross section for processes like $e^+ e^- \leftrightarrow \nu\bar{\nu}$ is: $\langle\sigma v\rangle \simeq 0.2 G_F^2 T^2$ (here $G_F \simeq 1.1 \times 10^{-5}\,\text{GeV}^{-2}$ is the Fermi coupling constant), and the number density of targets $n \simeq T^3$, so that $\Gamma \simeq 0.2 G_F^2 T^5$. Equating this to H it follows that

$$T_f \simeq (30 m_{pl}^{-1} G_F^{-2})^{1/3} \tag{1.23}$$

$$\simeq \text{few}\,\text{MeV}, \tag{1.24}$$

i.e., neutrinos freeze out before $e^\pm$ annihilations and do not share in subsequent entropy transfer. For $T \lesssim$ few MeV, neutrinos are decoupled and $T_\nu \propto R^{-1}$, while the entropy density in $e^\pm$ pairs and γ's $\propto R^{-3}$. Using the fact that before $e^\pm$ annihilation the entropy density of the $e^\pm$ pairs and γ's is: $s \propto (7 g_{e^\pm}/8 + g_\gamma)T^3 = 5.5 T^3$ and that after $e^\pm$ annihilation $s \propto g_\gamma T^3 = 2T^3$, it follows that after the $e^\pm$ annihilations

$$T_\nu/T = [g_\gamma/(g_\gamma + 7 g_{e^\pm}/8)]^{1/3}$$
$$= (4/11)^{1/3}. \tag{1.25}$$

Similarly, the temperature at the time of primordial nucleosynthesis T_i of a species which decouples at an arbitrary temperature T_d can be calculated:

$$T_i/T = [(g_\gamma + \frac{7}{8}(g_{e^\pm} + N_\nu g_{\nu\bar{\nu}}))/g_{*d}]^{1.3}$$
$$\simeq (10.75/g_{*d})^{1/3} \qquad (\text{for } N_\nu = 3). \tag{1.26}$$

Here $g_{*d} = g_*(T_d)$ is the number of species in equilibrium when the species in question decouples. Species which decouple at a temperature $30\,\text{MeV} \simeq m_\mu/3 \lesssim T \lesssim$ few $100\,\text{MeV}$ do not share in the entropy release from $\mu^\pm$ annihilations, and $T_i/T \simeq 0.91$; the important factor for limits based upon primordial nucleosynthesis $(T_i/T)^4 \simeq 0.69$. Species which decouple at temperatures $T_d \gtrsim \Lambda_{QCD}$, the temperature of the quark/hadron transition $\simeq$ few $100\,\text{MeV}$, do not share in the entropy transfer when the quark-gluon plasma $[g_* \simeq g_\gamma + g_{Gluon} + \frac{7}{8}(g_{e^\pm} + g_{\mu^\pm} + g_{\nu\bar{\nu}} + g_{u\bar{u}} + g_{d\bar{d}} + g_{s\bar{s}} + \ldots) \gtrsim 62]$ hadronizes, and $T_i/T \simeq 0.56$; $(T_i/T)^4 \simeq 0.10$.

'Hot' relics

Consider a stable particle species X which decouples at a temperature $T_f \gg m_x$. For $T < T_f$ the number density of $X's, n_x$, just decreases as R^{-3} as the Universe expands. In the absence of entropy production the entropy density s also decreases as R^{-3}, and hence the ratio n_x/s ($\propto$ number of X's per comoving volume) remains constant. At freeze-out

$$n_x/s = \frac{(g_{xeff}\varsigma(3)/\pi^2)}{(2\pi^2 g_{*d}/45),}$$
$$\simeq \frac{0.278 g_{xeff}}{g_{*d},} \tag{1.27}$$

where $g_{xeff} = g_x$ for a boson or $3g_x/4$ for a fermion, $g_{*d} = g_*(T_d)$, and $\varsigma(3) = 1.20206\ldots$. Today $s \simeq 7.04 n_\gamma$, so that the number density and mass density of X's are

$$n_x \simeq (2g_{xeff}/g_{*d})n_\gamma, \tag{1.28}$$
$$\Omega_x = \rho_x/\rho_c \simeq 7.6(m_x/100eV)(g_{xeff}/g_{*d})h^{-2}T_{2.7}^3. \tag{1.29}$$

Note, that if the entropy per comoving volume s has increased since the X decoupled, e.g., due to entropy production in a phase transition, then these values are decreased by the same factor that the entropy increased. As discussed earlier, Ωh^2 must be $\lesssim 0(1)$, implying that for a *stable* particle species

$$m_x/100eV \lesssim 0.13 g_{*d}/g_{xeff}; \tag{1.30}$$

for a neutrino species: $T_d \simeq$ few MeV$, g_{*d} \simeq 10.75, g_{xeff} = 2 \times (3/4)$, so that $n_{\nu\bar\nu}/n_\gamma \simeq 3/11$ and m_ν must be $\lesssim 96eV$. Note that for a species which decouples very early (say $g_{*d} = 200$), the mass limit (1.7 keV for $g_{xeff} = 1.5$) which $\propto g_{*d}$ is much less stringent.

Constraint (1.24) obviously does not apply to an unstable particle with $\tau < 10 - 15$ billion yrs. However, any species which decays radiatively is subject to other very stringent constraints, as the photons from its decays can have various unpleasant astrophysical consequences, e.g., dissociating D, distorting the microwave background, 'polluting' various diffuse photon backgrounds, etc. The astrophysical/cosmological constraints on the mass/lifetime of an unstable neutrino species and the photon spectrum of the Universe are shown in Figs. 1.8, 1.9.

'Cold' relics

Consider a stable particle species which is still coupled to the primordial plasma ($\Gamma > H$) when $T \simeq m_x$. As the temperature falls below m_x, its equilibrium abundance is given by

$$n_x/n_\gamma \simeq \left(\frac{g_x}{2\varsigma(3)}\right)\left(\frac{\pi}{8}\right)^{1/2}\left(\frac{m_x}{T}\right)^{3/2}\exp\left(-\frac{m_x}{T}\right), \tag{1.31}$$

$$n_x/s \simeq 0.14\left(\frac{g_x}{g_*}\right)\left(\frac{m_x}{T}\right)^{3/2}\exp\left(-\frac{m_x}{T}\right), \tag{1.32}$$

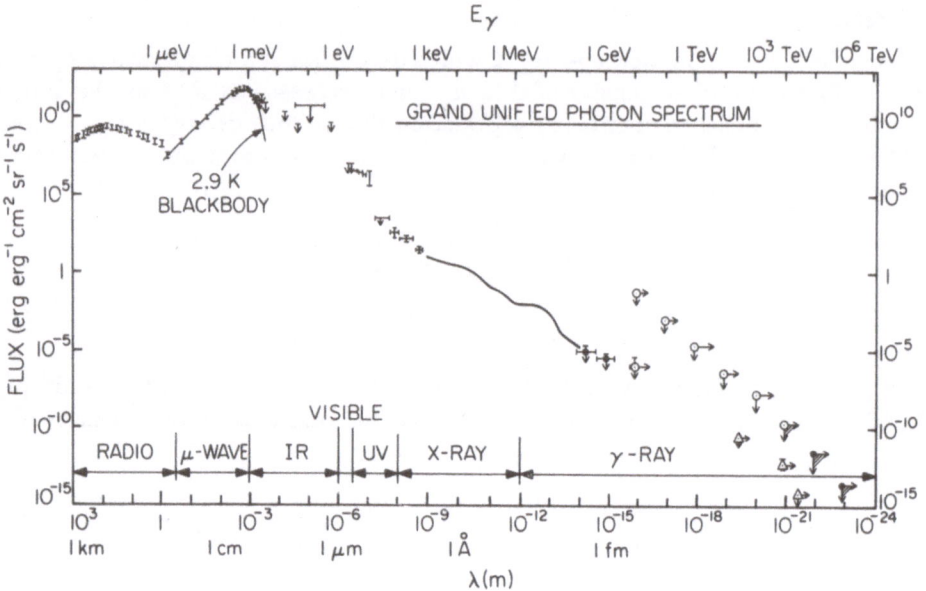

Figure 1.8: The diffuse photon spectrum of the Universe from $\lambda = 1$ km to 10^{-24} m. Vertical arrows indicate upper limits; horizontal arrows indicate integrated flux, i.e., Flux $(> E)$.

and in order to maintain an equilibrium abundance X's must diminish in number (by annihilations since by assumption the X is stable). So long as $\Gamma_{ann} \simeq n_x \langle \sigma v \rangle_{ann} \gtrsim H$ the equilibrium abundance of X's is maintained. When $\Gamma_{ann} \simeq H$, when $T = T_f$, the X's 'freeze-out' and their number density n_x decreases only due to the volume increase of the Universe, so that for $T \lesssim T_f$:

$$n_x/s \simeq (n_x/s)|_{T_f}. \tag{1.33}$$

The equation for freeze-out $(\Gamma_{ann} \simeq H)$ can be solved approximately, giving

$$\frac{m_x}{T_f} \simeq \ln[0.04(\sigma v)_o m_x m_{pl} g_x g_*^{-1/2}]$$

$$+ (\frac{1}{2} - n) \ln\{[0.04(\sigma v)_o m_x m_{pl} g_x g_*^{-1/2}]\},$$

$$\simeq 39 + \ln[(\sigma v)_o m_x] + (\frac{1}{2} - n) \ln[39 + \ln[(\sigma v)_o m_x], \tag{1.34}$$

$$\frac{n_x}{s} \simeq \frac{5\{\ln[0.04(\sigma v)_o m_x m_{pl} g_x g_*^{-1/2}]\}^{1+n}}{[(\sigma v)_o m_x m_{pl} g_*^{1/2}]},$$

$$\simeq \frac{4 \times 10^{-19}\{39 + \ln[(\sigma v)_o m_x]\}^{1+n}}{[(\sigma v)_o m_x g_*^{1/2}]} \tag{1.35}$$

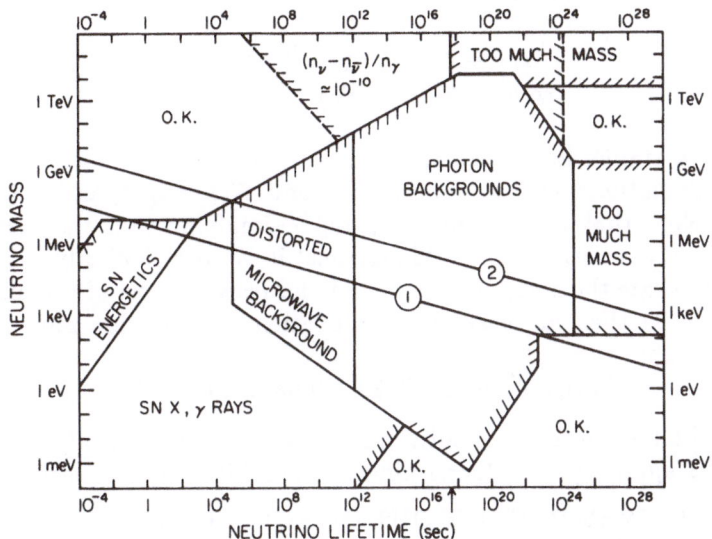

Figure 1.9: Summary of astrophysical/cosmological constraints on neutrino masses/lifetimes[38]. Lines 1 and 2 represent mass/lifetime relationships: $\tau = a \times 10^{-6} \sec(m_\mu/m_\nu)^5$, for $a = 1$, 10^{12}.

where $\langle \sigma v \rangle_{ann}$ is taken to be $(\sigma v)_o (T/m_x)^n$ ($n = 0$ corresponds to s-wave annihilation, $n = 1$ to p-wave, etc.). In the second form of each equation $g_x = 2, g_* \simeq 100$, and all dimensional quantities are to be measured in GeV units.

[The 'correct way' to solve for n_x/s is to integrate the Boltzmann equation which governs the X abundance, $d/dt(n_x/s) = -\langle \sigma v \rangle s[(n_x/s)^2 - (n_{xeq}/s)^2]$. This has been done carefully in ref. 35 (also see ref. 36), and the 'freeze-out' approximation used in Eqns(1.33–1.35) is found to be reasonably good. As discussed in ref. 35 a more accurate analytic approximation gives

$$\frac{m_x}{T_f} \simeq \ln[0.04(n+1)(\sigma v)_o m_x m_{pl} g_x g_*^{-1/2}]$$

$$- (n + \frac{1}{2}) \ln\{\ln[0.04(n+1)(\sigma v)_o m_x m_{pl} g_x g_*^{-1/2}]\},$$

$$\frac{n_x}{s} \simeq \frac{3.79(n+1)(m_x/T_f)^{n+1}}{[(\sigma v)_o m_x m_{pl} g_*^{1/2}]}.$$

This analytic approximation for n_x/s agrees with the numerical results for n_x/s to better than 5%.]

As an example, consider a 'heavy' (Dirac) neutrino species ($m_x \gg$ MeV), for which $\langle \sigma v \rangle \simeq 0(1)m_x^2 G_F^2$. In the absence of annihilations this species would decouple at $T \simeq$ few MeV which is $\ll m_x$, and so the X will become a 'cold relic'. Using

Eqns(1.34, 1.35), we find that today:

$$n_x/s \simeq 5 \times 10^{-9}(m_x/\text{GeV})^{-3}, \tag{1.36}$$

$$\Omega_x h^2 \simeq 2(m_x/\text{GeV})^{-2}, \tag{1.37}$$

implying that a stable, heavy neutrino species must be *more* massive than a few GeV. [This calculation was first done by Lee and Weinberg[37], and independently by Kolb[37].] Note that $\rho_x \propto n_x m_x \propto (\sigma v)_o^{-1}$—implying that the more weakly-interacting a particle is, the more 'dangerous' it is cosmologically. If a particle species is to saturate the mass density bound and provide most of the mass density today ($\Omega_x h^2 \simeq 1$) then its mass and annihilation cross section must satisfy the relation:

$$(\sigma v)_o g_*^{1/2} \simeq 10^{-10}\{39 + \ln[m_x(\sigma v)_o]\}^{1+n} \tag{1.38}$$

where as usual all dimensional quantities are in GeV units. Note also that the relic abundance of a species ($\Omega_x h^2$) 'determines' its annihilation cross section!

1.4. PARTICLES WHICH DECAY OUT-OF-EQUILIBRIUM

Consider a particle species 'X', whose abundance per comoving volume, n_X/s, has frozen out at a value $n_X/s = r$, due to the ineffectiveness of its annihilations. If the species X is not stable, then eventually it will decay when the age of the Universe is about equal to its lifetime, $\tau \equiv \Gamma^{-1}$. In the process of decaying X particles will produce entropy, thereby increasing the entropy per comoving volume, S, of the Universe.

Suppose the X particles become NR before they decay. Until they decay their energy density relative to that of relativisitc particles steadily increases, $\rho_X/\rho_{rad} \propto R(t)$. At the temperature of order $r m_X$ the energy density in X's begins to exceed that in radiation. If this occurs before they decay, then when they decay, X decays will significantly increase the entropy of the Universe.

The equations which govern this process are

$$\dot{\rho} = -3H\rho_X - \Gamma\rho_X, \tag{1.39}$$
$$(\Rightarrow \rho_X = \rho_{X0}(R/R_0)^{-3}\exp[-\Gamma(t - t_0)])$$
$$\dot{S} = (2\pi^2 g_*/45)^{1/3}\Gamma R^4 \rho_X/S^{1/3}, \tag{1.40}$$
$$H^2 = 8\pi G(\rho_{rad} + \rho_X)/3. \tag{1.41}$$

Eqn(1.39) is the usual exponential decay law for particle decays (where I am assuming that the X's are NR). Eqn(1.40) follows from the first law of thermodynamics, $dS = dQ/T$, where dQ, the heat released due to decays per comoving volume, is $R^3\rho_X dt$, and I have assumed that the decay products are relativistic and rapidly thermalize so that $S = 4\rho_{rad}/3T = (2\pi^2 g_*/45)T^3$. [If g_* is constant, then Eqn(1.40) can also be written as: $\dot{\rho}_{rad} = -4H\rho_{rad} + \Gamma\rho_X$.] Eqn(1.41) is the usual Friedmann equation and I have assumed that only X's and R particles contribute significantly to the energy density.

This set of equations is straightforward to solve. Let me discuss the case where the entropy release is significant (final entropy/initial entropy $\equiv S_f/S_0 \gg 1$). In this case the temperature of the Universe decreases as $g_*^{-1/3}R^{-1}$ until the energy released by decays per expansion time becomes comparable to $\rho_{rad}(\Gamma t\rho_X \simeq \rho_{rad})$. After this time the energy density in particles produced by X decays exceeds that in the original radiation component and $\rho_{rad} \propto t^{-1}$, $T \propto R^{-3/8}$. The temperature at the decay epoch ($t \simeq \Gamma^{-1}$; for $t \geq \Gamma^{-1}$, the X's rapidly disappear) is

$$T_{decay} \simeq 0.6g_*^{-1/4}(\Gamma m_{pl})^{1/2},$$

which is a factor of order $g_*^{1/12}(rm_x)^{1/3}/(\Gamma m_{pl})^{1/6}$ greater than what it would have been in the absence of X decays. Contrary to intuition, however, the Universe is not suddenly 'heated up' at $t \simeq \tau = \Gamma^{-1}$; rather, due to the exponential decay law, the entropy produced by X decays is steadily released, resulting in the Universe cooling more slowly, $T \propto R^{-3/8}$ (compared to the usual R^{-1}). The increase in entropy per comoving volume is given by

$$S_f/S_0 = 1.83g_*^{1/4}(rm_x)/(\Gamma m_{pl})^{1/2}, \tag{1.42}$$

where $g_* \simeq$ the value of g_* when $t \simeq \Gamma^{-1}$.

The equations for particles which decay out of equilibrium will arise in many interesting situations, including the reheating of the new inflationary Universe. A more detailed discussion of decaying particles is given in ref. 102 and in Lecture 3.

LECTURE 2:
BARYOGENESIS

I'll begin by briefly summarizing the evidence for the baryon asymmetry of the Universe and the seemingly insurmountable problems that render baryon symmetric cosmologies untenable. For a more detailed discussion of these I refer the reader to Steigman's review of the subject[39]. For a review of recent attempts to reconcile a symmetric Universe with both baryogenesis and the observational constraints, I refer the reader to Stecker[40].

2.1 EVIDENCE FOR A BARYON ASYMMETRY

Within the solar system we can be very confident that there are no concentrations of antimatter (e.g., antiplanets). If there were, solar wind particles striking such objects would be the strongest γ-ray sources in the sky. Also, NASA has yet to lose a space probe because it annihilated with antimatter in the solar system.

Cosmic rays more energetic than 0(0.1 GeV) are generally believed to be of "extrasolar" origin, and thereby provide us with samples of material from throughout the galaxy (and possibly beyond). The ratio of antiprotons to protons in the cosmic rays is about 3×10^{-4}, and the ratio of anti-^4He to ^4He is less than 10^{-5} (ref. 41). Antiprotons are expected to be produced as cosmic-ray secondaries (e.g. $p + p \rightarrow 3p + \bar{p}$) at about the 10^{-4} level. At present both the spectrum and total flux of cosmic-ray antiprotons are at variance with the simplest model of their production as secondaries. A number of alternative scenarios for their origin have been proposed including the possibility that the detected $\bar{p}$'s are cosmic rays from distant antimatter galaxies. Although the origin of these $\bar{p}$'s remains to be resolved, it is clear that they do not provide evidence for an appreciable quantity of antimatter in our galaxy. [For a review of antimatter in the cosmic rays we refer the reader to ref. 41.]

The existence of both matter and antimatter galaxies in a cluster of galaxies containing intracluster gas would lead to a significant γ-ray flux from decays of π°'s produced by nucleon-antinucleon annihilations. Using the observed γ-ray background flux as a constraint, Steigman[39] argues that clusters like Virgo, which is at a distance ≈ 20 Mpc ($\approx 10^{26}$ cm) and contains several hundred galaxies, must not contain both matter and antimatter galaxies.

Based upon the above-mentioned arguments, we can say that if there exist equal quantities of matter and antimatter in the Universe, then we can be absolutely certain they are separated on mass scales greater than $1 M_\odot$, and reasonably certain they are separated on scales greater than (1-100) $M_{galaxy} = 10^{12} - 10^{14} M_\odot$. As I will discuss below, this fact is virtually impossible to reconcile with a symmetric cosmology.

It has often been pointed out that we derive most of our direct knowledge of the large-scale Universe from photons, and since the photon is a self-conjugate particle we obtain no clue as to whether the source is made of matter or antimatter. Neutrinos, on the other hand, are not self-conjugate, and can in principle reveal information about the matter-antimatter composition of their source. Large neutrino detectors may someday provide direct information about the matter-antimatter composition of the Universe on the largest scales.

Baryons account for only a tiny fraction of the particles in the Universe, the 3K-microwave photons being the most abundant species (yet detected). The number density of 3K photons is $n_\gamma = 399T_{2.7}^3$ cm^{-3}. The baryon density is not nearly as well determined. Luminous matter (baryons in stars) contribute at least 0.01 of closure density ($\Omega_{lum} > 0.01$), and as discussed in Lecture 1 the age of the Universe requires that Ω_{tot} (and Ω_b) must be $< 0(2)$. These direct determinations place the baryon-to-photon ratio $\eta \equiv n_b/n_\gamma$ in the range 3×10^{-11} to 6×10^{-8}. As I also discussed in Lecture 1 the yields of big-bang nucleosynthesis depend directly on η, and the production of amounts of D, ^3He, ^4He, and ^7Li that are consistent with their present measured abundances restricts η to the narrow range (4-7) $\times 10^{-10}$.

Since today it appears that $n_b \gg n_{\bar{b}}, \eta$ is also the ratio of net baryon number to photons. The number of photons in the Universe has not remained constant, but has increased at various epochs when particle species have annihilated (e.g. $e^\pm$ pairs at T ≈ 0.5 MeV). Assuming the expansion has been isentropic (i.e., no significant entropy production), the entropy per comoving volume ($\equiv S \propto sR^3$) has remained constant. The "known entropy" is presently about equally divided between the 3K photons and the three cosmic neutrino backgrounds (e, μ, τ). Taking this to be the present entropy, the ratio of baryon number to entropy is

$$n_B/s \simeq \eta/7 \simeq (6 - 10) \times 10^{-11}, \tag{2.1}$$

where $n_B \equiv n_b - n_{\bar{b}}$ and η is taken to be in the range $(4-7) \times 10^{-10}$. So long as the expansion is isentropic and baryon number is at least effectively conserved this ratio remains constant and is what I will refer to as the baryon number of the Universe. As discussed earlier the net baryon number per comoving volume is $\propto n_B/s$.

Although the matter-antimatter asymmetry appears to be "large" today (in the sense that $n_B \approx n_b \gg n_{\bar{b}}$), the fact that $n_B/s \simeq 10^{-10}$ implies that at very early times the asymmetry was "tiny" $(n_B \ll n_b)$. To see this, let us assume for simplicity that nucleons are the fundamental baryons. Earlier than 10^{-6} s after the bang the temperature was greater than the mass of a nucleon. Thus nucleons and antinucleons should have been about as abundant as photons, $n_N \approx n_{\bar{N}} \approx n_\gamma$. The entropy density s is $\approx g_* n_\gamma \approx g_* n_N \approx 0(10^2) n_N$. The constancy of $n_B/s \approx 0(10^{-10})$ requires that for $t < 10^{-6}s$, $(n_N - n_{\bar{N}})/n_N (\approx 10^2 n_B/s) \approx 0(10^{-8})$. During its earliest epoch, the Universe was nearly (but not quite) baryon symmetric.

2.2 THE TRAGEDY OF A SYMMETRIC COSMOLOGY

Suppose that the Universe was initially locally baryon symmetric. Earlier than 10^{-6} s after the bang nucleons and antinucleons were about as abundant as photons. For $T < 1$ GeV the equilibrium abundance of nucleons and antinucleons is $(n_N/n_\gamma)_{EQ} \approx (m_N/T)^{3/2} \exp(-m_N/T)$, and as the Universe cooled the number of nucleons and antinucleons would decrease tracking the equilibrium abundance as long as the annihilation rate $\Gamma_{ann} \approx n_N \langle \sigma v \rangle_{ann} \approx n_N m_\pi^{-2}$ was greater than the expansion rate H. At a temperature T_f annihilations freeze out $((\Gamma_{ann}/H)|_{T_f} \simeq 1)$, nucleons and antinucleons being so rare they can no longer find each other to annihilate. Using Eqn(1.34) we can compute T_f: $T_f \simeq 0(20\,\text{MeV})$. Because of the incompleteness of the annihilations, residual nucleon and antinucleon to photon ratios (given by Eqn(1.35)) $n_{\bar{N}}/n_\gamma = n_N/n_\gamma \simeq 10^{-18}$ are "frozen in". Even if the matter and antimatter could subsequently be separated, n_N/n_γ is a factor of 10^8 too small. To avoid 'the annihilation catastrophe', matter and antimatter must be separated on large scales before $t \approx 3 \times 10^{-3}s(T \approx 20\,\text{MeV})$. I'll consider two possible mechanisms.

Statistical fluctuations:

One possible mechanism for doing this is statistical (Poisson) fluctuations. The comoving volume that encompasses our galaxy today contains $\simeq 10^{12} M_\odot \approx 10^{69}$ baryons and $\simeq 10^{79}$ photons. Earlier than 10^{-6} s after the bang this same comoving volume contained $\simeq 10^{79}$ photons and $\simeq 10^{79}$ baryons and antibaryons. In order to avoid the annihilation catastrophe, this volume would need an excess of baryons over antibaryons of $\simeq 10^{69}$, but from statistical fluctuations one would expect $N_b - N_{\bar{b}} \approx 0(N_b^{1/2}) \simeq 3 \times 10^{39}$—a mere 29.5 orders of magnitude too small!

A Causal 'Mystery Interaction'

Clearly, statistical fluctuations are of no help, so consider a hypothetical interaction that separates matter and antimatter. In the standard cosmology the distance over which light signals (and hence causal effects) could have propagated since the bang (the horizon distance) is finite and $\simeq 2t$. When $T \approx 20 \, \text{MeV}(t \simeq 10^{-3} \, \text{s})$ causally-coherent regions contained only about $10^{-5} M_\odot$. Thus, in the standard cosmology causal processes could have only separated matter and antimatter into lumps of mass $\lesssim 10^{-5} M_\odot \ll M_{galaxy} \approx 10^{12} M_\odot$. [In the next lecture I will discuss inflationary scenarios; in these scenarios it is possible that the Universe is globally symmetric, while asymmetric locally (within our observable region of the Universe). This is possible because inflation removes the causality constraint.]

It should be clear that the two observations, $n_b \gg n_{\bar{b}}$ on scales at least as large as $10^{12} M_\odot$ and $n_b/n_\gamma \approx (4-7) \times 10^{-10}$, effectively render all baryon-symmetric cosmologies untenable. A viable pre-GUT cosmology needed to have as an initial condition a tiny baryon number, $n_B/s \simeq (6-10) \times 10^{-11}$—a very curious initial condition at that!

2.3. THE INGREDIENTS NECESSARY FOR BARYOGENESIS

More than a decade ago Sakharov[42] suggested that an initially baryon-symmetric Universe might dynamically evolve a baryon excess of $0(10^{-10})$, after which baryon-antibaryon annihilations would destroy essentially all of the antibaryons, leaving the one baryon per 10^{10} photons that we observe today. In his 1967 paper Sakharov outlined the three ingredients necessary for baryogenesis: (a) B-nonconserving interactions; (b) a violation of both C and CP; (c) a departure from thermal equilibrium.

It is clear that B(baryon number) must be violated if the Universe begins baryon symmetric and then evolves a net B. In 1967 there was no motivation for B non-conservation. After all, the proton lifetime is at least 35 orders of magnitude longer than that of any unstable elementary particle—pretty good evidence for B conservation. Of course, grand unification provides just such motivation, and proton decay experiments are likely to detect B nonconservation in the next decade if the proton lifetime is $\lesssim 10^{34}$ years.

Under C (charge conjugation) and CP (charge conjugation combined with parity), the baryon number of a state (B) changes sign. Thus a state that is either C or CP invariant must have $B = 0$. If the Universe begins with equal amounts of matter and antimatter, and without a preferred direction (as in the standard cosmology), then its initial state is both C and CP invariant. Unless both C and CP are violated, the Universe will remain C and CP invariant as it evolves, and thus cannot develop a net baryon number even if B is not conserved. Both C and CP violations are needed to provide an arrow to specify that an excess of matter be produced. C is maximally violated in the weak interactions, and both C and CP are violated in the K° - $\bar{K}^\circ$ system. Although a fundamental understanding of CP violation is still lacking at present, GUTs can (and must) accommodate CP violation. It would be very surprising if CP violation only occurred in the K° -

$\bar{K}^\circ$ system and not elsewhere in the theory also (including the B-nonconserving sector). In fact, without miraculous cancellations the CP violation in the neutral kaon system will give rise to CP violation in the B-nonconserving sector at some level.

The necessity of a departure from thermal equilibrium is a bit more subtle. It is a straightforward exercise to show that CPT and unitary alone are sufficient to guarantee that equilibrium particle phase space distributions are given by: $f(p) = [\exp(\mu/T + E/T) \pm 1]^{-1}$. In equilibrium, processes like $\gamma + \gamma \leftrightarrow b + \bar{b}$ imply that $\mu_b = -\mu_{\bar{b}}$, while processes like (but not literally) $\gamma + \gamma \leftrightarrow b + b$ require that $\mu_b = 0$. Since $E^2 = p^2 + m^2$ and $m_b = m_{\bar{b}}$ by CPT, it follows that in thermal equilibrium, $n_b \equiv n_{\bar{b}}$. [Note, the number density $n = \int d^3 p f(p)/(2\pi)^3$.]

Because the temperature of the Universe is changing on a characteristic timescale H^{-1}, thermal equilibrium can only be maintained if the rates for reactions that drive the Universe to equilibrium are much greater than H. Departures from equilibrium have occurred often during the history of the Universe. For example, because the rate for $\gamma + \text{matter} \rightarrow \gamma' + \text{matter}'$ is $\ll H$ today, matter and radiation are not in equilibrium, and nucleons do not all reside in ^{56}Fe nuclei (thank God!).

2.4. THE STANDARD SCENARIO: OUT-OF-EQUILIBRIUM DECAY

The basic idea of baryogenesis has been discussed by many authors[43-48]. The model that incorporates the three ingredients discussed above and that has become the "standard scenario" is the so-called out-of-equilibrium decay scenario. I now describe the scenario in some detail.

Denote by "X" a superheavy ($\gtrsim 10^{14}$ GeV) boson whose interactions violate B conservation. X might be a gauge or a Higgs boson (e.g., the XY gauge bosons in SU(5), or the color triplet component of the 5 dimensional Higgs). [Scenarios in which the X particle is a superheavy fermion have also been suggested.] Let its coupling strength to fermions be $\alpha^{1/2}$, and its mass be M. From dimensional considerations its decay rate $\Gamma_D = \tau^{-1}$ should be

$$\Gamma_D \approx \alpha M. \tag{2.2}$$

At the Planck time ($\simeq 10^{-43}$ s) let us assume that the Universe is baryon symmetric ($n_B/s = 0$), with all fundamental particle species (fermions, gauge and Higgs bosons) present with equilibrium distributions. At this epoch $T \simeq g_*^{-1/4} m_{pl} \simeq 3 \times 10^{18}$ GeV $\gg M$. (Here I have taken $g_* \simeq 0(100)$; in minimal SU(5) $g_* \approx 160$.) At the Planck time X, $\bar{X}$ bosons should be very relativistic (as $T \gg M$) and up to statistical factors about as abundant as photons: $n_x = n_{\bar{x}} \approx n_\gamma$. Nothing of importance occurs until $T \approx M$.

For $T \lesssim M$ the equilibrium abundance of X, $\bar{X}$ bosons relative to photons is

$$X_{EQ} \simeq (M/T)^{3/2} \exp(-M/T), \tag{2.3}$$

where $X \equiv n_x/n_\gamma$ is just the number of X, $\bar{X}$ bosons per comoving volume. In order for X, $\bar{X}$ bosons to maintain an equilibrium abundance as T falls below M, they must be able to diminish in number rapidly compared to $H = |\dot{T}/T|$. The most important process in this regard is decay; other processes (e.g., annihilation) are higher order in α and self-limiting. If $\Gamma_D \gg H$ for T = M, then X, $\bar{X}$ bosons can adjust their abundance (by decay) rapidly enough so that X "tracks" the equilibrium value. In this case thermal equilibrium is maintained and no asymmetry is expected to evolve.

More interesting is the case where $\Gamma_D < H \approx 1.66 g_*^{1/2} T^2/m_{pl}$ when T = M, or equivalently $M > g_*^{-1/2}\alpha 10^{19}\text{GeV}$. In this case, X, $\bar{X}$ bosons are not decaying on the expansion timescale $(\tau > t)$ and so remain as abundant as photons $(X \simeq 1)$ for $T \lesssim M$; hence they are overabundant relative to their equilibrium number. This overabundance (indicated with an arrow in Fig. 2.1) is the departure from thermal equilibrium. Much later, when $T \ll M$, $\Gamma_D \approx H$ (i.e. $t \approx \tau$), and X, $\bar{X}$ bosons begin to decrease in number as a result of decays. To a good approximation they decay freely since the fraction of fermion pairs with sufficient center-of-mass energy to produce an X or $\bar{X}$ is $\simeq \exp(-M/T) \ll 1$, which greatly suppresses inverse decay processes $(\Gamma_{ID} \approx \exp(-M/T)\Gamma_D \ll H)$. Fig. 2.1 summarizes the time evolution of X; Fig. 2.2 shows the relationship of the various rates $(\Gamma_D, \Gamma_{ID},$ and $H)$ as a function of $M/T(\propto t^{1/2})$.

Now consider the decay of X and $\bar{X}$ bosons: suppose X decays to channels 1 and 2 with baryon numbers B_1 and B_2, and branching ratios r and (1-r). Denote the corresponding quantities for $\bar{X}$ by $-B_1, -B_2, \bar{r},$ and $(1 - \bar{r})$ [e.g., $1 = (\bar{q}\bar{q}), 2 = (q\ell), B_1 = -2/3,$ and $B_2 = 1/3$]. The mean net baryon number of the decay products of the X and $\bar{X}$ are, respectively, $B_x = rB_1 + (1 - r)B_2$ and $B_{\bar{x}} = -\bar{r}B_1 - (1 - \bar{r})B_2$. Hence the decay of an X, $\bar{X}$ pair on average produces a baryon number ε,

$$\varepsilon \equiv B_x + B_{\bar{x}} = (r - \bar{r})(B_1 - B_2). \tag{2.4}$$

If $B_1 = B_2$, or $r = \bar{r}$, $\varepsilon = 0$. If $B_1 = B_2$ X could have been assigned a baryon number B_1, and B would not be violated by X, $\bar{X}$ bosons.

It is simple to show that $r = \bar{r}$ unless both C and CP are violated. Let $\bar{X} =$ the charge conjugate of X, and $r_\uparrow, r_\downarrow, \bar{r}_\uparrow, \bar{r}_\downarrow$ denote the respective branching ratios in the upward and downward directions. [For simplicity, I have reduced the angular degree of freedom to up and down.] The quantities r and $\bar{r}$ are branching ratios averaged over angle: $r = (r_\uparrow + r_\downarrow)/2, \bar{r} = (\bar{r}_\uparrow + \bar{r}_\downarrow)/2$ and $\varepsilon = (r_\uparrow - \bar{r}_\uparrow + r_\downarrow - \bar{r}_\downarrow)/2$. If C is conserved, $r_\uparrow = \bar{r}_\uparrow$ and $r_\downarrow = \bar{r}_\downarrow$, and $\varepsilon = 0$. If CP is conserved $r_\uparrow = \bar{r}_\downarrow$ and $r_\downarrow = \bar{r}_\uparrow$, and once again $\varepsilon = 0$.

When the X, $\bar{X}$ bosons decay $(T \ll M, t \approx \tau)$, $n_x = n_{\bar{x}} \approx n_\gamma$. Therefore, the net baryon number density produced is $n_B \approx \varepsilon n_\gamma$. The entropy density $s \approx g_* n_\gamma$, and so the baryon asymmetry produced is $n_B/s \approx \varepsilon/g_* \approx 10^{-2}\varepsilon$.

Recall that the condition for a departure from equilibrium to occur is: $(\Gamma_D/H)|_{T=M} < 1$ or $M > g_*^{-1/2}\alpha m_{pl}$. If X is a gauge boson then $\alpha \approx 1/45$, and so M must be $\gtrsim 10^{16}\text{GeV}$. If X is a Higgs boson, then α is essentially arbitrary, although $\alpha \approx (m_f/M_W)^2\alpha_{gauge} \approx 10^{-3} - 10^{-6}$ if the X is in the same representation as the light Higgs bosons responsible for giving mass to the fermions

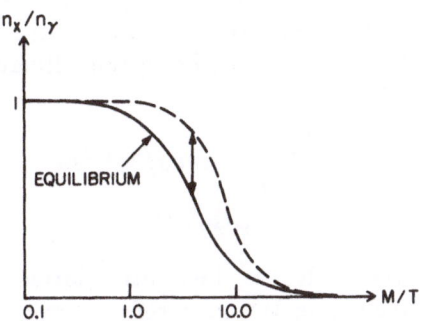

Figure 2.1: The abundance of X bosons relative to photons. The broken curve shows the actual abundance, while the solid curve shows the equilibrium abundance.

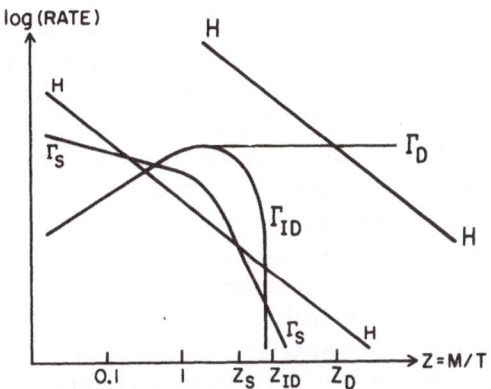

Figure 2.2: Important rates as a function of $z = M/T$. H is the expansion rate, Γ_D the decay rate, Γ_{ID} the inverse decay rate, and Γ_s the $2 \leftrightarrow 2$ $\not\!\!B$ scattering rate. Upper line marked H corresponds to case where $K \ll 1$; lower line the case where $K > 1$. For $K \ll 1$, X's decay when $z = z_D$; for $K > 1$, freeze out of IDs and S occur at $z = z_{ID}$ and z_s.

(here m_f = fermion mass, M_W = mass of the W boson ≈ 83 GeV). It is apparently easier for Higgs bosons to satisfy this mass condition than it is for gauge bosons. If $M > g_*^{-1/2} \alpha m_{pl}$, then only a modest C, CP-violation ($\varepsilon \approx 10^{-8}$) is necessary to explain $n_B/s \approx (6 - 10) \times 10^{-11}$. As I will discuss below ε is expected to be larger for a Higgs boson than for a gauge boson. For both these reasons a Higgs boson is the more likely candidate for producing the baryon asymmetry.

2.5 NUMERICAL RESULTS

Boltzmann equations for the evolution of n_B/s have been derived and solved numerically in refs. 49, 50. They basically confirm the correctness of the qualitative picture discussed above, albeit, with some important differences. The results can best be discussed in terms of

$$K \equiv \frac{\Gamma_D}{2H(M)} \simeq \alpha m_{pl}/3g_*^{1/2}M, \tag{2.5}$$

$$\simeq 3 \times 10^{17}\alpha \text{GeV}/M. \tag{2.6}$$

K measures the effectiveness of decays, i.e., rate relative to the expansion rate. K also measures the effectiveness of B-nonconserving processes in general because the decay rate characterizes the rates in general for B nonconserving processes, for $T \lesssim M$ (when all the action happens):

$$\Gamma_{ID} \simeq (M/T)^{3/2} \exp(-M/T) \, \Gamma_D, \tag{2.7}$$
$$\Gamma_s \simeq A\alpha(T/M)^5 \, \Gamma_D, \tag{2.8}$$

where Γ_{ID} is the rate for inverse decays (ID), and Γ_S is the rate for $2 \leftrightarrow 2$ B nonconserving scatterings (S) mediated by X. [A is a numerical factor which depends upon the number of scattering channels, etc, and is typically 0(100-1000).]

[It is simple to see why $\Gamma_s \propto \alpha(T/M)^5\Gamma_D \propto \alpha^2 T^5/M^4$. $\Gamma_s \simeq n\langle\sigma v\rangle$; $n \simeq T^3$ and for $T < M$, $\langle\sigma v\rangle \propto \alpha T^2/M^4$. Note, in some supersymmetric GUTs, there exist fermionic partners of superheavy Higgs which mediate B-nonconservation (and also lead to dim-5 $\not{B}$ operators). In this case $\langle\sigma v\rangle \propto \alpha^2/M^2$ and $\Gamma_s \simeq A\alpha(T/M)^3\Gamma_D$, and $2 \leftrightarrow 2$ $\not{B}$ scatterings are much more important.]

The time evolution of the baryon asymmetry $(n_B/s$ vs $z = M/T \propto t^{1/2})$ and the final value of the asymmetry which evolves are shown in Figs. 2.3 and 2.4 respectively. For $K < 1$ all B nonconserving processes are ineffective (rate $< H$) and the asymmetry which evolves is just ε/g_* (as predicted in the qualitative picture). For $K_c > K > 1$, where K_c is determined by

$$K_c(\ln K_c)^{-2.4} \simeq 300/A\alpha, \tag{2.9}$$

$2 \leftrightarrow 2$ B nonconserving scatterings 'freeze-out' before ID's do and can be ignored. Equilibrium is maintained to some degree (by Ds and IDs), however a sizeable asymmetry still evolves

$$n_B/s \simeq (\varepsilon/g_*)0.3K^{-1}(\ln K)^{-0.6}. \tag{2.10}$$

This is the surprising result: for $K_c > K \gg 1$, equilibrium is not well maintained and a significant n_B/s evolves, whereas the qualitative picture would suggest that for $K \gg 1$ no asymmetry should evolve. For $K > K_c$, S are very important, and the n_B/s which evolves becomes exponentially small:

$$n_B/s \simeq (\varepsilon/g_*)(AK\alpha)^{1/2} \exp[-4(AK\alpha)^{1/4}/3]. \qquad (2.11)$$

[In supersymmetric models which have dim-5 $\not{B}$ operators, $K_c(\ln K_c)^{-1.2} \simeq 18/A\alpha$ and the analog of Eqn(2.9) for $K > K_c$ is: $n_B/s \simeq (\varepsilon/g_*)A\alpha K \exp[-2(A\alpha K)^{1/2}]$.]

For the XY gauge bosons of SU(5) $\alpha \simeq 1/45, A \simeq$ few $\times 10^3$, and $M \simeq$ few $\times 10^{14} GeV$, so that $K_{XY} \simeq 0(30)$ and $K_c \simeq 100$. The asymmetry which could evolve due to these bosons is $\simeq 10^{-2}(\varepsilon_{XY}/g_*)$. For a color triplet Higgs $\alpha_H \simeq 10^{-3}$ (for a top quark mass of 40 GeV) and $A \simeq$ few $\times 10^3$, leading to $K_H \simeq 3 \times 10^{14}$ GeV$/M_H$ and $K_c \simeq$ few $\times 10^3$. For $M_H \lesssim 3 \times 10^{14}$ GeV, $K_H < 1$ and the asymmetry which could evolve is $\simeq \varepsilon_H/g_*$.

2.6. VERY OUT-OF-EQUILIBRIUM DECAY

If the X boson decays very late, when $M \gg T$ and $\rho_x > \rho_{rad}$, the additional entropy released in its decays must be taken into account. This is very easy to do. Before the Xs decay, $\rho = \rho_x + \rho_{rad} \simeq \rho_x = Mn_x$. After they decay $\rho_x \simeq \rho_{rad} = \pi^2 g_* T_{RH}^4/30 \simeq 0.75 s T_{RH}$ (s, T_{RH} are the entropy density and temperature after the X decays). As usual assume that on average each decay produces a mean net baryon number ε. Then the result in n_B/s produced is

$$\frac{n_B}{s} = \frac{\varepsilon n_x}{s}, \qquad (2.12)$$

$$\simeq 0.75\varepsilon(T_{RH}/M). \qquad (2.13)$$

[Note, I have assumed that when the X's decay $\rho_x \gg \rho_{rad}$ so that the initial entropy can be ignored compared to entropy produced by the decays; this assumption guarantees that $T_{RH} \lesssim M$. I have also assumed that $T \ll M$ so that IDs and S processes can be ignored. Finally, note that how the X's produce a baryon number of ε per X is *irrelevant*; it could be by $X \to$ q's l's, or equally well by $X \to \phi$'s $\to$ q's l's ($\phi =$ any other particle species).]

Note that the asymmetry produced depends upon the ratio T_{RH}/M and not T_{RH} itself—this is of some interest in inflationary scenarios in which the Universe does not reheat to a high enough temperature for baryogenesis to proceed in the standard way (out-of-equilibrium decays). For reference T_{RH} can be calculated in terms of $\tau_x \simeq \Gamma^{-1}$; when the Xs decay $(t \simeq \tau_x, H \simeq t^{-1} \simeq \Gamma)$: $\Gamma^2 = H^2 = 8\pi\rho_x/3m_{pl}^2$. Using the fact that $\rho_x \simeq g_*(\pi^2/30)T_{RH}^4$ it follows that

$$T_{RH} \simeq g_*^{-1/4}(\Gamma m_{pl})^{1/2} \qquad (2.14)$$

2.7. THE C, CP VIOLATION ε

The crucial quantity for determining n_B/s is ε—the C, CP violation in the superheavy boson system. Lacking 'The GUT', ε cannot be calculated precisely, and hence n_B/s cannot be predicted, as, for example, the ^4He abundance can be.

54

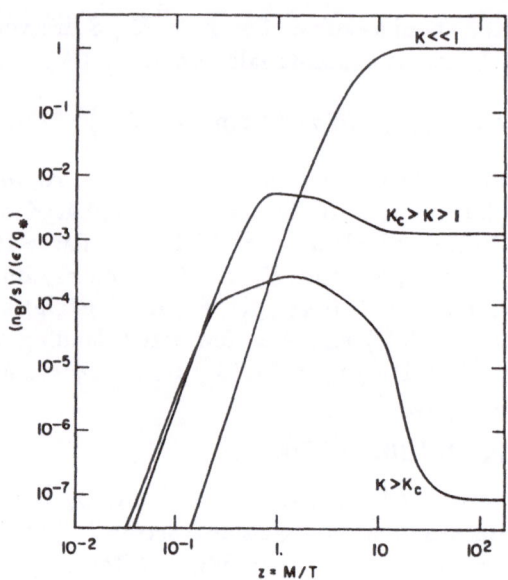

Figure 2.3: Evolution of n_B/s as a function of $z = M/T(\sim t^{1/2})$. For $K \ll 1$, n_B/s is produced when the X, $\bar{X}$ bosons decay out-of-equilibrium ($z \gg 1$). For $K_c > K > 1$, $n_B/s \propto z^{-1}$ (due to IDs) until the IDs freeze out ($z \simeq 10$). For $K > K_c$ $2 \leftrightarrow 2$ $\not{B}$ scatterings are important, and n_B/s decreases very rapidly until they freeze out ($z \simeq z_S$).

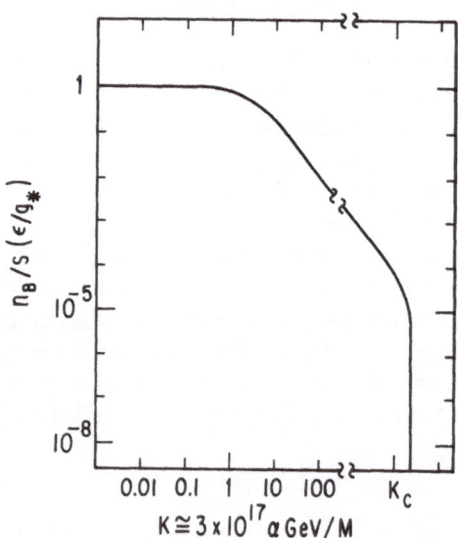

Figure 2.4: The final baryon asymmetry (in units of ε/g_*) as a function of $K \simeq 3 \times 10^{17} \alpha$ GeV/M. For $K \lesssim 1$, n_B/s is independent of K and $\simeq \varepsilon/g_*$. For $K_c > K > 1$, n_B/s decreases slowly, $\propto 1/(K(\ln K)^{0.6})$. For $K > K_c$ (when $2 \leftrightarrow 2$ $\not{B}$ scatterings are important), n_B/s decreases exponentially with $K^{1/4}$.

The quantity $\varepsilon \propto (r - \bar{r})$; at the tree graph (i.e., Born approximation) level $r - \bar{r}$ must vanish. Non-zero contributions to $(r - \bar{r})$ arise from higher order loop corrections due to Higgs couplings which are complex[47,51-55]. For these reasons, it is generally true that:

$$\varepsilon_{Higgs} \lesssim 0(\alpha^N) \sin \delta \tag{2.15}$$

$$\varepsilon_{gauge} \lesssim 0(\alpha^{N+1}) \sin \delta, \tag{2.16}$$

where α is the coupling of the particle exchanged in loop (i.e., $\alpha = g^2/4\pi$), $N \geq 1$ is the number of loops in the diagram which make the lowest order, non-zero contributions to $(r - \bar{r})$, and δ is the phase of some complex coupling. The C, CP violation in the gauge boson system occurs at 1 loop higher order than in the Higgs because gauge couplings are necessarily real. Since $\alpha \lesssim \alpha_{gauge}$, ε is at most $0(10^{-2})$—which is plenty large enough to explain $n_B/s \simeq 10^{-10}$. Because K for a Higgs is likely to be smaller, and because C, CP violation occurs at lower order in the Higgs boson system, the out-of-equilibrium decay of a Higgs is the more likely mechanism for producing n_B/s. [No additional cancellations occur when calculating $(r - \bar{r})$ in supersymmetric theories, so these generalities also hold for supersymmetric GUTs.]

In minimal SU(5)—one $\underline{5}$ and one $\underline{24}$ of Higgs, and three families of fermions, N = 3. This together with the smallness of the relevant Higgs couplings implies that $\varepsilon_H \lesssim 10^{-15}$ which is not nearly enough[47,51,52]. With 4 families the relevant couplings can be large enough to obtain $\varepsilon_H \simeq 10^{-8}$ – if the top quark and fourth generation quark/lepton masses are $0(M_W)$ (ref. 53). By enlarging the Higgs sector (e.g., by adding a second $\underline{5}$ or a $\underline{45}$), $(r - \bar{r})$ can be made non-zero at the 1-loop level, making $\varepsilon_H \simeq 10^{-8}$ easy to achieve.

In more complicated theories, e.g., E6, S0(10), etc., $\varepsilon \simeq 10^{-8}$ can also easily be achieved. However, to do so restricts the possible symmetry breaking patterns. Both E6 and SO(10) are C-symmetric, and of course C-symmetry must be broken before ε can be non-zero. In general, in these models ε is suppressed by powers of M_C/M_G where $M_C(M_G)$ is the scale of C(GUT) symmetry breaking, and so M_C cannot be significantly smaller than M_G.

It seems very unlikely that ε can be related to the parameters of the $K^\circ - \bar{K}^\circ$ system, the difficulty being that not enough C, CP violation can be 'fed up' to the superheavy boson system. It has been suggested that ε could be related to the electric dipole moment of the neutron[54].

Although baryogenesis is nowhere near being on the same firm footing as primordial nucleosynthesis, we now at least have for the first time a very attractive framework for understanding the origin of $n_B/s \simeq 10^{-10}$. A framework which is so attractive, that in the absence of observed proton decay, the baryon asymmetry of the Universe is probably the best evidence for some kind of quark/lepton unification.

There are several very interesting aspects of baryogenesis which I have not mentioned. Affleck and Dine have discussed a very interesting mechanism for generating a baryon asymmetry at low temperatures in supersymmetric models[56]. Very recently, Kuzmin, Rubakov, and Shaposnikov have pointed out that the rate for B nonconservation through the electroweak anomaly (B+L is anomalous in the standard SU(2) × U(1) theory) can be large ($\Gamma \gg H$) at temperatures $\simeq 100 - 200$ GeV, and can strongly damp any baryon and lepton asymmetry with $B - L = 0$. (In SU(5) any baryon/lepton asymmetry produced must necessarily have $B - L = 0$ since $B - L$ is conserved in SU(5); for larger gauge groups, e.g., SO(10), E6, etc., $B - L$ is not necessarily conserved, and $B - L \simeq 0(B, L)$ also typically evolves.) B nonconservation via the electroweak anomaly and its cosmological implications are discussed in detail in refs. 57, 58. [In writing up this lecture I have borrowed freely and heavily from the review on baryogenesis written by myself and E. W. Kolb (ref. 55), and refer the interested reader there for a more thorough discussion of the details of baryogenesis.]

LECTURE 3:
TOWARD THE INFLATIONARY PARADIGM

3.1. OVERVIEW

Guth's inflationary Universe scenario has revolutionized our thinking about the very early Universe. The inflationary scenario offers the possibility of explaining a handful of very fundamental cosmological facts—the homogeneity, isotropy, and flatness of the Universe, the origin of density inhomogeneities and the origin of the baryon asymmetry, while at the same time avoiding the monopole problem. It is based upon microphysical events which occurred early ($t \leq 10^{-34}$ sec) in the history of the Universe, but well after the planck epoch ($t \geq 10^{-43}$ sec). While Guth's original model was fundamentally flawed, the variant based on the slow-rollover transition proposed by Linde, and Albrecht and Steinhardt (dubbed 'new inflation') appears viable. Although old inflation and the earliest models of new inflation were based upon first order phase transitions associated with spontaneous-symmetry breaking (SSB), it now appears that the inflationary transition is a much more generic phenomenon, being associated with the evolution of a weakly-coupled scalar field which for some reason or other was initially displaced from the minimum of its potential. Models now exist which are based on a wide variety of microphysics: SSB, SUSY/SUGR, compactification of extra dimensions, R^2 gravity, induced gravity, and some random, weakly-coupled scalar field. While there are several models which successfully implement the inflation, none is particularly compelling and all seem somewhat *ad hoc*. The common distasteful feature of all the successful models is the necessity of a small dimensionless number in the model—usually in the form of a dimensionless coupling of order 10^{-15}. And of course, all inflationary scenarios rely upon the assumption that vacuum energy (or equivalently a cosmological term) was once dynamically very significant, whereas today there exists every evidence that it is not (although we have no understanding why it is not). For these reasons

I have entitled this lecture *Toward the Inflationary Paradigm*. I have divided my lecture into the following sections: Shortcomings of the standard cosmology; New inflation—the slow–rollover transition; Scalar field dynamics; Origin of density inhomogeneities; Specific models, I. Interesting failures; Lessons learned—prescription for successful inflation; Two models that work; The Inflationary paradigm; Loose ends; and Inflation confronts observation.

3.2. SHORTCOMINGS OF THE STANDARD COSMOLOGY

The standard cosmology is very successful—it provides us with a reliable framework for describing the history of the Universe as early as 10^{-2} sec after the bang (when the temperature was about 10 MeV) and perhaps as early as 10^{-43} sec after the bang (see Fig. 1.5). In sum, the standard cosmology is a great achievement. [There is nothing in our present understanding of physics that would indicate that it is incorrect to extrapolate the standard cosmology back to times as early as 10^{-43} sec—the fundamental constituents of matter, quarks and leptons, are point-like particles and their known interactions should remain 'weak' up to energies as high as 10^{19} GeV—justifying the dilute gas approximation made in writing $\rho_r \propto T^4$. (This fact was first pointed out by Collins and Perry[59]). However, at times earlier than 10^{-43} sec, corresponding to temperatures greater than 10^{19} GeV, quantum corrections to general relativity—a classical theory, should become very significant.]

However, it is not without its shortcomings. There are a handful of very important and fundamental cosmological facts which, while it can accommodate, it in no way elucidates. I will briefly review these puzzling facts.

(1–2) Large-scale Isotropy and Homogeneity

The observable Universe (size $\simeq H^{-1} \simeq 10^{28}$cm $\simeq 3000$ h^{-1} Mpc) is to a high degree of precision isotropic and homogeneous on the largest scales, say > 100Mpc. [Of course, our knowledge of the Universe outside our past light cone is very limited; see ref. 60.] The best evidence for the isotropy and homogeneity is provided by the uniformity of the cosmic background temperature (see Fig. 1.2): $(\delta T/T) < 10^{-3}$ (10^{-4} if the dipole anisotropy is interpreted as being due to our motion relative to the cosmic rest frame). Large-scale density inhomogeneities or anisotropic expansion would result in temperature fluctuations of comparable magnitude (see refs. 61, 62). The smoothness of the observed Universe is puzzling if one wishes to understand it as being due to causal, microphysical processes which operated during the early history of the Universe. Our Hubble volume today contains an entropy of about 10^{88}. At decoupling ($t \simeq 6 \times 10^{12}(\Omega h^2)^{-1/2}$ sec, $T \simeq 1/3$eV), the last epoch when matter and radiation were known to be interacting vigorously and particle interactions might have been able to smooth the radiation, the entropy within the horizon was only about 8×10^{82}; that is, the comoving volume which contains the presently-observable Universe, then was comprised of about 2×10^5 causally-distinct regions. How is it that they came to be homogeneous? Put another way, the particle horizon at decoupling only subtends an angle of about $1/2°$ on the sky today—how is it that the cosmic background temperature is so uniform on angular scales much greater than this?

The standard cosmology can accommodate these facts—after all the FRW cosmology is exactly isotropic and homogeneous, but at the expense of very special initial data. In 1973 Collins and Hawking[63] showed that the set of initial data which evolve to a Universe which globally is as smooth as ours has measure zero (provided that the strong and dominant energy conditions are always satisfied).

(3) Small-scale Inhomogeneity

As any real astronomer will gladly testify, the Universe is very lumpy—stars, galaxies, clusters of galaxies, superclusters, etc. Today, the density contrast on the scale of galaxies is: $\delta\rho/\rho \simeq 10^5$. The fact that the microwave background radiation is very uniform even on very small angular scales ($\ll 1°$) indicates that the Universe was smooth even on the scale of galaxies at decoupling. [The relationship between the angle on the sky and mass contained within the corresponding length scale at decoupling is: $\theta \simeq 1'(M/10^{12}M_\odot)^{1/3}\Omega^{-1/3}h^{1/3}$.] On small angular scales: $\delta T/T \simeq c(\delta\rho/\rho)_{dec}$, where the numerical constant $c \simeq 10^{-1} - 10^{-2}$ (see ref. 62 for further details). Whence came the structure which today is so conspicuous?

Once matter decouples from the radiation and is free of the pressure support provided by the radiation, any density inhomogeneities present will grow via the Jeans (or gravitational instability)—in the linear regime, $\delta\rho/\rho \propto R(t)$. [If the mass density of the Universe is dominated by a collisionless particle species, e.g., a light, relic neutrino species or relic axions, density perturbations in these particles can begin to grow as soon as the Universe becomes matter-dominated.] In order to account for the present structure, density perturbations of amplitude $\sim 10^{-3}$ or so at decoupling are necessary on the scale of galaxies. The standard cosmology sheds no light as to the origin or nature (spectrum and type—adiabatic or isothermal) of the primordial density perturbations so crucial for understanding the structure observed in the Universe today. [For a review of the formation of structure in the Universe according to the gravitational instability picture, see ref. 64.]

(4) Flatness (or Oldness) of the Universe

The observational data suggest that

$$0.01 \leq \Omega \leq \text{few}.$$

Ω is related to both the expansion rate of the Universe and the curvature radius of the Universe:

$$\Omega = 8\pi G\rho/3H^2 \equiv H_{crit}^2/H^2, \tag{3.1}$$

$$|\Omega - 1| = (H^{-1}/R_{curv})^2, \tag{3.2}$$

The fact that Ω is not too different from unity today implies that the present expansion rate is close to the critical expansion rate and that the curvature radius of the Universe is comparable to or larger than the Hubble radius. As the Universe expands Ω does not remain constant, but evolves away from 1

$$\Omega = 1/(1 - x(t)), \tag{3.3}$$

$$x(t) = (k/R^2)/(8\pi G\rho/3), \tag{3.4}$$

$$x(t) \propto \begin{cases} R(t)^2 & \text{radiation} - \text{dominated} \\ R(t) & \text{matter} - \text{dominated} \end{cases}$$

That Ω is still of order unity means that at early times it was equal to 1 to a very high degree of precision:

$$|\Omega(10^{-43} \text{ sec}) - 1| \simeq O(10^{-60}),$$
$$|\Omega(1 \text{ sec}) - 1| \simeq O(10^{-16}).$$

This in turn implies that at early times the expansion rate was equal to the critical rate to a high degree of precision and that the curvature of the Universe was much, much greater than the Hubble radius. If it were not, i.e., suppose that $|(k/R^2)/(8\pi G\rho/3)| \simeq O(1)$ at $t \simeq 10^{-43}$ sec, then the Universe would have collapsed after a few Planck times ($k > 0$) or would have quickly become curvature-dominated, ($k < 0$), in which case $R(t) \propto t$ and $t(T = 3K) \leq 10^{-11}$ sec! Why was this so?

The so-called flatness problem has sometimes been obscured by the fact that it is conventional to rescale $R(t)$ so that $k = -1$, 0, or +1, making it seem as though there are but three FRW models. However, that clearly is not the case; there are an infinity of models, specified by the curvature radius $R_{curv} = R(t)|k|^{-1/2}$ at some given epoch, say the planck epoch. Our model corresponds to one with a curvature radius that exceeds its initial Hubble radius by 30 orders-of-magnitude. Again, this fact can be accommodated by FRW models, but the extreme flatness of our Universe is in no way explained by the standard cosmology. [The flatness problem and the naturalness of the $k = 0$ model have been emphasized by Dicke and Peebles[65].]

(5) Baryon Number of the Universe

There is ample evidence (see ref. 66) for the dearth of antimatter in the observable Universe. That fact together with the baryon-to-photon ratio ($\eta \simeq 4 - 7 \times 10^{-10}$) means that our Universe is endowed with a net baryon number, quantified by the baryon number-to-entropy ratio

$$n_B/s \simeq (6 - 10) \times 10^{-11},$$

which in the absence of baryon number non-conserving interactions or significant entropy production is proportional to the constant net baryon number per comoving volume which the Universe has always possessed. Until five or so years ago this very fundamental number was without explanation. Of course it is now known that in the presence of interactions that violate B, C, and CP a net baryon asymmetry will evolve dynamically. Such interactions are predicted by Grand Unified Theories (or GUTs) and 'baryogenesis' is one of the great triumphs of the marriage of grand unification and cosmology. [See ref. 67 for a review of grand unification.] If the baryogenesis idea is correct, then the baryon asymmetry of the Universe is subject to calculation just as the primordial Helium abundance is. Although the idea is very attractive and certainly appears to be qualitatively correct, a precise calculation of the baryon number-to-entropy ratio cannot be performed until *The* Grand Unified Theory is known. [Baryogenesis is discussed in Lecture 2 and is reviewed in ref. 68.]

(6) The Monopole Problem

If the great success of the marriage of GUTs and cosmology is baryogenesis, then the great disappointment is 'the monopole problem'. 't Hooft-Polyakov monopoles[69] are a generic prediction of GUTs. In the standard cosmology (and for the simplest GUTs) monopoles are grossly overproduced during the GUT symmetry-breaking transition, so much so that the Universe would reach its present temperature of 3K at the very tender age of 30,000 yrs! [For a detailed discussion of the monopole problem, see refs. 70, 71.] Although the monopole problem initially seemed to be a severe blow to the Inner Space/Outer Space connection, as it has turned out it provided us with a valuable piece of information about physics at energies of order 10^{14} GeV and the Universe at times as early as 10^{-34} sec—the standard cosmology and the simplest GUTs are definitely incompatible! As it turned out, it was the search for a solution to the monopole problem which in the end led Guth to come upon the inflationary Universe scenario[72,73].

(7) The Smallness of the Cosmological Constant

With the possible exception of supersymmetry/supergravity and superstring theories, the absolute scale of the scalar potential $V(\phi)$ is not specified (here ϕ represents the scalar fields in the theory, be they fundamental or composite). A constant term in the scalar potential is equivalent to a cosmological term (the scalar potential contributes a term $V g_{\mu\nu}$ to the stress energy of the Universe[74]). At low temperatures (say temperatures below any scale of spontaneous symmetry-breaking) the constant term in the potential receives contributions from all the stages of SSB—chiral symmetry breaking, electroweak SSB, GUT SSB, etc. The observed expansion rate of the Universe ($H = 100h$ km sec^{-1}Mpc^{-1}) limits the total energy density of the Universe to be

$$\rho_{TOT} \leq O(10^{-46} \text{GeV}^4).$$

Making the seemingly very reasonable assumption that all stress energy self-gravitates (which is dictated by the equivalence principle) it follows that the vacuum energy of our $SU(3) \times U(1)$ vacuum must be less than 10^{-46}GeV^4. Compare this to the scale of the various contributions to the scalar potential: $0(M^4)$ for physics associated with a symmetry breaking scale of M

$$V_{today}/M^4 \leq \rho_{TOT}/M^4 \leq \begin{cases} 10^{-122} & M = m_{pl} \\ 10^{-102} & M = 10^{14}\text{GeV} \\ 10^{-56} & M = 300\text{GeV} \\ 10^{-46} & M = 1\text{GeV} \end{cases}$$

At present there is no explanation for the vanishingly small value of the energy density of our very unsymmetrical vacuum. It is easy to speculate that a fundamental

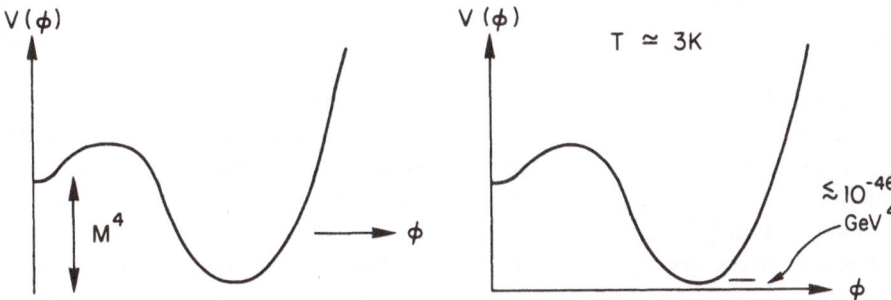

Figure 3.1: In gauge theories the vacuum energy is a function of one or more scalar fields (here denoted collectively as ϕ); however, the absolute energy scale is not set. Vacuum energy behaves like a cosmological term; the present expansion rate of the Universe constrains the value of the vacuum energy today to be $\leq 10^{-46}$ GeV^4.

understanding of the smallness of the cosmological constant will likely involve an intimate link between gravity and quantum field theory.

Today we can be certain the vacuum energy is small and plays a minor role in the dynamics of the expansion of the Universe (compared to the potential role that it could play). If we accept this as an empirical determination of the absolute scale of the scalar potential $V(\phi)$, it then follows that the energy density associated with an expectation value of ϕ near zero is enormous—of order M^4 (see Fig. 3.1) and therefore could have played an important role in the dynamics of the very early Universe. Accepting this *empirical determination* of the zero of vacuum energy— which is a very great leap of faith, is the starting point for inflation. In fact, the rest of the journey is downhill.

All of these cosmological facts can be accommodated by the standard model, but seemingly at the expense of highly special initial data (the possible exception being the monopole problem). Over the years there have been a number of attempts to try to understand and/or explain this apparent dilemma of initial data. Inflation is the most recent attempt and I believe shows great promise. Let me begin by briefly mentioning the earlier attempts:

* Mixmaster Paradigm–Starting with a solution with a singularity which exhibits the features of the most general singular solutions known (the so-called mixmaster model) Misner and his coworkers hoped that they could show that particle viscosity would smooth out the geometry. In part because horizons still effectively exist in the mixmaster solution 'the chaotic cosmology program' has

proven unsuccessful (for further discussion see refs. 75).

* Nature of the Initial Singularity–Penrose[76] explored the possibility of explaining the observed smoothness of the Universe by restricting the kinds of initial singularities which are permitted in Nature (those with vanishing Weyl curvature). In a sense his approach is to postulate a law of physics governing allowed initial data.

* Quantum Gravity Effects–The first two solutions involve appealing to classical gravitational effects. A number of authors have suggested that quantum gravity effects might be responsible for smoothing out the space-time geometry (deWitt[77]; Parker[78]; Zel'dovich[79]; Starobinskii[80]; Anderson[81]; Hartle and Hu[82]; Fischetti et al.[83]). The basic idea being that anisotropy and/or inhomogeneity would drive gravitational particle creation, which due to back reaction effects would eliminate particle horizons and smooth out the geometry. Recently, Hawking and Hartle[84] have advocated the Quantum Cosmology approach to actually compute the initial state. All of these approaches necessarily involve events at times $\lesssim 10^{-43}$ sec and energy densities $\gtrsim m_{pl}^4$.

* Anthropic Principle–Some (see, e.g., refs. 85) have suggested (or in some cases even advocated) 'explaining' many of the puzzling features of the Universe around us (and in some cases, even the laws of physics!) by arguing that unless they were as they are intelligent life would not have been able to develop and observe them! Hopefully we will not have to resort to such an explanation.

The approach of inflation is somewhat different from previous approaches. Inflation (at least from my point-of-view) is based upon well-defined and reasonably well-understood microphysics (albeit, some of it very speculative). That microphysics is:

* Classical Gravity (general relativity), at least as an effective, low-energy theory of gravitation.

* 'Modern Particle Physics'–grand unification, supersymmetry/supergravity, field theory limit of superstring theories, etc. at energy scales $\lesssim m_{pl}$

As I will emphasize, in all viable models of inflation the inflationary period (at least the portion of interest to us) takes place well after the planck epoch, with the energy densities involved being far less than m_{pl}^4 (although semi-classical quantum gravity effects might have to be included as non-renormalizable terms in the effective Lagrangian). Of course, it could be that a resolution to the cosmological puzzles discussed above involves both 'modern particle physics' and quantum gravitational effects in their full glory (as in a fully ten dimensional quantum theory of strings).

I will not take the time or the space here to review the historical development of our present view of inflation; I refer the interested reader to the interesting paper on this subject by Lindley[86]. It suffices to say that Guth's very influential paper of 1981 was the 'shot heard 'round the world' which initiated the inflation revolution,[73] and that Guth's doomed original model (see Guth and Weinberg[87]; Hawking et al[87]) was revived by Linde's[88] and Albrecht and Steinhardt's[89] variant, 'new inflation'. I will focus all of my attention on the present status of the 'slow-rollover' model of Linde[88] and Albrecht and Steinhardt[89].

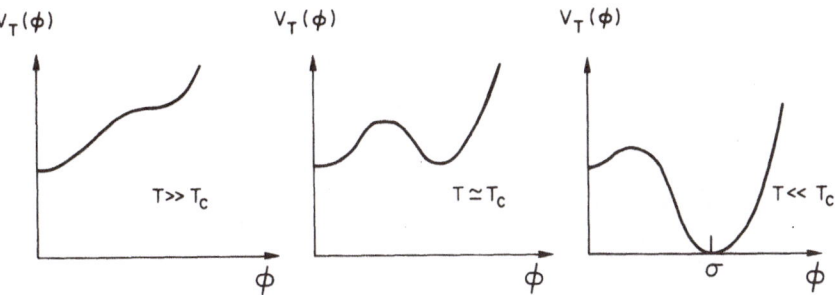

Figure 3.2: The finite temperature effective potential as a function of T (schematic). The Universe is usually assumed to start out in the high temperature, symmetric minimum ($\phi = 0$) of the potential and must eventually evolve to the low tempera- ture, asymmetric minimum ($\phi = \sigma$). The evolution of ϕ from $\phi = 0$ to $\phi = \sigma$ can prove to be very interesting—as in the case of an inflationary transition.

3.3. NEW INFLATION—THE SLOW-ROLLOVER TRANSITION

The basic idea of the inflationary Universe scenario is that there was an epoch when the vacuum energy density dominated the energy density of the Universe. During this epoch $\rho \simeq V \simeq$ constant, and thus $R(t)$ grew exponentially ($\propto \exp(Ht)$), allowing a small, causally-coherent region (initial size $\leq H^{-1}$) to grow to a size which encompasses the region which eventually becomes our presently-observable Universe. In Guth's original scenario[73], this epoch occurred while the Universe was trapped in the false ($\phi = 0$) vacuum during a strongly, first-order phase transition. In new inflation, the vacuum-dominated, inflationary epoch occurs while the region of the Universe in question is slowly, but inevitably, evolving toward the true, SSB vacuum. Rather than considering specific models in this section, I will try to discuss new inflation in the most general context. For the moment I will however assume that the epoch of inflation is associated with a first-order, SSB phase transition, and that the Universe is in thermal equilibrium before the transition. As we shall see later new inflation is more general than these assumptions. But for definiteness (and for historical reasons), let me begin by making these assumptions.

Consider a SSB phase transition characterized by an energy scale M. For $T \geq T_c \simeq 0(M)$ the symmetric ($\phi = 0$) vacuum is favored, i.e., $\phi = 0$ is the global minimum of the finite temperature effective potential $V_T(\phi)$ (=free energy density). As T approaches T_c a second minimum develops at $\phi_i = 0$, and at $T = T_c$, the two minima are degenerate. At temperatures below T_c the SSB ($\phi = \sigma$) minimum is the global minimum of $V_T(\phi)$ (see Fig. 3.2). However, the Universe does not instantly make the transition from $\phi = 0$ to $\phi = \sigma$; the details and time required are a question of dynamics. [The scalar field ϕ is the order parameter for the SSB transition under discussion; in the spirit of generality ϕ might be a gauge singlet field or might have nontrivial transformation properties under the gauge group, possibly even responsible for the SSB of the GUT.] Once the temperature of the

64

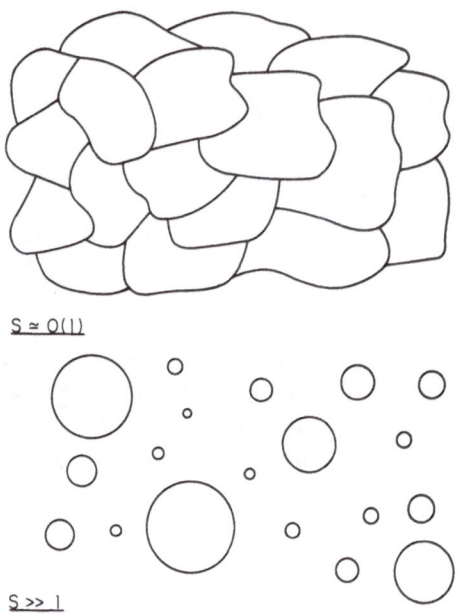

S ≃ O(1)

S ≫ 1

Figure 3.3: If the tunneling action is large $(S \gg 1)$, barrier penetration will proceed via bubble nucleation, while in the case that it becomes small $(S \simeq O(1))$, the Universe will fragment into irregularly-shaped fluctuation regions. The very large scale (scale ≫ bubble or fluctuation region) structure of the Universe is determined by whether $S \simeq O(1)$—in which case the Universe is comprised of irregularly-shaped domains, or $S \gg O(1)$—in which case the Universe is comprised of isolated bubbles.

Universe drops below $T_c \simeq O(M)$, the potential energy associated with ϕ being far from the minimum of its potential, $V \simeq V(0) \simeq M^4$, dominates the energy density in radiation ($\rho_r < T_c^4$), and causes the Universe to expand exponentially. During this exponential expansion (known as a deSitter phase) the temperature of the Universe decreases exponentially causing the Universe to supercool. The exponential expansion continues so long as ϕ is far from its SSB value. Now let's focus on the evolution of ϕ.

Assuming a barrier exists between the false and true vacua, thermal fluctuations and/or quantum tunneling must take ϕ across the barrier. The dynamics of this process determine when and how the process occurs (bubble formation, spinodal decomposition, etc.) and the value of ϕ after the barrier is penetrated. If the action for bubble nucleation remains large, $S_b \gg 1$, then the barrier will be overcome by the nucleation of Coleman-deLuccia bubbles;[90] on the other hand if the action for bubble nucleation becomes of order unity, then the Universe will undergo spinodal decomposition, and irregularly-shaped fluctuation regions will form (see Fig. 3.3; for a more detailed discussion of the barrier penetration process see refs. 89–91). For definiteness suppose that the barrier is overcome when the temperature is T_{MS} and that after the barrier is penetrated the value of ϕ is ϕ_0. From this point the journey to the true vacuum is downhill (literally). For the moment let us assume that the evolution of ϕ is adequately described by semi-classical equations of motion:

$$\ddot{\phi} + 3H\dot{\phi} + \Gamma\dot{\phi} + V' = 0, \qquad (3.5)$$

where ϕ has been normalized so that its kinetic term in the Lagrangian is $1/2\partial_\mu\phi\partial^\mu\phi$, and prime indicates a derivative with respect to ϕ. The subscript T on V has been dropped; for $T \ll T_c$ the temperature dependence of V_T can be neglected and the zero temperature potential ($\equiv V$) can be used. The $3H\dot{\phi}$ term acts like a frictional force, and arises because the expansion of the Universe 'redshifts away' the kinetic energy of $\phi(\propto R^{-3})$. The $\Gamma\dot{\phi}$ term accounts for particle creation due to the time-variation of ϕ[refs. 92, 93]. The quantity Γ is determined by the particles which couple to ϕ and the strength with which they couple ($\Gamma^{-1} \simeq$ lifetime of a ϕ particle). As usual, the expansion rate H is determined by the energy density of the Universe:

$$H^2 = 8\pi G\rho/3, \qquad (3.6)$$

$$\rho \simeq 1/2\dot{\phi}^2 + V(\phi) + \rho_r, \qquad (3.7)$$

where ρ_r represents the energy density in radiation produced by the time variation of ϕ. [For $T_{MS} \ll T_c$ the original thermal component makes a negligible contribution to ρ.] The evolution of ρ_r is given by

$$\dot{\rho}_r + 4H\rho_r = \Gamma\dot{\phi}^2, \qquad (3.8)$$

where the $\Gamma\dot{\phi}^2$ term accounts for particle creation by ϕ.

In writing Eqns(3.5–3.8) I have implicitly assumed that ϕ is spatially homogeneous. In some small region (inside a bubble or a fluctuation region) this will be a good approximation. The size of this smooth region will turn out to be unimportant; take it to be of order the 'Physics Horizon', H^{-1}—certainly, it is not likely to be larger. Now follow the evolution of ϕ within the small, smooth patch of size H^{-1}.

If $V(\phi)$ is sufficiently flat somewhere between $\phi = \phi_0$ and $\phi = \sigma$, then ϕ will evolve very slowly in that region, and the motion of ϕ will be 'friction-dominated' so that $3H\dot{\phi} \simeq -V'$ (in the slow growth phase particle creation is not important[94]). If V is sufficiently flat, then the time required for ϕ to transverse the flat region can be long compared to the expansion timescale H^{-1}; for definiteness say, $\tau_\phi = 100H^{-1}$. During this slow growth phase $\rho \simeq V(\phi) \simeq V(\phi = 0)$; both ρ_r and $1/2\dot{\phi}^2$ are $\ll V(\phi)$. The expansion rate H is then just

$$H \simeq (8\pi V(0)/3m_{pl}^2)^{1/2}$$
$$\simeq O(M^2/m_{pl}), \qquad (3.9)$$

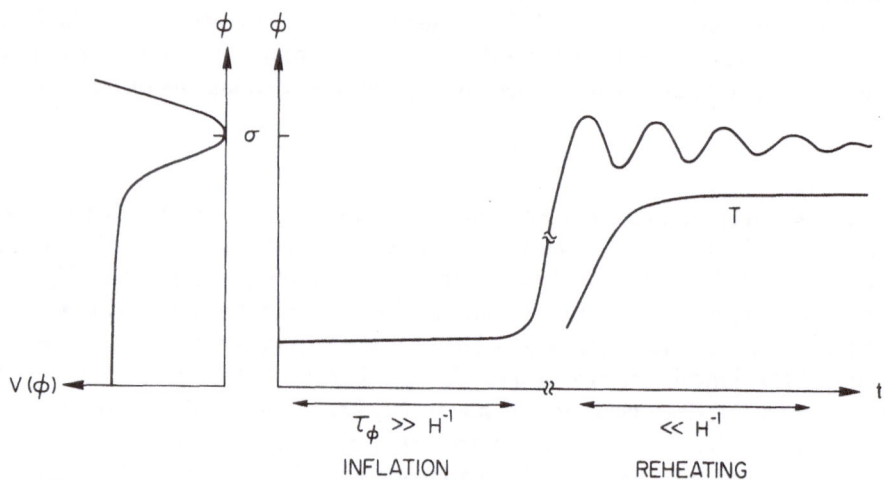

Figure 3.4: Evolution of ϕ and the temperature inside the bubble or fluctuation region (schematic). Early on ϕ evolves slowly (relative to the expansion timescale), then as the potential steepens ϕ evolves rapidly (on the expansion timescale). The oscillations of ϕ are damped by particle creation, which leads to the reheating of the bubble or fluctuation region.

where $V(0)$ is assumed to be of order M^4. While $H \simeq$ constant, R grows exponentially: $R \propto \exp(Ht)$; for $\tau_\phi = 100H^{-1}$, R expands by a factor of e^{100} during the slow rolling period, and the physical size of the smooth region increases to $e^{100}H^{-1}$.

As the potential steepens, the evolution of ϕ quickens. Near $\phi = \sigma$, ϕ oscillates around the SSB minimum with frequency m_ϕ : $m_\phi^2 \simeq V''(\sigma) \simeq O(M^2) \gg H^2 \simeq M^4/m_{pl}^2$. As ϕ oscillates about $\phi = \sigma$ its motion is damped both by particle creation and the expansion of the Universe. If $\Gamma^{-1} \ll H^{-1}$, then coherent field energy density $(V + 1/2\dot{\phi}^2)$ is converted into radiation in less than an expansion time $(\Delta t_{RH} \simeq \Gamma^{-1})$, and the patch is reheated to a temperature $T \simeq 0(M)$—the vacuum energy is efficiently converted into radiation ('good reheating'). On the other hand, if $\Gamma^{-1} \gg H^{-1}$, then ϕ continues to oscillate and the coherent field energy redshifts away with the expansion: $(V + 1/2\dot{\phi}^2) \propto R^{-3}$—the coherent energy behaves like non-relativistic matter. Eventually, when $t \simeq \Gamma^{-1}$ the energy in radiation begins to dominate that in coherent field oscillations, and the patch is reheated to a temperature $T \simeq (\Gamma/H)^{1/2}M \simeq (\Gamma m_{pl})^{1/2} \ll M$ ('poor reheating'). The evolution of ϕ is summarized schematically in Fig. 3.4. In the next section I will discuss the all-important scalar field dynamics in great detail.

For the following discussion let us assume 'good reheating' ($\Gamma \gg H$). After reheating the patch has a physical size $e^{100}H^{-1}$ ($\simeq 10^{17}$cm for $M \simeq 10^{14}$GeV), is at a temperature of order M, and in the approximation that ϕ was initially constant throughout the patch, the patch is exactly smooth. From this point forward the region evolves like a radiation-dominated FRW model. How have the cosmological conundrums been 'explained'?

First, *the homogeneity and isotropy*; our observable Universe today ($\simeq 10^{28}$cm) had a physical size of about 10cm ($= 10^{28}$cm $\times 3K/10^{14}$GeV) when T was 10^{14}GeV—thus it lies well within one of the smooth regions produced by the inflationary epoch. Put another way, inflation has resulted in a smooth patch which contains an entropy of order $(10^{17}$cm$)^3 \times (10^{14}$GeV$)^3 \simeq 10^{134}$, which is much, much greater than that within the presently-observed Universe ($\simeq 10^{88}$). Before inflation that same volume contained only a very small amount of entropy, about $(10^{-23}$cm$)^3(10^{14}$GeV$)^3 \simeq 10^{14}$. The key to inflation then is the highly non-adiabatic event of reheating (see Fig. 3.5). The very large-scale cosmography depends upon the state of the Universe before inflation and how inflation was initiated (bubble nucleation or spinodal decomposition); see ref. 96 for further discussion.

Since we have assumed that ϕ is spatially constant within the bubble or fluctuation region, after reheating the patch in question is precisely uniform, and at this stage *the inhomogeneity puzzle* has not been solved. Inflation *has* produced a smooth manifold on which small fluctuations can be impressed. Due to deSitter space produced quantum fluctuations in ϕ, ϕ is not exactly uniform even in a small patch. Later, I will discuss the density inhomogeneities that result from the quantum fluctuations in ϕ.

The flatness puzzle involves the smallness of the ratio of the curvature term to the energy density term. This ratio is exponentially smaller after inflation: $x_{\text{after}} \simeq e^{-200} x_{\text{before}}$ since the energy density before and after inflation is $0(M^4)$, while k/R^2 has exponentially decreased (by a factor of e^{200}). Since the ratio x is reset to an exponentially small value, the inflationary scenario predicts that today Ω should be $1 \pm 0(10^{-BIG\#})$.

If the Universe is reheated to a temperature of order M, a *baryon asymmetry* can evolve in the usual way, although the quantitative details may be different[55,94]. If the Universe is not efficiently reheated ($T_{RH} \ll M$), it may be possible for n_B/s to be produced directly in the decay of the coherent field oscillations[92-95] (which behave just like NR ϕ particles); this possibility will be discussed later. In any case, it is absolutely necessary to have baryogenesis occur after reheating since any baryon number (or any other quantum number) present before inflation is diluted by a factor of $(M/T_{MS})^3 \exp(3H\tau_\phi)$—the factor by which the total entropy increases. Note that if C, CP are violated spontaneously, then ϵ (and n_B/s) could have a different sign in different patches—leading to a Universe which on the very largest scales ($\gg e^{100}H^{-1}$) is baryon symmetric.

Since the patch that our observable Universe lies within was once (at the beginning of inflation) causally-coherent, the Higgs field could have been aligned throughout the patch (indeed, this is the lowest energy configuration), and thus there is likely to be ≤ 1 monopole within the entire patch which was produced as a topological defect. *The glut of monopoles* which occurs in the standard cosmology does not occur. [The production of other topological defects (such as domain walls, etc.) is avoided for similar reasons.] Some monopoles will be produced after reheating

68

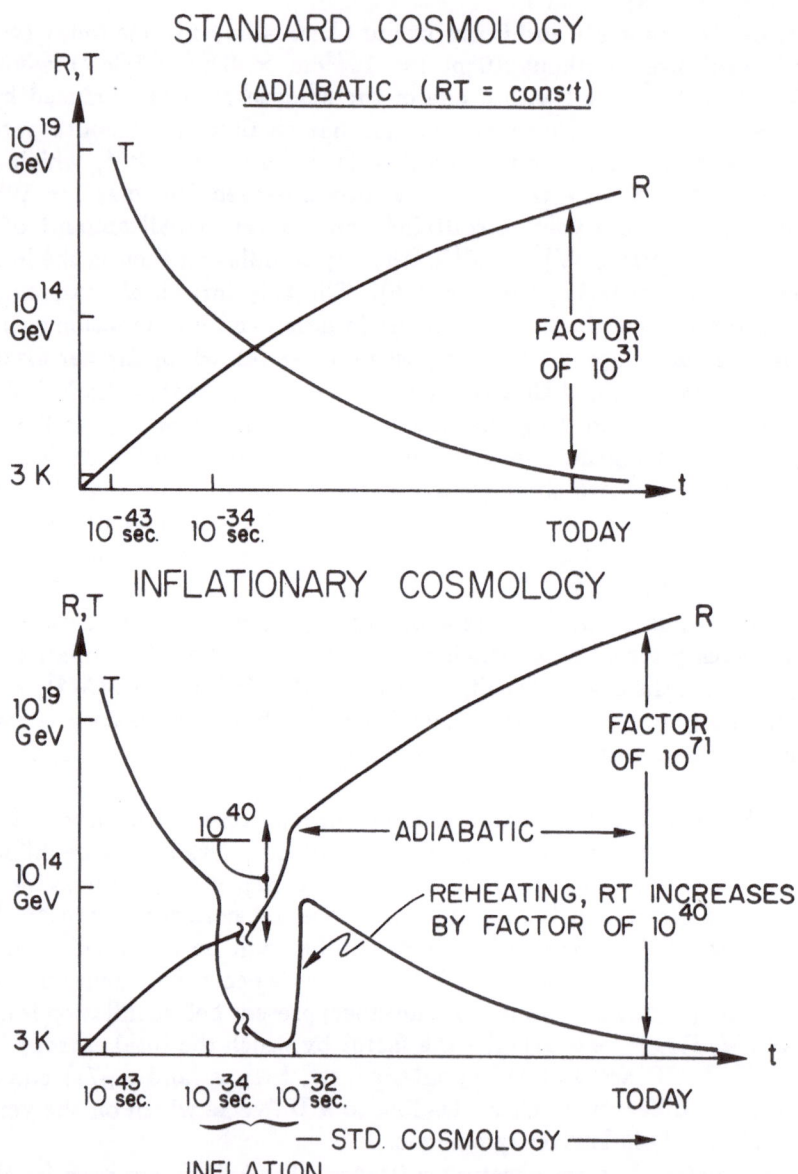

Figure 3.5: Evolution of the scale factor R and temperature T of the Universe in the standard cosmology and in the inflationary cosmology. The standard cosmology is always adiabatic ($RT \simeq$ const), while the inflationary cosmology undergoes a highly non-adiabatic event (reheating) after which it is adiabatic.

in rare, very energetic particle collisions[97]. The number produced is both exponentially small and exponentially uncertain. [In discussing the resolution of the monopole problem I am tacitly assuming that the SSB of the GUT is occurring during the SSB transition in question, or that it has already occurred in an earlier SSB transition; if not then one has to worry about the monopoles produced in the subsequent GUT transition.] Although monopole production is intrinsically small in inflationary models, the uncertainties in the number of monopoles produced are exponential and of course, it is also possible that monopoles might be produced as topological defects in a subsequent phase transition[98] (although it may be difficult to arrange that they not be overproduced).

Finally, the inflationary scenario sheds no light upon *the cosmological constant puzzle*. Although it can potentially successfully resolve all of the other puzzles in my list, inflation is, in some sense, a house of cards built upon the cosmological constant puzzle.

3.4. SCALAR FIELD DYNAMICS

The evolution of the scalar field ϕ is key to understanding new inflation. In this section I will focus on the semi-classical dynamics of ϕ. Later, I will return to the question of the validity of the semi-classical approach. Much of what I will discuss here is covered in more detail in ref. 99.

Stated in the most general terms, the current view of inflation is that it involves the dynamical evolution of a very weakly-coupled scalar field (hereafter referred to as ϕ) which is, for one reason or another, initially displaced from the minimum of its potential (see Fig. 3.6). While it is displaced from its minimum, and is slowly-evolving toward that minimum, its potential energy density drives the rapid (exponential) expansion of the Universe, now known as inflation.

The usual assumptions which are made (often implicitly) in order to analyze the scalar field dynamics inflation are:

$\star$ A FRW spacetime with scale factor $R(t)$ and expansion rate

$$H^2 \equiv (\dot{R}/R)^2 = 8\pi\rho/3m_{pl}^2 - k/R^2 \tag{3.10}$$

where the energy density is assumed to be dominated by the stress energy associated with the scalar field (in any case, other forms of stress energy rapidly redshift away during inflation and become irrelevant).

$\star$ The scalar field ϕ is spatially constant (at least on a scale $\gtrsim H^{-1}$) with initial value $\phi_i \neq \sigma$, where $V(\sigma) = V'(\sigma) = 0$.

$\star$ The semi-classical equation of motion for ϕ provides an accurate description of its evolution; more precisely,

$$\phi(t) = \phi_{cl}(t) + \Delta\phi_{QM}$$

where the quantum fluctuations (characterized by size $\Delta\phi_{QM} \simeq H/2\pi$) are assumed to be a small perturbation to the classical trajectory $\phi_{cl}(t)$. From

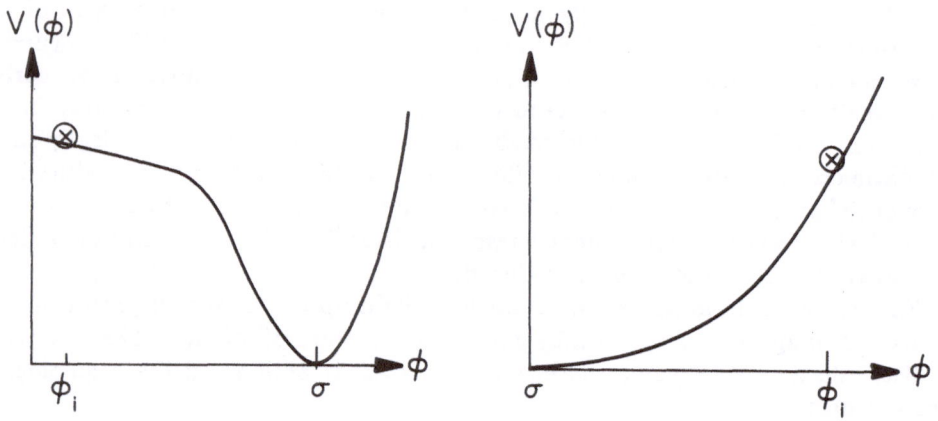

Figure 3.6: Stated in the most general terms, inflation involves the dynamical evolution of a scalar field which was initially displaced from the minimum of its potential, be that minimum at $\sigma = 0$ or $\sigma \neq 0$.

this point forward I will drop the subscript 'cl'. I will return later to these assumptions to discuss how they have been or can be relaxed and/or justified. Consider a classical scalar field (minimally coupled) with lagrangian density given by

$$\mathcal{L} = -\frac{1}{2}\partial_\mu\phi\partial^\mu\phi - V(\phi). \tag{3.11}$$

For now I will ignore the interactions that ϕ must necessarily have with other fields in the theory. As it will turn out they must be weak for inflation to work, so that this assumption is a reasonable one. The stress-energy tensor for this field is then

$$T_{\mu\nu} = -\partial_\mu\phi\partial_\nu\phi - \mathcal{L}g_{\mu\nu} \tag{3.12}$$

Assuming that in the region of interest ϕ is spatially-constant, $T_{\mu\nu}$ takes on the perfect fluid form with energy density and pressure given by

$$\rho = \frac{1}{2}\dot{\phi}^2 + V(\phi)(+(\nabla\phi)^2/2R^2), \tag{3.13a}$$

$$p = \frac{1}{2}\dot{\phi}^2 - V(\phi)(-(\nabla\phi)^2/6R^2), \tag{3.13b}$$

where I have included the spatial gradient terms for future reference. [Note, once inflation begins the spatial gradient terms decrease rapidly, $(\nabla\phi)^2/R^2 \propto R^{-2}$, for

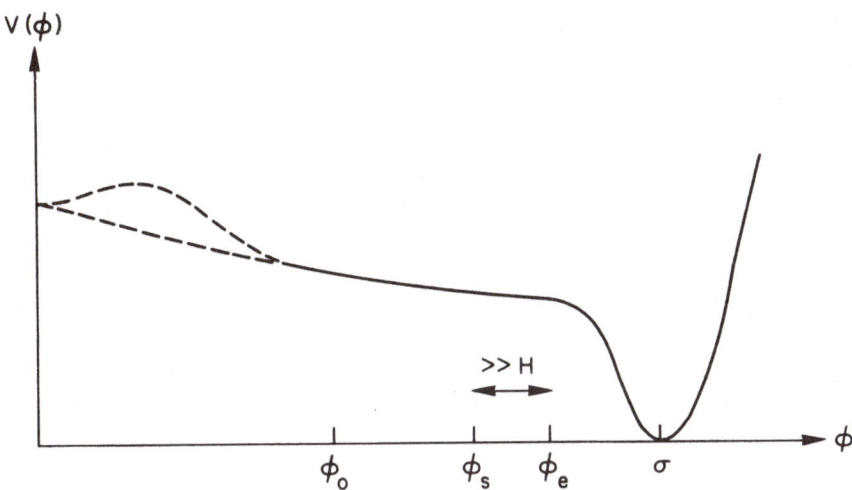

Figure 3.7: Schematic plot of the potential required for inflation. The shape of the potential for $\phi \ll \sigma$ determines how the barrier between $\phi = 0$ and $\phi = \sigma$ (if one exists) is penetrated. The value of ϕ after barrier penetration is taken to be ϕ_0; the flat region of the potential is the interval $[\phi_s,\ \phi_e]$.

wavelengths $\gtrsim H^{-1}$, and quickly become negligible.] That the spatial gradient term in ϕ be unimportant is crucial to inflation; if it were to dominate the pressure and energy density, then $R(t)$ would grow as t (since $p = -\frac{1}{3}\rho$) and not exponentially.

The equations of motion for ϕ can be obtained either by varying the action or by using $T^{\mu\nu}_{;\nu} = 0$. In either case the resulting equation is:

$$\ddot{\phi} + 3H\dot{\phi}(+\Gamma\dot{\phi}) + V'(\phi) = 0. \tag{3.14}$$

I have explicitly included the $\Gamma\dot{\phi}$ term which arises due to particle creation. The $3H\dot{\phi}$ friction term arises due to the expansion of the Universe; as the scalar field gains momentum, that momentum is redshifted away by the expansion.

This equation, which is analogous to that for a ball rolling with friction down a hill with a valley at the bottom, has two qualitatively different regimes, each of which has a simple, approximate, analytic solution. Fig. 3.7 shows schematically the potential $V(\phi)$.

(1) The slow-rolling regime

In this regime the field rolls at terminal velocity and the $\ddot{\phi}$ term is negligible. This occurs in the interval where the potential is very flat, the conditions for sufficient flatness being[95]:

$$|V''| \leq 9H^2, \tag{3.15a}$$

$$|V'm_{pl}/V| \leq (48\pi)^{1/2}, \tag{3.15b}$$

Condition (3.15a) usually subsumes condition (3.15b), so that condition (3.15a) generally suffices. During the slow-rolling regime the equation of motion for ϕ reduces to

$$\dot{\phi} \simeq -V'/3H. \tag{3.16}$$

During the slow-rolling regime particle creation is exponentially suppressed[94] because the timescale for the evolution of ϕ (which sets the energy/momentum scale of the particles created) is much greater than the Hubble time (which sets the physics horizon), i.e., any particles radiated would have to have wavelengths much larger than the physics horizon, which results in the exponential suppression of particle creation during this epoch. Thus, the $\Gamma\dot{\phi}$ term can be neglected during the 'slow roll'.

Suppose the interval where conditions (3.15a,b) are satisfied is $[\phi_s, \phi_e]$, then the number of e-folds of expansion which during the time ϕ is evolving from $\phi = \phi_s$ to $\phi = \phi_e$ ($\equiv N$) is

$$N \equiv \int H dt \simeq -3 \int_{\phi_s}^{\phi_e} H^2 d\phi/V'(\phi) \simeq -(8\pi/m_{pl}^2) \int V(\phi)d\phi/V'(\phi). \tag{3.17}$$

[Note that $R_e/R_s \equiv \exp(N)$ since $\dot{R}/R \equiv H$.] Taking H^2/V' to be roughly constant over this interval and approximating V' as $\simeq \phi V''$ (which is approximately true for polynominal potentials) it follows that

$$N \approx 3H^2/V'' \geq 3.$$

If there is a region of the potential where the evolution is friction-dominated, then N will necessarily be greater than 1 (by condition (3.15a)).

(2) Coherent field oscillations

In this regime

$$|V''| \gg 9H^2,$$

and ϕ evolves rapidly, on a timescale $\ll$ the Hubble time H^{-1}. Once ϕ reaches the bottom of its potential, it will oscillate with an angular frequency equal to $m_\phi \equiv V''(\sigma)^{1/2}$. In this regime it proves useful to rewrite Eqn(3.14) for the evolution of ϕ as

$$\dot{\rho}_\phi = -3H\dot{\phi}^2 - \Gamma\dot{\phi}^2. \tag{3.18}$$

where

$$\rho_\phi \equiv 1/2\dot{\phi}^2 + V(\phi).$$

Once ϕ is oscillating about $\phi = \sigma$, $\dot{\phi}^2$ can be replaced by its average over a cycle

$$< \dot{\phi}^2 >_{cycle} = \rho_\phi,$$

and Eqn(3.18) becomes

$$\dot{\rho}_\phi = -3H\rho_\phi - \Gamma\rho_\phi \tag{3.19}$$

which is nothing else but the equation for the evolution of the energy density of zero momentum, massive particles with a decay width Γ.

Referring back to Eqn(3.13) we can see that the cycle average of the pressure (i.e., space-space components of $T_{\mu\nu}$) is zero—as one would expect for NR particles. The coherent ϕ oscillations are in every way equivalent to a very cold condensate of ϕ particles. The decay of these oscillations due to quantum particle creation is equivalent to the decay of zero-momentum ϕ particles.

The complete set of semi-classical equations for the reheating of the Universe is

$$\dot{\rho}_\phi = -3H\rho_\phi - \Gamma\rho_\phi, \tag{3.20a}$$

$$\dot{\rho}_r = -4H\rho_r + \Gamma\rho_\phi, \tag{3.20b}$$

$$H^2 = 8\pi G(\rho_r + \rho_\phi)/3, \tag{3.20c}$$

where $\rho_r = (\pi^2/30)g_* T^4$ is the energy density in the relativistic particles produced by the decay of the coherent field oscillations. [I have tacitly assumed that the decay products of ϕ rapidly thermalize; Eqn(3.20b) is correct whether or not the decay products thermalize, so long as they are relativistic.] The evolution for the energy density in the scalar is easy to obtain

$$\rho_\phi = M^4(R/R_e)^{-3}\exp[-\Gamma(t-t_e)], \tag{3.21}$$

where I have set the initial energy equal to M^4, the initial epoch being when the scalar field begins to evolve rapidly (at $R = R_e$, $\phi = \phi_e$, and $t = t_e$).

From $t = t_e$ until $t \simeq \Gamma^{-1}$, the energy density of the Universe is dominated by the coherent sloshings of the scalar field ϕ, set into motion by the initial vacuum energy associated with $\phi \ll \sigma$. During this phase

$$R(t) \propto t^{2/3}$$

that is, the Universe behaves as if it were dominated by NR particles—which it is!

Interestingly enough it follows from Eqn(3.20) that during this time the energy density in radiation is actually decreasing ($\rho_r \propto R^{-3/2}$—see Fig. 3.8). [During the first Hubble time after the end of inflation ρ_r does increase.] However, the all important entropy per comoving volume is increasing

$$S \propto R^{15/8}.$$

When $t \simeq \Gamma^{-1}$, the coherent oscillations begin to decay exponentially, and the entropy per comoving volume levels off—indicating the end of the reheating epoch. The temperature of the Universe at this time is,

$$T_{RH} \simeq g_*^{-1/4}(\Gamma m_{pl})^{1/2} \tag{3.22}$$

If Γ^{-1} is less than H^{-1}, so that the Universe reheats in less than an expansion time, then all of the vacuum is converted into radiation and the' Universe is reheated to a temperature

$$T_{RH} \simeq g_*^{-1/4}M \quad (\text{if } \Gamma \geq H) \tag{3.22'}$$

the so-called case of good reheating.

To summarize the evolution of the scalar field ϕ: early on ϕ evolves very slowly, on a timescale $\gg$ the Hubble time H^{-1}; then as the potential steepens (and $|V''|$ becomes $> 9H^2$) ϕ begins to evolve rapidly, on a timescale $\ll$ the Hubble time H^{-1}. As ϕ oscillates about the minimum of its potential the energy density in these oscillations dominates the energy density of the Universe and behaves like NR matter ($\rho_\phi \propto R^{-3}$); eventually when $t \simeq \Gamma^{-1}$, these oscillations decay exponentially, 'reheating' the Universe to a temperature of $T_{RH} \simeq g_*^{-1/4}(\Gamma m_{pl})^{1/2}$ (if $\Gamma > H$, so that the Universe does not e-fold in the time it takes the oscillations to decay, then $T_{RH} \simeq g_*^{-1/4}M$). Saying that the Universe reheats when $t \simeq \Gamma^{-1}$ is a bit paradoxical as the temperature has actually been *decreasing* since shortly after the ϕ oscillations began. However, the fact that the temperature of the Universe was actually once greater than T_{RH} for $t < \Gamma^{-1}$ is probably of no practical use since the entropy per comoving volume increases until $t \simeq \Gamma^{-1}$—by a factor of $(M^2/\Gamma m_{pl})^{5/4}$, and any interesting objects that might be produced (e.g., net baryon number, monopoles, etc.) will be diluted away by the subsequent entropy production. By any reasonable measure, T_{RH} is the reheat temperature of the Universe. The evolution of ρ_ϕ, ρ_r, and S are summarized in Fig. 3.8.

Armed with our detailed knowledge of the evolution of ϕ we are ready to calculate the precise number of e-folds of inflation necessary to solve the horizon and flatness problems and to discuss direct baryon number production. First consider the requisite number of e-folds required for sufficient inflation. To solve the homogeneity problem we need to insure that a smooth patch containing an entropy of at least 10^{88} results from inflation. Suppose the initial bubble or fluctuation region has a size $H^{-1} \simeq m_{pl}/M^2$—certainly it is not likely to be significantly larger than this. During inflation it grows by a factor of $\exp(N)$. Next, while the Universe is dominated by coherent field oscillations it grows by a factor of

$$(R_{RH}/R_e) \simeq (M^4/T_{RH}^4)^{1/3},$$

where T_{RH} is the reheat temperature. Cubing the size of the patch at reheating (to obtain its volume) and multiplying its volume by the entropy density ($s \approx T_{RH}^3$), we obtain

$$S_{patch} \simeq e^{3N} m_{pl}^3/(M^2 T_{RH}).$$

Insisting that S_{patch} be greater than 10^{88}, it follows that

$$N \geq 53 + \frac{2}{3}\ln(M/10^{14}\text{GeV}) + \frac{1}{3}\ln(T_{RH}/10^{10}\text{GeV}). \tag{3.23}$$

Varying M from 10^{19}GeV to 10^8GeV and T_{RH} from 1GeV to 10^{19}GeV this lower bound on N only varies from 36 to 68.

The flatness problem involves the smallness of the ratio

$$x = (k/R^2)/(8\pi G\rho/3)$$

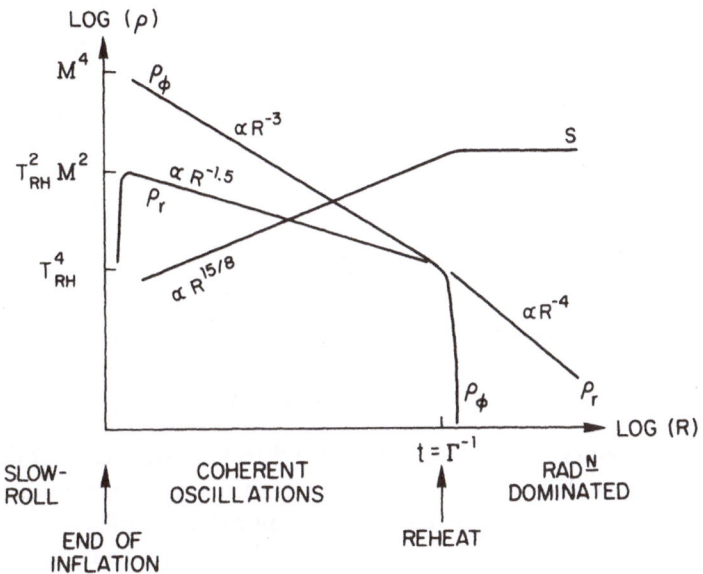

Figure 3.8: The evolution of ρ_ϕ, ρ_r, and S during the epoch when the Universe is dominated by coherent ϕ-oscillations. The reheat temperature $T_{RH} \simeq g_*^{-1/4}(\Gamma m_{pl})^{1/2}$. The maximum temperature achieved after inflation is actually greater, $T_{max} \simeq (T_{RH} M)^{1/2}$.

required at early times. Taking the pre-inflationary value of x to be x_i and remembering that

$$x(t) \propto \begin{cases} R^{-2} & \rho = \text{cons't} \\ R & \rho \propto R^{-3} \\ R^2 & \rho \propto R^{-4} \end{cases}$$

it follows that the value of x today is

$$x_{today} = x_i e^{-2N}(M/T_{RH})^{4/3}(T_{RH}/10eV)^2(10eV/3K).$$

Insisting that x_{today} be at most of order unity implies that

$$N \geq 53 + \ln(x_i) + \frac{2}{3}\ln(M/10^{14}\text{GeV}) + \frac{1}{3}\ln(T_{RH}/10^{10}\text{GeV})$$

—up to the term $\ln(x_i)$, precisely the same bound as we obtained to solve the homogeneity problem. Solving the isotropy problem depends upon the initial anisotropy present; during inflation isotropy decreases exponentially (see refs. 100).

Finally, let's calculate the baryon asymmetry that can be directly produced by the decay of the ϕ particles themselves. Suppose that the decay of each ϕ particle results in the production of net baryon number ϵ. This net baryon number might be produced directly by the decay of a ϕ particle (into quarks and leptons) or indirectly through an intermediate state ($\phi \to X\bar{X}$; X, $\bar{X} \to$ quarks and leptons; e.g., X might be a superheavy, color triplet Higgs[101]). The baryon asymmetry produced per volume is then

$$n_B \simeq \epsilon n_\phi.$$

On the other hand we have

$$(g_* \pi^2/30) T_{RH}^4 \simeq n_\phi m_\phi.$$

Taken together it follows that[93,102]

$$n_B/s \simeq (3/4)\epsilon T_{RH}/m_\phi. \tag{3.24}$$

This then is the baryon number per entropy produced by the decay of the ϕ particles directly. If the reheat temperature is not very high, baryon number non-conserving interactions will not subsequently reduce the asymmetry significantly. Note that the baryon asymmetry produced only depends upon the *ratio* of the reheat temperature to the ϕ particle mass. This is important, as it means that a very low reheat temperature can be tolerated, so long as the ratio of it to the ϕ particle mass is not too small.

3.5. ORIGIN OF DENSITY INHOMOGENEITIES

To this point I have assumed that ϕ is precisely uniform within a given bubble or fluctuation region. As a result, each bubble or fluctuation region resembles a perfectly isotropic and homogeneous Universe after reheating. However, because of deSitter space produced quantum fluctuations, ϕ cannot be exactly uniform, even within a small region of space. It is a well-known result that a massless and non-interacting scalar field in deSitter space has a spectrum of fluctuations given by (see, e.g., ref. 103)

$$(\Delta\phi)^2 \equiv (2\pi)^{-3} k^3 |\delta\phi_k|^2 = H^2/16\pi^3, \tag{3.25}$$

where

$$\delta\phi = (2\pi)^{-3} \int d^3k \delta\phi_k^{-ikx}, \tag{3.26}$$

and $\vec{x}$ and $\vec{k}$ are comoving quantities. This result is applicable to inflationary scenarios as the scalar field responsible for inflation must be very weakly-coupled and nearly massless. [That Universe is not precisely deSitter during inflation, i.e., $\rho + p = \dot{\phi}^2 \neq 0$, does not affect this result significantly; this point is addressed in ref. 104.] These deSitter space produced quantum fluctuations result in a calculable spectrum of adiabatic density perturbations. These density perturbations were first calculated by the authors of refs. 105-108; they have also been calculated by the authors of refs. 109 who addressed some of the technical issues in more detail. All the calculations done to date arrive at the same result. I will briefly describe the calculation in ref. 108; my emphasis here will be to motivate the result. I refer the reader interested in more details to the aforementioned references.

It is conventional to expand density inhomogeneities in a Fourier expansion

$$\delta\rho/\rho \;=\; (2\pi)^{-3} \int \delta_k e^{-ikx} d^3k. \tag{3.27}$$

The physical wavelength and wavenumber are related to comoving wavelength and wave number, λ and k, by

$$\lambda_{ph} \;=\; (2\pi/k)R(t) \equiv \lambda R(t),$$
$$k_{ph} \;=\; k/R(t).$$

The quantity most people refer to as $\delta\rho/\rho$ on a given scale is more precisely the RMS mass fluctuation on that scale

$$(\delta\rho/\rho)_k \;\equiv\; <(\delta M/M)^2>_k \;\simeq\; \Delta_k^2 \;\equiv\; (2\pi)^{-3}k^3|\delta_k|^2, \tag{3.28}$$

which is just related to the Fourier component δ_k on that scale.

The cosmic scale factor is often normalized so that $R_{today} = 1$; this means that given Fourier components are characterized by the physical size that they have today (neglecting the fact that once a given scale goes non-linear $(\delta\rho/\rho \gtrsim 1)$ objects of that size form bound 'lumps' that no longer participate in the universal expansion and remain roughly constant in size). The mass (in NR matter) contained within a sphere of radius $\lambda/2$ is

$$M(\lambda) \;\simeq\; 1.5 \times 10^{11} M_\odot (\lambda/\mathrm{Mpc})^3 \Omega h^2.$$

Although physics depends on physical quantities $(k_{ph},\ \lambda_{ph},\ \mathrm{etc.})$, the comoving labels k, M, and λ are the most useful way to label a given component as they have the affect of the expansions already scaled out.

I want to emphasize at the onset that the quantity $\delta\rho/\rho$ is not gauge invariant (under general coordinate transformations). This fact makes life very difficult when discussing Fourier components with wavelengths larger than the physics horizon (i.e., $\lambda_{ph} \gtrsim H^{-1}$). The gauge non-invariance of $\delta\rho/\rho$ is not a problem when $\lambda_{ph} \lesssim H^{-1}$, as the analysis is essentially Newtonian. The usual approach is to pick a convenient gauge (e.g., the synchronous gauge where $g_{oo} = -1$, $g_{oi} = 0$) and work very carefully (see refs. 110, 111). The more elegant approach is to focus on gauge-invariant quantities; see ref. 112. I will gloss over the subtleties of gauge invariance in my discussion, which is aimed at motivating the correct answer.

The evolution of a given Fourier component (in the linear regime—$\delta\rho/\rho \ll 1$) separates into two qualitatively different regimes, depending upon whether or not the perturbation is inside or outside the physics horizon $(\simeq H^{-1})$. When

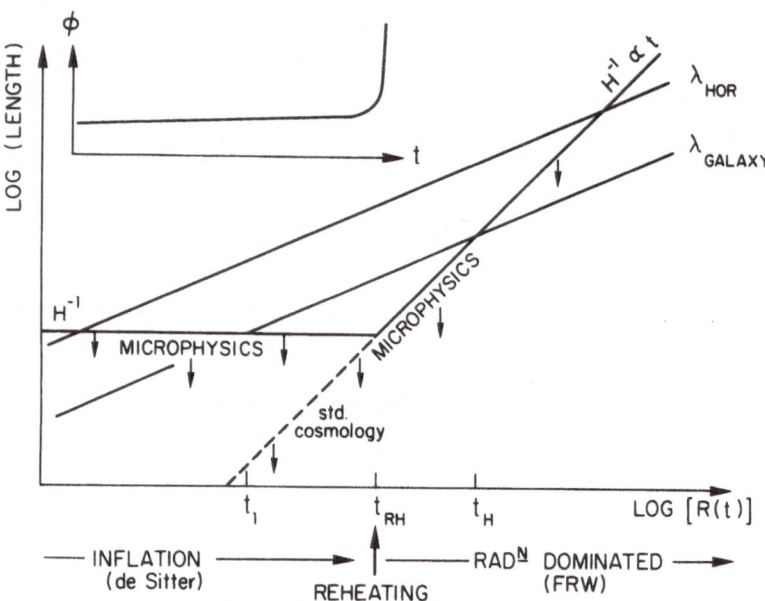

Figure 3.9: The evolution of the physical size of galactic- and (present) horizon-sized perturbations ($\lambda_{ph} \propto R$) and the size of the physics horizon H^{-1}. Causally-coherent microphysics operates only on scales $\leq H^{-1}$. In the standard cosmology a perturbation crosses the horizon but once as $H^{-1} \propto R^n (n > 1)$, making it impossible for microphysics to create density perturbations at early times. In the inflationary cosmology a perturbation crosses the horizon twice (since $H^{-1} \simeq$ const during inflation), and so microphysics can produce density perturbations at early times.

$\lambda_{ph} \leq H^{-1}$, microphysical processes can affect its evolution—such processes include: quantum mechanical effects, pressure support, free-streaming of particles, 'Newtonian gravity', etc. In this regime the evolution of the perturbation is very dynamical. When a perturbation is outside the physics horizon, $\lambda_{ph} \geq H^{-1}$, microphysical processes do not affect its evolution; in a very real sense its evolution is kinematic—it evolves as a wrinkle in the fabric of space–time.

In the standard cosmology, a given Fourier component crosses the horizon only once, starting outside the horizon and crossing inside at a time (see Fig. 3.9)

$$t \simeq (M/M_\odot)^{2/3} \text{ sec}$$

(valid during the radiation-dominated epoch). For this reason it is not possible to create adiabatic (more precisely, curvature) perturbations by causal microphysical processes which operate at early times[111,112]. In the standard cosmology, if adiabatic perturbations are present, they must be present *ab initio*. The smallness of the particle horizon at early times relative to the comoving volume occupied by the observable Universe today strikes again!

[It is possible for microphysical processes to create isothermal, more precisely isocurvature, perturbations. Once such perturbations cross inside the horizon they are characterized by a spectrum

$$(\delta\rho/\rho) \propto (M/M_H)^{-1/2}$$

or steeper. Here M_H is the horizon mass when the perturbations were created. Thus the earlier the processes operate, the smaller the perturbations are on interesting scales. By an appropriate choice of gauge it is possible to view these isothermal perturbations as adiabatic perturbations with a very steep spectrum, $\delta\rho/\rho \propto M^{-7/6}$; however, as must be the case, they cross the horizon with the amplitude mentioned above. For more details, see refs. 111, 112.]

Because the distance to the physics horizon remains approximately constant during inflation, the situation is very different in the inflationary Universe. All interesting scales start inside the horizon, cross outside the horizon and once again come inside the horizon (at the usual epoch); see Fig. 3.9. This means that causal microphysical processes can set up curvature perturbations on astrophysically-interesting scales. [This point seems to have been first appreciated by Press[113].]

Consider the evolution of a given Fourier component k. Early during the inflationary epoch $\lambda_{ph} \lesssim H^{-1}$, and quantum fluctuations in ϕ give rise to density perturbations on this scale. As the scale passes outside the horizon, say at $t = t_1$, microphysical processes become impotent, and $\delta\rho/\rho$ freezes out at a value,

$$\begin{aligned}(\delta\rho/\rho)_k &\simeq O(\dot{\phi}H\Delta\phi/M^4), \\ &\simeq O(\dot{\phi}H^2/M^4),\end{aligned} \tag{3.29}$$

as the scale leaves the horizon. From that point forward, the QM fluctuation is assumed to 'freeze in' and thereafter evolve classically. Note in the approximation that H and $\dot{\phi}$ are constant during the inflationary epoch the value of $\delta\rho/\rho$ as the perturbation leaves the horizon is independent of k. This scale independence of $\delta\rho/\rho$ when perturbations cross outside the horizon is of course traceable to the time translation invariance of deSitter space.

While outside the horizon the evolution of a perturbation is kinematical, independent of scale, and gauge dependent. There is a gauge independent quantity $(\equiv \varsigma)$ which remains constant while the perturbation is outside the horizon, and which at horizon crossing $(t = t_1$ and $t_H)$ is given by

$$\varsigma|_{\text{horizon crossing}} \simeq \delta\rho/(\rho + p),$$

$$\varsigma(t_1) = \varsigma(t_H)$$

$$\Rightarrow [\delta\rho/(\rho + p)]_{t=t_H} \simeq [\delta\rho/(\rho + p)]_{t=t_1}. \tag{3.30}$$

(see refs. 108 and 114 for more details). When the perturbation crosses back inside the horizon: $(\rho + p) = n\rho(n = 4/3-$ radiation-dominated; $n = 1$, matter-dominated) so that up to a numerical factor $(\delta\rho/\rho)|_{t_H} \simeq [\delta\rho/(\rho + p)]|_{t_H}$. During inflation, however, $\rho + p = \dot{\phi}^2 \ll \rho \simeq M^4$ so that $(\delta\rho/\rho)|_{t_1} \simeq (\dot{\phi}^2/M^4)[\delta\rho/(\rho + p)]|_{t_1}$. Note, $M^4/\dot{\phi}^2$ is typically a very large number. Eqns(3.29, 3.30) then imply

$$(\delta\rho/\rho)_H \equiv (\delta\rho/\rho)_{t=t_H} \simeq (M^4/\dot{\phi}^2)(\delta\rho/\rho)_{t_1} \simeq H^2/\dot{\phi}, \qquad (3.31)$$

Note that in the approximation that $\dot{\phi}$ and H are are constant during inflation the amplitude of $\delta\rho/\rho$ at horizon crossing $(= (\delta\rho/\rho)_H)$ is independent of scale. This fact is traceable to the time-translation invariance of the nearly-deSitter inflationary epoch and the scale-independent evolution of $(\delta\rho/\rho)$ while the perturbation is outside the horizon. The so-called scale-invariant or Zel'dovich spectrum of density perturbations was first discussed, albeit in another context, by Harrison[115] and Zel'dovich[116]. Scale-invariant adiabatic density perturbations are a generic prediction of inflation. [Because H and $\dot{\phi}$ are not precisely constant during inflation, the spectrum is not quite scale-invariant. However the scales of astrophysical interest, say $\lambda \simeq 0.1\text{Mpc} - 100\text{Mpc}$, cross outside the horizon during a very short interval, $\Delta N \simeq 6.9$, during which H, $\dot{\phi}$, and ϕ are very nearly constant. For most models of inflation the deviations are not expected to be significant; for further discussion see refs. 117, 118.] Although the details of structure formation are not presently sufficiently well understood to say what the initial spectrum of perturbations must have been, the Zel'dovich with an amplitude of about $10^{-4} - 10^{-5}$ is certainly a viable possibility.

Before moving on, let me be very precise about the amplitude of the inflation-produced adiabatic density perturbations. Perturbations which re-enter the horizon while the Universe is still radiation-dominated $(\lambda \leq \lambda_{eq} \simeq 13h^{-2}\text{Mpc})$, do so as a sound wave in the photons and baryons with amplitude

$$(\delta\rho/\rho)_H \equiv k^{3/2}|\delta_k|/(2\pi)^{3/2} \simeq H^2/(\pi^{3/2}\dot{\phi}) \qquad (3.32a)$$

Perturbations in non-interacting, relic particles (such as massive neutrinos, axions, etc.), which by the equivalence principle must have the same amplitude at horizon crossing, do not oscillate, but instead grow slowly $(\propto \ln R)$. By the epoch of matter-radiation equivalence they have an amplitude of 2–3 times that of the initial baryon–photon sound wave, or

$$(\delta\rho/\rho)_{MD} \simeq (2-3)(\delta\rho/\rho)_H \simeq (2-3)H^2/(\pi^{3/2}\dot{\phi}) \qquad (3.32b)$$

It is this amplitude which must be of order $10^{-5} - 10^{-4}$ for successful galaxy formation.

Perturbations which re-enter the horizon when the Universe is already matter-dominated (scales $\lambda \geq \lambda_{eq} \simeq 13h^{-2}\text{Mpc}$) do so with amplitude

$$(\delta\rho/\rho)_H \simeq k^{3/2}|\delta_k|/(2\pi)^{3/2} \simeq (H^2/10)/(\pi^{3/2}\dot{\phi}) \qquad (3.33)$$

Once inside the horizon they continue to grow (as $t^{2/3}$ since the Universe is matter-dominated).

When the structure formation problem is viewed as an initial data problem, it is the spectrum of density perturbations at the epoch of matter domination which is the relevant input spectrum. The shape of this spectrum has been carefully computed by the authors of ref. 119. Roughly speaking, on scales less than λ_{eq} the spectrum is almost flat, varying as $\lambda^{-3/4} \propto M^{-1/4}$ for scales around the galaxy scale ($\simeq$ 1Mpc). On scales much greater than λ_{eq}, $(\delta\rho/\rho) \propto \lambda^{-2} \propto M^{-2/3}$ (in the synchronous gauge where adiabatic perturbations grow as t^n; $n = 2/3$ matter dominated, $n = 1$ radiation dominated. Since these scales have yet to re-enter the horizon they have not yet achieved their horizon-crossing amplitude).

In order to compute the amplitude of the inflation-produced adiabatic density perturbations we need to evaluate $H^2/\dot\phi$ when the astrophysically-relevant scales crossed outside the horizon. Recall, in the previous section we computed when the comoving scale corresponding to the present Hubble radius crossed outside the horizon during inflation—up to 'ln terms' $N \simeq 53$ or so e-folds before the end of inflation, cf., Eqn.(3.23). The present Hubble radius corresponds to a scale of about 3000Mpc; therefore the scale λ must have crossed the horizon $\ln(3000\text{Mpc}/\lambda)$ e-folds later:

$$N_\lambda \simeq N_{HOR} - 8 + \ln(\lambda/\text{Mpc}) \simeq 45 + \ln(\lambda/\text{Mpc})$$
$$+ \frac{2}{3}\ln(M/10^{14}\text{GeV}) + \frac{1}{3}\ln(T_{RH}/10^{10}\text{GeV}).$$

Typically $H^2/\dot\phi$ depends upon N_λ to some power[117]; since N_λ only varies logarithmically ($\Delta N/N \simeq 0.14$ in going from 0.1Mpc to 3000Mpc), the scale dependence of the spectrum is almost always very minimal.

As mentioned earlier, a generic prediction of the inflationary Universe is that today Ω should be equal to one to a high degree of precision. Equivalently, that means

$$|(k/R^2)/(8\pi G\rho/3)| \ll 1$$

since

$$\Omega = 1/[1 - (k/R^2)/(8\pi G\rho/3)].$$

Therefore one might conclude that an accurate measurement of Ω would have to yield 1 to extremely high precision. However, because of the adiabatic density perturbations produced during inflation that is not the case. Adiabatic density fluctuations correspond to fluctuations in the local curvature

$$\delta\rho/\rho \simeq \delta(k/R^2)/(8\pi G\rho/3)$$

This means that should we be able to very accurately probe the value of Ω (equivalently the curvature of space) on the scale of our Hubble volume, say by using the Hubble diagram, we would necessarily obtain a value for Ω which is dominated by the curvature fluctuations on the scale of the present horizon,

$$\Omega_{obs} \simeq 1 + \delta(k/R^2)/(8\pi G\rho/3) \simeq 1 \pm O(10^{-4} - 10^{-5}),$$

and so we would obtain a value different from 1 by about a part in 10^4 or 10^5.

Finally, let me briefly mention that isothermal density perturbations can also arise during inflation. [Isothermal density perturbations are characterized by $\delta\rho = 0$, but $\delta(n_i/n_\gamma) \neq 0$ in some components. They correspond to spatial fluctuations in the local pressure due to spatial fluctuations in the local equation of state.] Such perturbations can arise from the deSitter produced fluctuations in other quantum fields in the theory.

The simplest example occurs in the axion-dominated Universe[120,121,122]. Suppose that Peccei-Quinn symmetry breaking occurs before or during inflation. Until instanton effects become important ($T \simeq$ few $100\,\text{MeV}$) the axion field $a = f_a\theta$ is massless and θ is in general not aligned with the minimum of its potential: $\theta = \theta_1 \neq 0$ (I have taken the minimum of the axion potential to be $\theta = 0$). Once the axion develops a mass (equivalently, its potential develops a minimum) θ begins to oscillate; these coherent oscillations correspond to a condensate of very cold axions, with number density $\propto \theta^2$. [For further discussion of the coherent axion oscillations see refs. 123–125.] During inflation deSitter space produced quantum fluctuations in the axion field gave rise to spatial fluctuations in θ_1:

$$\delta\theta \simeq \delta a/f_a \simeq H/f_a$$

Once the axion field begins to oscillate, these spatial fluctuations in the axion field correspond to fluctuations in the local axion to photon ratio

$$\delta(n_a/n_\gamma)/(n_a/n_\gamma) \simeq 2\delta\theta/\theta_1 \simeq 2H/(f_a\theta_1)$$

More precisely

$$(\delta n_a/n_a) = k^{3/2}|\delta a(k)|/(2\pi)^{3/2} = H/(2\pi^{3/2}f_\lambda\theta_1), \tag{3.34}$$

where f_λ is the expectation value of f_a when the scale λ leaves the horizon (in some models the expectation value of the field which breaks PQ symmetry evolves as the Universe is inflating so that f_λ can be $< f_a$). It is possible that these isothermal axion fluctuations can be important for galaxy formation in an axion-dominated, inflationary Universe[121].

3.6. SPECIFIC MODELS–PART I. INTERESTING FAILURES

(1) 'Old Inflation'

By old inflation I mean Guth's original model of inflation. In his original model the Universe inflated while trapped in the $\phi = 0$ false vacuum state. In order to inflate enough the vacuum had to be very metastable; however, that being the case, the bubble nucleation probability was necessarily small—so small that the bubbles that did nucleate never percolated, resulting in a Universe which resembled swiss cheese more than anything else[87]. The interior of an individual bubble was not suitable for our present Universe either. Because he was not considering flat potentials, essentially all of the original false vacuum energy resided in bubble walls rather than in vacuum energy inside the bubbles themselves. Although individual bubbles would grow to a very large size given enough time, their interiors would contain very little entropy (compared to the 10^{88} in our observed Universe). In sum, the Universe inflated all right, but did not 'gracefully exit' from inflation back to a radiation-dominated Universe—close, Alan, but no cigar!

(2) Coleman–Weinberg SU(5)

The first model of new inflation studied was the Coleman–Weinberg SU(5) GUT[88,89]. In this model the field which inflates is the 24-dimensional Higgs which also breaks SU(5) down to SU(3) × SU(2) × U(1). Let ϕ denote its magnitude in the SU(3) × SU(2) × U(1) direction. The one-loop, zero-temperature Coleman–Weinberg[126] potential is

$$V(\phi) = 1/2B\sigma^4 + B\phi^4\{\ln(\phi^2/\sigma^2) - 1/2\},$$
$$B = 25\alpha_{GUT}^2/16 \simeq 10^{-3} \tag{3.35}$$
$$\sigma \simeq 2 \times 10^{15}\text{GeV}$$

Due to the absence of a mass term $(m^2\phi^2)$, the potential is very flat near the origin (SSB arises due to one-loop radiative corrections[126]); for $\phi \ll \sigma$:

$$V(\phi) \simeq B\sigma^4/2 - \lambda\phi^4/4$$
$$\lambda \simeq |4B\ln(\phi^2/\sigma^2)| \simeq 0.1 \tag{3.36}$$

The finite temperature potential has a small temperature dependent barrier [height $O(T^4)$] near the origin [$\phi \simeq O(T)$]. The critical temperature for this transition is $O(10^{14} - 10^{15}\text{GeV})$, however the $\phi = 0$ vacuum remains metastable. When the temperature of the Universe drops to $O(10^9\text{GeV})$ or so, the barrier becomes low enough that the finite temperature action for bubble nucleation drops to order unity and the $\phi = 0$ false vacuum becomes unstable[89]. In analogy with solid state phenomenon it is expected that at this the temperature of the Universe will undergo 'spinodal decomposition', i.e., will break up into irregularly shaped regions within which ϕ is approximately constant (so-called fluctuation regions). Approximately the potential by Eqn(3.36). It is easy to solve for the evolution of ϕ in the slow-rolling regime [$|V''| \leq 9H^2$ for $\phi^2 \leq \phi_e^2 \simeq \sigma^2(\pi\sigma^2/m_{pl}^2|\ln(\phi^2/\sigma^2)|)$]

$$(H/\phi)^2 \simeq \frac{2\lambda}{3}N(\phi), \tag{3.37}$$

$$H^2 \simeq \frac{4\pi}{3}\frac{B\sigma^4}{m_{pl}^2}, \tag{3.38}$$

where $N(\phi) \equiv \int_\phi^{\phi_e} H dt$ is the number of e-folds of inflation the Universe undergoes while ϕ evolves from ϕ to ϕ_e. Clearly, the number of e-folds of inflation depends upon the initial value of $\phi(\equiv \phi_0)$; in order to get sufficient inflation ϕ_0 must be $\simeq O(H)$. Although one might expect ϕ_0 to be of this order in the fluctuation regions since $H \simeq 5 \times 10^9 \text{GeV} \simeq$ (temperature at which the $\phi = 0$ false vacuum loses its metastability), there is a fundamental difficulty. In using the semi-classical equations of motion to describe the evolution of ϕ one is implicitly assuming

$$\phi \simeq \phi_{cl} + \Delta\phi_{QM},$$
$$\Delta\phi_{QM} \ll \phi_{cl}$$

The deSitter space produced quantum fluctuations in ϕ are of order H. More specifically, it has been shown that[127,128]

$$\Delta\phi_{QM} \simeq (H/2\pi)(Ht)^{1/2}$$

Therein lies the difficulty—in order to achieve enough inflation the initial value of ϕ must be of the order of the quantum fluctuations in ϕ. At the very least this calls into question the semiclassical approximation.

The situation gets worse when we look at the amplitude of the adiabatic density perturbations:

$$(\delta\rho/\rho)_H \simeq (H^2/\pi^{3/2}\dot{\phi}) \tag{3.39}$$

$$\simeq (3/\pi^{3/2})(H^3/\lambda\phi^3) \tag{3.40}$$

$$(\delta\rho/\rho)_H \simeq (2/\pi)^{3/2}(\lambda/3)^{1/2}N^{3/2} \tag{3.41}$$

For galactic-scale perturbations $N \simeq 50$, implying that $(\delta\rho/\rho)_H \simeq 30$! Again, its clear that the basic problem is traceable to the fact that during inflation $\phi \leq H$.

The decay width of the ϕ particle is of order $\alpha_{GUT}\sigma \simeq 10^{13}\text{GeV}$ which is much greater than H (implying good reheating), and so the Universe reheats to a temperature of order 10^{14}GeV or so.

From Eqns(3.37, 3.41) it is clear that by reducing λ both problems could be remedied—however $\lambda \lesssim 10^{-13}$ is necessary[108]. Of course, as long as the inflating field is a gauge non-singlet λ is set by the gauge coupling strength. From this interesting failure it is clear that one should focus on weakly-coupled, gauge singlet fields for inflation.

(3) Geometric Hierarchy Model

The first model proposed to address the difficulty mentioned above, was a supersymmetric GUT[129,130]. In this model ϕ is a scalar field whose potential at tree level is absolutely flat, but due to radiative corrections develops curvature. In the model ϕ is also responsible for the SSB of the GUT. The potential for ϕ is of the form

$$V(\phi) \simeq \mu^4[c_1 - c_2 \ln(\phi/m_{pl})] \tag{3.42}$$

where $\mu \simeq 10^{12}\text{GeV}$ is the scale of supersymmetry breaking, and c_1 and c_2 are constants which depend upon details of the theory. This form for the potential is only valid away from its SSB minimum ($\sigma \simeq m_{pl}$) and for $\phi \gg \mu$. The authors presume that higher order effects will force the potential to develop a minimum for $\phi \simeq m_{pl}$. Since $V' \propto \phi^{-1}$ the potential gets flatter for large ϕ—which already sounds good.

The inflationary scenario for this potential proceeds as follows. The shape of the potential is not determined near $\phi = 0$; depending on the shape ϕ gets to some initial value, say $\phi = \phi_0$, either by bubble nucleation or spinodal decomposition. Then it begins to roll. During slow roll which begins when $|V''| \simeq 9H^2$ and $\phi_s \simeq (c_2/24\pi c_1)^{1/2}m_{pl}$,

$$H^2 \simeq \frac{8\pi}{3m_{pl}^2}c_1\mu^4 \qquad (3.43a)$$

$$(1 - \phi^2/m_{pl}^2) \simeq (c_2/4\pi c_1)N(\phi) \qquad (3.43b)$$

$$(\delta\rho/\rho)_H \simeq (H^2/\pi^{3/2}\dot{\phi}), \qquad (3.44a)$$

$$\simeq 8(8/3)^{1/2}(c_1^{3/2}/c_2)\mu^2\phi/m_{pl}^3. \qquad (3.44b)$$

Note that during the slow roll $(\phi \geq \phi_s)$

$$\frac{\phi}{H} \geq \frac{\phi_s}{H} \simeq \frac{c_2^{1/2}}{c_1} \frac{1}{8\pi} \frac{m_{pl}^2}{\mu^2},$$

$$\simeq 10^{13}c_2^{1/2}/c_1 \gg 1,$$

thereby avoiding the difficulty encountered in the Coleman–Weinberg where $\phi \leq H$ was required to inflate. For $c_1 \simeq O(1)$, $c_2 \simeq 10^{-8}$—acceptable values in the model, $(\delta\rho/\rho)_H \simeq 10^{-5}$ and $N(\phi_s) \simeq 4\pi c_1/c_2 \simeq 10^9$. The number of e-folds of inflation is very large—10^9. This is quite typical of the very flat potentials required to achieve $(\delta\rho/\rho) \simeq 10^{-4} - 10^{-5}$.

Now for the bad news. In this model ϕ is very weakly coupled—it only couples to ordinary particles through gravitational strength interactions. Its decay width is

$$\Gamma \simeq O(\mu^6/m_{pl}^5), \qquad (3.45)$$

which is much less than H (implying poor reheating) and leads to a reheat temperature of

$$T_{RH} \simeq O[(\Gamma m_{pl})^{1/2}], \qquad (3.46a)$$

$$\simeq O(\mu^3/m_{pl}^2), \qquad (3.46b)$$

$$\simeq 10\,\text{MeV}. \qquad (3.46c)$$

Such a reheat temperature safely returns the Universe to being radiation-dominated before primordial nucleosynthesis, and produces a smooth patch containing an enormous entropy—for $c_2 \simeq 10^{-8}$, $c_1 \simeq 1$, $S_{patch} \simeq (m_{pl}^3/\mu^2 T_{RH})e^{3N} \simeq 10^{35}\exp(3 \times 10^9)$, but does not reheat it to a high enough temperature for baryogenesis. Poor reheating is a problem which plagues almost all potentially viable models of inflation. Achieving $(\delta\rho/\rho)_H \lesssim 10^{-4}$ requires the scalar potential to be very flat, which necessarily means that ϕ is very weakly-coupled, and therefore $T_{RH}(\propto \Gamma^{1/2})$ tends to be very low.

(4) CERN SUSY/SUGR Models[131]

Early on members of the CERN theory group recognized that supersymmetry might be of use in protecting the very small couplings necessary in inflationary potentials from being overwhelmed by radiative corrections. They explored a variety of SUSY/SUGR models[131] (and dubbed their brand of inflation 'primordial inflation'). In the process, they encountered a difficulty which plagues almost all supersymmetric models of inflation based upon minimal supergravity theories.

It is usually assumed that at high temperatures the expectation value of an inflating field is at the minimum of its finite temperature effective potential (near $\phi = 0$); then as the Universe cools it becomes trapped there, and then eventually slowly evolves to the low temperature minimum (during which time inflation takes place). In SUSY models $\langle\phi\rangle_T$ is not necessarily zero at high temperatures. In fact in essentially all of their models $\langle\phi\rangle_T > 0$ and the high temperature minimum smoothly evolves into the low temperature minimum (as shown in Fig. 3.10)[132]. As a result the Universe in fact would never have inflated!

There are two obvious remedies to this problem: (i) arrange the model so that $\langle\varphi\rangle_T \le 0$ (as shown in Fig. 3.10), then ϕ necessarily gets trapped near $\phi = 0$; or (ii) assume that ϕ is never in thermal equilibrium before the phase transition so that ϕ is not constrained to be in the high temperature minimum of its finite temperature potential at high temperatures. Variants of the CERN models[131] based on these two remedies have been constructed by Ovrut and Steinhardt[133] and Holman, Ramond, and Ross[134]

3.7. LESSONS LEARNED–A PRESCRIPTION FOR SUCCESSFUL NEW INFLATION

The unsuccessful models discussed above have proven to be very useful in that they have allowed us to 'write a prescription' for the kind of potential that will successfully implement inflation[117]. The following prescription incorporates these lessons, together with other lessons which have been learned (sometimes painfully). As we will see all but the last of the prescribed features, that the potential be part of a sensible particle physics model, are relatively easy to arrange.

(1) The potential should have an interval which is sufficiently flat so that ϕ evolves slowly (relative to the expansion timescale H^{-1})—that is, flat enough so that a slow–rollover transition ensues. As we have seen, that means an interval

$$[\phi_s, \ \phi_e]$$

where

$$|V''| \le 9H^2,$$
$$|V'm_{pl}/V| \le (48\pi)^{1/2}.$$

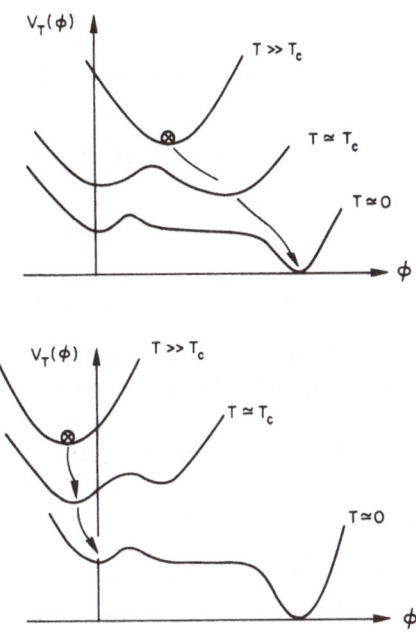

Figure 3.10: In SUSY/SUGR models $\langle\varphi\rangle_T$ is not necessarily equal to zero. If $\langle\varphi\rangle_T >$ 0, then there is the danger that $\langle\varphi\rangle_T$ smoothly evolves into the zero temperature minimum of the potential, thereby eliminating the possibility of inflation (upper figure). A sure way of preventing this is to design the potential so that $\langle\varphi\rangle_T \le 0$ (lower figure).

(2) The length of the interval where ϕ evolves slowly should be much greater than $H/2\pi$, the scale of the quantum fluctuations, so that the semi-classical approximation makes sense. [Put another way the interval should be long enough so that quantum fluctuations do not quickly drive ϕ across the interval.] Quantitatively, this calls for

$$|\phi_e - \phi_s| \gg (H\Delta t)^{1/2}(H/2\pi)$$

where Δt is the time required for ϕ to evolve from $\phi = \phi_s$ to $\phi = \phi_e$. [More precisely, the semi-classical change in ϕ in a Hubble time, $\Delta\phi_{Hubble} \simeq -V'm_{pl}^2/8\pi V$, should be much greater than the increase in $\langle\phi^2\rangle_{QM}^{1/2} \simeq H/2\pi$, due to the addition of another quantum mode; see ref. 108.]

(3) In order to solve the flatness and homogeneity problems the time required for ϕ to roll from $\phi = \phi_s$ to $\phi = \phi_e$ should be greater than about 60 Hubble times

$$N \equiv \int_{\phi_s}^{\phi_e} H\, dt \simeq \int_{\phi_s}^{\phi_e} 3H^2 d\phi/(-V') \simeq 3H^2/V'' \geq 60$$

The precise formula for the minimum value of N is given in Eqn(3.23).

(4) The scalar field ϕ should be smooth in a sufficiently large patch (say size L) so that the energy density and pressure associated with the $(\nabla\phi)^2$ term is negligible:

$$1/2(\nabla\phi)^2 \simeq (\phi_0/L)^2 \ll V(\phi_0) \simeq M^4.$$

(Otherwise the $(\nabla\phi)^2$ term will dominate ρ and p, so that $R(t) \propto t$—that is, inflation does not occur.) Usually this condition is easy to satisfy, as all it requires is that

$$L \gg \phi_0/M^2 \simeq (\phi_0/m_{pl})H^{-1};$$

since ϕ_0 is usually $\ll m_{pl}$, $(\phi_0/m_{pl})H^{-1} \ll H^{-1}$—that is, ϕ only need be smooth on a patch comparable to the physics horizon H^{-1}. [I will discuss a case where it is not easy to satisfy—Linde's chaotic inflation.] Once inflation does begin, any initial inhomogeneities in ϕ are rapidly smoothed by the exponential expansion.

(5a) In order to insure a viable scenario of galaxy formation (and microwave anisotropies of an acceptable magnitude) the amplitude of the adiabatic density perturbations must be of order $10^{-5} - 10^{-4}$. [In a Universe dominated by weakly-interacting relic particles such as neutrinos or axions, $(\delta\rho/\rho)_{MD}$ must be a few $\times 10^{-5}$.] This in turn results in the constraint

$$\text{few} \times 10^{-5} \simeq (\delta\rho/\rho)_{MD} \simeq (2-3)(\delta\rho/\rho)_H \simeq (2-3)(H^2/\pi^{3/2}\dot{\phi})_{Galaxy},$$

$$(H^2/\dot{\phi})_{Galaxy} \simeq 10^{-4}$$

In general, this is by far the most difficult of the constraints (other than sensible particle physics) to satisfy and leads to the necessity of extremely flat potentials. I should add, if one has another means of producing the density perturbations necessary for galaxy formation (e.g., cosmic strings or isothermal perturbations), then it is sufficient to have

$$(\dot{H}^2/\dot{\phi})_{Galaxy} \lesssim 10^{-4}$$

(5b) Isothermal perturbations produced during inflation, e.g., as discussed for the case of an axion-dominated Universe, also lead to microwave anisotropies and possibly structure formation. The smoothness of the microwave background dictates that

$$(\delta\rho/\rho)_{ISO} \lesssim 10^{-4}$$

while if they are to be relevant for structure formation

$$(\delta\rho/\rho)_{ISO} \simeq 10^{-5} - 10^{-4}$$

In the case of isothermal axion perturbations this is easy to arrange to have $(\delta\rho/\rho)_{ISO} \ll 10^{-4}$ unless the scale of PQ symmetry is larger than about 10^{18} GeV.

(6a) Sufficiently high reheat temperature so that the Universe is radiation-dominated at the time of primordial nucleosynthesis ($t \simeq 10^{-2} - 10^2$ sec, $T \simeq 10$ MeV $- 0.1$ MeV). Only in the case of poor reheating is T_{RH} likely to be anywhere as low as 10 MeV, in which case $T_{RH} \simeq (\Gamma m_{pl})^{1/2}$ and the condition that T_{RH} be ≥ 10 MeV then implies

$$\Gamma \geq 10^{-23} \text{GeV} \simeq (6.6 \times 10^{-2} \text{ sec})^{-1}$$

(6b) The more stringent condition on the reheat temperature is that it be sufficiently high for baryogenesis. If baryogenesis proceeds in the usual way[68], the out-of-equilibrium decay of a supermassive particle whose interactions violate B, C, P conservation, then T_{RH} must be greater than about 1/10 the mass of the particle whose out-of-equilibrium decays are responsible for producing the baryon asymmetry. Assuming that this particle couples to ordinary quarks and leptons, its mass must be greater than 10^9 GeV or so to insure a sufficiently longlived proton, implying that the reheat temperature must be greater than about 10^8 GeV (at the very least). On the other hand if the baryon asymmetry can be produced by the decays of the ϕ particles themselves, then

$$n_B/s \simeq (0.75)(T_{RH}/m_\phi)\epsilon$$

and a very low reheat temperature may be tolerable

$$T_{RH} \simeq 10^{-10}\epsilon^{-1}m_\phi$$

where as usual ϵ is the net baryon number produced per ϕ-decay.

(7) If ϕ is not a gauge singlet field, as in the case of the original Coleman-Weinberg SU(5) model, one must be careful that 'ϕ rolls in the correct direction'. It was shown that for the original Coleman-Weinberg SU(5) models ϕ might actually begin to roll toward the SU(4) $\times$ U(1) minimum of the potential even though the global minimum of the potential was the SU(3) $\times$ SU(2) $\times$ U(1) minimum[135]. This is because near $\phi = 0$ the SU(4) $\times$ U(1) direction is usually the direction of steepest descent. This is the so-called problem of 'competing phases'. As mentioned earlier, the extreme flatness required to obtain sufficiently small density perturbations probably precludes the possibility that ϕ is a gauge non-singlet, so the problem of competing phases does not usually arise. [Although in SUSY/SUGR models ϕ is often complex and one has to make sure that it rolls in the correct direction.]

(8) In addition to the scalar density perturbations discussed earlier, tensor or gravitational wave perturbations also arise (these correspond to tensor perturbations in

the metric $g_{\mu\nu})^{136}$. The amplitude of these perturbations is easy to estimate. The energy density in a given gravitational wave mode (characterized by wavelength λ) is

$$\rho_{GW} \simeq m_{pl}^2 h^2/\lambda^2 \tag{3.47}$$

where h is the dimensionless amplitude of the wave. As each gravitational wave mode crosses outside the horizon during inflation deSitter space produced fluctuations lead to

$$(\rho_{GW})_{\lambda \simeq H^{-1}} \simeq H^4, \text{ or } h \simeq H/m_{pl}. \tag{3.48}$$

While outside the horizon the dimensionless amplitude h remains constant (h behaves like a minimally coupled scalar field), and so each mode enters the horizon with a dimensionless amplitude

$$h \simeq H/m_{pl}. \tag{3.49}$$

Gravitational wave perturbations with wavelength of order the present horizon lead to a quadrupole anisotropy in the microwave temperature of amplitude h. The upper limit to the quadrupole anisotropy of the microwave background ($\delta T/T \lesssim 10^{-4}$) leads to the constraint

$$H/m_{pl} \leq 10^{-4},$$
$$M \leq O(10^{17}\text{GeV}).$$

In turn this leads to a constraint on the reheat temperature (using $g_* \simeq 10^3$)

$$T_{RH} \leq g_*^{-1/4}M \leq \text{ few} \times 10^{16}\text{GeV}$$

(9) One has to be mindful of various particles which may be produced during the reheating process. Of particular concern are stable or very long-lived, NR particles (including other scalar fields which may be set into oscillation and thereafter behave like NR matter). Since $\rho_{NR}/\rho_R \propto R(t)$ and today $\rho_{NR}/\rho_R \simeq 3 \times 10^4$ or so one has to be careful that ρ_{NR}/ρ_R is sufficiently small at early times

$$\rho_{NR}/\rho_R \leq \begin{cases} 3 \times 10^4 & \text{today} \\ 10^{-8} & T = 1\text{GeV} \\ 10^{-18} & T = 10^{10}\text{GeV} \end{cases}$$

which is not always easy—just ask any experimentalist about suppressing some effect by 18 orders-of-magnitude!

Of particular concern in supersymmetric models are gravitinos which, if produced, can decay shortly after nucleosynthesis and photodissociate the light elements produced (particularly D and ^7Li)^{137}. [In fact, the constraint that gravitinos not be overproduced during the reheating process leads to the very restrictive bound: $T_{RH} \leq 10^9\text{GeV}$ or so.] In supersymmetric models where SUSY breaking is done ala Polonyi^{138}, the Polonyi field can be set into oscillation^{139} and these oscillations which behave like NR matter can come to dominate the energy density of

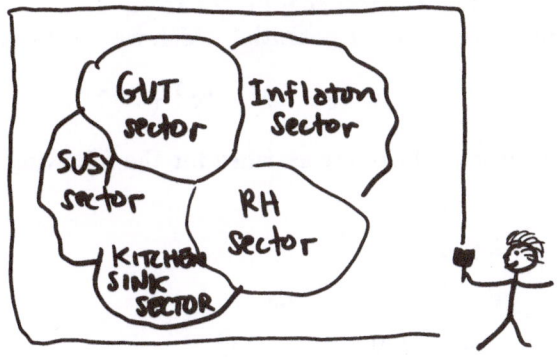

Figure 3.11: Constraint (11) in 'The Prescription for Successful Inflation'.

the Universe too early (leading to a Universe which if far too youthful when it cools to 3K) or even worse decay into dread gravitinos! In sum, one has to be mindful of the weakly-interacting, longlived particles which may be produced during reheating as they may eventually lead to an energy crisis.

(10) In SUSY/SUGR models where the scalar field responsible for inflation is in thermal equilibrium before the inflationary transition, one has to make sure that $\langle\varphi\rangle_T$ does not smoothly evolve into the zero temperature minimum of the potential. A sure way of doing this is to arrange to have

$$\langle\varphi\rangle_T \leq 0$$

this is the so-called thermal constraint[132].

(11) Last (in my probably incomplete list) but certainly not least, the scalar potential necessary for successful inflation should be but a part of a 'sensible, perhaps even elegant, particle physics model' (see Fig. 3.11). We do not want cosmology to be the tail that wags the dog!

These conditions are spelled out in more detail in ref. 117. In general they lead to a potential which is 'short and squat' and has a dimensionless coupling of order 10^{-15} somewhere. In order that radiative corrections not spoil the flatness, it is all but mandatory that ϕ be a gauge singlet field which couples very weakly to other fields in the theory.

[Suppose that ϕ has a nice, flat potential which will successfully implement inflation and has a ϕ^4 term whose coefficient $\lambda \simeq O(10^{-15})$ (as is usually the case). Now suppose that ϕ couples to another scalar field ψ or to a fermion field f through terms like: $\lambda'\psi^2\phi^2$ and $hff\phi$. One-loop corrections to the $\lambda\phi^4$ term in the scalar potential arise due to the coupling of ϕ to ψ of f: $(\lambda'^2\ln + h^4\ln)\phi^4$. In order that they not spoil the flatness of $V(\phi)$ by these 1-loop corrections, the couplings λ' and h must be small: $\lambda' \lesssim \ln^{-1/2}\lambda^{1/2}$; $h \lesssim \ln^{-1/4}\lambda^{1/4}$.]

To give an idea of the kind of potential which we are seeking consider

$$V = V_0 - a\phi^2 - b\phi^3 + \lambda\phi^4$$

The constraints discussed above are satisfied for the following sets of parameters

$$
SET\ 1 \begin{cases}
\lambda \leq 4 \times 10^{-16} \\
b \simeq 4 \times 10^7 \lambda^{3/2} m_{pl} \\
a \leq H^2/40 \simeq 10^{28}\lambda^3 m_{pl}^2 \\
\sigma \simeq 3 \times 10^7 \lambda^{1/2} m_{pl} \\
M \simeq V_0^{1/4} \simeq 3 \times 10^7 \lambda^{3/4} m_{pl} \\
\quad \simeq \lambda^{1/4}\sigma
\end{cases}
$$

$$
SET\ 2 \begin{cases}
V = \lambda(\phi^2 - \sigma^2)^2 \quad (b = 0,\ a = 2\lambda\sigma^2,\ V_0 = \lambda\sigma^4) \\
\sigma/m_{pl} = 1/2,\ 2,\ 3,\ 10 \\
\lambda = 2 \times 10^{-44},\ 5 \times 10^{-20},\ 10^{-15},\ 2 \times 10^{-15},\ 3 \times 10^{-16} \\
M \simeq \lambda^{1/4}\sigma
\end{cases}
$$

3.9. TWO SIMPLE MODELS THAT WORK

To date a handful of models that satisfy the prescription for successful inflation have been constructed[133,134,140−142,144,147−150]. Here, I will discuss two particularly simple and illustrative models. The first is an SU(5) GUT model proposed by Shafi and Vilenkin[141] and refined by Pi[142]. [Note, there is nothing special about SU(5); it could just as well be an E6 model.] I will discuss Pi's version of the model. In her model the inflating field $\vec{\phi}$ is a complex gauge singlet field whose potential is of the Coleman-Weinberg form[126]

$$V(\phi) = B[\phi^4 \ln(\phi^2/\sigma^2) + (\sigma^4 - \phi^4)/2]/4 \tag{3.50}$$

where $\phi = |\vec{\phi}|$ and B arises due to 1-loop radiative corrections from other scalar fields in the theory and is set to be $O(10^{-14})$ in order to successfully implement inflation. (Note, in Eqn(3.50) I have only explicitly shown the part of the potential relevant for inflation.) Since the 1-loop corrections due to other fields in the model are of order $(\lambda^2 \ln)\phi^4$ (λ is a typical quartic coupling, e.g., $\lambda\phi^2\psi^2$) the dimensionless couplings of ϕ to other fields in the theory must be of order 10^{-7} or so. In her model, $\vec{\phi}$ is the field responsible for Peccei–Quinn symmetry breaking; the vacuum expectation value of $|\vec{\phi}|$ breaks the PQ symmetry and the argument of $\vec{\phi}$ is the axion degree of freedom. In addition, the vacuum expectation value of $|\vec{\phi}|$ induces SU(5) SSB as it leads to a negative mass-squared term for the 24-dimensional Higgs in the theory which breaks SU(5) down to SU(3) × SU(2) × U(1). In order to have the correct SU(5) breaking scale, the vacuum expectation value of $|\vec{\phi}|$ must be of order 10^{18}GeV. In addition to the usual adiabatic density perturbations there are isothermal axion fluctuations of a similar magnitude[121]. The model reheats (barely) to a high enough temperature for baryogenesis. So far the model successfully implements inflation, albeit at the cost of a very small number ($B \simeq 10^{-14}$) whose origin is not explained and whose value is not stabilized (e.g., by supersymmetry).

The second model is a SUSY/SUGR model proposed by Holman, Ramond, and Ross[134] which is based on a very simple superpotential. They write the superpotential for the full theory as

$$W = I + S + G \qquad (3.51)$$

where I, S, G pieces are the inflation, SUSY, and GUT sectors respectively. For the I piece of the superpotential they choose the very simple form

$$I = (\Delta^2/M)(\varphi - M)^2, \qquad (3.52)$$

where $M = m_{pl}/(8\pi)^{1/2}$, Δ is an intermediate scale, and ϕ is the field responsible for inflation. Their potential has one free parameter: the mass scale Δ. This superpotential leads to the following scalar potential

$$\begin{aligned}V_I(\phi) &= \exp(|\phi|^2/M^2)[|\partial I/\partial\phi + \phi^* I/M^2|^2 - 3|I|^2/M^2] \\ &= \Delta^4 \exp(\phi^2/M^2)[\phi^6/M^6 - 4\phi^5/M^5 + 7\phi^4/M^4 - 4\phi^3/M^3 - \phi^2/M^2 + 1].\end{aligned}$$

Expanding the exponential one obtains

$$V_I(\phi) = \Delta^4(1 - 4\phi^3/M^3 + 6.5\phi^4/M^4 - 8\phi^5/M^5 + ...), \qquad (3.53a)$$
$$V_I' = \Delta^4(-12\phi^2/M^3 + 26\phi^3/M^4 - 40\phi^4/M^5 + ...) \qquad (3.53b)$$

It is sufficient to keep just the first two terms in $V_I(\phi)$ to solve the equations of motion

$$\phi/M \simeq [12(N(\phi) + 1/3)]^{-1} \qquad (3.54a)$$
$$H^2/\dot{\phi} \simeq (12\sqrt{3})^{-1}(\Delta/M)^2(\phi/M)^{-2} \simeq (12/\sqrt{3})(\Delta/M)^2 N^2, \qquad (3.54b)$$

By choosing $\Delta/M \simeq 9 \times 10^{-5}$ density perturbations of an acceptable magnitude result (and about 2×10^6 e-folds of inflation!). Taking $\Delta/M \simeq 9 \times 10^{-5}$ corresponds to an intermediate scale in the theory of about $\Delta \simeq 2 \times 10^{14}$GeV—a very suggestive value.

The ϕ field couples to other fields in the theory only by gravitational strength interactions and

$$\Gamma \simeq m_\varphi^3/M^2 \simeq \Delta^6/M^5, \tag{3.55}$$

where $m_\varphi^2 \simeq 8e\Delta^4/M^2$.

The resulting reheat temperature is

$$T_{RH} \simeq (\Gamma m_{pl})^{1/2} \simeq (\Delta/M)^3 M \simeq 10^6 \text{GeV}. \tag{3.56}$$

The baryon asymmetry in this model is produced directly by ϕ-decays ($\phi \to H_3\bar{H}_3$; $H_3\bar{H}_3 \to q's \; l's$; H_3 = color triplet Higgs

$$n_B/s \simeq (0.75\epsilon)T_{RH}/m_\phi$$
$$\simeq 10^{-1}\epsilon(\Delta/M)$$

Since $10^{-1}\Delta/M \simeq 10^{-5}$, a C, CP violation of about $\epsilon \simeq 10^{-5}$ is required to explain the observed baryon asymmetry of the Universe ($n_B/s \simeq 10^{-10}$).

Their model satisfies all the constraints for successful inflation except the thermal constraint. They argue that the thermal constraint is not relevant as the interactions of the ϕ field are too weak to put it into thermal equilibrium at early times. They therefore must take the initial value of ϕ ($\equiv \phi_0$) to be a free parameter and assume that in some regions of the Universe ϕ_0 is sufficiently far from the minimum so that inflation occurs ($\phi_0 \lesssim 10^{-3}M$). This model is somewhat ad hoc in that it contains a special sector of the theory whose sole purpose is inflation. Once again the model contains a small dimensionless coupling (the coefficient of the ϕ^4-term $\simeq 3 \times 10^{-16}$) or equivalently, a small mass ratio

$$(\Delta/M)^4 \simeq 10^{-16}$$

Since the model is supersymmetric that small number is stabilized against radiative corrections. Although the small ratio is not explained in their model, its value when expressed as a ratio of mass scales suggests that it might be related to one of the other small dimensionless numbers in particle physics (which also beg explanation)

$$(m_{GUT}/m_{pl}) \simeq 10^{-4}$$
$$(m_W/m_{pl}) \simeq 10^{-17}$$
$$g_e \simeq m_e/300\text{GeV} \simeq 10^{-6}$$

While neither of these models is particularly compelling and both have been somewhat contrived to successfully implement inflation, they are at the very least 'proof of existence' models which demonstrate that it is possible to construct a simple model which satisfies all the know constraints. Fair enough!

3.10. TOWARD THE INFLATIONARY PARADIGM

Guth's original model of inflation was based upon a strongly, first order phase transition associated with SSB of the GUT. The first models of new inflation were based upon Coleman-Weinberg potentials, which exhibit weakly-first order transitions. It now appears that the key feature needed for inflation is a very flat potential and that even potentials which lead to second order transitions (i.e., the $\phi = 0$ state is never metastable) will work just as well[143]. Most of the models for inflation now do not involve SSB, at least directly, they just involve the evolution of a scalar field which is initially displaced from the minimum of its potential. [There is a downside to this; in many models inflation is a sector of the theory all by itself.] Since the fields involved are very weakly coupled, thermal corrections can no longer be relied upon to set the initial value of ϕ. Inflation has become much more than just a scenario—it has become an early Universe paradigm!

On the horizon now are models which inflate, but are even more far removed from the original idea of a strongly-first order, GUT SSB phase transition; I'll discuss three of them here. Inflation—that is the rapid growth of our three spatial dimensions, appears to be a very generic phenomenon associated with early Universe microphysics.

(1) Chaotic Inflation

Linde[144] has proposed the idea that inflation might result from a scalar field with a very simple potential, say

$$V(\phi) = \lambda\phi^4$$

which due to 'chaotic initial conditions' (which thus far have not been well-defined) is displaced from the minimum of its potential—in this case $\phi = 0$ (see Fig. 3.12). With the initial condition $\phi = \phi_0$ this potential is very easy to analyze:

$$N(\phi_0) \equiv \int_{\phi_0}^{0} H dt \simeq \pi(\phi_0/m_{pl})^2, \tag{3.57}$$
$$(\delta\rho/\rho)_H \simeq (H^2/\dot\phi) \simeq (32/3)^{1/2}\lambda^{1/2}N(\phi)^{3/2}.$$

In order to obtain density perturbations of the proper amplitude ($\delta\rho/\rho \simeq 10^{-4}$) λ must be very small

$$\lambda \simeq 10^{-14}$$

—as usual! In order to obtain sufficient inflation, the initial value of ϕ must be

$$N(\phi_0) \simeq \pi(\phi_0/m_{pl})^2 \geq 60$$
$$\Rightarrow \phi_0 \geq 4.4 m_{pl}$$

Both of these two conditions are rather typical of potentials which successfully implement inflation. However, when one talks about truly chaotic initial conditions one wonders if a large enough patch exists where ϕ is approximately constant. Remember the key constraint is that the gradient energy density be small compared to the potential energy

$$(\nabla\phi)^2/2 \ll \lambda\phi_0^4$$

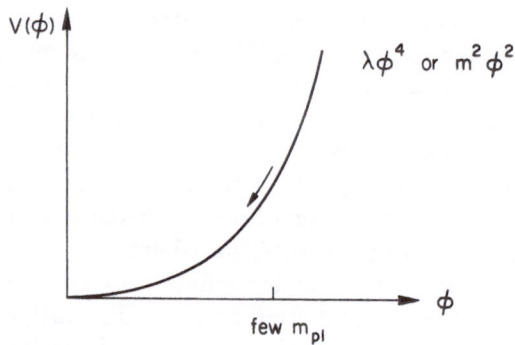

Figure 3.12: A potential for 'chaotic inflation'. In Linde's chaotic inflation, due to initial conditions, ϕ is displaced from the minimum of its potential ($\phi = 0$) and inflation occurs as it evolves to $\phi = 0$.

Labeling the typical dimension of the patch L, the above requirement translates to

$$L \gg \lambda^{-1/2}(m_{pl}/\phi_0)m_{pl}^{-1} \simeq 2(\phi_0/m_{pl})H^{-1}$$

which requires that L be rather large compared to the Hubble radius, therefore seeming to require rather special initial conditions. Still the simplicity of Linde's idea is very appealing.

[Note that the potential $V = \frac{1}{2}m^2\phi^2$ (corresponding to a massive scalar field) is also suitable for inflation. In this case

$$N(\phi_0) \simeq 2\pi(\phi_0/m_{pl})^2, \tag{3.58a}$$

$$(\delta\rho/\rho)_H \simeq H^2/\dot{\phi} \simeq 4(\pi/3)^{1/2}(m/m_{pl})N. \tag{3.58b}$$

Sufficient inflation requires: $\phi_0 \gtrsim 3m_{pl}$, and density perturbations of an acceptable magnitude requires: $(m/m_{pl}) \simeq 10^{-4}/(4N) \simeq 4 \times 10^{-7}$. This potential has been analyzed by I. Moss (private communication) and L. Jensen (private communication), and more recently by the authors of refs. 145.]

(2) Induced Gravity Inflation

Consider the Ginzburg-Landau theory of induced gravity based upon the effective Lagrangian[146]

$$\mathcal{L} = -\epsilon\phi^2 R/2 - \partial_\mu\phi\partial^\mu\phi/2 - \lambda(\phi^2 - v^2)^2/8, \tag{3.59}$$

where ϵ, λ are dimensionless couplings, R is the Ricci scalar, and $v \equiv \epsilon^{-1/2}(8\pi G)^{-1/2}$. The equation of motion for ϕ is

$$\ddot{\phi} + 3H\dot{\phi} + \dot{\phi}^2/\phi + [V' - 4V/\phi]/(1 + 6\epsilon) = 0 \tag{3.60a}$$

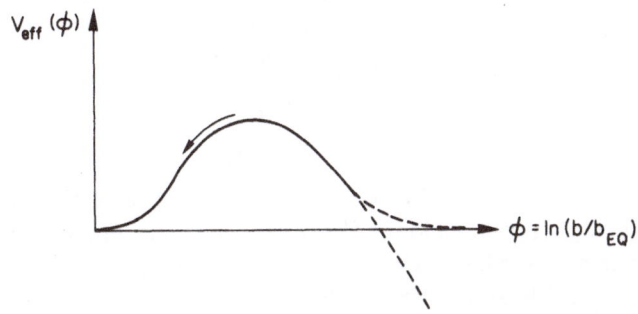

Figure 3.13: In theories with additional spatial dimensions there must be an effective potential associated with the size of the extra dimensions (shown here schematically). One might expect that very early on ($t \leq 10^{-43}$ sec) the size of the extra dimensions is displaced from its equilibrium value ($\equiv b_{eq}$), due to finite temperature corrections, initial conditions, or whatever. It is speculated that inflation might occur as the size of the extra dimensions evolves to its equilibrium value, thereby solving both the usual cosmological puzzles and the puzzle of why the extra dimensions are so small compared to our three familiar spatial dimensions.

supplemented by

$$H^2[1 + (2\dot{\phi}/\phi)/H] = (3\epsilon\phi^2)^{-1}[\dot{\phi}^2/2 + V(\phi)] \qquad (3.60b)$$

Successful inflationary scenarios can be constructed for $\phi_0 \ll v$ and for $\phi_0 \gg v$ ($\phi_0 =$ the initial value of ϕ), so long as $\epsilon \leq 10^{-2}$ and $\lambda \simeq O(10^{-12} - 10^{-16})$[147,148]. The small dimensionless coupling constant required in the scalar potential is by now a very familiar condition.

(3) The Compactification Transition

Ever increasing numbers of physicists are pursuing the idea that unification of the forces may require additional spatial dimensions (or as the optimist would say, unification of the forces is evidence for extra dimensions!), e.g., Kaluza-Klein theories, supergravity theories, and most recently, superstring theories. We know experimentally that these extra dimensions must be very small ($\ll 10^{-17}$cm) and indeed in most theories the extra dimensions form a compact manifold of characteristic size 10^{-34}cm or so. If our space-time is truly more than four dimensional, then we have yet another problem to add to our list of puzzling cosmological facts—the extreme smallness of the extra spatial dimensions, some $62 \simeq log(10^{28}\text{cm}/10^{-34}\text{cm})$ or so orders of magnitude smaller than the more familiar three spatial dimensions. The possible use of inflation to explain this largeness problem has not escaped the attention of researchers in this field.

In these theories there is a natural candidate for the 'inflating field' (which is also automatically a gauge singlet)—the radius of the extra dimensions. If there are extra dimensions there must be some dynamics which determine their present, equilibrium size ($\equiv b_{eq}$), and in principle one should be able to construct an effective potential associated with the size of the extra dimensions

$$V_{eff} = V(\phi),$$
$$\phi = \ln(b/b_{eq}).$$

(see Fig. 3.13). [The substitution $\phi = \ln(b/b_{eq})$ results in the usual kinetic term for ϕ when the higher dimensional Einstein equations are written down.] If the extra dimensions are displaced from their equilibrium value—an idea which seems not at all unreasonable since very early on ($t < 10^{-43}$ sec) one might expect all the dimensions to be on equal footing, then while they are evolving to their equilibrium value ($\phi = 0$) the Universe will be endowed with a large potential energy (and may very well inflate), thereby explaining the largeness of our three spatial dimensions as well as the usual cosmological puzzles. Inflationary models involving the compactification transition have already been investigated and the results are encouraging[149].

3.11. LOOSE ENDS

Inflation offers the possibility of making the present state of the Universe (on scales as large as our Hubble radius) insensitive to the initial data for the Universe. Since we have little hope of ever knowing what the initial data were this is a very attractive proposition. It has by no means yet achieved that lofty goal. There are a number of loose ends (and perhaps even a loose thread which may unravel the entire tapestry). I will briefly mention a few of them here.

(1) 'Who is ϕ?'

Inflationary models exist in which the scalar field ϕ: effects SSB of the GUT[140,141,142], effects SSB of SUSY[133], induces Newton's constant (in a Landau-Ginzburg model of induced gravity)[147,148], is $\sim \ln(r_X/r_{XEQ})$ (where r_X is the radius of compactified extra dimensions) in theories with extra dimensions which become compactified[149], is $\propto$ (scalar curvature)$^{1/2}$ in R^2 theories of gravity,[150] is just some 'random' scalar field[144], or is merely in the theory to effect inflation.[131,134] Given the number of different kinds of inflationary scenarios which exist, it seems as though inflation is generic to early Universe microphysics, occurring whenever a weakly-coupled scalar field finds itself displaced from the minimum of its potential. Clearly, a key question at this point is just how 'the inflation sector' of the theory fits into the Big Picture!

(2) What Determines the Initial Value of ϕ?

One thing is certain, and that is that ϕ must be very weakly-coupled, as quantified by its small dimensionless coupling constant. Because of this fact, it is almost certain that ϕ was not initially in thermal contact with the rest of the Universe and so the initial value of $\phi(\equiv \phi_0)$ is unlikely to be determined by thermal considerations (in the earliest models of new inflation, ϕ_0 was determined by thermal considerations, however these models resulted in density perturbations of an unacceptably large amplitude). At present ϕ_0 must be taken as initial data. Some have argued that ϕ_0 might be determined in an anthropic-like way, as regions of the Universe where ϕ_0 is sufficiently far displaced from equilibrium will undergo inflation and eventually occupy most of the physical volume of the Universe. Perhaps the wave-function of the Universe approach will shed some light on the initial distribution of the scalar field ϕ. Or it could be that due to 'as-of-yet unknown dynamics' ϕ was indeed in thermal equilibrium at a very early epoch. It goes without saying that it is crucial that ϕ be initially displaced from its minimum.

(3) Validity of the Semi-Classical Equations of Motion for ϕ

While it may seem perfectly plausible that ϕ evolves according to its semi-classical equations of motion, the validity of this assumption has troubled inflation-ists from the 'dawn of new inflation'. While a full quantum field theory treatment of inflation is very difficult and has not been effected, a number of specific issues have been addressed. Several authors have studied the role of inhomogeneities in ϕ, and have found that for the very weakly-coupled fields one is dealing with, mode coupling is not important and the individual modes are quickly smoothed by the exponential expansion of their physical wavelengths[151]. I already mentioned the necessity of having ϕ smooth over a sufficiently large region so that the gradient terms in the stress energy do not dominate.

The effect of quantum fluctuations on the evolution of ϕ has been studied in some detail by Guth and Pi[143], Fischler et al.[152], Linde[127], Vilenkin and Ford[128], Semenoff and Weiss[153], and Evans and McCarthy[154]. The basic conclusion that one draws from the work of these authors is that the use of the semi-classical equations of motion is valid so long as $\phi_{cl} \gg \Delta\phi_{QM} \simeq N^{1/2}H/2\pi$, which is almost always satisfied for the very flat potentials of interest to inflationists (at least for the last 50 or so e-folds which affect our present Hubble volume). [More precisely, the semi-classical change in ϕ in a Hubble time, $\Delta\phi_{Hubble} \simeq -V'/3H^2 \simeq -V'm_{pl}^2/(8\pi V)$, should be much greater than the increase in $< \phi^2 >_{QM}^{1/2}, \simeq H/2\pi$, due to the addition of another quantum mode; see Bardeen et al.[108].] At present the validity of the semi-classical equations of motion seems to be reasonably well established.

(4) No Hair Conjectures

While inflation has been touted from the very beginning as making the present state of the Universe insensitive to the initial spacetime geometry, not much has been done to justify this claim until very recently. As I mentioned earlier, inflation is nearly always analyzed in the context of a flat, FRW cosmological model, making such a claim somewhat dubious. However, it has now been shown that all of the homogeneous models (with the exception of the highly-closed models) undergo infla-tion, isotropize and remain isotropic to the present epoch providing that the model would have inflated the requisite 60 or so e-folds in the absence of anisotropy[155].

The proof of this result involves three parts. First, Wald[156] demonstrated that all homogeneous models with a positive cosmological term asymptotically approach deSitter (less the aforementioned highly-closed models which recollapse before the cosmological term becomes relevant). Wald's result follows because all forms of 'anisotropy energy density' decrease with increasing proper volume element, whereas the cosmological term remains constant, and so eventually triumphs. Of course, inflationary models do not, in the strictest sense, have a cosmological term, rather they have a positive vacuum energy as long as the scalar field is displaced from the minimum of its potential. Thus the dynamics of the scalar field comes into play: does ϕ stay displaced from the minimum of its potential long enough so that the vacuum energy comes to dominate? Due to the presence of anisotropy the expansion rate is *greater* than if there were only vacuum energy density, and so the friction felt by ϕ as it trys to roll (the $3H\dot{\phi}$ term) is greater and it takes ϕ longer to evolve to its minimum than without anisotropy. For this reason the Universe does become vacuum dominated before the vacuum energy disappears, and in fact the Universe inflates slightly longer in the presence of anisotropy (one or two e-folds)[157]. Finally, is the anisotropy reduced sufficiently so that the Universe today is still nearly isotropic? As it turns out, the requisite 60 or so e-folds needed to solve the other conundrums reduces the growing modes of anisotropy sufficiently to render them small today.

Allowing for inhomogeneous initial spacetimes makes matters much more difficult. Jensen and Stein-Schabes[158] and Starobinskii[159] have proven the analogue of Wald's theorem for spacetimes which are negatively-curved. Jensen and Stein-Schabes[158] have gone on to conjecture that spacetimes which have sufficiently large regions of negative curvature will undergo inflation, resulting in a Universe today which although not globally homogeneous, at least contains smooth volumes as large as our current Hubble volume.

Does this improve the situation that Collins and Hawking[63] discussed in 1973? While the work of Jensen and Stein-Schabes[158] seems to indicate that many inhomogeneous spacetimes undergo inflation and even leads one to speculate that the measure of the set of initial spacetimes which eventually inflate is non-zero, it is not possible to draw a definite conclusion without first defining a measure on the space of initial data. In fact, as Penrose[76] pointed out there is at least one way of defining the measure such that this is not the case. Consider the set of all Cauchy data at the present epoch; intuitively it is clear that those spacetime slices which are highly irregular are the rule, and those which are smooth in regions much larger than our current Hubble volume are the exception. Defining the measure today, it seems very reasonable that the smooth spacetime slices are a set of measure zero. Now evolve the spacetimes back to some initial epoch (for example $t = 10^{-43}$ sec). Using the seemingly very reasonable measure defined today and the mapping back to 'initial' spacetimes, one could argue that the set of initial data which inflate is still of measure zero. While I believe that this argument is technically correct, I also believe that it is silly. First, upon close examination of all of those initial spacetimes which led to spacetimes today without smooth regions as large as our present Hubble volume, one would presumably find that the scalar field responsible for inflation would be very close to the minimum of its potential (in order that they not inflate)–not a very generic initial condition. Secondly, if one adopts the point-of-view of an evolving Universe which has an 'initial epoch' (and not everyone does), then there is a preferred epoch at which one would define a measure–the 'initial epoch,' and at that epoch I believe any reasonably defined measure would lead to the set of initial spacetimes which inflate being of non-zero measure.

Although it is not possible yet to claim rigorously that inflation has resolved the problem of the seemingly special initial data required to reproduce the Universe we see today (at least within our Hubble volume), I think that any fairminded person would admit that it has improved the situation dramatically. Extrapolating from the solid results that exist, it seems to me that starting with a general inhomogeneous spacetime, there will exist regions which undergo inflation and which today are much larger than our present Hubble volume, thereby accounting for the smooth region we find ourselves in. From a more global perspective, one might expect that on scales $\gg H^{-1}$ the Universe would be highly irregular. [The evolution of a model universe which is isotropic and homogeneous except for one spherically-symmetric region of false vacuum (where $\phi \neq \sigma$) has been studied by the authors of ref. 160. The results are interesting in that they begin to address the problem of general initial conditions. The vacuum-dominated bubble becomes causally detached from the rest of the spacetime, becoming a 'child Universe' spawned by inflation.]

(5) The Present Vanishingly Small Value of the Cosmological Constant

Inflation has shed no light on this difficult and very fundamental puzzle (nor has anything else for that matter!). In fact, since inflation runs on vacuum energy so to speak, the fate of inflation hinges upon the resolution of this puzzle. For example, suppose there were a grand principle that dictated that the vacuum energy of the Universe is always zero, or that there were an axion-like mechanism which operated and ensured that any cosmological constant rapidly relaxed to zero; either would be a disaster to inflation, shorting out its source of power–vacuum energy. [Another possibility which has received a great deal of attention recently is the possibility that deSitter space might be quantum mechanically unstable[161]–of course, if its lifetime were at least 60 some e-folds that would not necessarily adversely affect inflation.]

3.12. INFLATION CONFRONTS OBSERVATION

No matter how appealing a theory may be, it must meet and pass the test of experimental verification. Experiment and/or observation is the final arbiter. One of the few blemishes on early Universe physics is the lack, thus far, of experimental/observational tests of the many beautiful and exciting predictions. That situation is beginning to change as the field starts to mature. Inflation is one of the early Universe theories which is becoming amenable to verification or falsification. Inflation makes the following very definite predictions (postdictions?):

(1) $\Omega = 1.0$ (more precisely, $R_{curv} = R(t)|k|^{-1/2} = H^{-1}/|\Omega - 1|^{1/2} \gg H^{-1}$)

(2) Harrison-Zel'dovich spectrum of constant curvature perturbations (and possibly isocurvature perturbations as well) and tensor mode gravitational wave mode perturbations

The prediction of $\Omega = 1.0$ together with the primordial nucleosynthesis constraint on the baryonic contribution, $0.014 \lesssim \Omega_B h^2 \lesssim 0.035 \lesssim 0.15$ (ref. 9), suggests that most of the matter in the Universe must be nonbaryonic. The simplest and most plausible possibility is that it exists in the form of relic WIMPs (Weakly-Interacting Massive Particles, e.g., axions, photinos, neutrinos; for a review, see ref. 162). Going a step further, these two original predictions then lead to testable consequences:

(3) $H_0 t_0 = 2/3$ (providing that the bulk of the matter in the Universe today is in the form of NR particles)

The observational data on both H_0 and t_0 are far from being definitive: $H_0 \simeq 40 - 100 \mathrm{km\ sec^{-1} Mpc^{-1}}$ and $t_0 \simeq 12 - 20\mathrm{Gyr}$, implying only that $H_0 t_0 \simeq 0.5 - 2.0$.

(4) $\Omega = 1.0$

All of the dynamical observations suggest that the fraction of critical density contributed by matter which is clumped on scales $\lesssim 10 - 30\mathrm{Mpc}$ is only about: $\Omega_{\lesssim 30} \simeq 0.2 \pm 0.1$ (± 0.1 is not meant to be a formal error estimate, but indicates the spread in the observations) (see refs. 11). If inflation is not to be falsified, that leaves but two options: (1) the observations are somehow misleading or wrong; or (2) there exists a component of energy density which is smoothly distributed on scales $\lesssim 10 - 30\mathrm{Mpc}$ (and therefore would not be reflected in the dynamical determinations). Candidates for the smooth component include: relic, light neutrinos, which by virtue of the large length scale ($\lambda_\nu \simeq 13h^{-2}\mathrm{Mpc}$) on which neutrino perturbations are damped by freestreaming, would likely still be smooth on these scales; relic relativistic particles produced by the recent decay of an unstable WIMP species;[163] a relic cosmological term;[164] 'failed galaxies,' referring to a population of galaxies which have the same mix of dark matter to baryons, but are more smoothly distributed and are too faint to observe (at least thus far);[165] relic population of light strings—either fast moving non-intercommuting strings or a tangled network of non-Abelian strings[166]. All of these smooth component scenarios have testable consequences[167]—their predictions for $H_0 t_0$ differ from 2/3; the growth of perturbations is different; the evolution of the cosmic scale factor $R(t)$ is different from the matter-dominated model and various kinematic tests (magnitude-redshift, angular size-redshift, lookback time-redshift, proper volume element-redshift, etc.) can in principle differentiate between them.

(5) Microwave Fluctuations

Both the scalar and tensor metric perturbations lead to fluctuations in the CMBR on large angular scales ($\gg 1°$). On such large scales causal microphysical processes (such as reionization) cannot have erased the primordial fluctuations, and so if ever present, they must still be there. The scalar perturbations (if they have anything to do with structure formation) must be of amplitude $\gtrsim$ few $\times 10^{-6}$, which is within a factor of 10 or less of the current upper limits on these scales.

(6) Two Detailed Stories of Structure Formation

The simplest possibility, namely that most of the mass density is in relic WIMPs ($\Omega_{WIMP} = 1.0 - \Omega_B \simeq 0.9$) leads to two very detailed scenarios of structure formation: hot dark matter (the case where the dark matter is neutrinos) and cold dark matter (essentially any other WIMP as dark matter). At present, the numerical simulations of these scenarios are sufficiently definite that it is possible to falsify them–and in fact, both of these simplest scenarios have difficulties (see the recent review by White[168]). In the hot dark matter case it is forming galaxies early enough. The large-scale structure which evolves in this case (voids, superclusters, froth) qualitatively agrees with what is observed; however, in order to get agreement with the galaxy-galaxy correlation function, galaxies must form very recently (redshifts $\lesssim 1$) in contradiction to all the galaxies (redshifts as large as 3.2) and QSO's (redshifts as large as 4.0) which are seen at redshifts $\gtrsim 1$.

With cold dark matter the simulations can nicely reproduce galaxy clustering, most of the observed properties of galaxies (masses and densities, rotation curves, etc)[169]. However the simulations do not seem to be able to produce sufficient large-scale structure. In particular, they fail to account for the amplitude of the cluster-cluster correlation function (by a factor of about 3), large amplitude, large-scale peculiar velocities, and voids. [In fairness I should mention that our knowledge of large-scale structure of the Universe is still very fragmentary, with the first moderate sized ($\sim 10^4$), 3-dimensional surveys having just recently been completed.] In order to account for $\Omega = 1.0$, galaxy formation must be biased (i.e., only density-averaged peaks greater than some threshold, typically $2 - 3\sigma$, are assumed to evolve into galaxies which we see today, the more typical 1σ peaks resulting in 'failed galaxies' for some reason or another; see ref. 165).

[The situation with respect to large scale structure is becoming more interesting every moment. Several groups have now reported large-amplitude ($600 - 1000$km sec^{-1}) peculiar velocities on large scales ($\sim 50h^{-1}$Mpc) (Burstein et al.[170]; Collins et al.[171]). Such large peculiar velocities are very difficult, if not impossible, to reconcile with either hot or cold dark matter (or even smooth component models) and the Zel'dovich spectrum (see ref. 172). If these data hold up they may pose an almost insurmountable obstacle to any scenario with the Zel'dovich spectrum of density perturbations. The frothy structure observed in the galaxy distribution by de Lapparent et al.[173], galaxies distributed on the surfaces on large ($\sim 30h^{-1}$Mpc), empty bubbles, although somewhat more qualitative, also seems difficult to reconcile with cold dark matter.]

There are a number of observations/experiments which can and will be done in the next few years and which should really put the inflationary scenario to the test. They include improved sensitivity measurements of the CMBR anisotropy. The microwave background anisotropies predicted in the hot dark matter scenario are very close to the observational upper limits on angular scales of both 5 or so arcminutes and $\gtrsim$ few degrees[62]. With cold dark matter, the predictions are a factor of $3 - 10$ away from the observational limits (for the isocurvature spectrum, the quadrupole upper limit may actually rule out this possibility; see, Efstathiou and Bond[174]). An improvement in sensitivity to microwave anisotropies of the order of $3 - 10$ could either begin to confirm one of the scenarios or rule them both out, and is definitely within the realm of experimental reality (Wilkinson in ref. 8).

The relic WIMP hypothesis for the dark matter can also be tested. While it was once almost universally believed that all WIMP dark matter candidates were, in spite of their large abundance, essentially impossible to detect because of the feebleness of their interactions, a number of clever ideas have recently been suggested (and are being experimentally implemented) for detecting axions[175], photinos, sneutrinos, heavy neutrinos, etc.[176] Results and/or limits will be forth coming soon. With the coming online of the Tevatron at Fermilab, the SLC at SLAC, and hopefully the SSC it is possible that one of the candidates may be directly produced in the lab. Experiments to detect neutrino masses in the eV mass range also continue.

A geometric measurement of the curvature of the Universe (which uses the dependence of the comoving volume element as a function of redshift) has recently been made by Loh and Spillar[177]. Their preliminary results indicate $\Omega = 0.9^{+0.7}_{-0.5}$ (95% confidence) (for a matter-dominated model). This technique appears to have great cosmological leverage and looks very promising (especially the value!)—far more promising than the traditional approach of determining the density of the Universe through the deceleration parameter q_0.

Another area with great potential for improvement is 3d surveys of the distribution of galaxies. The largest redshift surveys at present contain only a few 1000 galaxies, yet have been very tantalizing, indicating evidence of voids and froth-like structure to the galaxy distribution[173]. The large, automated surveys which are likely to be done in the next decade could very well lead to a quantum leap in our understanding of the large scale features of the Universe and help to provide hints as to how they evolved.

The peculiar velocity field of the Universe is potentially a very valuable and direct probe of the density field of the Universe:

$$|\delta v_k| = |\dot{\delta}_k/k| \quad (= (\lambda H/2\pi)\delta_k \ for \ \Omega = 1) \tag{3.61}$$

$$(\delta v/c)_\lambda \simeq (\lambda/10^4 h^{-1}\text{Mpc})(\delta\rho/\rho)_\lambda \tag{3.62}$$

where δ_k and δv_k are the $k-th$ Fourier components of $\delta\rho/\rho$ and $\delta v/c$ respectively. The very recent measurements which indicate large amplitude peculiar velocities on scales of $\sim 50h^{-1}\text{Mpc}$ are surprising in that they indicate substantial power on these scales, and are problematic to almost every scenario of structure formation. Should they be confirmed they will provide a very acute test of structure formation in inflationary models.

Of course, theorists are very accommodating and have already started suggesting alternatives to the simplest scenarios for structure formation. As I mentioned earlier, scenarios with a smooth component to the energy density have been put forward to solve the Ω problem. Cosmic strings present a radically different approach to structure formation with their non-gaussian spectrum of density fluctuations (for further discussion see refs. 183). [It is interesting to note that cosmic strings of the right 'weight' ($G\mu \simeq 10^{-6}$ or so, where μ is the string tension) seem to be somewhat incompatible with inflation, as they must necessarily be produced after inflation and require reheating to a temperature $\gtrsim \mu^{1/2} \simeq 10^{16}\text{GeV}$ which seems difficult.] Somewhat immodestly I mention a proposal Silk and I recently made: 'double inflation'[178]. While the Harrison-Zel'dovich spectrum is a beautiful prediction both because of its geometric simplicity and its definiteness, it may well be in conflict with observation because it does not seem to allow enough power on large scales to account for the recent observations of froth and large amplitude peculiar velocities. In the variant we have proposed there are two (or more) episodes of inflation, with the final episode lasting only about 40 e-folds or so, so that the amplitudes of perturbations on large scales are set by the first episode and those on small scales by the second episode. That there might be multiple episodes of inflation seems quite plausible given the number of different microphysical scenarios which result in inflation. Arranging the most recent episode to last for only 40 or so e-folds so that some of the scales within our present Hubble volume crossed outside the horizon during an earlier episode of inflation is a more formidable task—but not an impossible or implausible one! If this can be arranged then it is possible to have very large amplitude perturbations on small scales (of order 10^{-1}) and larger than usual amplitude perturbations on large scales (nearly saturating the large scale microwave limits), thereby providing enough power for the large scale structure which the recent redshift surveys and peculiar velocity measurements indicate. The large amplitude perturbations on small scales allow for very early galaxy formation (and reionization of the Universe, thereby erasing the CMBR fluctuations on small angular scales). If the second episode of inflation proceeds via the nucleation of bubbles, they might directly explain the froth-like structure recently reported by de Lapparent et al.[173]

3.13. EPILOGUE

Despite the absence of a compelling model which successfully implements the inflationary paradigm, inflation remains a very attractive means of accounting for a number of very fundamental cosmological facts by microphysics that we have some understanding of: namely, scalar field dynamics at sub-planck energies. The lack of a compelling model at present must be viewed in the light of the fact that at present we have no compelling, detailed model for the 'Theory of Everything' and the fact that despite vigorous scrutiny there has yet to be a No-Go Theorem for inflation unearthed. It is my belief that the undoing of inflation (if it should come) will involve observations and not theory. At the very least The Inflationary Paradigm is still worthy of further consideration–and I hope that I have convinced you of that fact!

Due to space/time limitations my review of inflation has necessarily been incomplete, for which I apologize. I refer the interested reader to the more complete reviews by Linde[179]; by Abbott and Pi[180]; by Steinhardt[181]; by Brandenberger[182]; by Bonometto and Masiero[185]; and by Blau and Guth[184]. My prescription for successfully implementing inflation borrows heavily from the paper written by Steinhardt and myself[117]. This work was supported in part by the DoE (at Chicago) and by my Alfred P. Sloan Fellowship.

REFERENCES

1. M.S. Turner, in *Architecture of the Fundamental Interactions at Short Distances*, eds. P. Ramond and R. Stora (North-Holland, Amsterdam, 1987); S. Bonometto and A. Masiero, *La Vista del Nuovo Cimento* **9**, 3 (1986); *Inner Space/Outer Space*, eds. E.W. Kolb et al. (University of Chicago Press, Chicago, 1986); *Proceedings of the First Jerusalem Winter School*, eds. T. Piran and S. Weinberg (World Scientific, Singapore, 1986); *Proceedings of the Erice School on Gauge Theory and the Early Universe*, eds. P. Galeotti and D. Schramm (Reidel, Dordrecht, 1987); K. Olive, in *Grand Unification With and Without Supersymmetry* (World Scientific, Singapore, 1984); A. Dolgov and Ya.B. Zel'dovich, *Rev.Mod.Phys.* **53**, 1 (1981); G. Steigman, *Ann.Rev.Nucl.Part.Sci.* **29**, 313 (1979); E. Kolb in *Proceedings of the 1986 Santa Cruz TASI*, ed. H. Haber (World Scientific, Singapore, 1987).

2. M. B. Green and J. H. Schwarz, *Nucl.Phys.* **B181**, 502 (1981); **B198**, 252 (1982); **B198**, 441 (1982); *Phys.Lett.* **109B**, 444 (1982); M. B. Green, J. H. Schwarz, and L. Brink, *Nucl.Phys.* **B198**, 474 (1982); J. H. Schwarz, *Phys. Rep.* **89**, 223 (1982); M. B. Green and J. H. Schwarz, *Phys.Lett.* **149B**, 117 (1984); **151B**, 21 (1985); *Nucl.Phys.* **B243**, 475 (1984); L. Alvarez-Gaume and E. Witten, *Nucl.Phys.* **B243**, 475 (1984); G. Chapline and N. Manton, *Phys.Lett.* **120B**, 105 (1983); D. Gross, J. Harvey, E. Martinec, and R. Rohm, *Phys.Rev.Lett.* **54**, 502 (1984); P. Candelas, G. Horowitz, A. Strominger, and E. Witten, *Nucl.Phys.* **B256**, 46 (1985); M. B. Green, *Nature* **314**, 409 (1985); E. Kolb, D. Seckel, and M. S. Turner, *Nature* **314**, 415 (1985); K. Huang and S. Weinberg, *Phys.Rev.Lett.* **25**, 895 (1970).

3. M. Srednicki, *Nucl.Phys.* **B202**, 327 (1982); D. V. Nanopoulos and K. Tamvakis, *Phys.Lett.* **110B**, 449 (1982); J. Ellis, J. S. Hagelin, D. V. Nanopoulos, K. Olive, and M. Srednicki, *Nucl.Phys.* **B238**, 453 (1984); A. Salam and J. Strathdee, *Ann. Phys. (NY)* **141**, 316 (1982); P. G. O. Freund, *Nucl.Phys.* **B209**, 146 (1982); P. G. O. Freund and M. Rubin, *Phys.Lett.* **37B**, 233 (1980); E. Kolb and R. Slansky, *Phys.Lett.* **135B**, 378 (1984); C. Kounnas *et al.*, *Grand Unification With and Without Supersymmetry* (World Scientific, Singapore, 1984); Q. Shafi and C. Wetterich, *Phys.Lett.* **129B**, 387 (1983); P. Candelas and S. Weinberg, *Nucl.Phys.* **B237**, 397 (1984); R. Abbott, S. Barr, and S. Ellis, *Phys.Rev.* **D30**, 720 (1984); E. Kolb, D. Lindley, and D. Seckel, *Phys.Rev.* **D30**, 1205 (1984).

4. S. Djorgovski, H. Spinrad, P. McCarthy, and M. Strauss, *Astrophys.J.* **299**, L1 (1985).

5. B.A. Peterson *et al.*, *Astrophys.J.* **260**, L27 (1982); D. Koo, *Astron.J.* **90**, 418 (1985); C. Hazard, R-G. McMahon, and W.L.W. Sargent, *Nature* **322**, 38 (1986); *Nature* **322**, 40 (1986); S.J. Warren, et al., *Nature* **325**, 131 (1987).

6. P.J.E. Peebles, *Physical Cosmology* (Princeton Univ. Press, 1971); S. Weinberg, *Gravitation and Cosmology* (Wiley, NY, 1972), chapter 15; *Physical Cosmology*, eds. J. Audouze, R. Balian, and D.N. Schramm (North-Holland, Amsterdam,

1980).

7. J. Peterson, P. Richards, and T. Timusk, *Phys.Rev.Lett.* **55**, 332 (1985); G.F. Smoot et al., *Astrophys.J.* **291**, L23 (1985); D. Meyer and M. Jura, *Astrophys.J.* **276**, L1 (1984); D. Woody and P. Richards, *Astrophys.J.* **248**, 18 (1981).

8. D. Wilkinson, in *Inner Space/Outer Space*, eds. E. Kolb et al. (Univ. of Chicago Press, Chicago, 1986).

9. J. Yang, M.S. Turner, G. Steigman, D.N. Schramm, and K. Olive, *Astrophys.J.* **281**, 493 (1984); A. Boesgaard and G. Steigman, *Ann.Rev.Astron.Astrophys.* **23**, 319 (1985).

10. J. Huchra, and A. Sandage and G. Tammann, in *Inner Space/Outer Space*, eds. E. Kolb et al. (Univ. of Chicago Press, Chicago, 1986); J. Gunn and B. Oke, *Astrophys.J.* **195**, 255 (1975); J. Kristian, A. Sandage, and J. Westphal, *Astrophys.* **221**, 383 (1978); R. Buta and G. deVaucouleurs, *Astrophys.J.* **266**, 1 (1983); W.D. Arnett, D. Branch, and J.C. Wheeler, *Nature* **314**, 337 (1985).

11. S. Faber and J. Gallagher, *Ann.Rev.Astron.Astrophys.* **17**, 135 (1979); V. Trimble, *Ann.Rev.Astron.Astrophys.* **25**, in press (1987).

12. I. Iben, Jr., *Ann.Rev.Astron.Astrophys.* **12**, 215 (1974); A. Sandage, *Astrophys.J.* **252**, 553 (1982); D.N. Schramm, in *Highlights of Astronomy*, ed. R. West (Reidel, Dordrecht, 1983), vol.6; I. Iben and A. Renzini, *Phys.Rep.* **105**, 329 (1984).

13. G. Gamow, *Phys.Rev.* **70**, 572 (1946).

14. R. A. Alpher, J. W. Follin, and R. C. Herman, *Phys.Rev.* **92**, 1347 (1953).

15. P. J. E. Peebles, *Astrophys. J.* **146**, 542 (1966); R. V. Wagoner, W. A. Fowler, and F. Hoyle, *Astrophys. J.* **148**, 3(1967).

16. A. Yahil and G. Beaudet, *Astrophys. J.* **206**, 26 (1976).

17. R. V. Wagoner, *Astrophys. J.* **179**, 343 (1973).

18. K. Olive, D. N. Schramm, G. Steigman, M. S. Turner, and J. Yang, *Astrophys. J.* **246**, 557 (1981).

19. D. A. Dicus *et al.*, *Phys.Rev.* **D26**, 2694 (1982).

20. J. Yang, M. S. Turner, G. Steigman, D. N. Schramm, and K. Olive, *Astrophys. J.* **281**, 493 (1984).

21. R. Epstein, J. Lattimer, and D. N. Schramm, *Nature* **263**, 198 (1976).

22. M. Spite and F. Spite, *Astron. Astrophys.* **115**, 357 (1982).

23. D. Kunth and W. Sargent, *Astrophys. J.* **273**, 81 (1983); J. Lequeux *et al.*, *Astron. Astrophys.* **80**, 155 (1979).

24. S. Hawking and R. Tayler, *Nature* **209**, 1278 (1966); J. Barrow, *Mon. Not. r. Astron. Soc.* **175**, 359 (1976); K. Thorne, *Astrophys.J.* **148**, 51 (1967).

25. R. Matzner and T. Rothman, *Phys.Rev.Lett.* **48**, 1565 (1982).

26. Y. David and H. Reeves, in *Physical Cosmology*, see ref. 6; N. Terasawa and K. Sato, *Astrophys.J.* **294**, 9 (1985).

27. A. Boesgaard and G. Steigman, *Ann.Rev.Astron.Astrophys.* **23**, 319 (1985).

28. S. Faber and J. Gallagher, *Ann. Rev. Astron. Astrophys.* **17**, 135 (1979); V. Trimble, *Ann.Rev.Astron.Astrophys.* **25**, in press (1987); *Dark Matter in*

the Universe (IAU 117), eds. J. Kormendy and G. Knapp (Reidel, Dordrecht, 1987).

29. M. Davis and P. J. E. Peebles, *Ann. Rev. Astron. Astrophys.* **21**, 109 (1983).

30. M.S. Turner, in *Dark Matter in the Universe*, p.445, see ref. 28.

31. V. F. Shvartsman, *JETP Lett.* **9**, 184 (1969).

32. G. Steigman, D. N. Schramm, and J. Gunn, *Phys.Rev.Lett.* **43**, 202 (1977); J. Yang, D. N. Schramm, G. Steigman, and R. T. Rood, *Astrophys. J.* **227**, 697 (1979); J. Barrow and J. Morgan, *Mon.Not.r.Astron.Soc.* **202**, 393 (1983); K.A. Olive, D.N. Schramm, and G. Steigman, *Nucl.Phys.* **B180**, 497 (1981); A.S. Szalay, *Phys.Lett.* **101B**, 453 (1981); E.W. Kolb and R.J. Scherrer, *Phys.Rev.* **D25**, 1481 (1982); E.W. Kolb, M.S. Turner, and T.P. Walker, *Phys.Rev.* **D34**, 2197 (1986); R.J. Scherrer and M.S. Turner, *Astrophys.J.*, in press (1987); M.H. Reno and D. Seckel, *Phys.Rev.D*, in press (1987).

33. G. Arnison *et al.*, *Phys.Lett.* **126B**, 398 (1983).

34. D. N. Schramm and G. Steigman, *Phys.Lett.* **141B**, 337 (1984); D. Cline, D.N. Schramm, and G. Steigman, *Comments on Nucl.Part.Phys.*, in press (1987).

35. R. J. Scherrer and M. S. Turner, *Phys.Rev.* **D33**, 1585 (1986); **D34**, 3263(E) (1986).

36. S. Wolfram, *Phys.Lett.* **82B**, 65 (1979); G. Steigman, *Ann. Rev. Nucl. Part. Sci.* **29**, 313 (1979).

37. B. Lee and S. Weinberg, *Phys.Rev.Lett.* **39**, 169 (1977); E. W. Kolb, Ph.D. thesis (Univ. of Texas, 1978); also see, E. Kolb and K. Olive, *Phys.Rev.* **D33**, 1202 (1986).

38. M. S. Turner, in *Proceedings of the 1981 Int'l. Conf. on ν Phys. and Astrophys.*, eds. R. Cence, E. Ma, and A. Roberts, 1, 95 (1981).

39. G. Steigman, *Ann. Rev. Astron. Astrophys.* **14**, 339 (1976).

40. F. Stecker, *Ann. NY Acad. Sci.* **375**, 69 (1981).

41. T. Gaisser, in *Birth of the Universe*, eds. J. Audouze and J. Tran Thanh Van (Editions Frontiers: Gif-sur-Yvette, 1982).

42. A. Sakharov, *JETP Lett.* **5**, 24 (1967).

43. M. Yoshimura, *Phys.Rev.Lett.* **41**, 281; **42**, 746(E) (1978).

44. D. Toussaint, S. Treiman, F. Wilczek, and A. Zee, *Phys.Rev.* **D19**, 1036 (1979).

45. S. Dimopoulos and L. Susskind, *Phys.Rev.* **D18**, 4500 (1978).

46. A. Ignatiev, N. Krasnikov, V. Kuzmin, and A. Tavkhelidze, *Phys.Lett.* **76B**, 486 (1978).

47. J. Ellis, M. Gaillard, and D. V. Nanopoulos, *Phys.Lett.* **80B**, 360; **82B**, 464(E) (1979).

48. S. Weinberg, *Phys.Rev.Lett.* **42**, 850 (1979).

49. E. Kolb and S. Wolfram, *Nucl.Phys.* **172B**, 224; *Phys.Lett.* **91B**, 217 (1980); J. Harvey, E. Kolb, D. Reiss, and S. Wolfram, *Nucl.Phys.* **201B**, 16 (1982).

50. J. N. Fry, K. A. Olive, and M. S. Turner, *Phys.Rev.* **D22**, 2953; 2977; *Phys.Rev.Lett.* **45**, 2074 (1980).

51. D. V. Nanopoulos and S. Weinberg, *Phys.Rev.* **D20**, 2484 (1979).

52. S. Barr, G. Segrè and H. Weldon, *Phys.Rev.* **D20**, 2494 (1979).

53. G. Segrè and M. S. Turner, *Phys.Lett.* **99B**, 339 (1981).

54. J. Ellis, M. Gaillard, D. Nanopoulos, and S. Rudaz, *Phys.Lett.* **99B**, 101 (1981).

55. E. W. Kolb and M. S. Turner, *Ann.Rev.Nucl.Part.Sci.* **33**, 645 (1983).

56. I. Affleck and M. Dine, *Nucl.Phys.* **B249**, 361 (1985).

57. V.A. Kuzmin, V.A. Rubakov, and M.E. Shaposhnikov, *Phys.Lett.* **155B**, 36 (1985).

58. F. Accetta, P. Arnold, E.W. Kolb, L. McLerran, and M.S. Turner, *Phys.Rev.D*, submitted (1987); E.W. Kolb and M.S. Turner, *Mod.Phys.Lett.A*, in press (1987).

59. C.B. Collins and M.J. Perry, *Phys.Rev.Lett.* **34**, 1353 (1975).

60. G.F.R. Ellis, *Ann.NY.Acad.Sci.* **336**, 130 (1980).

61. R. Sachs and A. Wolfe, *Astrophys.J.* **147**, 73 (1967).

62. N. Vittorio and J. Silk, *Astrophys.J.* **285**, L39 (1984); J.R. Bond and G. Efstathiou, *Astrophys.J.* **285**, L44 (1984); J. Silk, in *Inner Space/Outer Space*, eds. E.W. Kolb et al. (University of Chicago Press, Chicago, 1986).

63. C.B. Collins and S. Hawking, *Astrophys.J.* **180**, 317 (1973).

64. See, e.g., P.J.E. Peebles, *Large-Scale Structure of the Universe* (Princeton Univ. Press, Princeton, 1980); G. Efstathiou and J. Silk, *Fund.Cosmic Phys.* **9**, 1 (1983); S.D.M. White, in *Inner Space/Outer Space*, eds. E. Kolb et al. (Univ. of Chicago Press, Chicago, 1986).

65. R.H. Dicke and P.J.E. Peebles, in *General Relativity: An Einstein Centenary Survey*, eds. S.W. Hawking and W. Israel (Cambridge University Press, Cambridge, 1979).

66. G. Steigman, *Ann.Rev.Astron.Astrophys.* **14**, 339 (1976).

67. G.G. Ross, *Grand Unified Theories* (Benjamin/Cummings, Menlo Park, 1984).

68. E.W. Kolb and M.S. Turner, *Ann.Rev.Nucl.Part.Sci.* **33**, 645 (1983).

69. G. 't Hooft, *Nucl.Phys.* **B79**, 276 (1974); A.M. Polyakov, *JETP Lett.* **20**, 194 (1974).

70. T.W.B. Kibble, *J.Phys.* **A9**, 1387 (1976); J. Preskill, *Phys.Rev.Lett.* **43**, 1365 (1979); Ya. B. Zel'dovich and M. Yu. Khlopov, *Phys.Lett.* **79B**, 239 (1978).

71. J. Preskill, *Ann.Rev.Nucl.Part.Sci.* **34**, 461 (1984); M.S. Turner, in *Monopole '83*, ed. J. Stone (Plenum Press, NY, 1984).

72. A. Guth and S-H. H. Tye, *Phys.Rev.Lett.* **44**, 631 (1980).

73. A. Guth, *Phys.Rev.* **D23**, 347 (1981).

74. S. Bludman and M. Ruderman, *Phys.Rev.Lett.* **38**, 255 (1977); Ya. B. Zel'dovich, *Sov.Phys.Uspekhi* **11**, 381 (1968).

75. C.W. Misner, *Astrophys. J.* **151**, 431 (1968); in *Magic Without Magic*, ed. J. Klauder (Freeman, San Francisco, 1972); R. Matzner and C.W. Misner, *Astrophys. J.* **171**, 415 (1972).

76. R. Penrose, in *General Relativity: An Einstein Centenary Survey*, eds. S.W. Hawking and W. Israel (Cambridge University Press, Cambridge, 1979).

77. B. deWitt, *Phys.Rev.* **90**, 357 (1953).

78. L. Parker, *Nature* **261**, 20 (1976).

79. Ya.B. Zel'dovich, *JETP Lett.* **12**, 307 (1970).

80. A.A. Starobinskii, *Phys.Lett.* **91B**, 99 (1980).

81. P. Anderson, *Phys.Rev.* **D28**, 271 (1983).

82. J.M Hartle and B.-L Hu, *Phys.Rev.* **D20**, 1772 (1979).

83. M.V. Fischetti, J. Hartle, and B.-L. Hu, *Phys.Rev.* **D20**, 1757 (1979).

84. J.M. Hartle and S.W. Hawking, *Phys.Rev.* **D28**, 2960 (1983).

85. B.J. Carr and M.J. Rees, *Nature* **278**, 605 (1979); J.D. Barrow and F. Tipler, *The Anthropic Cosmological Principle* (Oxford University Press, Oxford, 1986).

86. D. Lindley, Fermilab preprint (1985).

87. A. Guth and E. Weinberg, *Nucl.Phys.* **B212**, 321 (1983); S. Hawking, I. Moss, and J. Stewart, *Phys.Rev.* **D26**, 2681 (1982).

88. A. Linde, *Phys.Lett.* **108B**, 389 (1982).

89. A. Albrecht and P.J. Steinhardt, *Phys.Rev.Lett.* **48**, 1220 (1982).

90. S. Coleman and F. de Luccia, *Phys.Rev.* **D21**, 3305 (1980); S. Coleman, *Phys.Rev.* **D15**, 2929 (1977); C.G. Callan and S. Coleman, *Phys.Rev.* **D16**, 1762 (1977).

91. S.W. Hawking and I.G. Moss, *Phys.Lett.* **110B**, 35 (1982); L. Jensen and P.J. Steinhardt, *Nucl.Phys.* **239B**, 176 (1984); K. Lee and E.J. Weinberg, *Nucl.Phys.* **B267**, 181 (1986).

92. A. Albrecht, P.J. Steinhardt, M.S. Turner, and F. Wilczek, *Phys.Rev.Lett.* **48**, 1437 (1982).

93. L. Abbott, E. Farhi, and M. Wise, *Phys.Lett.* **117B**, 29 (1982).

94. A. Dolgov and A. Linde, *Phys.Lett.* **116B**, 329 (1982).

95. P.J. Steinhardt and M.S. Turner, *Phys.Rev.* **D29**, 2162 (1984).

96. J.R. Gott, *Nature* **295**, 304 (1982); J.R. Gott and T.S. Statler, *Phys.Lett.* **136B**, 157 (1984).

97. M.S. Turner, *Phys.Lett.* **115B**, 95 (1982); J. Preskill, in *The Very Early Universe*, eds. G. Gibbons, S. Hawking, and S. Siklos (Cambridge Univ. Press, Cambridge, 1983); G. Lazarides, Q. Shafi, and W.P. Trower, *Phys.Rev.Lett.* **49**, 1756 (1982).

98. G. Lazarides and Q. Shafi, in *The Very Early Universe, ibid*; K. Olive and D. Seckel, in *Monopole '83*, ed. J. Stone (Plenum Press, NY, 1984).

99. M.S. Turner, *Phys.Rev.* **D28**, 1243 (1983).

100. J.D. Barrow and M.S. Turner, *Nature* **292**, 35 (1981); J.D. Barrow and D.H. Sonoda, *Phys.Rep.* **139**, 1 (1986).

101. D. Nanopoulos, K. Olive, and M. Srednicki, *Phys.Lett.* **127B**, 30 (1983).

102. R.J. Scherrer and M.S. Turner, *Phys.Rev.* **D31**, 681 (1985).

103. T. Bunch and P.C.W. Davies, *Proc.Roy.Soc.London* **A360**, 117 (1978); also see, G. Gibbons and S. Hawking, *Phys.Rev.* **D15**, 2738 (1977).

104. L. Abbott and M. Wise, *Nucl.Phys.* **B244**, 541 (1984).

105. S. Hawking, *Phys.Lett.* **115B**, 295 (1982).

106. A.A. Starobinskii, *Phys.Lett.* **117B**, 175 (1982).

107. A. Guth and S.-Y. Pi, *Phys.Rev.Lett.* **49**, 1110 (1982).

108. J. Bardeen, P. Steinhardt, and M.S. Turner, *Phys.Rev.* **D28**, 679 (1983).

109. R. Brandenberger, R. Kahn, and W. Press, *Phys.Rev.* **D28**, 1809 (1983); W. Fischler, B. Ratra, and L. Susskind, *Nucl.Phys.* **B259**, 730 (1985); R. Brandenberger, *Rev.Mod.Phys.* **57**, 1 (1985).

110. E.M. Lifshitz and I.M. Khalatnikov, *Adv.Phys.* **12**, 185 (1963); E.M. Lifshitz, *Zh.Ek.Teor.Fiz.* **16**, 587 (1946).

111. W. Press and E.T. Vishniac, *Astrophys.J.* **239**, 1 (1980).

112. J.M. Bardeen, *Phys.Rev.* **D22**, 1882 (1980).

113. W. Press, in *Cosmology and Particles*, eds. J. Audouze et al. (Editions Frontieres, Gif-sur-Yvette, 1981), p. 137.

114. J.A. Frieman and M.S. Turner, *Phys.Rev.* **D30**, 265 (1984); R. Brandenberger and R. Kahn, *Phys.Rev.* **D29**, 2172 (1984).

115. E. Harrison, *Phys.Rev.* **D1**, 2726 (1970).

116. Ya.B. Zel'dovich, *Mon.Not.Roy.Astron.Soc.* **160**, 1p (1972).

117. P.J. Steinhardt and M.S. Turner, *Phys.Rev.* **D29**, 2162 (1984).

118. M.S. Turner and E.T. Vishniac, in preparation (1985).

119. J. Bond and A. Szalay, *Astrophys.J.* **276**, 443 (1983); P.J.E. Peebles, *Astrophys.J.* **263**, L1 (1982); J. Bond, A. Szalay, and M.S. Turner, *Phys.Rev.Lett.* **48**, 1036 (1982).

120. M. Axenides, R. Brandenberger, and M.S. Turner, *Phys.Lett.* **120B**, 178 (1983); P.J. Steinhardt and M.S. Turner, *Phys.Lett.* **129B**, 51 (1983).

121. D. Seckel and M.S. Turner, *Phys.Rev.* **D32**, 3178 (1985).

122. A.D. Linde, *JETP Lett.* **40**, 1333 (1984); *Phys.Lett.* **158B**, 375 (1985).

123. J. Preskill, M. Wise, and F. Wilczek, *Phys.Lett.* **120B**, 127 (1983); L. Abbott and P. Sikivie, *Phys.Lett.* **120B**, 133 (1983); M. Dine and W. Fischler, *Phys.Lett.* **120B**, 137 (1983).

124. M.S. Turner, F. Wilczek, and A. Zee, *Phys.Lett.* **125B**, 35, 519(E) (1983); J. Ipser and P. Sikivie, *Phys.Rev.Lett.* **50**, 925 (1983).

125. M.S. Turner, *Phys.Rev.* **D33**, 889 (1986).

126. S. Coleman and E. Weinberg, *Phys.Rev.* **D7**, 1888 (1973).

127. A.D. Linde, *Phys.Lett.* **116B**, 335 (1982).

128. A. Vilenkin and L. Ford, *Phys.Rev.* **D26**, 1231 (1982).

129. S. Dimopoulos and S. Raby, *Nucl.Phys.* **B219**, 479 (1983).

130. A. Albrecht, S. Dimopoulos, W. Fischler, E. Kolb, S. Raby, and P. Steinhardt, *Nucl.Phys.* **B229**, 528 (1983).

131. J. Ellis, D. Nanopoulos, K. Olive, and K. Tamvakis, *Phys.Lett.* **118B**, 335 (1982); D.V. Nanopoulos, K. Olive, M. Srednicki, and K. Tamvakis, *Phys.Lett.* **123B**, 41 (1983); J. Ellis, K. Enqvist, D. Nanopoulos, K. Olive, and M. Srednicki, *Phys.Lett.* **152B**, 175 (1985); K. Enqvist and D. Nanopoulos, *Phys.Lett* **142B**, 349 (1984); C. Kounnas and M. Quiros, *Phys.Lett.* **151B**, 189 (1985); for a review of the CERN SUSY/SUGR models see D.V. Nanopoulos, *Comments on Astrophysics* **X**, 219 (1986).

132. B. Ovrut and P.J. Steinhardt, *Phys.Lett.* **133B**, 161 (1983); L. Jensen and K. Olive, *Nucl.Phys.* **B263**, 731 (1986).

112

133. B. Ovrut and P.J. Steinhardt, *Phys.Rev.Lett.* **53**, 732 (1984); *Phys.Lett.* **147B**, 263 (1984).

134. R. Holman, P. Ramond, and G.G. Ross, *Phys.Lett.* **137B**, 343 (1984); C.D. Coughlan, R. Holman, P. Ramond, and G.G. Ross, *Phys.Lett.* **140B**, 44 (1984); **158B**, 47 (1985).

135. J. Breit, S. Gupta, and A. Zaks, *Phys.Rev.Lett.* **51**, 1007 (1983).

136. A.A. Starobinskii, *JETP Lett.* **30**, 682 (1979); V. Rubakov, M. Sazhin, and A. Veryaskin, *Phys.Lett.* **115B**, 189 (1982); R. Fabbri and M. Pollock, *ibid* **125B**, 445 (1983); L. Abbott and M. Wise, *Nucl.Phys.* **B244**, 541 (1984).

137. J. Ellis, J. Kim, and D. Nanopoulos, *Phys.Lett.* **145B**, 181 (1984); L.L. Krauss, *Nucl.Phys.* **B227**, 556 (1983); M.Yu. Khlopov and A.D. Linde, *Phys.Lett.* **138B**, 265 (1984).

138. J. Polonyi, Budapest preprint KFKI 1977-93, unpublished (1977).

139. C. Coughlan, W. Fischler, E. Kolb, S. Raby, and G.G. Ross, *Phys.Lett.* **131B**, 54 (1983).

140. S. Gupta and H.R. Quinn, *Phys.Rev.* **D29**, 2791 (1984).

141. Q. Shafi and A. Vilenkin, *Phys.Rev.Lett.* **52**, 691 (1984).

142. S.-Y. Pi, *Phys.Rev.Lett.* **52**, 1725 (1984).

143. A. Guth and S.-Y. Pi, *Phys.Rev.* **D32**, 1899 (1985).

144. A.D. Linde, *Phys.Lett.* **129B**, 177 (1983).

145. V.A. Belinsky, L.P. Grischuk, I.M. Khalatnikov, and Ya.B. Zel'dovich, *Phys.Lett.* **155B**, 232 (1985); T. Piran and R.M. Williams, *ibid* **163B**, 331 (1985).

146. S. Adler, *Rev.Mod.Phys.* **54**, 729 (1982); L. Smolin, *Nucl.Phys.* **B160**, 253 (1979); A. Zee, *Phys.Rev.Lett.* **42**, 417 (1979).

147. F. Accetta, D. Zoller, and M.S. Turner, *Phys.Rev.* **D31**, 3046 (1985).

148. B.L. Spokoiny, *Phys.Lett.* **147B**, 39 (1984).

149. Q. Shafi and C. Wetterich, *Phys.Lett.* **129B**, 387 (1983); **152B**, 51 (1985).

150. A.A. Starobinskii, *Phys.Lett.* **91B**, 99 (1986); M.B. Mijic, M.S. Morris, and W.-M. Suen, *Phys.Rev.* **D34**, 2934 (1986).

151. A. Albrecht and R. Brandenberger, *Phys.Rev.* **D31**, 1225 (1985); *ibid* **D32**, 1280 (1985).

152. W. Fischler, B. Ratra, and L. Susskind, *Nucl.Phys.* **B259**, 730 (1985).

153. G. Semenoff and N. Weiss, *Phys.Rev.* **D31**, 689 (1985).

154. M. Evans and J. McCarthy, *Phys.Rev.* **D31**, 1799 (1985).

155. M.S. Turner and L. Widrow, *Phys.Rev.Lett.* **57**, 2237 (1986); L. Jensen and J. Stein-Schabes, *Phys.Rev.* **D34**, 931 (1986).

156. R.M. Wald, *Phys.Rev.* **D28**, 2118 (1983).

157. G. Steigman and M.S. Turner, *Phys.Lett.* **128B**, 295 (1983).

158. L. Jensen and J. Stein-Schabes, *Phys.Rev.*, in press (1987).

159. A.A. Starobinskii, *JETP Lett.* **37**, 66 (1983).

160. S.K. Blau, E.I. Guendelman, and A.H. Guth, *Phys.Rev.D*, in press (1987); K. Sato, M. Sasaki, H. Kodama, and K. Maeda, *Phys.Lett.* **108B**, 103 (1982); *Prog.Theor.Phys.* **65**, 1443 (1981); *ibid* **68**, 1979 (1982); *ibid* **66**, 2287 (1981).

161. N. Myhrvad, *Phys.Lett.* **132B**, 308 (1983); E. Mottola, *Phys.Rev.* **D31**, 754(1985); **D33**, 2136; L. Parker, *Phys.Rev.Lett.* **50**, 1009 (1983); L. Ford, *Phys. Rev.* **D31**, 710 (1985); P. Anderson, *Phys.Rev.* **D32**, 1302 (1985); J. Traschen and C.T. Hill, *Phys. Rev.* **D33**, 3519 (1986).

162. M.S. Turner, in *Dark Matter in the Universe*, eds. J Kormendy and G. Knapp (Reidel, Dordrecht, 1987), p. 445.

163. M.S. Turner, G. Steigman, and L.L. Krauss, *Phys.Rev.Lett.* **52**, 2090 (1984); D.A. Dicus, E.W. Kolb, and V. Teplitz, *Phys.Rev.Lett.* **39**, 168 (1977); K.A. Olive, D. Seckel, and E.T. Vishniac, *Astrophys.J.* **292**, 1 (1985).

164. P.J.E. Peebles, *Astrophys.J.* **284**, 439 (1984); M.S. Turner, et al. in ref. 111.

165. N. Kaiser, *Astrophys.J.* **273**, L17 (1983); J.M. Bardeen, J. Bond, N. Kaiser, and A. Szalay, *Astrophys.J.* **304**, 15 (1986).

166. A. Vilenkin, *Phys.Rev.Lett.* **53**, 1016 (1984).

167. J.C. Charlton and M.S. Turner, *Astrophys.J.* **313**, 495 (1987).

168. S.D.M. White, in *Inner Space/Outer Space*, eds. E.W. Kolb et al. (University of Chicago Press, Chicago, 1986).

169. G. Blumenthal, S. Faber, J. Primack, and M. Rees, *Nature* **311**, 517 (1984); M. Davis, G. Efstathiou, C. Frank, and S.D.M. White, *Astrophys.J.* **292**, 371 (1985).

170. D. Burstein, et al., in *Galaxy Distances and Deviations from Universal Expansion*, eds. B. Madore and R. Tully (Reidel, Dordrecht, 1987), p. 123.

171. C.A. Collins, et al., *Nature* **320**, 506 (1986).

172. N. Vittorio and M.S. Turner, *Astrophys.J.* **316**, in press (1987).

173. V. deLapparent, M. Geller, and J. Huchra, *Astrophys.J.* **302**, L1 (1986).

174. G. Efstathiou and J.R. Bond, *Mon.Not.r.Astron.Soc.* **218**, 103 (1986).

175. P. Sikivie, *Phys.Rev.Lett.* **51**, 1415 (1983).

176. M. Goodman and E. Witten, *Phys.Rev.* **D31**, 3059 (1985).

177. E. Loh and E. Spillar, *Astrophys.J.* **307**, L1 (1986); *Phys.Rev.Lett.* **57**, 2865.

178. J. Silk and M.S. Turner, *Phys.Rev.* **D35**, 419 (1986); M.S. Turner, et al., *Astrophys.J.*, in press (1987).

179. A. Linde, *Rep.Prog.Phys.* **47**, 925 (1984); *Comments on Astrophysics* **10**, 229 (1985); *Prog.Theo.Phys. (Suppl.)* **85**, 279 (1985).

180. *The Inflationary Universe*, eds. L. Abbott and S.-Y. Pi (World Publishing, Singapore, 1986).

181. P.J. Steinhardt, *Comments on Nucl.Part.Phys.* **12**, 273 (1984).

182. R. Brandenberger, *Rev.Mod.Phys.* **57**, 1 (1985).

183. A. Vilenkin, *Phys.Rep.* **121**, 263 (1985); A. Albrecht and N. Turok, *Phys.Rev.Lett.* **54**, 1868 (1985); N. Turok, *Phys.Rev.Lett.* **55**, 1801 (1985); R. Scherrer, *Astrophys.J.*, in press (1987); A. Melott and R. Scherrer, *Nature*, in press (1987).

184. S.K.Blau and A.H. Guth, in *300 Years of Gravitation*, eds. S.W. Hawking and W. Israel (Cambridge University Press, Cambridge, 1987).

185. S. Bonometto and A. Masiero, *La Rivista del Nuovo Cimento* **9**, 3 (1986).

SUPERSYMMETRY AND THE EARLY UNIVERSE

Graciela Gelmini[*]
Lyman Laboratory of Physics
Harvard University
Cambridge, MA 02138

Introduction

There are deep connections between supersymmetry and cosmology. Supersymmetry, actually super-gravity or superstring theories,[1] offers the hope of unifying all particle interactions. $N=1$ supersymmetry at "low" (with respect to the Planck scale) energies may produce a good phenomenology of particle physics while solving the gauge hierarchy problem in Grand Unified Theories (GUT). A viable supergravity model must also have an acceptable cosmology. For example, it must produce inflation, which is required not only to solve cosmological problems of the standard Big Bang model, such as the horizon, flatness/oldness, entropy problems and the production of density fluctuations. Inflation is also required to solve internal problems of the elementary particle models, for example to dilute the number of monopoles (or walls) in GUT's or the number of gravitinos in minimal supergravity. An important constraint on supersymmetric models is to avoid the "entropy crisis" on (Polyonyi" problem). This is a strong constraint on fields coupled only gravitationally to matter (such as the gravitino or fields in the "hidden sector" of supergravity models) which on decay generate too large entropy at late times (e.g., after nucleosynthesis). On the other hand cosmology may need supersymmetry not only to provide the Theory of Everything. Inflation may require supersymmetry to produce very flat potentials, needed to get small enough density perturbations during inflation and maybe, to have inflation at all (at least in semiclassical descriptions).

In the following we will present the "standard" supersymmetric candidates for dark matter. Then we will see cosmological problems generated by gravitinos and by the "hidden sector" of supergravity models (that is the "Polonyi" problem or "entropy crisis"). Inflation produced by the "hidden sector" of supergravity and the open problems in this approach is our next point. The so-called "thermal constraint", that is the request that at high temperatures the universe approaches the false vacuum of inflation, lead to different types of models, those which ignore it, those in which inflation is produced by two coupled scalar fields, those so-called "no scale" models. These last two types of models have an entropy crisis.

Supersymmetric dark matter candidates

One of the most appealing properties of supersymmetry is that it solves technically the gauge hierarchy problem in Grand Unified Theories. In those theories there are two very different scales of masses in the potential for the Higgs fields; the electroweak scale, $10^2 GeV$, and the GUT scale, $10^{14} - 10^{16}$ GeV. The hierarchy is not maintained by radiative corrections which add up to the small scale terms of the order of the large scale. Supersymmetry does not say *why* certain numbers in the original potential are much smaller than others in the first place, but once they are small supersymmetry insures that they remain small after radiative corrections, stabilizing the hierarchy. This is the "set and forget" policy or "technical" solution to the gauge hierarchy problem. Flat potentials require small couplings which must also be stabilized against radiative corrections. The gauge hierarchy solution and flat potentials for inflation are due to the absence of some divergencies in supersymmetric models. While supersymmetry is exact or broken by soft terms, there are not quadratic divergencies, and fewer logarithmic ones.[1,2] A loop of fermions has opposite sign than a loop of bosons and if the masses and quantum numbers of

W. G. Unruh and G. W. Semenoff (eds.), The Early Universe, 115–124.
© *1988 by D. Reidel Publishing Company.*

fermions and bosons are identical both loops cancel exactly. Supersymmetry ensures that there are bosons and fermions, supersymmetric partners of each other, that differ only in the spin. A supersymmetric generator, Q has spin 1/2. Applied to a boson state $|b>$ gives a fermion $|f>$ and vice versa, $Q|b> = |f>$ and $Q|f> = |b>$. In $N=1$ supersymmetry there is only one generator Q, thus (f, b) form a supersymmetric multiplet. The minimal number of independent degrees of freedom in a fermion (a Majorana or Weyl fermion) is two. Thus the minimal supersymmetric multiplet involves a complex scalar field.

The solution to the gauge hierarchy implies that (in $N=1$ supersymmetry) there is one "sparticle" for every particle we know. We also know that supersymmetry is not a good symmetry now, since we know of no degenerate boson and fermion. The difference in mass between them must, however, be at most of the electroweak order, then, radiative corrections to the electroweak scale are of the same order of the bare quantity. That is

$$m_{SPARTICLES} < 1 \; TeV \tag{1}$$

There is another rule that early model builders discovered, *no known particle can be the sparticle of any other*. Thus the sparticles must all be new particles!

Particle		Sparticle	
$S=1/2$:	fermion (e, ν, μ,q) (example: electron)	$S=0$:	sfermion ($\tilde{e}$, $\tilde{\nu}$, $\tilde{\mu}$ $\tilde{q}$) (ex: selectron)
$S=0,1$:	boson (γ, H, gluon) (ex: photon, Higgs)	$S=1/2$:	bosino ($\tilde{\gamma}$, $\tilde{H}$, $\tilde{g}$) (ex: photino, Higgsino)
$S=2$:	graviton	$S=3/2$:	gravitino

Experimentally, we know that for sleptons (selectrons, smuons, staus) $m_{\tilde{l}} \geq (20 \; to \; 30) \; GeV$ and for gluinos and squarks, $m_{\tilde{g}, \tilde{q}} \geq (40 \; to \; 50) \; GeV$. [3]

Supersymmetry requires plenty of new particles. Could any of them be a candidate for the dark matter? Most supersymmetric models have a multiplicative conserved quantum number called R-parity which distinguishes particles from sparticles: R(particle) = + (particle), R (sparticle) = - (particle).[4] The definition of the R-parity of a particle or sparticle is

$$R = (-1)^{3(B-L)+F} \tag{2}$$

where B is the baryon number, L the lepton number and F the fermion number, i.e., twice the spin S, $F = 2S$. If S, B and L are conserved then R-parity is also conserved. This means that 1) sparticles are produced in pairs, and 2) heavier sparticles decay into lighter ones, thus the lightest supersymmetric particle, L.S.P., is stable.

In most models R-parity is conserved. But in some, R-parity is not conserved, due to an explicit[5] or spontaneous[6] violation of the lepton number, and still in some other models no parity (Z_2) holds but a larger discrete symmetry Z_n with $n \geq 3$. [7] In these cases the L.S.P. is *not* stable, it would decay into lighter normal particles (the details are very model dependent).

Stable relics from the big bang can not have electromagnetic or strong interactions. If they had they would mix with conventional matter in galaxies, planets, etc. Unsuccessful searches for anomalous superheavy isotopes imply that the number of such relics $\tilde{x}$ per proton p is
$(n(\tilde{x})/n(p)) < 0(10^{-20} - 10^{-30})$ for masses $m_{\tilde{x}}$ between 4 GeV and 1.2 TeV.[8] This number is much smaller than the abundances computed on the basis of annihilation in the early universe, at least by a factor 10^{10} for $\tilde{x}$ particles with strong interactions and by 10^{14} for electromagnetically interacting $\tilde{x}$'s.[9]

In minimal low energy supergravity model,[10], those with minimal number of particles and sparticles at low energies, the possible relics are: a sneutrino $\tilde{\nu}$ ($s=0$), photinos $\tilde{\gamma}$ or Higgsinos $\tilde{H}$ ($s=1/2$), gravitinos ($s=3/2$). Could they account for the dark matter? That is, could it be $\rho=\rho_c$ (ρ_c is the critical density) or at least $\rho = 0.1 \, \rho_c$ (enough to form the haloes of galaxies)? The abundance of these relics is controlled by their annihilation in the early universe. When the rate of annihilations becomes smaller than the rate of expansion of the universe the number of particles per comoving value is fixed and remains constant. The interactions of $\tilde{v}$, $\tilde{\gamma}$, $\tilde{H}$ are similar to the weak interaction of neutrinos, thus the bounds are similar.[11] Higgsinos $\tilde{H}$ and photinos $\tilde{\gamma}$ are Majorana fermions (they coincide with their antiparticles). Due to the presence of two identical particles in the initial state, the annihilation of non-relativistic $\tilde{\gamma}$'s and $\tilde{H}$'s as well as of Majorana neutrinos, can occur only in a p-wave if the mass m_f of the two fermions in the final state is zero, or in s-wave with an amplitude proportional to $m_f \neq 0$.[12,13] This happens if chirality is conserved in the vertices (see footnote 1). Therefore the bounds depend on m_f. Ellis et al.,[13] found that the bound $\rho_{\tilde{H}} < \rho_c$ is fulfilled if $m_{\tilde{H}} \gtrsim m_b \simeq 4.5 \, GeV$, the mass of the bottom quark, in the models in which the annihilation through a Z^0 resonance is dominant (and $m_{\tilde{H}} \gtrsim m_t > 30 \, GeV$, the mass of the top quark, in the models in which Z^0 boson exchange is suppressed and the dominant contribution comes from the Yukawa couplings which are smaller, i.e., through sfermion exchange). Photino annihilation proceeds through sfermion $\tilde{f}$ exchange, thus the limits in this case depend on both m_f and $m_{\tilde{f}}$. The bound $\rho_{\tilde{\gamma}} \leq \rho_c$ is fulfilled if $m_{\tilde{\gamma}} \gtrsim 0.5$ or 1.8 or 5 GeV for $m_{\tilde{f}} \sim 20$ or 40 or 100 GeV respectively.[13] If we are interested in $\rho_{\tilde{\gamma}} \leq 0.1 \, \rho_c$ then $m_{\tilde{\gamma}} \gtrsim 1.8$ or 3 or 15 GeV for the same values of $m_{\tilde{f}}$. The sfermion mass appears in the propagator thus the annihilation cross-section decreases as $m_{\tilde{f}}$ increases. To keep the cross-section constant m_f must increase with $m_{\tilde{f}}$ and thus $m_{\tilde{\gamma}}$ correspondingly increases (to allow $\tilde{\gamma}$'s to annihilate into the final fermions).

Actually, $\tilde{H}$ and $\tilde{\gamma}$ are only two special cases of a continuum of possibilities for the lightest mass eigenstate of a four by four mass matrix, mixing four neutral fermions, two Higgsinos (there are necessarily two different Higgs multiplets in the supersymmetric electroweak model) and two gauginos. However in a more general treatment the results are dominated by these two special limits.[13]

Sneutrinos annihilate mainly through a $\tilde{Z}^0$ exchange process which is not p-wave suppressed, but whose amplitude is proportional to one of the components of the $\tilde{Z}^0$ mass, that can be identified with $m_{\tilde{\gamma}}$. Thus if $m_{\tilde{\gamma}}$ is large the annihilation cross-section is large and almost no sneutrinos remain today. To obtain a large density in $\tilde{v}$, such as $\rho_{\tilde{v}} \simeq \rho_c$ the $\tilde{\gamma}$ can not be very heavy, and the $\tilde{v}$ should be even lighter (since we want it to be the L.S.P.) Thus $m_{\tilde{v}} < m_{\tilde{\gamma}} \simeq 3 \, GeV$. Or for $\rho_{\tilde{v}} \simeq 0.1 \, \rho_c$ then $m_{\tilde{v}} < m_{\tilde{\gamma}} \simeq 10 \, GeV$. Thus, $\rho_{\tilde{v}}$ smaller than something does not imply any mass limit on $m_{\tilde{v}}$. It is if $\tilde{v}$'s are a significant contribution to the dark matter in our universe that there are upper bounds on $m_{\tilde{v}}$. [14]

Models inspired in superstrings are, in general, non-minimal, thus they may have other L.S.P. candidates which depend strongly on the model.[15]

If $\tilde{\gamma}$, $\tilde{H}$ or $\tilde{v}$'s are in the halo of our galaxy they may be detected either indirectly through their annihilation products or through the energy deposited in elastic collisions on nuclei. Examples of possible indirect detection signatures are: an excess of antiprotons in cosmic rays produced by annihilations of photinos in the halo;[16] if photinos and sneutrinos have masses larger than 6 GeV they would be trapped in the sun and then annihilate there,[17] producing an excess of neutrinos coming from the sun; if sneutrinos are heavier than 12 GeV they may be trapped in the earth and produce too many neutrinos, unless they are heavier than 20 GeV.[18] Particles lighter than those mentioned "evaporate", that is, the rate with which they acquire thermal velocities large enough to escape from the sun or earth is larger than the rate at which they are captured from the halo of our galaxy. Thus they never accumulate. While the scattering cross-sections of $\tilde{v}$ and of $\tilde{\gamma}$ on protons in the sun are similar, those on heavier nuclei in the earth are very different. The sneutrinos (as Dirac neutrinos) have vectorial couplings in the Z^0-mediated interactions. For a small momentum transfer q such that $(1/q) > R$, where R is the radius of the target nucleus, the interaction is coherent, the nucleus is seen like a point particle. For weak interactions the coherence factor in the cross-section is $(Z(1 - 4\sin^2\theta_W) - N)^2$, where Z and N are the number of protons and neutrons. Since $\sin^2\theta_W \simeq 0.23$ the contribution of protons is negligible. This is a large factor of order $10^3 - 10^4$. Photinos (as Majorana neutrinos) have only spin dependent interactions, in the nonrelativistic limit (particles in the halo or in the sun have velocities of order $10^{-3}c$), the cross-section is proportional to $J(J + 1)$, where J is the spin of the nucleus, and it is never much larger for a heavy nucleus than for a proton. Higgsinos scatter mainly through Yukawa couplings which are very small, thus they are practically undetectable.

The differences in scattering cross-sections are very relevant for direct detection.[19] The rate of collisions with nuclei are expected to be $(10^2$ to $10^4)/(kg$ day) for sneutrinos (and Dirac neutrinos) while it is expected to be $(10^{-1} - 10)/(kg$ day) for photinos, putting the first over and the second under usual background rates. There are already bounds on coherently interacting particles coming from a Ge spectrometer, but not on photinos.[20] When a nucleus recoils a part of its energy goes into ionization, which can be seen in a semiconductor detector, an ultralow background Ge spectrometer. Dirac neutrinos with mass between 20 GeV and 2 TeV have been excluded as main components of the halo.[20] A small area in a mass of sneutrinos - mass of photinos space has also been excluded. Most of the recoil energy of a nucleus goes, however, to the lattice, producing phonons which, when thermalized, increase the temperature of the target.[19] Ballistic phonons or small temperature increases may be detected in low temperature targets. The attempts to detect GeV or heavier dark matter particles constitute a rapidly expanding field.

The Gravitino Problem

Supersymmetry is broken at low energies. Old models based on global supersymmetry have semipositive definite scalar potentials $V \geq 0$. When $V = 0$ supersymmetry is conserved. With spontaneous breaking of global supersymmetry a Goldstino, a zero mass spin 1/2 particle, appears in a similar way as a Nambu-Goldstone boson a zero mass $S = 0$ particle, appears when a normal global symmetry is spontaneously broken.[1] Supergravity is local supersymmetry. In supergravity models the scalar potential can be negative, positive, or zero and supersymmetry can be broken while $V = 0$ (i.e., the cosmological constant can be zero). The spontaneous breaking of supergravity is signaled by the "super Higgs effect": the massless gravitino "eats" the Goldstino, i.e., it incorporates the degrees of freedom of the Goldstino, and becomes massive, $m_{3/2} \neq 0$. This effect is equivalent to the Higgs effect in normal gauge theories in which the gauge bosons "eat" the Nambu-Goldstone bosons and become massive, when the gauge symmetry is spontaneously broken.[1] Supergravity models are completely defined by two functions of all the fields: a real function called the Kähler potential G and an analytic function, $f_{\alpha\beta}$, of the gauge fields[1,21]. The gravitino mass when supergravity is broken is $m_{3/2} = e^{G/2}$. The potential (in units of $M = M_{PLANCK}/\sqrt{8\pi} = 2.4 \times 10^{18}$ GeV $= 1$) is

$$V = e^{G}[G_i[(G'')^{-1}]^i_j \, G^j - 3] + f_{\alpha\beta}^{-1} D^{\alpha} D^{\beta} \tag{3}$$

where G_i is a first derivative of G, and G'' is the second derivative of G, and D^{α} depend on the gauge generators T^{α} and all the fields with gauge quantum numbers. D^{α} is zero in a minimum of V where all gauge non-singlet fields have zero vacuum expectation values. Minimal models are based in the simplifying choice of the second derivatives of G, G^i_j, and $f_{\alpha\beta}$ as delta functions: $G^i_j = \delta^i_j$ and $f_{\alpha\beta} = \delta_{\alpha\beta}$. Let us call ϕ_i, $i = 1,2,...$, the scalar fields, so $G_i = \partial G/\partial\phi_i$, $G^i = \partial G/\partial\phi_i^*$, ($\phi_i^*$ are the complex conjugate fields). The most general form of G in minimal models is, then,

$$G = \sum_i |\phi_i|^2 + \ln|W(\phi)|^2 \tag{4}$$

where $W(\phi)$ is an analytic function of dimension three of all the ϕ_i. When the low energy limit, $M \to \infty$, is taken, these models have a remarkable property. Masses equal to $m_{3/2}$ appear for the sfermions. As we have seen above, to solve the gauge hierarchy problem sfermion masses must be of the order of the electroweak scale. Therefore $m_{3/2}$ must be of 0 (100 GeV). How is it possible to obtain such a value? Since $m_{3/2} = e^{G/2} \simeq W(\phi)$ and $W(\phi)$ has dimension three when some of the ϕ_i fields acquire non-zero vacuum expectation values $<\phi> \neq 0$ and break supergravity, $W(\phi) \simeq <\phi>^3$ and

$$m_{3/2} \simeq \frac{<\phi>^3}{M^2} . \tag{5}$$

Therefore some of the $<\phi_i>$ need to be

$$<\phi> = (M^2 m_{3/2})^{1/3} \simeq 10^{13} \, GeV \quad . \tag{6}$$

This is a special new scale. A "hidden sector", weakly coupled to conventional matter was proposed in order to separate the special fields that break supergravity from the others. Fields in the hidden sector only couple to matter fields gravitationally, i.e., with couplings $(1/M)$. Thus, the function W is splitt into two pieces,

$$W = W_{OBSERVABLE}(y_i) + W_{HIDDEN}(z) \tag{7}$$

where y_i are the matter fields and z the hidden fields. In the simplest cases just one z field is enough. The first example was given by Polonyi:[22]

$$W_{HIDDEN}(z) = \mu^2(z + \beta M) \tag{8}$$

where $\mu = 10^{10} \, GeV$ and β is adjusted so that $V=0$ at the minimum, $\beta = 2 - \sqrt{3}$. The value of z at the minimum is $z = (\sqrt{3} - 1)M$, then $m_{3/2} = \mu^2 \exp[(\sqrt{3} - 1)^2/2]/M$ as wanted. The mass of z at the minimum is also $m_z \simeq \mu^2/M \simeq m_{3/2} \simeq 10^2 \, GeV$. Thus the potential is very flat.

There are non-minimal supergravity models in which G is chosen so that the first term of V is always zero. The simplest [23] of such functions G, involves just one field z in the hidden sector

$$G = -3 \ln(z + z^*) \tag{9}$$

(z^* is the complex conjugate of z). This choice gives origin to a flat potential $V \equiv 0$ at tree level. Models of this type, in which matter fields are incorporated, are called "no scale models"[24] because the spontaneous breaking scales of gauge theories are generated by radiative corrections. In some of these models the masses of the sfermions result at low energies to be a function of the gravitino mass $m_{3/2}$. Thus $m_{3/2}$ can have practically any value.[25] "Superstring inspired" models are non-minimal.

The gravitinos are coupled to matter only gravitationally, with coupling constant $1/M$. Thus, unless they are the L.S.P., they decay with a long lifetime which, on dimensional grounds, is of the order

$$\tau \simeq \frac{M^2}{m_{3/2}^3} \simeq 10^8 \sec \left[\frac{10^2 GeV}{m_{3/2}} \right]^3 \quad . \tag{10}$$

For $m_{3/2} = 10 \, MeV$, τ is equal to the life of the universe now. Gravitinos with $m_{3/2} < 10 \, MeV$ are stable and if $m_{3/2} > 1 \, keV$ their density is unacceptably large, $\rho_{3/2} > \rho_c$. Gravitinos with mass $m_{3/2} > 10 \, MeV$ have decayed by now but unless $m_{3/2} > 10^4 \, GeV$ they decayed after nucleosynthesis releasing a large entropy, which is also unacceptable.[26] Therefore the number of gravitinos is unacceptably large if

$$1 \, keV < m_{3/2} < 10^4 \, GeV \tag{11}$$

This is the so-called "gravitino problem". The problem is severe for minimal supergravity models, where $m_{3/2} = 0(100 \, GeV)$. The only way of decreasing the number of gravitinos is to dilute them through inflation: the density decreases exponentially when the volume increases exponentially.[27] There is a further problem. Gravitinos are regenerated, as everything else, in the reheating after inflation. Thus, bounds are imposed on the reheating temperature T_{RH} achieved after the inflation. For stable gravitinos $\rho_{3/2} < \rho_c$ imply [28]

$$m_{3/2} < 10^2 \, GeV \left[\frac{1.3 \times 10^{12} \, GeV}{T_{RH}} \right] \tag{12}$$

For unstable gravitinos, the possible photodissociation of light elements by photons produced in the decays of gravitinos imply[29]

$$T_{RH} < 2.5 \times 10^8 \ GeV \left[\frac{10^2 \ GeV}{m_{3/2}} \right] . \tag{13}$$

The solution to the gravitino problem requires quite low reheating temperatures in minimal models. ($m_{3/2} = 10^2 \ GeV$) where new ways of generating the baryon asymmetry at $T \leq 10^8 \ GeV$ have to be considered. Otherwise it requires non minimal models where $m_{3/2}$ can have any value.

The "Polonyi" Problem or "Entropy Crisis"

The potentials for fields in the hidden sector are very flat, such as in the Polonyi case, Eq. 8. Since fields in the hidden sector only interact gravitationally their decay occurs very late, $\tau \simeq M^2/m_z^3$ as in Eq. 10. In early times, at high temperatures T, the field z lies away from the minimum. When T decreases and supergravity is broken, z oscillates around a minimum at $z \simeq 10^{10} \ GeV$ where the curvature is only $m_z^2 = (10^2 \ GeV)^2$. The field z oscillates coherently until it decays. When it decays the energy stored in the oscillations, which behaves like non-relativistic matter, is released. The entropy generated is large enough as to dilute the baryon number and the temperature is not enough to generate the baryon asymmetry again. This is the "Polonyi problem" or "entropy crisis",[30,31] one of the major unsolved problems of most supergravity models for inflation. This problem can not be solved by inflation generated by other fields different from z. There is no dilution of the potential energy stored in z, the oscillations restart at the end of inflation. This is a very severe problem. In order to get an acceptable picture the amplitude of the oscillations should be $\leq 10^{-20}$ and $\leq 10^{-12}$ of the value it has in minimal and in no scale models, respectively.[31]

Supergravity Inflationary Models

Already in the original models for new inflation Coleman-Weinberg potentials failed because the scalar field responsible for inflation (hereafter referred to as the *inflaton*) had too large coupling constants. It was not possible to obtain density perturbations small enough $\delta\rho/\rho < 10^{-4}$. The inflaton must be very weakly coupled.[32] Gauge singlet fields in the hidden sector of supergravity models are very weakly coupled. Thus they appear as natural candidates to be the inflaton.[33] That potentials for hidden fields are very flat implies a poor reheating. This is a general characteristic of these models where $T_{RH} \simeq 10^6$ to $10^9 \ GeV$. New mechanisms to generate a baryon asymmetry at such low T are needed. These flat potentials guarantee that the objections of Mazenko, Unruh and Wald[34] do not apply and a semiclassical approximation is valid[32,35]

One of the open questions in these models is that of initial conditions, i.e., how is it that the inflaton field was localized in the false vacuum before inflation? The inflaton is so weakly coupled that it does not achieve thermal equilibrium at any temperature below the planck scale. This is why Linde claims that supergravity inflationary models only makes sense in the framework of chaotic initial conditions.[36] The other possibility is to *assume* that the inflaton was in equilibrium at the planck scale, $T \geq T_{PLANCK}$, before becoming a free field. Thus thermal distributions are maintained at $T < T_{PLANCK}$, and thermal effective potentials can be used. Then the existence of a barrier between the false and true minima at high temperatures can be requested. Could we have identified the Polonyi field, z, with the inflaton field, ψ? In this case the "entropy crisis" would be identified with the "reheating" after inflation, ceasing to cause a problem at later times. This is not possible in minimal supergravity models because the inflaton mass m_ψ can not be as small as m_z. While $m_z \simeq 10^2 \ GeV$ the inflaton mass needed for the reheating is at least $m_\psi \simeq 10^{10} \ GeV$. However in minimal models with just one inflaton field, ψ can not be *different* than z either.[37] The "thermal constraint"[38] is the requirement that the minimum of the effective temperature corrected potential at high temperatures approaches the false minimum. It has been shown that,[37] in minimal models with *one* inflaton field ψ, the thermal constraint is satisfied only if the true minimum in ϕ breaks supersymmetry, thus if $\psi \equiv z$. So ψ can not be either equal to z or different from z. This contradiction has three ways out:

1) Ignore the thermal constraint. After all, the validity of the effective temperature corrected potential depends on the unjustified assumption of thermal equilibrium at $T > T_{PLANCK}$ where all descriptions break down. This is the choice of Holman, Ramond and Ross.[39] They use a minimal model with a superpotential separated in three sectors: $W = I + S + G$, where I, responsible for inflation and S, responsible for the supersymmetry breaking, are hidden sectors. G contains the matter fields. $I = I(\psi)$ depends on only one inflaton. They use chaotic initial conditions or a second scalar field coupled to the inflaton just to create a thermal barrier around the false vacuum. The model may not be elegant but it works.

2) Use minimal models with two inflaton fields. This was the choice of Ovrut and Steinhardt.[40,41] One field drives inflation initially and the other generates the reheating. The first problem of this model was that, even if the dependence of the potential on the real parts of the fields was properly adjusted, the potential was unstable in the imaginary directions. Gravitational corrections at one loop were used[41] to correct this feature. However it has been shown[42] that this model has a severe unsolved entropy crisis problem. Thus it does not work.

3) Use non-minimal supergravity models. No scale models with one inflaton field proposed and studied by the CERN group[33,43] have an unsolved entropy crisis also. Models inspired in superstrings result in no scale models with two fields called generally S and T. Superstring models are originally in 10-dimensions, 6 of which are compactified. The T field is associated with the fluctuations in the overall size of the compactified six-dimensional manifold, and $S = T^3/\phi$ where ϕ is a dilaton field. It has been shown that S can not be used as inflaton. The T field breaks supergravity, it acts as a Polonyi field and generates an entropy crisis if $m_T < 10^{10}$ GeV [46]. As a solution Maeda, Pollock and Vayonakis[47] try to identify this T field with the inflaton (to solve the entropy crisis, which becomes the reheating, and to obtain inflation).

We are far from the Theory of Everything. Many models remain still to be formulated and tested.

Acknowledgements. I am pleased to thank Pierre Binnètruy and Robert Brandenberger for many instructive conversations.

* Address from October 1986: The Enrico Fermi Institute, University of Chicago, Chicago, IL 60637. On leave of absence from Department of Physics, University of Rome II, via Orazio Raimondo, Rome, Italy 00173. Work partly supported by the U.S. Department of Energy, Grant No. DE AC02 82ER-40073 and NSF Grant PHY-82-15249.

References and Footnotes

Footnote 1. In the annihilations of $\tilde{\gamma}$'s, $\tilde{H}$'s and Majorana ν's the vertices conserve chirality. If the masses in the final state are zero, chirality and helicity coincide, and the spin of the final state is $S_f = 1$. Therefore, the total angular momentum is $J_f = s_f + l_f \geq 1$, where l_f is the final state orbital angular momentum. Thus $J_i = J_f \geq 1$, where i refers to the initial state. The initial wave function must be totally antisymmetric (for identical particles) thus $(-1)^{l_i + S_i + 1} = (-1)$. Thus if $l_i = 0$ we should have $S_i = 0$ but $J_i = l_i + S_i$ must be 1 or larger. Therefore it must be $l_i \geq 1$. If the mass of the final state fermions is $m_f \neq 0$, a mass insertion can change the helicity of a final fermion and $S_f = 0$ is possible. Thus $J_f \geq 0$ allows for $l_i = 0$, but the amplitude is proportional to m_f.

1. Some reviews about supersymmetry, supergravity and superstrings are
   H. P. Nilles "Supersymmetry, Supergravity and Particle Physics", Physics rep. **110**, 1 (1984).
   J. Ellis "Supersymmetry, Supergravity and Superstring Phenomenology" CERN-TH 4255/85, lectures at the 28th Scottish Universities Summer School in Physics, Edinburgh, Scotland, July 28th to August27 27th, 1985.
   Proceedings of the workshop "Unified String Theories" ITP-Santa Barbara, 1985, ed. M. Green and D. Gross, World Scientific.

2. J. Wess and B. Zumino, Phys. Lett. **49B**, 52 (1974). J. Illipoulos and B. Zumino, Nucl. Phys. **B76**, 310 (1974); S. Ferrara, J. Illipoulos and B. Zumino, Nucl. Phys. **B77**, 413 (1974).

3. See, for example, J. Ellis in Ref. 1, D. Cline in proceedings of the 13th Texas Symposium on Relativistic Astrophysics, Chicago, 14-19 December 1986.

4. G. Farrar and P. Fayet, Phys. Lett. **76B**, 575 (1978); G. R. Farrar and S. Weinberg, Phys. Rev. **D27**, 2732 (1983).

5. L. Hall and M. Suzuki, Nucl. Phys. **B231**, 419 (1984).

6. G. G. Ross and J.W.F. Valle, Phys. Lett. **151B**, 375 (1985); J. Ellis, G. Gelmini, C. Jarlskog, G. G. Ross and J.W.F. Valle, Phys. Lett. **150B**, 142 (1985).

7. L. Hall, Harvard preprint HUTP-86/A057 talk at the "Quarks and Galaxies" Workshop, LBL, August 1986.

8. P.P. Smith and J.R.J. Bennett, Nucl. Phys. **B149**, 525 (1979); P. F. Smith et al., Nucl. Phys. **B206**, 333 (1982) and references therein.

9. S. Wolfram, Phys. Lett. **82B**, 65 (1979); C. B. Dover, T. K. Gaisser and G. Steigman, Phys. Rev. Lett. **42**, 1117 (1979).

10. L. Alvarez Gaumè, J. Polchinski and M. Wise, Nucl. Phys. **B221**, 495 (1983); J. Ellis, J. Hagelin, D. Nanopoulos and K. Tamvakis, Phys. Lett. **125B**, 275 (1983).

11. B. W. Lee and S. Weinberg, Phys. Rev. Lett. **39**, 165 (1977).

12. H. Goldberg, Phys. Rev. Lett. **50**, 1419 (1983).

13. J. Ellis, J. S. Hagelin, D. V. Nanopoulos, K. Olive and M. Srednicki, Nucl. Phys. **B238**, 453 (1984).

14. J. S. Hagelin, G. L. Kane, and S. Raby Nucl. Phys. **B241**, 638 (1984); L. Ibañez, Phys. Lett. **137B**, 160 (1984).

15. B. A. Campbell, J. Ellis, K. Enqvist, D. Nanopoulos, J. Hagelin and K. Olive Phys. Lett. **B173**, 270 (1986); B. Greene, K. Kirklin, P. Miron and G. G. Ross, Nucl. Phys. **B278**, 667 (1986, and Phys. Lett. **180B**, 69 (1986).

16. J. Silk and M. Srednicki, Phys. Rev. Lett. **53**, 624 (1984).

17. J. Silk, K. Olive and M. Srednicki, Phys. Rev. Lett. **55**, 257 (1985); M. Srednicki, K. Olive and J. Silk, Nucl. Phys. **B279**, 804 (1987); T. Gaisser, G. Steigman and S. Tilav, Phys. Rev. **D34**, 2206 (1986).

18. K. Freese, Phys. Lett. **B167**, 295 (1986); L. Krauss, M. Srednicki and F. Wilczek, Phys. Rev. **D33**, 2079 (1986).

19. For a recent review see P.F. Smith *Possible Experiments for Direct Detection of Particle Candidates for the Galactic Dark Matter*, Rutherford Lab preprint RAL-86-029 to be published in the proceedings of the 2nd ESO/CERN Symposium on *Cosmology, Astronomy and Fundamental Physis*, Garching, March 1986; and A. Drukier in this proceedings.

20. S. P. Ahlen, F. T. Avignone III, R. L. Brodzinski, A. K. Drukier, G. Gelmini and D. N. Spergel, *Limits on Cold Dark Matter Candidates from the Ultralow germanium Spectrometer*, Harvard Center for Astrophysics Preprint No. 2292, 1986.

21. E. Cremmer, B. Julia et al., Phys. Lett. **79B**, 28 (1978) and Nucl. Phys. **B147**, 105 (1979); E. Cremmer et al., Phys. Lett. **116B**, 231 (1982) and Nucl. Phys. **B212**, 413 (1983); J. Bagger and E. Witten, Phys. Lett. **115B**, 202 (1982) and **118B**, 103 (1982). J. Bagger, Nucl. Phys. **B211**, 302 (1983).

22. J. Polonyi, Budapest preprint KFKI-93 (1977).

23. E. Cremmer, S. Ferrara, C. Kounnas and D. Nanopoulos, Phys. Lett. **133B**, 61 (1983).

24. J. Ellis, A. Lahanas, D. Nanopoulos and K. Tamvakis, Phys. Lett. **134B**, 429 (1984).

25. J. Ellis, C. Kounnas and D. V. Nanopoulos, Phys. Lett. **143B**, 410 (1984); J. Ellis, K.D. Enqvist and D. V. Nanopoulos, Phys. Lett. **147B**, 99 (1984); J. Ellis, K. Enqvist, D. Nanopoulos and K. Tamvakis, Phys. Lett. **155B**, 381 (1985).

26. H. Pagels and J. Primack, Phys. Rev. Lett. **48**, 223 (1982); S. Weinberg, Phys. Rev. Lett. **48**, 1303 (1982).

27. J. Ellis, A. Linde and D. V. Nanopoulos, Phys. Lett., **118B**, 59 (1982).

28. J. Ellis, J. Kim and D. V. Nanopoulos, Phys. Lett. **145B**, 181 (1984).

29. J. Ellis, D. V. Nanopoulos and S. Sarkar, Nucl. Phys. **B259**, 179 (1985); D. Lindley, Ap. J. **294**, 1 (1985).

30. G. D. Coughlan, W. Fischler, E. Kolb, S. Raby and G. G. Ross, Phys. Lett. **131B**, 59 (1983).

31. A. S. Goncharov, A. D. Line and M. I. Vysotsky, Phys. Lett. **147B**, 279 (1984).

32. See M. Turner in these proceedings.

33. J. Ellis, D. V. Nanopoulos, K. A. Olive and K. Tamvakis, Nucl. Phys. **B221**, 524 (1983); D. V. Nanopoulos et al., Phys. Lett. **123B**, 41 (1983); D. V. Nanopoulos et al., Phys. Lett. **127B**, 30 (1983).

34. G. Mazenko, W. Unruh and R. Wald, Phys. Rev. **D31**, 273 (1985), see W. Unruh in these proceedings.

35. A. Albrecht and R. Brandenberger, Phys. Rev. **D31**, 1225 (1985); A. Albrecht, R. Brandenberger and R. Matzner, Phys. Rev. **D32**, 1280; G. D. Coughlan and G. G. Ross Phys. Lett. **157B**, 151 (1985); L. Jensen and K. Olive, Nucl. Phys. **B263**, 731 (1986).

36. A. D. Linde, Phys. Lett. **132B**, 317 (1983).

37. B. Binètruy and M. K. Gaillard, Nucl. Phys. **B254**, 388 (1985); P. Binètruy, proceedings 6th Workshop on Grand Unification (1985).

38. B. Ovrut and P. Steinhardt, Phys. Lett. **133**, 161 (1983); G. Gelmini, D. Nanopoulos, and K. Olive, Phys. Lett. **131B**, 53 (1983).

39. R. Holman, P. Ramond and G. G. Ross, Phys. Lett. **137B**, 343 (1984); G. D. Coughlan et al., Phys. Lett. **140B**, 44 (1984); G. D. Coughlan et al., Phys. Lett. **158B**, 401 (1985).

40. B. A. Ovrut and P. Steinhardt, Phys. Rev. Lett. **53**, 732 (1984) and Phys. Rev. **D30**, 2061 (1984).

41. P. Lindblom, B. Ovrut and P. Steinhardt, Phys. Lett. **B172**, 309 (1986).

42. O. Bertolami and G. G. Ross, Phys. Lett. **171B**, 46 (1986).

43. J. Ellis, K. Enqvist, D. Nanopoulos, K. Olive and M. Srednicki, Phys. Lett. **152B**, 175 (1985); G. Gelmini, C. Kounnas and D. Nanopoulos, Nucl. Phys. **B250**, 177 (1984); K. Enqvist and D. Nanopoulos, Phys. Lett. **142B**, 349 (1984); C. Kounnas and M. Quiros, Phys. Lett. **151B**, 189 (1985).

124

44. See for example, P. Binétruy, proceedings of the "Quarks and Galaxies Workshop" Berkeley, August, 1986, and references therein.

45. P. Binétruy and M. K. Gaillard, UCB-PTH-86/15 (1986).

46. G. German and G. G. Ross, Phys. Lett. **172**, 305 (1986).

47. K. Maeda, M. D. Pollock, and C. E. Vayonakis, Class. Quant. Grav. **3**: L89 (1986).

RELATIVISTIC COSMOLOGY

John D Barrow

Astronomy Centre
University of Sussex
Brighton BN1 9QH, U.K.

INTRODUCTION

These lectures do not aim to provide a complete survey of
relativistic cosmology. Rather, we shall confine attention to
a number of specific aspects of relativistic cosmology which
might be of interest to current investigations of high energy
physics in the early universe and to the wider audience of
relativists and astrophysicists at this School. In particular,
we shall focus upon features of general relativistic cosmology
that are of relevance to the inflationary universe theory and
the cosmological questions that it confronts[1,2]. A number of
new results will be described in the second half of the notes.
The detailed particle physics motivation for the existence of
an inflationary phase during the early history of the
expanding universe will be covered in other lectures.

A number of research problems arising from the subject
matter covered in the lectures have been added to supplement
these notes. These are signalled in the text by the symbol □
and are, as far as the author is aware, unsolved; relevant
associated references which might be helpful in their solution
have also been added where possible.

W. G. Unruh and G. W. Semenoff (eds.), The Early Universe, 125–201.
© 1988 by D. Reidel Publishing Company.

ORIENTATION

The aim of modern cosmology is to determine the structure of the Universe -- its age, shape, composition, history and so forth -- by observation and to understand those observations within the framework of a unified mathematical theory of gravitation and matter, locally and independently tested by experiment. This grandiose programme must come to terms with a number of unusual problems that are characteristic of cosmology:

The Universe is unique

This tautological observation is relevant to any deductions regarding the stability or generality of particular properties of the Universe with respect, say, to all the possibilities admitted in the initial data space of the Einstein equations. The fact that a particular property is not open dense[3] (*generic*) in the solution space of Einstein's equations need not be an argument against it being a property of the observed Universe.

There may also exist unique selection effects in operation if the Universe is infinite. For example, there is an old argument that exhaustively random infinite initial data sets possess the awkward property that any property that *can* arise, *must* arise somewhere infinitely often with probability unity[4-6].

There may also prove to exist fundamental problems[7,91] of principle when dealing with any quantum cosmological theory unless the many worlds interpretation of quantum mechanics is adopted, in which case what is meant by the uniqueness of the Universe must be more carefully defined. Other lecturers will discuss what is meant by "prediction" in a quantum cosmological model.

Non-local influences

The Einstein equations do not fix the topology of space yet this global feature of space-time can have very strong local effects upon current observables as well as upon quantum processes in the early universe. Although it is conventional to assume that the Universe possesses the natural spatial topology (S^3 or R^3) it may be far more probable for any universe "created out of nothing"[8] to possess one of the myriad of non-standard topologies. If extra spatial dimensions exist they are not expected to possess a natural topology[14].

Some cosmological boundary conditions may be necessary either at an initial singularity or at past infinity (the alternative -- that all timelike and null geodesics are closed, perhaps with periods $>> 10^{10}$ yrs is not appealing). The influence of initial conditions is occluded by the possibility that some portions of the initial singularity may be timelike (that is they lie in the causal future of other parts of it) so there is a breakdown of our usual picture of determinism.

The Universe may be significantly inhomogeneous over very large length scales. This is actually predicted by most inflationary universe scenarios[1].

Mach's Principle[9], if it true and can be suitably formulated, may reveal some global constraints upon the local structure of space-time. Even more speculatively, quantum non-locality may have some unexpected cosmological manifestation within a future theory of quantum gravitation.

Horizons

The causal structure of space-time ensures[10] that astronomical observations are limited to that part of space-time which lies in or on our past null cone. Our direct observations are limited at present to redshifts of order[11] 3.8 (if quasar redshifts are entirely cosmological) and our earliest indirect information comes from primordial nucleosynthesis at redshifts ~ 10^{10}. We have no direct or indirect observational evidence of any sort regarding the structure of the Universe at earlier times than ~ 1s. The most exciting observational development in this respect would be the discovery of a primordial mini black hole or super-massive monopole (again?!).

How many spatial dimensions are there?

Recent developments in particle physics have resurrected
the old idea that gauge invariances in the observed three
spatial dimensions may be interpreted as coordinate
invariances of a gravitation theory with additional spatial
dimensions. All Kaluza-Klein[12,13] and superstring[14] theories
predict the existence of additional spatial dimensions. This
considerably expands the possible evolutionary histories for
the very early history of the Universe in ways that are as yet
not fully understood. The additional dimensions are always
assumed to be spatial although there seems to be no
fundamental reason for this restriction.

Variation of fundamental "constants"

The mathematical formulations of physical laws that we
have found most expedient necessarily contain certain
proportionality factors whose precise values are in general
not constrained by the solution of those equations. These
proportionality constants we call the "constants of Nature".
It is possible that those quantities we assume to be
fundamentally constant do vary on cosmological timescales[6].
Such a variation of the "constants" we observe in three
dimensional space is predicted to occur necessarily in
cosmological theories possessing uncompactified extra spatial
dimensions, but is strongly constrained by observation[15-17].

Unknown Physics

As yet, we possess no very strong-field tests of general
relativity and no high-energy tests of asymptotic freedom. The
structure of the very early universe will be strongly affected
by any deviation from these standard theories at high
energies: for example by the influence of quadratic
corrections to the Einstein gravitational lagrangian in
regions of high space-time curvature, or the breakdown of
quantum field theory at high energies. We may soon run out of
practical and affordable local tests of the theoretical ideas
that are necessary to model the early history of the Universe.
The widespread, and essentially teleological view that all the

"right" theories must have testable consequences that *we* will be able to check by experiment is no more than a hope. In particular, if the Universe is as close to the critical density today as some inflationary theories predict[18] then we may never be able to determine whether the Universe is open or closed[19].

Selection Effects

Astronomical observations of distant objects are beset by well-known evolutionary selection effects — for instance, because of light-travel time-delays, distant galaxies are now seen when they were younger and hence intrinsically different from nearby galaxies of similar appearance.

The truth of some theories of galaxy formation requires some quantitative assessment to be made of the (as yet) purely subjective impression of "filaments" in the clustering of galaxies[21].

If there is no law of Nature that fixes initial conditions uniquely, completely and self-consistently, then our own existence may act as an irreducible selection effect[6]. This would moderate our surprise at finding the Universe to possess properties that are unlikely *a priori*. These properties (for example the great size and age of the observed Universe) may possess no explanation other than that they are necessary for the evolution of biological complexity.

Unknown Matter Fields

We have only scant knowledge of the quantity of matter in the Universe[22] (the so called "missing matter problem") and the form that it takes. We do not know whether the initial and present stages of the expansion are dominated by the effects of non-classical fluids, examples of which are those described by the cosmological constant or vacuum strings. We do not know which energy conditions it is reasonable to expect the material content of the early universe to obey. For instance, violation of the once inviolate[23] strong energy condition ($\rho + 3p \geq 0$ for an isotropic perfect fluid having density ρ and pressure p) is now accepted as a necessary ingredient of inflationary universe theories.

☐ Examine the observable effects of infinite cosmic strings and inflation in a finite open universe with 3-torus topology.
☐ Determine whether it is more probable for a finite universe that is created 'out of nothing' to possess an exotic topology than the natural one (S^3).
☐ Find a probability measure for the space of cosmological initial data. Can you exploit the fact that chaotic cosmological models (like the Mixmaster models[33,56,132]) preserve a natural dynamical measure as the singularity is aproached?

HOW LITTLE COULD WE KNOW?

In order to create some perspective for later discussion of the popular standard Big Bang model described by an isotropic and homogeneous Friedman model we shall begin by summarizing what might be called 'the state of maximum ignorance'. Let us suppose that we are allowed to assume only that *gravitation is described by a metric theory*. That is, that there exists some space-time geometry with metric interval

$$ds^2 = g_{ab}\, dx^a dx^b \quad ; \quad 0 \leqslant a,b \leqslant 3 \qquad\qquad (1)$$

and so there exist geodesic equations determined by paths of stationary action between any two points in the space-time. However, lacking a gravitation theory we do not yet know how the geometry is coupled to the mass-energy content of space-time (or even if it is coupled to it).

Now, suppose that we are also allowed to assume that *all the familiar local (non-gravitational) physics holds*, that *we can ignore selection effects*, and that *our observations are perfectly accurate*. How much can we deduce about the structure of the Universe[24,38]?

In Figure 1 we show the observational situation that

confronts us in our attempts to map out the structure of space-time. We are able to make direct observations by collecting photons, neutrinos, gravitational waves and any other relativistic or massless particles travelling down *our* past null cone. These observations will be occluded by various known or unknown sources of opacity in the past. For example, in Figure 1 we have indicated how far back in redshift we can observe directly using photons, neutrinos and gravitational waves using the numbers appropriate in the standard Friedman model as a guide.

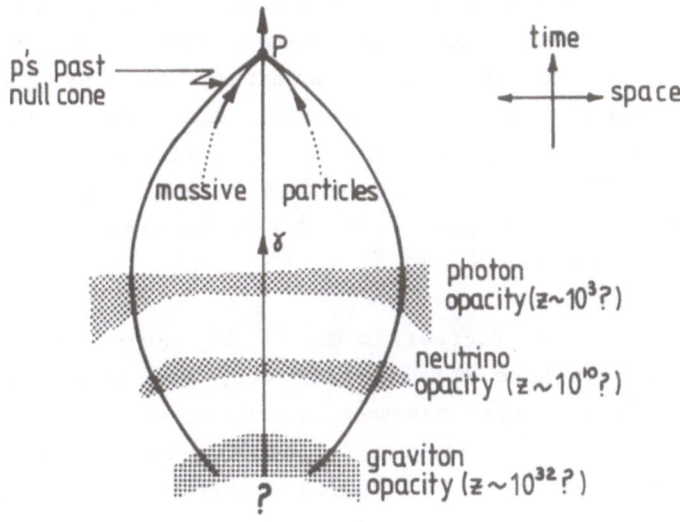

Fig. 1. The region of space-time to the causal past of an astronomer moving along the world-line γ and presently at p. The regions directly accessible to observation are the past null cone along which we receive massless and relativistic particles and gravitational waves and its interior, from which we receive massive particles like cosmic rays or primordial nuclei moving at subluminal velocities.

We can also make observations of the structure of the local region inside our past null cone by observing the motions of non-relativistic massive objects like cosmic rays,

planets, stars, galaxies, nuclei, massive neutrinos and so
forth. Notice that we can say absolutely nothing *in principle*
about the structure of space-time outside our past null cone
or about the past null cones of other observers or about our
future null cone, and without a theory of gravitation we
cannot predict the structure on the inside of our past null
cone from data observed on it.

The direct observations available to us[24,38] are *number
counts of celestial objects with solid angle, proper motions;
shapes, sizes and intensities of images; the spectra and
intensity distributions of radiation fields*. These do not
allow us, in fact, to deduce either the complete metric of
space-time or the full energy-momentum tensor of its material
content even down our own past null cone without the
introduction of unverifiable assumptions about its
structure[24,38]. No observations can demonstrate the
Cosmological Principle that the Universe is globally
homogeneous and isotropic.

This lack of progress is not really surprising. But it is
interesting to note that certain very strong conclusions are
possible under the minimal assumptions we have allowed
ourselves. In particular, it is possible to prove a variety of
singularity theorems[23]. A typical statement of the simplest of
which is that: if S is a Cauchy surface with extrinsic
curvature $X_a{}^a > C > 0$, with C constant, and the Ricci tensor
of space-time obeys the reconvergence condition $R_{ab}v^a v^b > 0$,
where v^a is a timelike vector field, then all inextendible
timelike geodesics terminate in the past at a physical
singularity. (It is straightforward to modify these
assumptions so that they refer to conditions in and on our
past null cone only). If we assume a theory of gravity, (for
example general relativity), then the Einstein equations allow
us to re-express the reconvergence condition on the Ricci
tensor as the *strong energy condition* on the energy-momentum
tensor, T_{ab}, of the matter; that is, for any timelike vector
field v^a,

$$[T_{ab} - \tfrac{1}{2}g_{ab} T_a{}^a]v^a v^b > 0 \qquad\qquad (2)$$

(Note that (2) is not a necessary condition for the formation
of a past singularity since the shear of the geodesics also
contributes positively to their convergence and can compensate
for the negativity of the combination (2).) Other theories of
gravity, for example those generated by adding terms quadratic
in the curvature invariants (R^2 or $R_{ab}R^{ab}$) to the Einstein
gravitational action can produce space-time singularities[39]
but the conditions on T_{ab} that are sufficient for them to do
so will generally be considerably more complicated than (2).

Suppose that we now widen our assumptions to
include[24-26,38]: *general relativity is the theory of
gravitation*. It then turns out that the observations we are
able to make (ignoring selection effects and practical
imperfections) are *necessary and sufficient to determine the
metric and the energy momentum tensor along our past null cone
until any singularity is reached*. This theorem, proved by Nel
and Stoeger[38], is a remarkable and unexpected result. It would
be interesting to know for what other theories of gravitation
it is true.

Now that we have assumed a theory of gravitation we have
some hyperbolic evolution equations which enable us to
determine some of the structure off the past null cone from
information on it. However, because of our inability to
observe outside the past null cone we could still never verify
that the Universe is or was homogeneous and isotropic, even if
it really is to some approximation[38]. It is interesting to
recall that if current inflationary models of the early
universe are correct then it is likely that the Universe is
extremely inhomogeneous beyond our horizon. The only way in
which we can make deductions about this region is by assuming
some form of Copernican or Cosmological Principle. The only
information available to us in the absence of such
unverifiable assumptions is that conveyed by the elliptic
constraint equations of general relativity. These allow us to
"feel" the gravitational effects of these regions even though
we cannot see them. The conservation equations prevent
anything happening there (half the Universe outside our

horizon disappearing, for example) which could give us acausal contradictions here and now.

The fact that we find we cannot base cosmology upon observations alone means that we must proceed as in other quantitative sciences by specifying some variables, in this case g_{ab} and T_{ab}, up to a set of free parameters. These parameters are then determined by fitting the model to the observations until a contradiction or a superior fit to another model is obtained. If we make the assumption of spatial homogeneity then all local information about the Universe is taken to be global information and the role of the past null cone structure upon what is directly observable is obscured.

▢ Give a rigorous statistical description of approximate spatial homogeneity and a resolution of the "closure problem" that arises. Relate it to the idea of 'almost' homogeneous groups of motions[42].

▢ Determine whether other theories of gravitation[43,50] (*eg* Brans-Dicke theory or general relativity plus quadratic or higher-order curvature corrections to the lagrangian[39]) allow our past null cone to be determined uniquely and completely by ideal observations.

▢ Prove singularity theorems in other theories of gravity subject to energy conditions different from the usual strong energy condition (2) by using their field equations to express the geodesic convergence condition as a condition on the stress-energy tensor.

▢ Prove a singularity theorem that makes use of Einstein's equations in an essential way (*i.e.* not simply to arrive at (2) as a restatement of the geodesic convergence condition $R_{ab}u^a u^b \geq 0$ or to deduce differentiability conditions).

▢ Show that observables plus general relativity do not allow our past null cone to be det rmined if selection effects and observational limitations are allowed for.

▢ Reanalyse the problem of determining our past null cone assuming that there exist N>3 spatial dimensions. In particular, evaluate the effect of having N even, and/or N not equal to 3 on the propagation of weak gravitational waves and other wave fields. Determine whether weak gravity waves can propagate both *inside* and on the past null cone in any of these cases[44].

NEWTONIAN GRAVITATION

Before looking in more detail at relativistic cosmological models it is instructive to compare the content of general relativity with that of Newtonian graviation, to which it reduces in the limit of slow motion and weak gravitational fields. The principal contrasts are summarized below:

Relativistic Cosmology	Newtonian Cosmology
10 field equations	1 field equation
10 potentials	1 potential
Non-linear equations	Linear equation
Intrinsically geometrical	Absolute space and time
Can cope with ∞ space	Requires finite space
All energies gravitate	Mass-density gravitates
Hyperbolic propagation	Instantaneous propagation
Singularities of space-time	Singularities in space
Horizons and black holes	No horizons or black holes
Gravitational waves	No gravitational waves

This list indicates the greater content of general relativistic cosmology compared with the Newtonian theory: Newtonian cosmological models possess relativistic analogues but not all general relativistic cosmologies possess a Newtonian analogue. We stress also the sense in which "general" relativity generalises special relativity: the latter is a theory of a space-time manifold R^4 endowed with the Minkowski metric η_{ab}, whereas the arena of general relativity is a general four-manifold M^4 endowed with a pseudo-Riemannian metric g_{ab}.

The field equations of general relativity are usually derived from the demand that the second-rank divergenceless energy-momentum tensor T_{ab} be determined by the most general divergenceless combination of second-rank tensors linear in the curvature. This prescription allows the presence of a cosmological constant in the Einstein equations although it had no known counterpart in Newtonian gravity. However, it is instructive to see[6] how the presence of a cosmological constant is inevitable in Newtonian theory and could even have been found by Newton.

Suppose we ask, as Newton did[58], for the most general form of the gravitational potential, $\Phi(r)$, such that the external potential of a spherical shell is identical to that of a point of equal mass to the shell located at its centre O. If the shell has surface density σ and radius a then at some point P located at a distance $r > a$ from the centre of the shell the potential due to the shell and due to a point mass M(a) at O will be equal if

$$M(a)\Phi(r) + 2\pi\sigma a\lambda(a) = 2\pi\sigma a r^{-1} \int_{r-a}^{r+a} x\Phi(x)dx \qquad (3)$$

where $\lambda(a)$ is a constant and $M(a) = 4\pi a^2\sigma$ is the mass of the shell. If we differentiate this integral equation it can be solved to give $\lambda(a) = 2Ba^2$ and the general solution of Newton's problem is

$$\Phi(r) = Ar^{-1} + \Lambda r^2 \; ; \; A, \Lambda \text{ constants} \qquad (4)$$

where $A = -GM(a)$ and Λ is what Einstein termed the *cosmological constant*. The appropriate field equation for $\Phi(r)$ is the modified Poisson equation

$$\nabla^2\Phi + \Lambda = 4\pi G\rho \qquad (5)$$

☐ Find the most general gravitational lagrangians describing a metric theory of gravity (assume it to be a function of the independent curvature invariants as a first hypothesis) for which the external gravitational field of a sphere can be replaced *(i)* by that of a point mass and, *(ii)* by a point of the same mass as the sphere.

☐ What is the most general gravitational lagrangian that reduces to the Poisson equation in the slow-motion and weak-field limit?

☐ What is the relativistic theory of gravitation in 2+1 dimensional space-time which has 2+1 dimensional Newtonian gravity as a limit? (It is *not* 2+1 dimensional general relativity[37,45]. Note: the logarithmic potential of 2+1 Newtonian gravity diverges at spatial infinity).

NEWTONIAN COSMOLOGY

The kinematical content of Newtonian cosmology[27,46] is most simply seen by considering the generalization of Hubble's law for the relative recesion velocities v of objects possessing relative separations r. If this expansion proceeds isotropically and homogeneously then we write

$$v = Hr \tag{6}$$

where the scalar function H is the usual Hubble parameter. Let us now extend (6) to a completely general expansion flow which proceeds at different rates in different directions. If the three orthogonal velocity components are v_α, $\alpha = 1,2,3$, along orthogonal directions r_α then we have, in general,

$$v_\alpha = H_{\alpha\beta} \, r_\beta \tag{7}$$

where the 3x3 matrix $H_{\alpha\beta}$ is the Hubble tensor of velocity gradients. We split $H_{\alpha\beta}$ into its symmetric ($\theta_{\alpha\beta}$) and antisymmetric ($\omega_{\alpha\beta}$) parts

$$H_{\alpha\beta} = \theta_{\alpha\beta} + \omega_{\alpha\beta} \tag{8}$$

and then decompose $\theta_{\alpha\beta}$ into its trace (H) and tracefree ($\sigma_{\alpha\beta}$) parts,

$$H_{\alpha\beta} = \sigma_{\alpha\beta} + H\delta_{\alpha\beta}/3 + \omega_{\alpha\beta} \tag{9}$$

where $\delta_{\alpha\beta}$ is the Kronecker delta symbol. Physically, the symmetric matrix $\sigma_{\alpha\beta}$ describes the shear distortion of the flow at constant volume, $\omega_{\alpha\beta}$ the vorticity and H the pure volume dilation. Observation indicates that in the present-day Universe the isotropic volume dilation of the Hubble flow exceeds any rotation or shear by at least[100,107] a factor of order 10^5.

The dynamics of Newtonian cosmology are given by the field equation (5) together with the continuity and momentum conservation equations

$$\dot{\rho} + \rho\nabla.v = 0 \tag{10}$$

$$\dot{v} + (v\cdot\nabla)v = -\nabla\Phi \tag{11}$$

If we define a vorticity vector ω_λ, vorticity scalar ω, and shear scalar σ by

$$\omega_\lambda = \tfrac{1}{2}\epsilon_{\lambda\mu\nu}\,\omega_{\mu\nu} \quad ; \quad \omega^2 = 2\omega_\mu\omega^\mu \quad ; \quad \sigma^2 = 2\sigma_{\alpha\beta}\sigma^{\alpha\beta} \tag{12}$$

then equations (7)-(9) and (12) reduce (5), (10) and (11) to

$$4\pi\rho a^3/3 = M \tag{13}$$

$$(a^2\omega_\lambda)^{\cdot} = a^2\omega_\mu\sigma_{\mu\lambda} \tag{14}$$

$$\ddot{a} - a(\Lambda - \tfrac{1}{2}\sigma^2 + \omega^2)/3 + GMa^{-2} = 0 \tag{15}$$

where the expansion scale factor $a(t)$ is defined by[50]

$$a^3(t) = \exp(\int H_{\alpha\alpha}\,dt) \tag{16}$$

If $\sigma = \omega = 0$ we obtain the zero-pressure Friedman universes of general relativity. Curiously, neither the Newtonian cosmological models[51], nor their perturbations[41,52] were analysed until after their general relativistic counterparts.

In general relativity there exists an additional non-Newtonian effect which can influence the expansion of the universe. This is the possibility of a spatial curvature anisotropy. The Newtonian analogue of the spatial curvature term in cosmological models is necessarily isotropic.

☐ Give a singularity theorem for Newtonian cosmology.
☐ Classify the singularities arising in irrotational Newtonian cosmological models using catastrophe theory[48,49] and find the general behaviour in the neighbourhood of the singularity.
☐ Adapt the classic work of Leray and others[40,33,47] on the global and local existence of solutions to the equations (10) and (11) of Newtonian hydrodynamics to prove results about the Newtonian and general relativistic cosmological problems.
☐ Determine the asymptotic behaviour of ever-expanding Newtonian cosmological models. Relate the answer to the general behaviour of general relativistic cosmological models with isotropic spatial curvature and to the asymptotic solution of the Newtonian N-body problem with potential of the form (4).
☐ Which Newtonian cosmological models have no relativistic analogue?
☐ Show that Newtonian black holes cannot exist.

GENERAL RELATIVISTIC COSMOLOGY

In order to appreciate the degree of generality of the particular solutions of general relativity that are known, let us consider the specification of the general solution of Einstein's equations for cosmological evolution from a spacelike hypersurface of constant time, S. If the metric is expressed in synchronous coordinates

$$ds^2 = dt^2 - g_{\alpha\beta}(x^\alpha, t)dx^\alpha dx^\beta \quad ; \quad 1 \leqslant \alpha, \beta \leqslant 3 \qquad (17)$$

and the stress tensor, T_{ab}, is that of a perfect fluid with pressure p, density ρ, equation of state $p(\rho)$ and normalized 4-velocity u_i, $0 \leqslant i,j \leqslant 3$,

$$T_{ab} = (\rho + p)u_a u_b - pg_{ab} \quad ; \quad u_a u^a = 1 \qquad (18)$$

then on S we can, in general, specify 6 $g_{\alpha\beta}$, 6 $\dot{g}_{\alpha\beta}$, $3u_\alpha$ and 1 $p(\rho)$ = 16 independently arbitrary functions of the three spatial variables x^α. However, 8 of these functions are not independent of the others by virtue of the four $R_0{}^a$ constraint equations and the four general coordinate covariances of the general relativity. This leaves a general solution, in the presence of perfect fluid, specified by 8 independent arbitrary functions of three spatial variables. In the vacuum case only 12-8 = 4 arbitrary functions are necessary to specify the general solution[28,29,30]. Note that the most general transformations of the time and space coordinates which leave the metric in the synchronous form (10) contain four arbitrary functions of the space coordinates.

If we have a gravitation theory that is derived from a gravitational lagrangian that contains terms *quadratic* in the curvature invariants then the resulting field equations are (except for a special case) of 4th order in g_{ab}. In vacuum, the initial data on a {t=constant} surface must now also include 6 components of $\dddot{g}_{\alpha\beta}$ and 6 of $\ddddot{g}_{\alpha\beta}$ and the general vacuum solution must be prescribed by a total of 24-4-4 = 16 independently arbitrary functions of 3 variables.

□ Determine what type of global information is required to patch together local approximations to the general solution of the Einstein equations and how it affects the assessment of generality[54-56].

□ What meaning can be associated with the function-counting assessment of generality when T_{ab} is not specified by an equation of state? How is the situation altered by requiring that the Second Law of thermodynamics hold?

□ Find a method of characterising the relative generality of cosmological solutions with timelike portions.

□ Determine the number of arbitrary functions necessary to characterise the general solution of a theory of gravity in N space-time dimensions whose gravitational lagrangian contains terms linear in the curvature and quadratic in the different possible independent quadratic curvature invariants.

□ Determine what under what conditions on T_{ab} and g_{ab} cosmological models exhibit chaotic behaviour[56] as $t \to 0$ and $t \to \infty$. Hence formulate a definition of generality based upon the number of arbitrary constants necessary to describe the *asymptotes* of a solution rather than the solution itself and compare this with the standard function-counting assessments of generality[57].

□ Determine the number of arbitrary functions necessary to characterize the most general relativistic cosmological models which possess a Newtonian analogue.

THE FRIEDMAN COSMOLOGICAL MODELS

The simplest cosmological solutions are the homogeneous and isotropic Friedman models. They possess the metric

$$ds^2 = dt^2 - a^2(t)[\ dr^2/(1-kr^2) + r^2(de^2 + \sin^2 e d\phi^2)]\quad(19)$$

The spatial geometry of each {t = constant} surface is that of a space of constant curvature. The sign of this curvature is determined by that of the constant k, which, by a coordinate transformation, $(r \to |k|^{\frac{1}{2}} r$ and $a \to |k|^{-\frac{1}{2}} a)$, may without loss of generality be set to the values 0, +1 or −1. The salient features of the three cases are as follows:

_k = 0:_

The spacelike surfaces of constant time are Euclidean. The spatial sections an the spatial volume are infinite if their topology is R^3. An identification of Cartesian coordinates (x,y,z) with (x + A, y + B, z + C) where A,B, and C are constants produces the 3-torus topology and the spatial volume is finite.

_k = −1:_

The spacelike surfaces of constant time possess constant negative curvature as can be seen explicitly by the transformation $r = sh\chi$ which brings (19) to the form

$$ds^2 = a^2(t)[\ dx^2 + sh^2\chi\ (d\theta^2 + sin^2\theta d\phi^2)]\qquad\qquad (20)$$

The spatial sections and volume are infinite if the natural R^3 topology is adopted. (In fact all the possible topologies have yet to be classified[62]). If we introduce a fourth pseudo-coordinate, u, along with the Cartesians x,y,z via

$$x = ash\chi cos\theta,\quad y = ash\chi cos\phi,\quad z = ash\chi sin\theta sin\phi,\quad u = ach\chi$$

then we see that

$$dx^2 + dy^2 + dz^2 - du^2 =$$
$$a^2(t)[\ dx^2 + sh^2\chi\ (d\theta^2 + sin^2\theta d\phi^2)]\qquad (21)$$

and the 3-geometry can be viewed as embedded in a 4-dimensional pseudo-Euclidean space. Remarkably, Clarke[61] has shown that *any* non-compact four-dimensional space-time can be embedded in the 89-dimensional Euclidean space, R^{89}, with metric signature $(-,-,+,....+)$.

<u>k = +1</u>:

The spatial sections and volume are finite although there is no boundary to the space. The space is one of constant positive curvature as can be seen by using the transformation $r = asin\chi$ in (20). This displays it to be a 3-sphere of radius a(t) which may be embedded in a 4-dimensional Euclidean space by introducing an analogous pseudo-coordinate to that used to obtain (21).

In each case we see that the 3-spaces of the Friedman space-times can be considered to be curved relative to an artificial flat 4-dimensional space whose curvature radius in its fourth dimension is just a(t). The speed of expansion of the universe into this fourth dimension is described by da/dt.
It is *not* an expansion into an external three-dimensional space.

The scale factor a(t) is determined by solving the two independent Einstein equations. In the presence of a *comoving* (that is $u_a = \delta_a{}^0$ so geodesic observers expanding with the universe retain constant values of their spatial coordinates (r,θ,ϕ)) perfect fluid stress (18), these equations are

$$\frac{\dot{a}^2}{a^2} = \frac{8\pi G\rho}{3} - \frac{k}{a^2} \tag{22}$$

$$\dot{\rho} + 3\dot{a}(\rho + p)/a = 0 \tag{23}$$

These equations have no content unless some restriction is placed on the behaviour of p and ρ, otherwise any a(t) solves (22) and (23) for some p and ρ. If we pick an equation of state

$$p = (\gamma - 1)\rho \quad ; \quad \gamma \text{ constant}, -1 < \gamma \leqslant 2, \tag{24}$$

then the solution of (22)-(23) will be characterised by just *two* independently arbitrary *constants* rather than the eight functions of the general solution described above.

There are two simple classes of solution to (22)-(24) which do not involve elliptic functions: when k = 0 we have the Einstein-de Sitter universes

$$a(t) \propto t^{2/3\gamma} \quad ; \quad \rho = 1/6\pi G\gamma^2 t^2 \tag{25}$$

In the case of blackbody radiation ($\gamma = 4/3$), the general solution has the simple form

$$a(t) \propto (At - kt^2)^{\frac{1}{2}} \quad ; \quad \rho = 3A^2/32\pi G(At-kt^2)^2 \tag{26}$$

where A is an arbitrary constant. When A = 0 and k = -1 we have the vacuum solution of Milne with a(t) $\propto$ t.

The schematic evolution of the Friedman universes for k = 0, +1 and -1 when $2/3 < \gamma \leqslant 2$ is shown in Figure 2. When $\gamma = 2/3$ there are no recollapsing closed universes and a(t) $\propto$ t for all k.

By differentiating (22) and using (23) one sees that

$$\ddot{a} = -4\pi G(\rho + 3p)a \qquad (27)$$

Hence, if the Friedman universe is expanding now ($\dot{a}/a > 0$) it must have had a = 0 and $\rho = \infty$ at a finite time in the past if $\rho + 3p \geq 0$.

It is instructive to consider the Friedman models in *conformal time* τ defined in terms of the comoving proper time t by

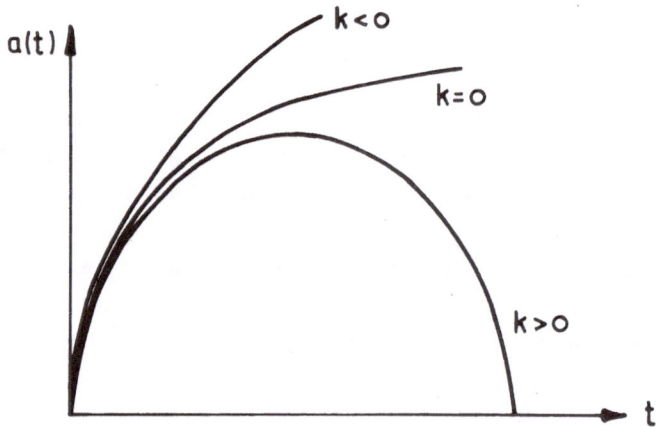

Fig. 2. The evolution of the scale-factor a(t) with respect to the comoving proper time t. The qualitative behaviour is determined by the sign of the curvature parameter k. In this time coordinate each member of the family of different closed universes with k > 0 has a different total lifetime.

$$dt = a(t)d\tau \; ; \; ' = d/d\tau \qquad (28)$$

If we now make the change of variable

$$y = a^{(3\gamma-2)/2} \qquad (29)$$

then the Friedman universes are *all* described by the simple
harmonic oscillator equation

$$y'' = -\tfrac{1}{4}k(3\gamma-2)^2 y \tag{30}$$

This neat result appears previously to have gone unnoticed but
has recently been exploited to simplify the treatment of some
quantum cosmological models[7,59,60].

Another interesting feature of the description in conformal
time is that for fixed y all closed universe have the
same conformal lifetime. If we choose $y(\tau=0) = 0 = a(t=0)$ then
this lifetime is $\Delta\tau = 2\pi/(3\gamma-2)$; see Figure 3. In t-time
closed universes do not all begin and end at the same moment.

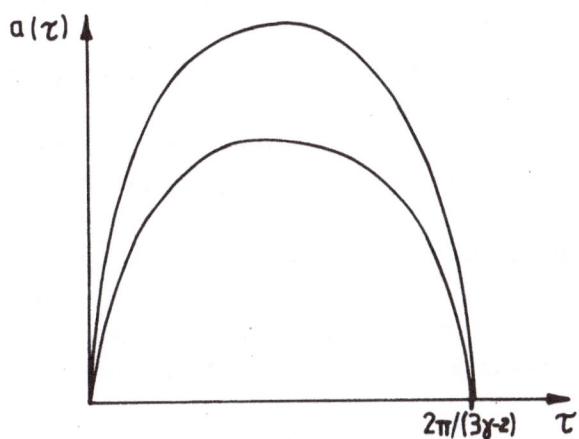

Fig. 3. The evolution of closed Friedman models in conformal
time. For fixed $y \neq 2/3$ they all have the same conformal
lifetime and attain their expansion maxima at the same moment
of conformal time.

In general the closed models are described by the simple
formulae

$$a(\tau) = a_{max} \{\sin[(3\gamma-2)\tau/2]\}^{2/(3\gamma-2)} \tag{31}$$

$$t(\tau) = a_{max} \int \{\sin[(3\gamma-2)x/2]\}^{2/(3\gamma-2)} \, dx \qquad (32)$$

where a_{max} is the value of $a(\tau)$ at the moment of maximum expansion, τ_{max} defined by $a'(\tau_{max}) = 0$, $a''(\tau_{max}) < 0$.

The conformal time coordinate also exhibits the causal structure of the Friedman models. For example, using (28) and the radial coordinate $x = \sin^{-1}(r/a)$ in the $k = +1$ model, the metric is

$$ds^2 = a^2(\tau)[\, d\tau^2 - dx^2 - \sin^2x(\, d\theta^2 + \sin^2\theta \, d\phi^2)] \qquad (33)$$

The equations of radial ($d\theta = d\phi = 0$) light-rays ($ds^2 = 0$) are therefore $dx = \pm \, d\tau$. So, after a conformal time τ an observer located at $\tau = 0$ can only see everything inside the radius $x = \tau$. This radius is called the *particle horizon at time τ*.

☐ Find a method of characterising the relative generality of cosmological solutions with timelike portions.

☐ Investigate the effect of the 'wedge' structure of space-time outside a vacuum string in a Friedman space-time with 3-torus topology.

☐ Examine the effects of inflation in a finite open universe with 3-torus topology.

☐ Find a technique for solving equations like (22) and (23) subject to inequalities on p and ρ rather than initial conditions.

☐ Investigate the behaviour of perturbations to the Friedman models with non-standard topologies.

☐ Examine the evolution of a spherical overdensity to the black hole state in a closed Friedman Universe using the conformal time coordinate. Investigate ways of discriminating between the black hole and the closed universe.

OBSERVABLE PARAMETERS

It is useful to pick the two mathematical quantities needed to completely specify a Friedman model so that they are directly observable (modulo selection effects). We define the

Hubble parameter

$$H = \dot{a}/a \tag{34}$$

and the deceleration parameter

$$q = -\ddot{a} a/ \dot{a}^2 \tag{35}$$

Their present values (subscript '0') are found to be

$$H_0 = 75 \pm 25 \text{ Kms}^{-1}\text{Mpc}^{-1} ; \quad q_0 < 2, \tag{36}$$

If $\Lambda = 0$ then differentiating (22) and evaluating today we have

$$H_0^2(2q_0 - 1) = ka_0^2 \tag{37}$$

Therefore $k \geq 0 \leftrightarrow q_0 \geq \frac{1}{2} \leftrightarrow \Omega_0 \geq 1$ where the density parameter Ω_0 is defined by

$$\Omega_0 = \rho_0/\rho_{cr} : \rho_{cr} = 3H_0^2/8\pi G \tag{38}$$

We note that the expansion scale factor $a(t)$ is not observable but it is convenient to measure times and distance measures in units of the present value of a. Accordingly we define the *redshift* $z > 0$ by

$$1 + z = a_0/a(t) \tag{39}$$

For example, the angular scale, θ_h, subtended by the particle horizon, defined above, at redshift z in a k = 0 Friedman universe is

$$2\sin(\theta_h/2) = [(1 + z)^{\frac{1}{2}} - 1]^{-1} \tag{40}$$

So, quasars at $z = 2$ separated by more than about 84° on the sky are causally disjoint, as are regions of the microwave background radiation separated by more than about 2° on its last scattering surface at $z = 1000$.

WHEN DO CLOSED UNIVERSES RECOLLAPSE?

One of the most interesting features of the inflationary universe picture is that it predicts[18] that the present density of the Universe should be within one part in a million of the critical density of the $k = 0$ Friedman model. If there is some quantum or superstring(?) constraint upon Nature which ensures that k must be zero ($\Omega = 1$ if $\Lambda = 0$) then this prediction has no content. However, if our Universe has $k \neq 0$ ($\Omega \neq 1$) then it is very difficult to understand why the two terms on the right-hand-side of equation (22) are still of the same order of magnitude after $a(t)$ has changed by thirty-two orders of magnitude since the Planck time $t_p = G^{-\frac{1}{2}} \sim 10^{-43}$s. The observed state of affairs requires k/a^2 to have been initially smaller than ρ by a huge factor $> 10^{60}$. Providing an explanation for this state of affairs is equivalent to explaining why the Universe has expanded for $10^{60}t_p \sim 10^{17}$ secs.

It was first realized by Guth[18] (see also Sato[63]) that a sufficiently long period of expansion with the equation of state of the matter obeying $p = -\rho$ so that $a(t) \propto \exp(H_0 t)$ with H_0 constant (or even $p = (\gamma-1)\rho$ with $0 \leq \gamma < 2/3$ and $a(t) \propto t^{2/3\gamma}$ in fact) can explain why the k/a^2 term in the Friedman equation is not enormously larger than the $8\pi G\rho/3 \sim a^{-3\gamma}$ term for large a.

The inflationary picture also predicts that there should exist a spectrum of constant curvature density fluctuations out to the maximum scale of the inflation[1]. As yet there is no natural way to obtain the observed amplitude of these fluctuations ($\delta\rho/\rho \sim 10^{-4}$). However, if the theory did generate fluctuations of this magnitude with constant amplitude on every scale as it enters the horizon then the level of density fluctuations is larger than the distance of the predicted mean density from the critical density. The possibility of such a situation arising in the post-inflationary phase of the Universe led Zeldovich and Grishchuk[64] to investigate whether it is possible for local unbound sub-regions of a closed universe to avoid hitting the final singularity. They, and subsequently Bonnor[65] and Hellaby and Lake[66], examined this problem in the context of spherically symmetric space-times containing zero pressure matter. They found that under these circumstances it is not possible for negatively-curved shells of material to avoid the final singularity. Barrow and Tipler[67] examined this question in greater generality to determine the conditions under which closed universes (defined as those possessing compact Cauchy surfaces) collapse to an all-encompassing future singularity. They showed that the existence of a maximal hypersurface is a necessary and sufficient condition for the existence of both an initial and a final all-encompassing "strong" singularity in a universe with a compact Cauchy surface satisfying the strong energy condition (2) and a generic condition to the effect that all geodesics feel some tidal gravitational field.

A spacelike hypersurface, S, is said to be a *maximal hypersurface* if $z_{;a}^a = 0$ everywhere on S, where z^a is the unit normal vector to S (this corresponds to the physical notion of the moment of maximum expansion). A "strong" singularity is one which, roughly speaking, crushes out of existence objects which hit it. Examples of "strong" singularities are strong curvature singularities and crushing singularities[68]. (These restrictions on the singularities are required only for the 'necessary' part of the singularity theorem: that is, a maximal hypersurface will give all-encompassing initial and final singularities, but only "strong" initial and final singularities will have a maximal hypersurface between them).

It has been shown[6,67,71] that only closed universes with spatial topology either S^3 or $S^2 \times S^1$ or more complicated hybrids formed by connected summations and special identifications of these two basic topologies actually admit maximal hypersurfaces so long as the differentiable structure of 4-dimensional space-time does not possess one of the infinite number of "exotic" differentiable structures of Freedman[69] and Donaldson[70] that pure mathematicians are currently so excited about (that is, we want the coordinate systems covering the 4-dimensional space-time to be only those generated by pulling-up the 3-dimensional systems covering the 3-spaces). Thus closed universes with other topologies (for example the 3-torus, T^3) cannot recollapse to an all-encompassing final singularity. However, it is still an open question whether or not all closed universes with S^3 and $S^2 \times S^1$ topologies do in fact recollapse to a singularity.

By "recollapsing universe", here, we mean a non-static space-time which has a maximal compact Cauchy hypersurface, and by "all-encompassing initial singularity" we mean that all inextendible timelike curves have a length less than a constant L to the past of any Cauchy surface, where L may depend on the Cauchy surface, but not on the particular curve. "All-encompassing final singularity" is defined analogously. As stated above, a closed recollapsing universe begins and ends in an all-encompassing singularity if the strong energy and a generic condition hold. However, a recollapsing universe is distinct from the Wheeler universe defined in Marsden and Tipler[68] and in Barrow and Tipler[67]. A Wheeler universe begins and ends in all-encompassing singularities, but it could be that the singularity might not be strong, and hence a Wheeler universe might not have a maximal hypersurface. But a Wheeler universe with a maximal hypersurface is a recollapsing universe, if the strong energy and a generic condition hold.

If an equation of state is not defined then it turns out that the conditions under which even closed Friedman universes recollapse are rather subtle. a recent investigation[72] shows that old theorems of Tolman and Ward[73] concerning this question are actually incorrect. Neither the conditions[27] {$\rho > 0$ and $p \geq 0$} nor {$\rho + p > 0$ and $\rho + 3p > 0$} are strong enough to guarentee that a closed Friedman universe recollapses. The problem is that the pressure may diverge *before* a maximal hypersurface is reached. For example[72], if we take

$$a(t) = 1 + t - 2(1 - t)^{3/2}/3 \quad ; \quad 0 < t < 1 \tag{41}$$

then we can calculate ρ and p from equations (22) and (23) for $k = +1$ and we find $p \to \infty$ as $t \to 1$ but ρ, a and à remain finite. Sufficient conditions to obtain recollapse are found to be {$\rho + 3p > 0$ and $|p| \leqslant C\rho$, C constant} together with a *matter regularity condition*: T_{ab} is regular except at space-time singularities. The conditions on T_{ab} for the recollapse of anisotropic closed universes are as yet unknown[72].

□ Prove or disprove the conjecture[72] that all globally hyperbolic closed vacuum universes with S^3 or $S^2 \times S^1$ spatial topology expand from an all-encompassing initial singularity to a maximal hypersurface, and recollapse to an all-encompassing final singularity.

□ Prove or disprove the conjecture[72] that all globally hyperbolic spatially homogeneous closed universes with S^3 or $S^2 \times S^1$ spatial topology and with stress-energy tensors which obey (i) the strong energy condition, (ii) the positive pressure criterion, (iii) the dominant energy condition, and (iv) the matter regularity condition expand from an all-encompassing initial singularity to a maximal hypersurface and recollapse to an all-encompassing final singularity.

□ Prove or disprove the conjecture[72] that all globally hyperbolic closed universes with S^3 or $S^2 \times S^1$ spatial topology and with stress-energy tensors which obey (i) the strong energy condition, and (ii) the positive pressure criterion, begin in an all-encompassing initial singularity and end in an all-encompassing final singularity.

□ Find conditions on T_{ab} for the closed Kantowski-Sachs[74,76] and Taub [86] axisymmetric Bianchi type IX universes to recollapse to a final singularity.

□ Examine the consequences of the discovery[69,70] that 4-dimensional manifolds admit an infinite number of distinct differentiable structures (for all other dimensions there is one unique differentiable structure only) for cosmological models. Does the special character of 4-dimensional manifolds provide any clue as to why the observed dimension of space-time is 4 in the context of Kaluza-Klein or superstring theories exhibiting dimensional compactification[116]?

SPATIALLY HOMOGENEOUS UNIVERSES

Physically speaking, we would regard as spatially homogeneous a universe in which all comoving observers record the same picture of cosmic history. Mathematically[79], this requires that for any pair of observers A and B there is a symmetry which allows the universe model considered as centred on A to be transformed into an identical model centred on B. Spaces of this type either fall into the Bianchi classification and possess a three-dimensional group of translation symmetries or into the exceptional case of Kantowski-Sachs[74], (found also by Kompanyeets and Chernov[75]), which possesses a four-dimensional group of motions with no simply transitive three-dimensional subgroup. The Kantowski-Sachs models are not considered here because their exact solutions have been found and fully studied elsewhere[76]. Henceforth any reference to spatially homogeneous cosmologies will refer to the Bianchi[77,139] types.

The Bianchi spaces allow us to identify spacelike hypersurfaces S(t) on which we may define at least three independent Killing vector fields ξ_A (where upper case Latin indices run from 1 to 3) which satisfy Killing's equations,

$$\xi_{m;n} + \xi_{n;m} = 0 \qquad (42)$$

The commutators of the ξ_A are determined by a set of structure constants, $C_{AB}{}^D$,

$$[\xi_A, \xi_B] = \xi_A \xi_B - \xi_B \xi_A = C_{AB}{}^D \xi_D \qquad (43)$$

The components of the metric, g_{ab}, are invariant under the isometry generated by infinitesimal translations along these Killing vector fields - the functional time-dependence of the metric is the same at all points. The Einstein equations relate the energy-momentum tensor T_{mn} to the first and second derivatives of g_{mn} and so, if g_{mn} is invariant under an isometry, all physical properties of T_{mn} also remain invariant. Since the Killing vector fields span each S(t) we have a spatially homogeneous space-time

The set of n Killing vectors possesses an n-dimensional group structure, G_n, characterised by the equivalence classes of structure constants, $C_{BC}{}^A$, and these can be used to classify all spatially homogeneous cosmologies. The possible group types possess a G_3, G_4 or a G_6, and in all but one exceptional (Kantowski-Sachs) case there is at least one G_3 subgroup. The classification of this subgroup leads to Bianchi's famous decomposition into types but we shall give the modified scheme developed by Estabrook, Wahlquist and Behr[78], and Ellis and MacCallum[79,80,81].

On any particular spacelike hypersurface, the Killing vector basis can be chosen so that the structure constants can be decomposed as

$$C_{BC}{}^A = \epsilon_{BCE} n^{EA} + \delta^A_C a_B - \delta^A_B a_C \tag{44}$$

where ϵ_{BCE} is the completely antisymmetric symbol ($\epsilon_{123} = 1$) and

$$n^{AB} = \text{diag}(n_1, n_2, n_3) \tag{45}$$

$$a^B = (a, 0, 0) \tag{46}$$

and all the $\{n_1, n_2, n_3, a\}$ can be normalized to ± 1 or 0. If an $n_2 n_3 = 0$, then n_2 and n_3 can be set equal to ± 1 and a to $\sqrt{|h|}$, where h is a parameter used in the group classification. The possible combinations of n_1 and a fix the Bianchi-Behr group types. They are divided into two classes, A and B, according to whether a is zero or non-zero, as shown in Table 1.

The isotropic Friedman models have G_6 symmetry groups with G_3 subgroups such that the zero curvature model can be thought of as a special case of Bianchi types I or VII_0, the open Friedman model as a special case of types V or VII_h and the closed model as a special case of type IX.

Table 1. The number of independent, arbitrary constants necessary to specify a Bianchi type universe on a spacelike hypersurface of constant time in vacuum (r) and with perfect fluid source (s). Typically, s = r + 4, except for some particular cases where geometrical constraints exclude particular degrees of freedom for the fluid motion and then s < r + 4. The Bianchi group dimension is p.

BIANCHI TYPE		p	ARBITRARY FUNCTIONS IN VACUUM, r	ARBITRARY FUNCTIONS WITH PERFECT FLUID, s
I		0	1	2
II		3	2	5
VI_0		5	3	7
VII_0	*Class A*	5	3	7
VIII		6	4*	8*
IX		6	4*	8*
. . . .				
IV		5	3	7
V		3	1	5
VI_h	*Class B*	6	4*	8*
VII_h		6	4*	8*
$VI_{-1/9}$		6	4*	7

In the Table we have indicated the relative degree[82] of generality of each type in two ways. First, the group dimension p gives the dimension of the orbit of $C_{BC}{}^A$ as a subset of all the 9 distinct $C_{BC}{}^A$ components. Because the Jacobi identities must be satisfied by the Killing vectors, that is

$$c^{ABC}[[\xi_A, \xi_B], \xi_C] = 0 \qquad (47)$$

these imply

$$C_{[AB}{}^D C_{C]D}{}^E = 0 \qquad (48)$$

and so,

$$n^{AB}a_B = 0, \qquad\qquad (49)$$

and hence the orbits of any particular group type are at most *six-dimensional*.

We are interested in the relative generality of particular Bianchi universe *solutions* of the field equations rather than their intrinsic group structure. The best way[83] of chacterising the generality of particular metrics is to count the independent parameters in the initial data set giving rise to the general solution of each group type for a specified T_{ab}. The quantity r in Table 1 is the number of arbitrary constants which appear in the *general vacuum solution* of each group type (in types VI_h and VII_h this count includes the parameter h).

The most general vacuum solutions are therefore those of types VII_h, VI_h, VIII, $VI_{-1/9}$ and IX, and are parametrised by four independent arbitrary constants. (The reason for the appearance of type $VI_{-1/9}$ which appears to be more special than VI_h is[82-84] that two of the Einstein constraint equations become null identities for the choice $h = -1/9$). We recall our earlier discussion following equations (17) and (18) in which the general inhomogeneous vacuum solution was characterised locally by four free *functions* of three spatial variables rather than four *constants*. The less general Bianchi types admit fewer independent shear modes ($\dot{g}_{ij}$ components); for example, the Bianchi type I vacuum solution is the famous Kasner metric[29,85] and is determined by only one free constant.

The column labelled s in Table 1 gives the number of independent arbitrary constants necessary to specify the general solution of each Bianchi type with a *perfect fluid* energy-momentum tensor. Recall that this required the independent specification of eight functions in the general inhomogeneous case. We expect 4 additional parameters to specify the general perfect fluid solution compared with the vacuum specification. In fact, we do not have $s = r + 4$ in all cases because the constraint equations place additional restrictions upon the generality of the T_{ab} allowed in types I and II. For example, if a perfect fluid is added to the vacuum type I (Kasner) solution the Bianchi I geometry requires the time-space Ricci components to be identically zero, hence T_{mn} is constrained to have $T_{o\alpha} = 0$, $\alpha = 1, 2, 3$ and so we must have comoving velocities with $u_a = \delta_a{}^0$. Therefore only one additional piece of data is required beyond the one-parameter vacuum datum: the density ρ. The perfect fluid $VI_{-1/9}$ is also not as general as perfect fluid VI_h, VII_h, VIII and IX solutions (unlike the situation in vacuum) because the double degeneracy of the R_{oi} constraint equations only occurs in vacuum or when the velocity field is comoving, $(u_a = \delta_a{}^0)$; in all other cases $s = r + 4$.

The *general* vacuum solutions[86] are known only for Bianchi types I (1 parameter), II (2 parameters) and V (1 parameter) in vacuum and only for type I (2 parameters) in the presence of perfect fluid obeying (18) and (24). Special vacuum solutions are known for type VI_0 (1 parameter), VI_h (2 parameter), $VI_{-1/9}$ (zero parameter), VII_h (2 parameter) along with various special perfect fluid solutions. The stability of these special solutions has been studied in detail recently[34,82,88].

☐ Classify the homogeneous spaces and thereby the possible spatially homogeneous cosmological models in N dimensions.
☐ Which Bianchi types admit the de Sitter universe?
☐ Use the approximate symmetry group construction of Spero[42] to classify the known inhomogeneous cosmological solutions.
☐ Produce an analogous classification of relative generality to that in Table 1 for the self-similar generalization of the Bianchi classification developed by Eardley[87].
☐ Study inhomogeneous perturbations of the known exact spatially homogeneous universe solutions.
☐ Extend the evaluation of the relative generality of the Bianchi universes given in Table 1 to the case where non-perfect fluid stress tensors are admitted. Does the hierarchy of generality remain the same?

THE MICROWAVE BACKGROUND AND THE DENSITY OF THE UNIVERSE

In this section we shall describe the observational consequences of small anisotropies in the Hubble expansion and curvature of the Universe. Such anisotropies *must* exist at some level because the Universe is not precisely homogeneous in space. We shall also see that this question turns out to have an unexpected connection with the observational problem of determining Ω_O, the present density of the Universe, (38).

There exist[19] a variety of theoretical prejudices regarding the likely value of Ω_O. They are worth commenting on since they all support a situation in which any observational test of whether the Universe is open or closed appears to be impossibly difficult:

Simplicity: $\Omega_O = 1$ *exactly*

This minimizes the number of free parameters in the isotropic Friedman model if we assume the spatial sections possess natural topology. It avoids awkward questions regarding the fine-tuning of the initial conditions. However in light of the fact that the Universe is not exactly a Friedman model and quantum gravitational fluctuations inevitably exist in the spatial curvature when the expansion emerges from the Planck era, we might want to refine this form of our belief in simplicity to something like...

Quantum Simplicity: $\Omega_O = 1 \pm$ (quantum correction)

In some theories the sign of this quantum correction can be specified. For example, if the Universe was created spontaneously out of a quantum vacuum[64,89], then we require $\Omega_O > 1$ if the topology is natural. However, the existence of so many more nonstandard topologies, (*e.g.*, the 3-torus), may make the quantum creation of universes with nonstandard topology out of "nothing" a more probable occurrence; however, such topologies require additional parameters to specify them. It is interesting to note that the Friedman model allows the largest number of possible topologies. One might even interpret the fact that our Universe possesses a density so close to the critical one as evidence for its original creation out of a quantum vacuum state.

Inflation.

Models for the evolution of the early universe which admit either a phase transition with very special properties[63], or the existence of hypothetical scalar fields with appropriately weak nonlinear self-interactions[1,90] or quadratic lagrangians[39] will undergo a short period of de Sitter expansion. This inflationary interlude in the expansion can explain the proximity of the observed Universe to the critical density and leads us to predict that the present value of Ω_0 will satisfy[18]

$$|\Omega_0 - 1| < 10^{-6} \tag{50}$$

This is just the condition that would have permitted sufficient inflation to resolve the horizon and monopole problems[63] (assuming, of course, that such problems exist following a quantum gravitational or superstring era). Inflation does not predict the sign of $|\Omega_0 - 1|$. In a universe obeying (50) with $\delta\rho/\rho \sim 10^{-4}$ on all scales local observations may never be a reliable guide to the global density[64].

Quantum Gravity

To deal with boundary terms in the gravitational action and associated path integrals, these models are formulated in closed universes[91]. Whether or not this restriction is fundamental is still unclear. It also appears possible[92,93] that quantum restrictions will lead to universes in which $|\Omega_0 - 1|$ is very small with high probability. One should bear in mind that mathematical cosmologists appear to have a natural prejudice for closed universes because it is that much easier to prove theorems concerning the properties of compact spaces.

Many of these theoretical prejudices lean towards universe models that will result in Ω_O lying tantalizingly close to unity. This appears to be a depressing state of affairs for observational cosmologists aiming to determine whether or not the Universe is open, since they are already faced with a complicated and ambiguous collection of astronomical data. Direct observations[94] of luminous galaxies lead to an estimate of $\Omega_0 \sim 0.007$, while[95,96] the dynamics of individual galaxies and those in groups and clusters limit the density clustered within them to $\Omega_O \sim 0.2$. The totality of observations of helium -4, helium -3, and deuterium place a limit on the total density of material in baryonic form of $\Omega_O \sim 0.15$, assuming these light elements were manufactured primarily by primordial nucleosynthesis[97]. The possibility that large quantities of dark material could exist, either in the form of black holes of a size that avoids the nucleosynthesis constraints or in an unclustered sea of nonbaryonic particles, makes it impossible to draw any definite conclusions regarding the total density of the Universe that are robust enough to exclude the theoretical prejudices outlined above.

If we forget the theoretical prejudices for a very small value of $|\Omega_O - 1|$, then the observational data still allows the dull solution that all dark matter is baryonic and resides in planetary-sized objects or faint stars. Attempts to resolve this "missing mass problem" by measuring the deceleration of the Hubble expansion are bedevilled by a number of notorious selection and evolutionary effects[31], and the lingering possibility of a nonzero cosmological constant[98] complicates the issue even further.

Despite the difficulties outlined above, we shall see there may exist[99,100] ways of using the structure of the microwave background radiation anisotropy over intermediate angular scales to determine the density of the Universe and decide whether or not a cosmological model is open or closed no matter how small the value of $|\Omega_O - 1|$.

MICROWAVE BACKGROUND OBSERVATIONS

Temperature fluctuations of dipole signature with an amplitude ~ 10^{-3} are the only ones to have been positively detected[101] in the microwave background radiation. Only upper limits exist on any possible quadrupole moment and these are summarized in Table 2

Table 2. Microwave background observations: limits on the amplitude of the quadrupole anisotropy in the microwave background at various wavelengths.

Upper Limit on $\Delta T/T_o$	Angular scale θ(deg)	Wavelength (cm)	Ref.
3.0×10^{-5}	10-90	1.2	102
4.4×10^{-5}	6	0.07	103
6.2×10^{-5}	90	0.03	104
7.0×10^{-5}	90	0.8	105
4.0×10^{-4}	6	0.8	105

We shall be concerned with the structure of the background radiation over "intermediate" angular scales exceeding about 10° and its associated temperature profile $T(\theta,\phi)$. These scales exceed those of fine scale associated with fluctuations from embryonic protoclusters, the effects of reheating and the thickness of the last-scattering surface.

CHARACTERISTIC MICROWAVE BACKGROUND PATTERNS

There have been many investigations into the form of the microwave background expected in anisotropic, homogeneous cosmological models[99,100,106-111]. We shall describe the simplest examples of each possibility for illustrative purposes.

First, we note that in general, the temperature anisotropy can contain three distinct pieces: a dipole term due to our motion relative to the universal frame in which the radiation is isotropic; a Doppler term, which need not be purely dipolar, due to any similar relative motion of the last scattering surface of the radiation; and, finally, an anisotropic distortion term due to the intrinsic expansion anisotropy of the Universe. This last distortion term can have three important forms in realistic cosmological models, depending on the density of the Universe and the form of the anisotropy:

Quadrupole

The simplest example is the axisymmetric Bianchi type I universe[109,99]. The metric is

$$ds^2 = dt^2 - X^2(t)[dx^2 + dy^2] - Y^2(t)dz^2 \qquad (51)$$

The geodesic equations can be integrated exactly and when the anisotropy level is low, the temperature anisotropy has the simple quadrupolar θ-variation,

$$\frac{\Delta T(\theta,\phi)}{T_0} \equiv \frac{T(\theta,\phi) - T_0}{T_0} \quad \alpha \; P_2(\cos \theta) \qquad (52)$$

where $P_2(\cos\theta)$ is a Legendre polynomial and T_0 is the mean radiation temperature. When the model is not axisymmetric, then the dependence with angle is given by the spherical harmonic $Y_{2m}(\theta,\phi)$. The solutions of the geodesic equations allow no convergence or divergence of geodesic separations between last scattering (L) and observation (O) at the present epoch. This behaviour is essentially Newtonian and also occurs in some very special anisotropic models with $\Omega_0 < 1$ and also in general closed models like Bianchi IX. (In these more complicated cases it may be possible for a dipole to exist in addition to a quadrupole whereas in the simple metric (51) the Ricci tensor satisfies $R_{0\alpha} = 0$, $\alpha = 1$, 2, and 3, and hence any perfect fluid must be comoving and there can be no dipole.

Hotspot

If an anisotropic universe is open ($\Omega_0 < 1$), then there can occur a new effect that was first pointed out by Novikov[112]. Various other authors[106,108,111] have studied theoretical aspects of this problem, and Barrow *et al*[99] and Bajtlik *et al*[113] have computed observable features in detail. The simplest example is an axisymmetric Bianchi-type-V model containing pressureless matter,

$$ds^2 = dt^2 - X^2(t)e^{2z\sqrt{1 - \Omega}_0}(dx^2 + dy^2) - Y^2(t)dz^2 \qquad (53)$$

This reduces to (51) when $\Omega_0 = 1$ and is the open Friedman universe in unusual coordinates when $X(t) = Y(t)$. This metric only solves the Einstein equations if the 3-velocity possesses a nonzero component:

$$u_\alpha = (0, 0, \pm 2\sigma_0(1 - \Omega_0)^{\frac{1}{2}}(1 + z)/\sqrt{3} H_0\Omega_0) \qquad (54)$$

where the present ratio of shear to Hubble rate, σ_0/H_0, is a constant parametrizing the amplitude of the anisotropy; and z is the red shift. In both (51) and (53) the scale factors can be normalized so that $X = Y$ at the red shift $z_L = 1000$ when the radiation first becomes collisionless. The resulting temperature profile observed on the sky today can be computed and has two components: a dipole, because of the velocity term (but this need not be present in a nonaxisymmetric model of this type), and a distorted quadrupole or "hotspot". This is illustrated in Figure 4 for various values of Ω_0.

As Ω_0 takes lower values, the quadrupole pattern is focussed into a region of diminishing angular scale. The diameter of the focussed quadrupole is roughly $2\Omega_0$ radians. If one follows the evolution of $T(\theta,\phi)$ with falling red shift, from z_L to the present, one finds that the focussing starts to become particularly effective at red shifts less than $\sim \Omega_0^{-1}$

become particularly effective at red shifts less than $\sim \Omega_0^{-1}$ when the negative curvature dominates the effect of the matter on the dynamics. The focussing effect can be seen explicitly in the geodesic behaviour. The relation between angles θ_0 and θ_L is

$$\tan\left(\frac{\theta_0}{2}\right) = \tan\left(\frac{\theta_0}{2}\right) \exp\left(\tau_L - \tau_0\right) \tag{55}$$

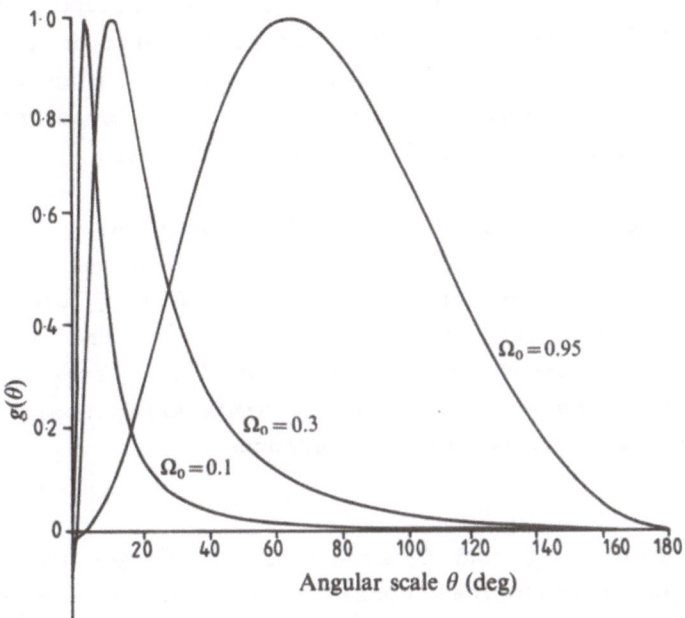

Fig.4. Quadrupole focussing. The variation of $\Delta T(\theta_0,\phi_0)/T_0 \propto g(\theta_0)$ with angular scale in type V universes for various values of Ω_0. The focussing effect increases as Ω_0 decreases.

where we have used the τ time coordinates defined by

$$\tau_0 = ch^{-1}(2\Omega_0^{-1} - 1), \quad \tau_L = ch^{-1}\left[\frac{2(\Omega_0^{-1} - 1)}{1 + z_L} + 1\right] \tag{56}$$

Hence, if $\Omega_0 < 1$, we have $\theta_0 < \theta_L$ and there is *focussing*.

The type V model we have used as a hot spot paradigm has isotropic spatial curvature and possesses a simple Newtonian analogue. The anisotropy modes in Bianchi type V models all decay as the universe expands. In order to give a detectable temperature anisotropy signal today, the anisotropy would have been so large at the epoch of primordial nucleosynthesis that the helium and deuterium abaundances produced then would have been in severe conflict with observation[115] However, this does not mean that the hotspot effect can never be physically relevant in our Universe; there exist other, more general, open homogeneous universes containing the hotspot effect in which the anisotropy level need not be so large in the distant past. Also, it transpires that the simple Bianchi type V example is closely related to the description of small-amplitude density inhomogeneities in open universes[19,114]. Notice also that the hotspot effect, although allowing one to infer the total density of the Universe directly if that density lies significantly below the critical value, is of no use if the density lies very close to the critical value, as suggested by many of the theoretical prejudices discussed above. To discriminate between open and closed universes when $|\Omega_0 - 1|$ is arbitrarily small, we have to examine general relativistic effects.

Spirals

The Bianchi I and V models are not the most general anisotropic cosmological models which contain the isotropic Friedman models as special cases as is evident from Table 1. We should therefore consider the type VII_0 and VII_h models which generalise them. In each case the generalisation is equivalent to adding circularly polarized homogeneous gravitational waves to the I and V models respectively. The VII_0 and VII_h models reduce to the I and V models in the limit when the wavelength of these gravitational waves goes to infinity. Each type VII universe can be parametrized by a measure of the wavelength of these gravitational wave modes. The most convenient is the Collins-Hawking[107] parameter, x, defined as

$$x = \frac{\text{wavelength of gravitational waves}}{\text{Hubble radius}} \qquad (58)$$

The effects of finite values of x on the propagation of the microwave background radiation from its last scattering red shift are intrinsically general relativistic because the presence of these gravitational waves produces a non-Newtonian effect: anisotropic spatial curvature. It was first noticed by Collins and Hawking[22], and recently studied in more detail by Barrow et al[88,100], that this curvature anisotropy produces a geodesic spiralling effect in the type VII models with $\Omega_0 \leqslant 1$.

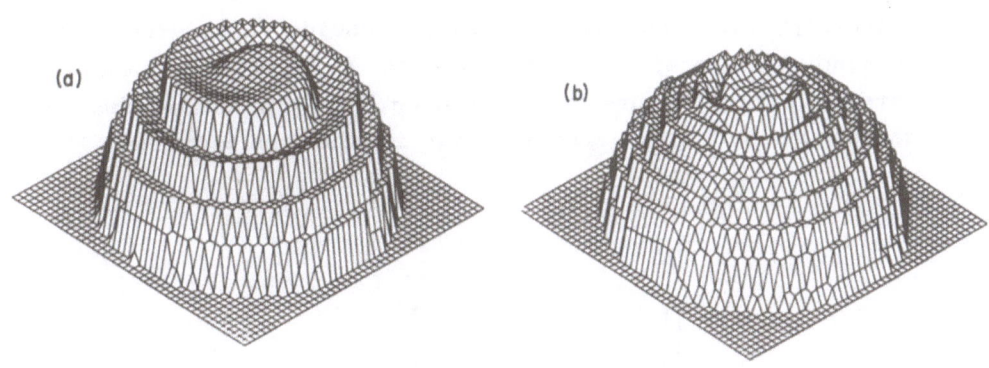

(a)

(b)

Fig.6. Spirals.(a) The temperature pattern, $T(\theta,\phi)$ predicted in a type VII_0 universe (with $\Omega_0 = 1$). The radial distance from the hemispheric centre gives the magnitude of $T(\theta_{obs}, \phi_{obs})$; θ runs over the top of the hemisphere whilst ϕ runs from $0 \to 2\pi$ around the circular base. The value of $x = 0.067$ and the anisotropy level has been set at the largest value compatible with the observations in Table 2. (b) As in (a) but for an open universe of type VII_h (with $\Omega_0 = 0.7$). The spiral plus quadrupole seen in (a) has been focussed in the θ angle into a spiral hotspot.

166

This effect disappears as the spatial curvature becomes flat, $\Omega_0 \to 1$. When $\Omega_0 < 1$, then instead of (52) we have, approximately,

$$\frac{\Delta T(\theta,\phi)}{T_0} \propto P_2 \left[\cos \left\{ 2 \tan^{-1}(2 \tan(\frac{\theta}{2}) \Omega_0^{-1}) \right\} \right] \tag{57}$$

Elsewhere, we have produced detailed temperature maps of microwave skies possessing the hotspot feature[99,100] and Lukash and Novikov[114] have also discussed the structure of inhomogeneous hotspots[19]. The angular scale of the hotspot feature emphasizes the need for complete sky coverage in observational searches for anisotropies. It is also important to note that the maximum amplitude of the temperature anisotropy in the hotspot will considerably exceed that outside the hotspot region; Figure 5. We note that the hotspot effect occurs only in open ($\Omega_0 < 1$) universes that contain anisotropies, although not in all open universes

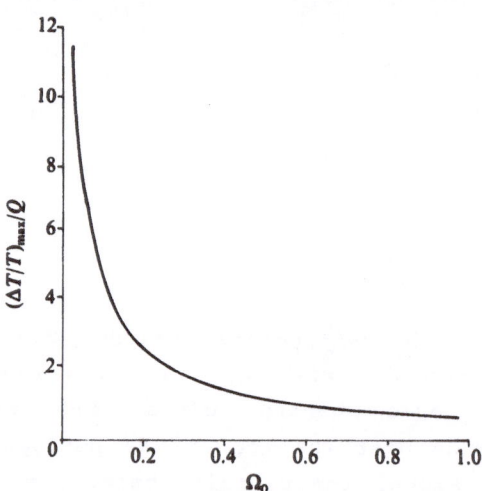

Fig.5. Hotspot contrast. The enhancement predicted at the hotspot's temperature maximum compared with the quadrupole over the rest of the sky. The enhancement disappears, along with the hotspot, as $\Omega_0 \to 1$.

This spiralling effect is superimposed upon a quadrupole in the flat (VII$_0$) case but the spiral plus quadrupole are both focussed by the hotspot effect in the open (VII$_h$) universe; see Figure 6. This effect does not occur in the closed, spatially homogeneous, cosmological models of Bianchi type IX. They have an anisotropic distortion term that is essentially quadrupolar with no spiralling in the ϕ angle.

The number of twists, N, of the full spiral patterns shown in Figure 6 is determined by x as[100]

$$N \sim \frac{2}{\pi x} \tag{59}$$

and is not influenced by Ω_0 to first order. Therefore, the presence of the spiral effect allows observers to discriminate between open and closed homogeneous universes: *closed homogeneous universes cannot exhibit the spiral efect no matter how close Ω_0 is to unity.* This appears to be the only discrete cosmological observable so far discovered that has such a property.

As $x \rightarrow \infty$, the spiralling disappears. Note that the geodesic spirals are left-handed in the observer angles ($\pi - \theta_0$, $\pi + \phi_0$). The solutions[22,100] for the geodesics reveal the spiral effect and generalize the solution (55)

$$\tan(\frac{\theta_0}{2}) = \tan(\frac{\theta_L}{2}) \exp [-(\tau_0 - \tau_L)\sqrt{h}] \tag{60}$$

$$\Phi_0 = \Phi_L + (\tau_0 - \tau_L) - \frac{1}{\sqrt{h}} \tag{61}$$
$$\times \ln\left[\sin^2 (\frac{\theta_L}{2}) + \cos^2 (\frac{\theta_L}{2}) \exp \{2(\tau_0 - \tau_L)\sqrt{h}\}\right]$$

where τ_0 and τ_L are still defined by (56). To obtain the Bianchi type I and V results, we take the limit $x \rightarrow \infty$ of the VII$_0$ and VII$_h$ models respectively with *fixed* Ω_0. The VII$_0$

behaviour is obtained from (60) and (61) by taking the limits
$h \to 0$ and $\Omega_0 \to 1$ with finite x. There is a simple relationship
between x, h, and the density parameter Ω_0

$$x = \left[\frac{h}{1 - \Omega_0} \right]^{\frac{1}{2}} \tag{62}$$

The angular behaviour of the temperature profile in
type-VII universes now picks up a nontrivial azimuthal
variation in ϕ owing to the spiral effect, and we find in the
simplest examples[100], that

$$\frac{\Delta T(\theta_0, \phi_0)}{T_0} \propto [A^2(\theta_0) + B^2(\theta_0)]^{\frac{1}{2}} \cos(\phi_0 + \tilde{\phi}) \tag{63}$$

where $\tilde{\phi}$ is given by

$$\cos \tilde{\phi} = \left[\left[\frac{\sigma_{12}}{\sigma} \right]_0 B(\theta_0) - \left[\frac{\sigma_{13}}{\sigma} \right]_0 A(\theta_0) \right] \tag{64}$$

$$\times (A^2(\theta_0) + B^2(\theta_0))^{-\frac{1}{2}}$$

Here, $(\sigma_{12}/\sigma)_0$ and $(\sigma_{13}/\sigma)_0$ are constants measuring the
amplitude of any anisotropy, and the functions $A(\theta)$ and $B(\theta)$
are approximately[100]

$$\left. \begin{array}{l} A(\theta) \\ B(\theta) \end{array} \right\} \sim \left. \begin{array}{l} \sin[(2 \cos \theta_0)x^{-1}] \\ \cos[(2 \cos \theta_0)x^{-1}] \end{array} \right\} \cdot f(x, \theta_0) \tag{65}$$

for some messy function $f(x, \theta_0)$. Since there is a complete
spiral turn every time $2\cos(\theta_0)x^{-1}$ is an integral multiple of
2π, we can see why (59) holds. For further details see ref.
(100).

□ Determine the detailed nature of the analogue of the spiral
effect in Bianchi type VI and VIII models.
□ Interpret the spiral effect and its analogues in other

Bianchi types by reference to the action of the three-dimensional group of motions acting on the hypersurfaces of homogeneity.

□ Investigate the behaviour of the geodesics in inhomogeneous perturbations of homogeneous models containing hotspots and spirals.

□ Formulate Fourier analysis on a negatively-curved space of constant curvature and expand finite wavelength density perturbations of open Friedman universes in terms of the Fourier components. Relate these Fourier components to exact Bianchi V universes.

□ By recalling that the velocity-space of special relativity is a space of constant negative curvature, relate the hotspot phenomenon to that of relativistic aberration.

□ Although Bianchi type VI and VIII universes cannot completely isotropise (because they do not contain an isotropic subcase) they can become arbitrarily isotropic. Determine the form of the background anisotropy in such cases. What is the analogue of the spiral effect generated by the hyperbolic rotations in the type VI_h group of motions?

□ Discover whether the discrete difference between the microwave background sky in closed and non-closed spatially homogeneous universes still exists when inhomogeneities are admitted.

□ What spatial topologies are allowed for each Bianchi type universe and how would a non-standard topology alter the observed microwave background patterns? Is it possible for local obseravtaions of the microwave background to determine the global topology of the Universe?

□ Determine new limits on the largest spatially homogeneous magnetic and electric fields allowed in the universe by calculating the limits imposed by the microwave background isotropy in Bianchi VII and IX models. Investigate the formal analogy between magnetic field stresses and vorticity.

□ Give a geometrical argument to explain why the hotspot phenomenon occurs in Class B but not Class A of the Bianchi classification of Table 1.

□ Find an inhomogeneous cosmological model which exhibits the hotspot or spiral-like effect on geodesic propagation.

OBSERVATIONAL LIMITS

A detailed analysis of the microwave background in spatially homogeneous cosmological models containing the Friedman universes as special isotropic subcases allows us to place limits upon the large scale rotation and shear distortion of the Universe. By using the observational limits on the quadrupole listed in Table 2, and assuming that the microwave photons were last-scattered at $z = 1000$, the calculations of Barrow et al[100] yield the following upper limits on the rotation and shear of the Universe:

Type IX (Closed universe, $\Omega_0 \sim 1$)

$$(\sigma/H)_0 < 2.6 \cdot 10^{-5}$$
$$;$$
$$(\omega/H)_0 < 3.9 \cdot 10^{-13}$$

(66)

Type VII (Open and Flat universes, $\Omega_0 \leq 1$)

$$\max \{(\omega/H)_0 , (\sigma/H)_0\}$$

$$< 5.10^{-9}\Omega_0^{-1}x^{-1}[1 + x^2(1 - \Omega_0)]^{\frac{1}{2}}[1 + 9x^2(1 - \Omega_0)]^{\frac{1}{2}} , \quad (67)$$

For $x < 1$, this upper limit reduces to $5.10^{-9}x^{-2}\Omega_0^{-1}$; for $x > 1$ it becomes $1.5.10^{-8}\Omega_0^{-1}$.

These limits considerably improve upon the ones given earlier by Collins and Hawking[107] but they are not as strong as those obtained from primordial nucleosynthesis[115] since the latter are , in effect, imposed at $z = 10^{10}$ rather than $z = 10^3$. We have not quoted limits for the Bianchi universes of lower generality (I and V) since these limits are obtainable from those above by specialization of parameters (see refs.(100) & (107) for full details). The limits (66) and (67) are remarkably strong and provide direct evidence for the isotropy of the Universe.

In Figure 7 we show two specific examples of numerical analyses leading to the upper limits imposed on the vorticity of the Bianchi VII universes by the dipole and quadrupole observations of the microwave background temperature isotropy[100].

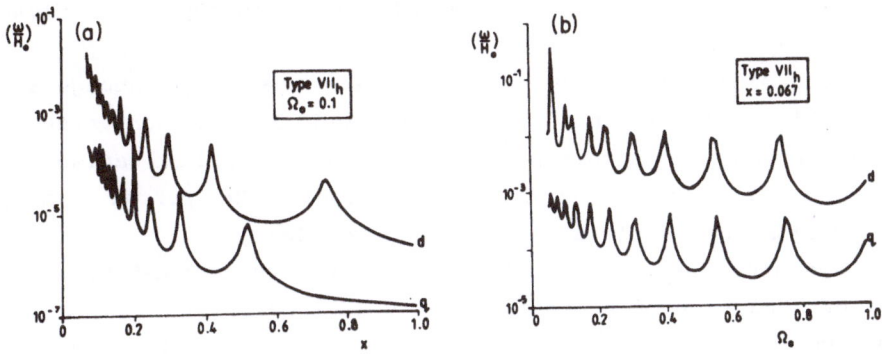

Fig.7. Upper limits[100] on the present vorticity to Hubble rate, $(\omega/H)_0$, compatible with present observations of the dipole (d) and quadrupole (q) observations of the microwave background radiation temperature isotropy: (a) in a type VII_h universe with $\Omega_0 = 0.1$ for various $x < 1$, and (b) in a type VII_h universe with $x = 0.067$ for various values of Ω_0.

The following features of Figures 7(a) & (b) are worth noting: the limits imposed by the dipole and quadrupole observations have an oscillatory character because of the spiral effect. The limits (66) and (67) result from averaging over these oscillations. The limits weaken with decreasing x and decreasing Ω in (a) and (b) because in each case the overall anisotropy pattern is becoming more poorly described by the first two *spherical* harmonics[100]. In (a) this is because of the pronounced spiral effect at small x, whereas in (b) it is because of the increasing hotspot focussing at small Ω_0.

☐ Determine whether the very strong limit (66) on any vorticity in the universe if it is closed, (66), is primarily an artefact of the restriction to spatially homogeneous cosmologies, or primarily indicative of the difficulty of fitting a vortex into the restrictive geometry of a closed universe.

☐ Determine limits on the largest spatially homogeneous magnetic and electric fields allowed in the Universe by calculating the accompanying microwave background anisotropy in type VII and IX models. Investigate the formal analogy between magnetic field stresses and vorticity.

☐ Determine the maximum shear and vorticity allowed in models of our Universe of Bianchi VI and VIII types.

☐ How do the presence of collisionless distributions of primordial neutrinos and gravitons in the Universe affect the limits derived for the rotation and shear of the Universe? Extend the calculations described to determine the relative anisotropy in photons, neutrinos and gravitons today (remember to include the effects of massive particle annihilation on the mean temperatures of the three collisionless components).

ISOTROPY AND HOMOGENEITY

Direct observations of the Hubble flow[117], radio source counts[118], the intensity distribution of the cosmic x-ray background[119] as well as the impressive temperature isotropy of the cosmic microwave background radiation discussed above, all witness to the high degree of isotropy in the expansion of the Universe back to the redshift at which the microwave photons were last scattered by electrons. Further, the excellent agreement between the predictions[97] of primordial nucleosynthesis in the standard 3 (or 4)-neutrino isotropic Friedman model of the early universe and the observed (or inferred) primordial abundances of helium-4, helium-3, deuterium and lithium-7 imply that the Universe was extremely isotropic at a redshift ~ 10^{10}. *Prior to this redshift we possess no direct or indirect observational evidence regarding the isotropy and homogeneity of the Universe.*

The isotropy of the microwave background radiation is the more surprising as regions of the microwave background separated by more than ~ 2° on its last scattering surface at z = 1000 do not appear to have been in causal communication with each other before the time when they were last scattered. How, then, did these widely separated, independent regions conspire to have the same temperature and radiation density today to better than one part in a thousand?

Traditionally this question has not been posed. Prior to the discovery of the microwave background radiation the only direct test of the isotropy of the Universe was the counting of galaxies in different solid angles around the sky, first carried out by Hubble[120]. Because the evidence gleaned by this technique is so meagre, and non-uniform, anisotropic solutions of general relativity are so hard to find cosmologists predicated their early observational and theoretical studies on the assumption of isotropy and homogeneity. Such models provided an excellent description of the present state of the Universe and hence the question of why large scale isotropy and homogeneity does exist did not arise prior to 1967. Rather, attention was focussed upon explaining the presence of the small deviations from perfect homogeneity: the heterogeneities that grew into stars and galaxies.

Soon after the microwave isotropy was first measured, Misner[121,122] stressed that it was the existence of the underlying isotropy and homogeneity that constituted the major mystery in need of explanation rather than that of the source of the small irregularities.

□ Determine whether thermodynamic aspects of classical and quantum gravitation favour the evolution of a universe from uniformity to irregularity or *vice versa*.

THE COSMOLOGICAL PRINCIPLE(S)

The power of assuming that the Universe is isotropic and spatially homogeneous to a first approximation was stressed primarily by Milne[123] and is known as the *Cosmological Principle*. It implies that the metric of space-time is that of a space of constant curvature. It is used in one form or another to break through the Geordian knot of determining the structure of the Universe by observation which we discussed at the beginning of these notes. It is most expedient to assume isotropy and homogeneity and then look for evidence that this assumption is false. However, this is awkward in that the Universe clearly cannot be *exactly* spatially homogeneous (SH) and isotropic (I); therefore we encounter a variety of weaker versions of the Cosmological Principle; for example:

(i) the Universe is SH and I "on the average"
(ii) the Universe is SH "on the average"
(iii) we are not at the centre of the Universe
(iv) we are at a typical position in the Universe
(v) the observed portion of the Universe is a fair sample.

Usually, it is stated that because we observe the microwave background to be extremely close to isotropy and assume the truth of the Copernican Principle (either in the form *(iii)* or *(iv)*), we cannot regard this observation as unique to our location. Hence the Universe must look isotropic about any point and so must be spatially homogeneous. However, we note that although a space that is *exactly* isotropic about every point must be spatially homogeneous[31], we do not have a theorem which tells us that a space which is *almost* isotropic about every point must be *almost homogeneous*. However, it is the latter result that is being assumed in practice.

A recent attempt has been made by Stoeger *et al*[124] to formulate the Cosmological Principle in a fashion that is observationally testable. It involves comparing variations in density ρ over space and determining an upper bound on the size of these variations over each length scale L. However, it would seem to be more useful to evaluate the *metric perturbations* to space-time induced by these density inhomogeneities. This evaluation would take into account the fact that a particular level of inhomogeneity on a large scale (*eg* that of galaxy clusters) is far more significant *vis a vis* the truth of the Cosmological Principle than is an inhomogeneity of the same amplitude over the scale of the solar system. In addition, all forms of the Cosmological Principle are used solely to justify the adoption of the Friedman *metric* and so one should evaluate the distortions to the metric implied by observations in order to test this assumption. The metric perturbation associated with a density perturbation $\delta\rho/\rho$ over a length scale L is of order

$$\delta g/g \sim (\delta\rho/\rho)(L/ct)^2 \tag{68}$$

where the c is the velocity of light and hence $ct \sim cH_0^{-1}$ is roughly the present size of the particle horizon. Similar expressions can be written down to give the metric perturbations associated with vortical and shear motions[126]. Finally, we note that it is the metric perturbation that is observed in any temperature anisotropy of the microwave background radiation[100,107].

□ Give as many inequivalent versions of the "Cosmological Principle" as you can and give examples of cosmological models which obey and do not obey each formulation you give.
□ Formulate the idea of an "almost symmetry" in a precise manner and determine whether a space that is almost isotropic everywhere is almost homogeneous.
□ Classify almost homogeneous spaces according to their "distance" from homogeneous ones.
□ Produce a stochastic version of the Einstein equations,

treating the metric as a random variable. Determine the form of the metric under the assumptions of statistical homogeneity and isotropy employed in studies of homogeneous turbulence.

☐ Give a precise version of a Cosmological Principle which can be applied to the real Universe and checked quantitatively against observation?

☐ Give examples of inhomogeneous distributions of matter which obey a density but not a metric criterion for homogeneity and *vice versa*.

CAN WE PROVE A COSMOLOGICAL PRINCIPLE?

Hoyle and Narlikar[125] were the first to appreciate the idea of explaining the uniformity of the Universe without appeal to special initial conditions. In 1963, before the discovery of the microwave background radiation they pointed out that one of the advantages of the steady state theory of the expanding universe was that the de Sitter solution is stable against the growth of anisotropic and inhomogeneous perturbations..."*any finite portion of the universe gradually loses its 'memory' of an initially imposed anisotropy or inhomogeneity...the universe attains the observed regularity irrespective of initial boundary conditions*". This idea was resurrected by Misner[121,122] in the context of the Big Bang models after the discovery of the background radiation isotropy in 1967.

Misner aimed to show that large scale properties of the Universe, like its isotropy, were the inevitable consequences of physical processes occurring within the Universe and could thus be predicted to exist in any sufficiently old Universe *independent of its initial conditions*. This was termed the *chaotic cosmology programme* and focussed initially upon finding dissipative processes (*eg* shock-wave damping[127], neutrino viscosity[121-2,128-9], and quantum particle production[130-1]) which could remove arbitrarily large amounts of initial anisotropy by the present, and in finding ways of

enlarging particle horizons[132] near the initial singularity so
that dissipative processes could have effect over regions that
would expand to extragalactic extent by today. In both these
searches the "chaotic cosmologists" were unsuccessful. No
solutions which allow horizon removal or enlargement with
significant probability were found[133,33]; the dissipation of
large amounts of anisotropy in the very early universe was
shown to give rise to a very large value of the entropy per
baryon, in excess of that observed today[134], if baryon number
was conserved (which it is expected to be over most of the
classical history of the Universe after ~ 10^{-30}s).

IS ISOTROPY A STABLE PROPERTY OF COSMOLOGICAL MODELS?

One way of evaluating the possibility of proving the
inevitability of observing a Cosmological Principle to hold
after 15 billion years of cosmic expansion is to evaluate
whether *isotropy* is a stable property of cosmological
solutions to Einstein's equations. This was attempted by
Collins and Hawking[135] in 1973. Restricting themselves to
spatially homogeneous cosmological models they investigated
whether the isotropic flat and open Friedman universes are
stable solutions of Einstein's equations as $t \to \infty$.

We recall for Table 1 that the most general class of
homogeneous universe models containing the open ($k = -1$)
Friedman models as a special case are those of Bianchi type
VII_h. The Bianchi type VII_0 is the most general class
including the flat ($k = 0$) Friedman model. The Bianchi type
VII equations can thus be linearized about the Friedman
solutions and the stability of the latter determined as $t \to \infty$.

Collins and Hawking[135] (CH) define conditions for
isotropisation of ever-expanding universes which include the
following two requirements:

(i) the ratio of the shear to the mean Hubble expansion rate,

σ/H, tends to *zero* as $t \to \infty$.

(ii) the energy-momentum tensor for the matter content of the universe obeys the *weak energy condition*:

$$T_{ab}u^a u^b \geqslant 0 \qquad\qquad (69)$$

the *dominant energy condition*:

$$T_{oo} \geqslant |T_{\alpha\beta}|, \quad \alpha,\beta = 1,2,3 \qquad\qquad (70)$$

and the *positive pressure criterion*:

$$\sum_{\alpha=1}^{3} T_{\alpha\alpha} \geqslant 0 \qquad\qquad (71)$$

It is then possible to prove[135] that no open set of spatially homogeneous initial data isotropises as $t \to \infty$. Only if we restrict the initial data to that which is *spatially flat*, and the equation of state of matter to be $p = 0$ to leading order, can it be proved that there exists an open neighbourhood of initial data which isotropises. Thus, it was argued that isotropic universes are a set of measure zero amongst cosmological solutions to Einstein's equations.

☐ Determine whether ever-expanding Newtonian cosmological models isotropise.

☐ Give a criterion to determine whether a closed universe isotropizes by defining a time coordinate which places the expansion maximum at infinity. Use your definition to determine whether the closed Friedman universe is a stable solution of Bianchi type IX.

☐ Extend the Collins and Hawking stability analysis of the Friedman model to the self-similar generalization of the Bianchi classification developed by Eardley.[87]

☐ Determine whether inhomogeneous perturbations alter the stability properties of isotropic universes as $t \to \infty$.

☐ Determine the conditions on g_{ab} and T_{ab} under which isotropy is a stable property of singular homogeneous cosmological models as t → 0.

☐ How large is the class of Bianchi type VI and VIII universes which can approach arbitrarily close to isotropy as t → ∞ ?

IS ISOTROPY REALLY UNSTABLE AND DOES IT MATTER ANYWAY?

Because the above result is widely misunderstood and because it is an important precursor to our discussion of recent attempts to prove cosmic "No Hair" theorems we shall now analyse it more carefully.[6,136,34]

First, we should appreciate that the energy conditions, (ii), ensure that as t → ∞ "matter does not matter" in open universes; that is, the Friedman stability problem reduces to the *vacuum* stability of the isotropic Milne model (eqn. (26) with A = 0). But our Universe can only have been vacuum (=curvature) -dominated since a redshift z ~ Ω_0^{-1} ~ 15 until the present. The observed low anisotropy of the real Universe is primarily a consequence of the evolution from the initial state (z = ∞ or z = 10^{32} ?) until z ~ Ω_0^{-1} −1 and the asymptotic behaviour in the curvature-dominated regime analysed by the CH stability theorems is essentially irrelevant to this.

In the case of the k = 0 Friedman model the matter is always important, but in order to obtain isotropisation it was necessary to require that the equation of state be p = 0. Isotropisation in the CH sense therefore only occurs in this Universe during the dust-dominated era from z ~ 10^4 to the present. However, the present isotropy level of the real Universe is determined primarily by the integrated effect of the anisotropy domination during the entire radiation era when z > 10^4. Again, the asymptotic stability result is irrelevant to the present isotropy level.

No asymptotic result regarding the anisotropy evolution as $t \to \infty$ can explain why the anisotropy level is lower than a particular value at any finite time. (For these reasons the CH stability analyses do not admit the Anthropic interpretation generally associated with them.[6,36,34])

Approach to a Family of Plane Waves

Notwithstanding the argument just presented, it is very instructive to investigate the $t \to \infty$ stability of the open Friedman universe a little further. The isotropisation criterion (i) of CH requires $\sigma/H \to 0$ as $t \to \infty$; that is, in mathematicians' language, *asymptotic stability* of the Friedman solution with respect to the type VII evolution equations. Collins and Hawking proved that this cannot occur in general when (ii) holds. However, it has since been shown[34,88] that $\sigma/H \to$ constant as $t \to \infty$; that is, the Friedman model is *stable*, although not asymptotically stable.

In physical problems asymptotic stability (perturbations decay to zero) is too strong a requirement to make — the solar system is not stable in this sense. The practical definition of stability is that perturbations be bounded and this is what we shall mean by *stability*.

We can in fact show[34,88] that spatially homogeneous perturbations of the ever-expanding Friedman model asymptote in general to a 2-parameter family of plane waves of Bianchi type VII_h described by a family of exact solutions to the Einstein equations first found by Lukash[137,138].

The Lukash solutions are spatially homogeneous vacuum solutions of Bianchi type VII_h described by the metric

$$ds^2 = dt^2 - g_{\alpha\beta}dx^\alpha dx^\beta, \qquad x^\alpha = (x,y,z) \tag{72}$$

where

$$g_{\alpha\beta} = \begin{bmatrix} a^2 e^{2z} (ch\mu + sh\mu cos\psi), & a^2 e^{2z} sh\mu sin\psi, & 0 \\ a^2 e^{2z} sh\mu sin\psi, & a^2 e^{2z} (ch\mu - sh\mu cos\psi), & 0 \\ 0, & 0, & c^2 \end{bmatrix} \quad (73)$$

where $a(t)$, $c(t)$, $\psi(t)$, $\mu(t)$ with k constant.

This metric describes two monochromatic circularly-polarized gravitational waves of wavelength $2\pi c(t)k^{-1}$, and amplitude $\mu(t)$ moving in the $\pm z$ direction ($=x^3$) on a space of constant negative curvature. Physically, this is like a Bianchi type V metric of constant negative curvature plus gravitational waves which create a spatial curvature anisotropy. The Lukash solution is a *2-parameter family of plane waves*[83] (all curvature invariants vanish) with

$$a(t) = t^{1/(1+\lambda^2)} \quad ; \quad c(t) = (1+\lambda^2) \, t$$

$$\psi(t) = \frac{k\ln t}{1+\lambda^2} \quad ; \quad \lambda = \frac{k}{2} sh\mu \quad (74)$$

where λ is an arbitrary constant.

When $k = \mu = 0$ (74) reduces to the isotropic Milne universe; and when $k \to 0$ with $\mu \neq 0$ we approach the axisymmetric Bianchi type V model. It is worth noting here that $k = x^{-1} = h^{-\frac{1}{2}}$, where x is the 'spiral parameter' of Collins and Hawking introduced earlier[58] and h is the type VII$_h$ label (see equation (62) with $\Omega_0 = 0$ in vacuum). Each Lukash model is labelled by 2-parameters (k,μ) compared with the 4 parameters required for the general type VII$_h$ vacuum solution. However, one can show that as $t \to \infty$ all vacuum and perfect fluid[34,88] VII$_h$ universes approach the 2-parameter family of Lukash metrics. In particular, perturbations of the open Friedman universe (whether large or small) of VII$_h$ type *all* approach the Lukash solutions and have $\mu \to$ constant, hence $\sigma/H \to$ constant as $t \to \infty$. This stability turns out to be non-trivial. If we start with a Lukash solution characterized by $\mu = \mu_0 =$ constant then as $t \to \infty$ it will evolve towards another member of the family of Lukash plane waves with $\mu = \mu_\infty$ = constant $\neq \mu_O$. Since the $\mu = o$ case is the isotropic Milne model, we see that this means that *the isotropic solution is stable but not asymptotically stable* (for full details see ref. 100) In fact, one has $\mu_\infty < \mu_O$ and the model evolves closer to isotropy as $t \to \infty$ whenever

$$y < \frac{2(4 + k^2 sh^2 \mu_O)}{12 + k^2 sh^2 \mu_O} \tag{75}$$

if the model contains perfect fluid with $p = (y - 1)\rho$ equation of state when $2/3 < y \leqslant 2$.

We can conclude that open isotropic universes are not unstable to spatially homogeneous perturbations as $t \to \infty$. However, the present observed isotropy cannot be a consequence of this asymptotic behaviour because we live at a finite time after the initial state, that is in a real sense far from the asymptotic region.

☐ Numerically integrate the null geodesics of the Lukash solutions to determine the exact nature of the spiral pattern produced. Compare it with the results presented earlier for the case of small anisotropy.

☐ Determine whether all Bianchi type VII$_h$ cosmological models approach the Lukash metrics in the presence of electromagnetic fields.

☐ Investigate inhomogeneous perturbations of the Lukash solutions and use them to analyse the general behaviour of cosmological solutions to Einstein's equations as $t \to \infty$.

☐ Since the Lukash solutions are plane waves one might expect there to be no quantum particle production in such a space-time. Determine whether this is so.

NO HAIR THEOREMS

The attempt to prove asymptotic stability of isotropic expansion by Collins and Hawking[135] imposed very restrictive energy conditions by demanding positive pressures, (71). This ensures that open universes approach vacuum solutions as t → ∞ as we have already mentioned, but it also excludes the presence of a positive cosmological constant, Λ. Since 1981 there has been growing interest[1] in the cosmological consequences of a finite period of "inflationary" early expansion history during which the dynamics were dominated by an effective cosmological constant. From the viewpoint of general relativity, inflation reduces to the possibility that some set of hypothetical matter fields in the early universe give rise to an energy momentum tensor of the form

$$T_{ab} \approx \Lambda \, g_{ab} \tag{76}$$

where Λ is a constant. By comparison with equation (18) we see this is equivalent to a perfect fluid with $p = -\rho = -\Lambda$, a point first made by McCrea[139]. If we examine the Friedman equation for an isotropic universe containing a perfect fluid with equation of state $p = (\gamma - 1)\rho$ and the stress T_{ab} in (76) we see that (22)-(23) give

$$\frac{\dot{a}^2}{a^2} = \frac{\rho}{3} - \frac{k}{a^2} + \frac{\Lambda}{3} \tag{77}$$

where $\rho \propto a^{-3\gamma}$. Hence as $a \to \infty$, for $\gamma > 0$, we see that

$$\frac{\dot{a}^2}{a^2} \to \frac{\Lambda}{3} \tag{78}$$

and $a \to \exp(t\sqrt{\Lambda/3})$ That is, the dynamics approach the de Sitter universe.

The 'No Hair' conjecture[140-1] is that this asymptotic approach to de Sitter is true in general even in the presence of anisotropies, inhomogeneities and other fluid stresses. It is important to discover under what circumstances this might be true since this will delineate the set of initial conditions which can be driven towards isotropy and homogeneity by inflation.

Various 'proofs' of versions of the No Hair conjecture by Hoyle & Narlikar[125], Barrow[142], Boucher & Gibbons[143-4], Starobinskii[145] and Wald[146] have appeared in the literature for particular situations (*eg* perturbations close to isotropy and homogeneity, spatial homogeneity). For example, Starobinskii shows that if $\Lambda > 0$, then as $t \to \infty$ the Einstein equations admit a series approximation of the form

$$ds^2 = dt^2 - [e^{2t\sqrt{\Lambda/3}}A_{\alpha\beta}(\mathbf{x}) + B_{\alpha\beta}(\mathbf{x}) + e^{t\sqrt{\Lambda/3}} C_{\alpha\beta}(\mathbf{x}) + \ldots]dx^{\alpha}dx^{B} \qquad (79)$$

where, in the presence of a perfect fluid with $2 < \gamma \leqslant 1$, *eight* components of the spatial functions, $A_{\alpha\beta}$, $B_{\alpha\beta}$ and $C_{\alpha\beta}$ are left arbitrary by the field equations. Deviations from the de Sitter state are seen to decay exponentially rapidly in time within the event horizon of a geodesic observer. Outside that horizon the inhomogeneity ($A_{\alpha\beta}(\mathbf{x})$) may still be large though. Inflation does not remove inhomogeneity: it simply dilutes it by expanding it to exponentially large length scales.

Other examples of No Hair "theorems", this time for ever-expanding spatially homogeneous models, but without the need for deviations from isotropy to be assumed small initially so that a series expansion is not required for an asymptotic analysis, are those of Wald[146], Jensen & Stein-Schabes[148], Turner & Widrow[149] and Moss & Sahni[147]. Exact inhomogeneous solutions which approach de Sitter in accord with (79) have also been found[150-1] They show, in particular, that if the universe contains an effective stress of the form (76) created by a self-interacting scalar field then as $t \to \infty$, (ever-expanding universes) the anisotropy falls-off exponentially rapidly and the de Sitter universe is approached so long as the remaining matter fields obey the *weak energy condition* (69) (*ie* the energy density of the scalar field is positive) and also the *strong energy condition* (2).

▢ What do non-geodesic (non-hypersurface orthogonal) observers see as $t \to \infty$ in the inflationary approach to de Sitter?

▢ What is the range of gravitation theories for which the No Hair conjecture holds in the form (79)?

INFLATION AND THE INITIAL VALUE PROBLEM

It is important to stress that even if the No Hair conjecture were true and all conceivable cosmological initial conditions asymptotically approached the de Sitter solution locally this would not provide an explanation of the present low isotropy of the Universe. There can exist no such explanation of the isotropy level in the Universe at a finite time[153].

There have been many claims that a finite period of de Sitter expansion has the attraction of simultaneously solving the 'flatness', 'horizon' problems as well as explaining why the universe is so old[1,18]. It is worth noting that these are in fact all the same mathematical problem under different names. It is not possible, in general, to solve one without automatically solving the others. The remaining independent cosmological problem that inflation aims to resolve -- that of the present isotropy of the universe -- has been the subject of numerous recent papers that focus upon displaying the wide range of cosmological initial conditions that inevitably evolve towards the isotropic de Sitter state.

Two points of principle need to be made about claims that inflation can explain the present isotropy of the Universe. (We shall assume that there do not arise any general relativistic or quantum gravitational processes which prevent inflation from actually occurring). Confine attention to spatially homogeneous but anisotropic cosmological models with everywhere spacelike surfaces of constant density so the Einstein equations are a set of ordinary differential equations ($\dot{x} = F(x)$) obeying the local Lifschitz and continuity conditions (so $\|\partial F / \partial x\|$ is continuous). They will contain some matter source, for example a scalar field, ϕ, with potential $V(\phi)$ able to drive inflation for a finite period. This constitutes a well-posed initial value problem which has the property that its solution at any time t is a continuous and unique function of conditions at any earlier time T < t. This makes it impossible *in principle* for inflation, or any other classical or quantum dissipative process operating in classical space-time, to explain the present isotropy of the Universe independently of cosmological initial conditions. We can always choose a cosmological model that today is more anisotropic than observation allows and evolve it backwards through any period of inflation to determine a set of pre-inflationary initial conditions which will therefore fail to evolve a universe consistent with observation today. At present we have no unique probability measure to apply to cosmological initial conditions and so it is not possible to say how 'large' the set of such counter-examples is.

It is clear that if one *first* chooses cosmological initial data then one can subsequently choose those constants of Nature which fix the amount of inflation so that by the present time any initial anisotropy will be less than any pre-set level, no matter how small. This is what is often done in the literature[147-9]. It does not conflict with our first principle above, which is equivalent to the statement that if the constants of Nature which determine the duration of the inflationary epoch are chosen *first* then it is always possible to subsequently choose cosmological initial data which cannot become more isotropic than some pre-chosen level by the present day. There is a tendency for particle physicists to choose the inflationary parameters *after* the initial data to solve the isotropy problem whereas, as in all other areas of cosmology and physics, we choose the fundamental constants *before* the initial data. If that is done, inflation cannot explain the isotropy of the Universe irrespective of the initial data.

INFLATION AND THE STRONG ENERGY CONDITION

Many authors have interpreted the above-mentioned No Hair 'theorems' as a proof of the effectiveness of inflation in explaining the present high level of isotropy in the Universe independent of initial conditions. For, it is claimed, no matter how anisotropic the initial conditions, there will be an exponentially rapid approach to the isotropic de Sitter solution within the event horizon of any geodesic observer during a period of inflation.

The No Hair conjecture is, roughly speaking, also equivalent to the statement that in the presence of a positive effective cosmological constant, the de Sitter space-time is a stable asymptotic solution of the Einstein equations. A variety[142-9] of mathematical demonstrations of such a result, subject to particular assumptions, have appeared in the literature. All contain, either implicitly or explicitly , one major technical weakness: *the assumption that the energy-momentum tensor, T_{ab} of the material content other than that of the cosmological constant or inflating matter field obeys the strong energy condition (2).*

The imposition of the strong energy condition is unsatisfactory because the positive cosmological constant arising from an effective stress like (76) necessarily violates the strong energy condition (2). Indeed, violation of the strong energy condition is a necessary condition for inflation or generalised inflation[152] to occur and resolve the flatness-horizon-age problem. It is therefore quite unreasonable to expect all the other matter sources near the Planck time to obey the condition (2). We shall now show that when the unreasonable strong energy condition is dropped the cosmological No Hair conjecture fails[153-4].

The Deflationary Universe

We consider a zero-curvature Friedman model containing a perfect fluid with pressure p and energy density ρ having equation of state (15) but we shall also assume the presence of a bulk viscosity $\eta = \alpha\rho$, with α constant. The total pressure is therefore given by p' where[31]

$$p' = (\gamma - 1)\rho - 3H\alpha\rho \tag{80}$$

The two essential field equations are, the Friedman equation

$$3H^2 = \rho \tag{81}$$

(where $H(t) = \dot{a}/a$ is the Hubble expansion rate and we have picked units with $8\pi G = 1$)), and the conservation equation ($'\cdot' = d/dt$)

$$\dot{\rho} + 3H(\rho + p') = 0 \tag{82}$$

These yield a single differential equation for $H(t)$:

$$2\dot{H} = 3H^2(3H\alpha - \gamma), \quad \rho = 3H_o^2 = \gamma^2/3\alpha^2 \tag{83}$$

There exists a special de Sitter solution of (83) with constant expansion rate

$$H = H_o = \gamma/3\alpha \tag{84}$$

This special solution is a stable attractor as $t \to -\infty$, but is *unstable* as $t \to +\infty$. This can be seen by solving (*83*) to obtain $H(t)$:

$$H(t) = H_o(1 + Aa^{3\gamma/2})^{-1}; \quad A \geq 0, \text{ constant} \tag{85}$$

Integrating (85), we obtain for the expansion scale factor, $a(t)$, of the Friedman metric

$$\ln a + Aa^{3\gamma/2} = 3H_o\gamma t/2 \tag{86}$$

We see that this cosmological model evolves from an initial de
Sitter state at t = - ∞ and actually *deflates* to the zero-
curvature Friedman state with a ∝ $t^{2/3\gamma}$ as t → + ∞. The
dominant energy condition, (70), $\rho + p' \geq 0$ is always obeyed
since H ≤ H_0, but the strong energy condition (2), which
reduces to $\rho + 3p' \geq 0$ is *violated* at early times.

This solution, which was first found by Murphy[155] in
another context as an example of a non-singular cosmological
model, reveals the restrictive role played by the strong
energy condition and shows that the de Sitter state need not
be stable. The stress (80) responsible for the instability
can just be viewed mathematically as a form of energy-momentum
tensor which de-stabilizes a de Sitter state that it has
initially created after the manner of chaotic inflation.
However, it is also physically motivated as a classical bulk
viscosity[31]. If we pick γ = 2 then the perfect fluid stress
becomes equivalent to a massless scalar field, ϕ, with p = ρ =
$\dot{\phi}^2/2$ and a dissipative coupling $-3\eta\dot{\phi}$. Classically, bulk
viscous stresses arise because the expansion of universe is
continually trying to pull the fluid out of thermal
equilibrium. They should vanish when the equation of state is
pure radiation. Bulk viscosities also appear as
phenomenological descriptions of quantum particle production
near the Planck time[156-7]. This correspondence is of
partiucular relevance when interpreting the above result in
the context of early universe inflation. Since it is
necessary for inflation to come to an end if the Universe is
to resume the Friedman-like expansion we now observe, there
must arise a violation of one of the energy conditions used to
prove the existing cosmic no hair theorems. It is the
particle production arising from the rapidly-varying coupling
to the inflaton field to other matter fields when it
oscillates about the global minimum of the potential that
gives rise to the subsequent decay of the inflaton field.
This decay is usually put into the equation of motion of the
inflaton field by hand. The model described by (85) and (86)
can be viewed as an exact description of this decay by
particle production.

Models of chaotic inflation[1], say with a massive scalar field, necessarily involve fluids with a *time-varying* equation of state spanning the domain $-\rho \leqslant p \leqslant \rho$. The state (86) is an exactly soluble example of a fluid with a time-varying equation of state.

If one chooses the viscosity coefficient, η, to be constant in (80) then a different type of *inflationary* solution to (80)-(82) exists. There is again a special de Sitter solution with $H = H_O = \eta/\gamma$, but it is now stable as $t \to \infty$ but unstable as $t \to -\infty$. The solution subject to $a(0) = 0$ is[158]

$$a^{3\gamma/2} = 2(e^{3\eta t/2} - 1)/3\eta \qquad (87)$$

This solution begins as $t \to 0$ in the Friedman state and evolves to de Sitter as $t \to \infty$. If one takes $\eta = \alpha \rho^n$ then solutions with $0 < n < 1/2$ display this inflationary behaviour typified by (87) but those with $n > 1/2$ have the *deflationary* behaviour of (85)-(86). The peculiar intermediate case $n = 1/2$ exhibits power-law inflation with

$$a \propto t^{2(\gamma - \alpha\sqrt{3})/3} \quad ; \quad \gamma > \alpha\sqrt{3} \qquad (88)$$

The deflationary example (86) of how the No Hair conjectures can fail when (2) is relaxed may be interesting in the context of the very first inflationary universe model which was suggested by Starobinskii[159]. This assumed an initial de Sitter state destabilised by 1-loop quantum effects and one suspects that these can therefore be modelled by a phenomenological bulk viscosity. In the model (86) one also has a natural and inevitable transition *out* of an initial de Sitter state. This deflationary evolution provides an interesting possible resolution of the "exit problem" that besets most inflationary models[1]. The fact that the model (86) is exactly soluble means also that it is possible to carry out detailed calculation of the generation of irregularities at the classical and quantum level. The initial de Sitter phase of the solution (86) should result in the generation[160] of density and graviton fluctuations with constant curvature spectra. The deviations from this state caused by the transition to the exact Friedman asymptote should leave traces in the gravitational-wave background.

We note in passing that the special de Sitter solution (84) is a gravitational equilibrium state of the field equations: it possesses an equilibrium Hawking temperature T_H = $H_o/2\pi$ seen by[140] geodesic observers. However, this state is not in local thermal equilibrium since entropy is being generated by the bulk viscosity at a constant rate to maintain the static de Sitter state (7). It would be interesting to study the unusual relationship between the gravitational and local thermodynamics in this situation.

Finally, one might wonder whether the deflationary behaviour (86) is stable when anisotropies are added to the spatially flat Friedman universe. The case of a Bianchi type I universe containing perfect fluid and bulk viscosity was examined by Belinskii and Khalatnikov[162] who were interested in determining whether the non-existence of a singularity in the isotropic solution (86) was stable. The field equations governing the type I evolution are given by (80) and (82) but (81) is generalised to include the shear anisotropy $\sigma^2 = \Sigma^2 a^{-6}$,

$$H^2 = \rho/3 + \Sigma^2 a^{-6} \quad ; \quad \Sigma \text{ constant} \tag{89}$$

where a is now the geometric-mean scale-factor of the three orthogonal directions of expansion.

The resulting evolution can be determined qualitatively by recourse to a phase plane portrait in H and ρ. It was claimed[162-3] that when $\Sigma \neq 0$ the initial de Sitter state is unstable to the formation of an anisotropic Kasner-like singularity with

$$a(t) \propto (\Sigma t)^{1/3}, \quad \rho \propto t^{-\gamma} \exp(-\alpha/t^2) \qquad (90)$$

as $t \to 0$ (we still have a $\propto t^{2/3\gamma}$ as $t \to \infty$). However, this asymptote is unphysical since it violates the dominant energy condition; i.e $\rho + p' \to (\gamma - \alpha t^{-1})\rho$ approaches zero from below as $t \to 0$. The phase plane is shown in Figure 8.

The dominant energy condition will be violated whenever $\dot{\rho} > 0$ and this requires the solution to satisfy

$$H \leq H_0 = \gamma/3\alpha \qquad (91)$$

The only solution trajectory that does not violate the constraint (91) is the isotropic solution (85).

In Figure 8 the exact inviscid Friedman asymptote is located at the origin ($H = 0$, $\rho = 0$) of the phase-plane whilst the de Sitter solution (84) is the critical point A. Isotropic solutions lie on the parabola $3H^2 = \rho$, whilst anisotropic ($\sigma^2 \neq 0$) solutions lie outside it. In general, in such portraits one can read-off where the dominant energy condition will be violated by considering (82), from which we see that $\rho + p'$ will be negative in regions of the phase space where $H > 0$ and ρ is increasing or where $H < 0$ and ρ is decreasing.

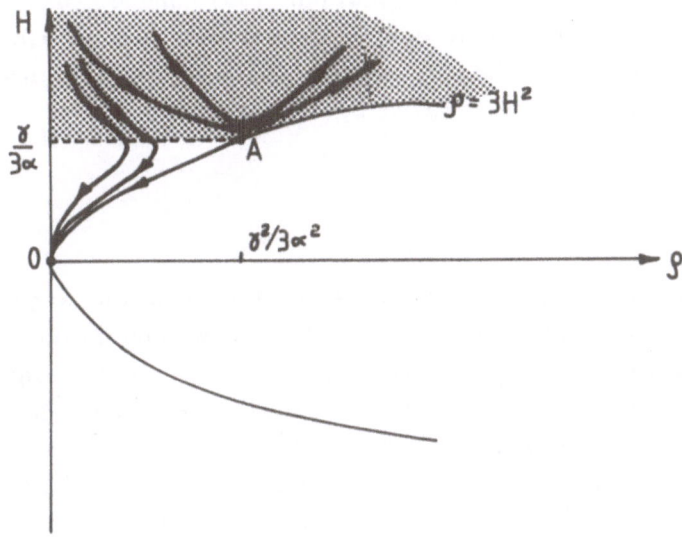

Fig. 8. The (H, ρ) phase portrait[162] for the Bianchi type I universes possessing perfect fluid $p = (\gamma - 1)\rho$ and a bulk viscous stress $\eta = \alpha\rho$. Arrows indicate the direction of increasing time. The shaded region is that in which the dominant energy condition $p' + \rho \geq 0$ is violated. The only solution trajectory that does not enter this physically forbidden region is the isotropic solution (86) which runs along the parabola $3H^2 = \rho$ from A, the de Sitter solution (84), to O the isotropic Friedman model with $a \propto t^{2/3\gamma}$. In accord with (89) no solution trajectories lie on the interior of this parabola.

It would be interesting to ascertain how general is the feature that for T_{ab} of a particular form, the dominant energy condition alone is sufficient to fix the solution uniquely. Parenthetically it is worth noting that the unphysical asymptote (90) in the type I viscous model possesses a Weyl curvature singularity (the shear $\sigma^2 = \Sigma^2 a^{-6} \to \infty$ as $t \to 0$) but $\rho \to 0$ as $t \to 0$. This illustrates a remark made in the opening pages of these notes, (following equation (2)), that the strong energy condition (2) is not necessary for the existence of a singularity. The solution (90) shows that both the strong and the dominant energy conditions can be violated and yet a singularity still arises because of the over-riding geodesic convergence created by the shear[164].

RESUMÉ

In these notes we have attempted to provide a guided tour through those areas of relativistic cosmology which should be of interest to those wanting to apply general relativity to current research areas in cosmology. We have deliberately avoided treating aspects of particle physics associated with the missing mass and galaxy formation problems since these have been given saturation coverage in other review articles and conference proceedings. We have focussed upon the following topics and issues: the relationship between observational and theoretical cosmology, the status of the Friedman universes with respect to the most general inhomogeneous and homogeneous solutions of Einstein's equations, and the use of these solutions to extract the maximum of information from observations concerning the extent to which the Universe is isotropic. The extremely high level of isotropy established by these analyses of the microwave background radiation leads us to seek an explanation for it. Hence the stability of the isotropic cosmological models under various assumptions was critically investigated. These investigations culminate naturally in an analysis of the effectiveness of 'inflation' as an explanation of the present level of isotropy. We showed that it is not possible for inflation to explain the structure of the Universe independently of initial conditions. Finally, we argued that attempts to prove the Cosmic No Hair conjecture should not use the strong energy condition and then demonstrated that when this condition is dropped the No Hair conjecture is false. A simple isotropic example involving a bulk viscous stress shows it to fail in the most dramatic way: an initial de Sitter state deflates into an Einstein-de Sitter universe.

Acknowledgements

I would like especially to thank G Galloway, R Juszkiewicz, R Matzner, A Ottewill, S Siklos, D Sonoda, J Stein-Schabes, F J Tipler and J Wainwright, with whom I have enjoyed working on some of the topics described herein. I have also benefitted from helpful discussions and correspondence with S Bhavsar, A Burd, B Carter, C B Collins, G F R Ellis, G Gibbons, J Hartle, B L Hu, W Kolb, V Lukash, M MacCallum, W McCrea, M Madsen, M Mejic, I Moss, J Narlikar, I Novikov, D Page, V Sahni, J Silk, R J Tayler and R Wald, and finally my thanks go to our long-suffering secretaries at Sussex.

REFERENCES

1. A. Linde, Rep. Prog. Phys., 47:925 (1984)

2. G.Gibbons, S.W. Hawking & S.T.C.Siklos, (Eds.) "The Very Early Universe", C.U.P., Cambridge (1983).

3. S.W.Hawking, Gen. Rel. & Gravn., 1: 393 (1971)

4. F.J.Tipler, Q. Jl. R. astron. Soc., 22:133 (1981).

5. G.F.R. Ellis & G.B.Brundrit, Q. Jl. R. astron. Soc., 20:37 (1979).

6. J.D.Barrow & F.J.Tipler, "The Anthropic Cosmological Principle", Oxford U.P., Oxford & N.Y. (1986).

7. F.J.Tipler, Phys. Rep., 137: 231 (1986).

8. A. Vilenkin, Phys. Lett., B117:25 (1982).

9. D.Raine & M. Heller, "The Science of Space-Time", Pachart, Arizona (1981).

10. W.Rindler, Mon. Not. R. astron. Soc., 116:662 (1956)

11. C.Hazard, R.G.McMahon & W. Sargent, Nature, 322:38 (1986).

12. T. Kaluza, Sber. preuss. Akad. Wiss. Phys. Math. Kl., 966 (1921).

13. O. Klein, Z. Physik., 37:895 (1926)

14. M. Green & J.Schwarz, Phys. Lett. B., 149: 117 (1984).

15. W. J. Marciano, Phys. Rev. Lett., 52: 489 (1984).

16. E.Kolb, M.Perry & T.P.Walker, Phys. Rev. D., 33:869 (1986)

17. J.D.Barrow, Phys. Rev. D, (in press) (1986)

196

18. A.Guth, Phys. Rev. D., 23:347 (1981).

19. J.D.Barrow, Can. J. Phys., 64:152 (1986).

20. E.Wampler, C.Gaskell, W.L.Burke & J.A.Baldwin, Astrophys. 276:403 (1984).

21. J.D.Barrow & S.P.Bhavsar, Q. Jl. R. astron. Soc., (in press) (1987)

22. J.D.Barrow, P.J.E.Peebles & D.W.Sciama, eds. "The Material Content of the Universe", The Royal Society, London (1986).

23. S.W. Hawking & G.F.R. Ellis, "The Large scale Structure of Space-Time", C.U.P., Cambridge, (1973).

24. J. Kristian & R.K. Sachs, Astrophys. J., 143: 379 (1966).

25. G. Daucourt, J. Phys. A., 16: 3507 (1983).

26. W.H.McCrea, Zeit. f. Ap., 18: 98 (1939).

27. G.F.R. Ellis, "General Relativity and Cosmology", Varenna Lectures in Physics, ed. R.K.Sachs, Acad. Press, London (1971).

28. A.Z.Petrov, "Einstein Spaces", Pergamon Press, Oxford (1969).

29. L.Landau & E.M. Lifshitz, "The Classical Theory of Fields" Pergamon, Oxford (1974).

30. S.T.C. Siklos, in "Relativistic Astrophysics and Cosmology", eds. E.Verdaguer & X.Fustero, World, Singapore (1984).

31. S.Weinberg, "Gravitation and Cosmology", Wiley, New York, (1972).

32. F.J.Tipler, Phys. Rev. D., 17:2521 (1978).

33. J.D.Barrow, Phys. Rep., 85:1 (1982).

34. J.D.Barrow & D.H.Sonoda, Phys. Rep., 139:1 (1986).

35. J.D.Barrow, in "Proceedings of the 1st International Conference on Phase Space", ed. A.Zachary, Plenum, NY, (1986).

36. A.Vilenkin, Phys. Rev. D., 24:2082 (1981).

37. A. Staruskiewicz, Acta Phys. Polon., 24:734 (1963).

38. G.F.R.Ellis, S.D.Nel, R.Maartens, W.R.Stoeger and A.P.Whiteman, Phys. Rep., 124:315 (1985).

39. J.D.Barrow & A.Ottewill, J. Phys. A., 16: 2757 (1983).

40. J.Leray, Acta Math., 63:193 (1934).

41. E.M.Lifshitz & I. Khalatnikov, Adv. Phys., 12:185 (1963).

42. A.Spero & R.Baierlein, J. Math. Phys., 19:1324 (1978)

43. C.Will, "Theory and Experiment in Gravitational Physics", C.U.P., Cambridge (1983).

44. J.D.Barrow, Phil. Trans. Roy. Soc., A310:337 (1983).

45. J.D.Barrow, A.Burd & D.Lancaster, Class. & Quantum Grav., 3:551 (1986).

46. P.Landsberg & D.Evans, "Mathematical Cosmology", O.U.P., Oxford (1977).

47. J.Marsden, "Applications of Global Analysis in Mathematical Physics", Publish or Perish Inc., Boston,(1974).

48. V.Arnold, "Catastrophe Theory", Springer, NY (1985).

49. V.Arnold, S.F.Shandarin & Y.B.Zeldovich, Geophys. Astrophys. Fluid. Dyn., 20:111 (1982).

50. J.V.Narlikar & A.K.Kembhavi, Fund. Cosmic Phys., 6:1 (1980).

51. E.Milne & W.H.McCrea, Quart. J. Math. Oxford Ser., 5:73 (1934).

52. W.Bonnor, Mon. Not. R. astron. Soc., 117:104 (1957).

53. E.M.Lifshitz, J.Phys. USSR 10:116 (1946).

54. J.D.Barrow & F.J.Tipler, Phys. Reports 56:371 (1979).

55. J.D.Barrow & F.J.Tipler, Phys. Lett. A.

56. V.Belinskii, E.M.Lifshitz & I.Khalatnikov, Sov. Phys. Usp., 13:745 (1971).

57. J.D.Barrow, Sussex Preprint (1986).

58. I. Newton, "Philosophiae naturalis principia mathematica" (1713), transl. A. Motte (revised F.Cajori), Univ. California Press, Berkeley, (1946).

59. J.D.Barrow, unpublished (1984).

60. J.D.Barrow & F.J.Tipler, in preparation (1986).

61. C.J.S. Clarke, Proc. Roy. Soc. A., 314:417 (1970).

62. G.F.R.Ellis, Gen. Rel. Gravn., 2:7 (1971).

63. K.Sato, Mon. Not. R. astron. Soc., 195:467 (1981).

64. Y.B.Zeldovich & L.Grishchuk, Mon. Not. R. astr. Soc., 207:23P (1984).

65. W.Bonnor, Class & Quantum Gravity, 2:781 (1985).

66. C.Hellaby & K.Lake, Astrophys. J., 290:381 (1985), errata 300:000 (1986).

67. J.D.Barrow & F.J.Tipler, Mon. Not. R. astr. Soc., 216:395 (1985).

68. J.E.Marsden & F.J.Tipler, Phys. Reports 66:109 (1980).

69. M.Freedman, J. Diff. Geometry, 17:357 (1982).

70. S.K.Donaldson, J. Diff. Geometry 18:269 (1983).

71. R.Schoen & S.-T. Yau, Manuscript Math., 28:159 (1979).

72. J.D.Barrow, G.Galloway & F.J.Tipler, Mon. Not. R. astr. Soc., (in press) (1986).

73. R.C.Tolman & M.Ward, Phys. Rev. 32:835 (1932).

74. R.Kantowski & R.K.Sachs, J. Math. Phys., 7:443 (1966).

75. A.S.Kompanyeets & A.S. Chernov, Sov. Phys. JETP 20:1303 (1964).

76. C.B.Collins, J. Math. Phys., 18:2116 (1977).

77. L.Bianchi, Mem. Soc. It., 11:267 (1898), reprinted in Opere IX, ed. A.Maxia, Editzioni Crenonese, Rome (1952).

78. F.B.Estabrook, H.D.Wahlquist & C.G.Behr, J. Math. Phys., 9:497 (1968).

79. M.A.H. MacCallum, Cargese Lectures in Physics Vol 6, ed. E.Schatzman, Gordon & Breach, NY (1973).

80. G.F.R. Ellis & M.A.H. MacCallum, Comm. Math. Phys., 12:108 (1969).

81. G.F.R.Ellis & M.A.H.MacCallum, Comm. Math. Phys., 19:31 (1970).

82. O.Bogoiavlenskii and S.P.Novikov, Russian Math. Surveys 31:31 (1971).

83. S.T.C.Siklos, Comm. Math. Phys., 58:255 (1978).

84. M.A.H.MacCallum, Lecture Notes in Physics 109; ed. M.Demianski, Springer, NY (1979).

85. E.Kasner, Am. J. Math., 43:217 (1921).

86. D.Kramer, E.Herlt, H.Stephani and M.A.H.MacCallum, "Exact Solutions of Einstein's Equations", C.U.P., Cambridge (1981).

87. D.Eardley, Comm. Math. Phys., 37:289 (1974).

88. J.D.Barrow & D.H.Sonoda, Gen. Rel. Gravn., 17:409 (1985).

89. A.Vilenkin, Phys. Rev. D 27:2848 (1983).

90. A.Linde, Lett. Nuovo Cim., 39:401 (1984).

91. J.Hartle & S.W.Hawking, Phys. Rev. D 28:2960 (1983).

92. J.V.Narlikar & T.Padmanabhan, Phys.Rep., 100:151 (1983).

93. S.W.Hawking, Nucl. Phys. B 239:257 (1984).

94. S.Faber & J.S.Gallagher, Ann. Rev. Astron. Astrophys., 17:135 (1979).

95. M.Davis & P.J.Peebles, Ann. Rev. Astron. Astrophys., 21:109 (1983).

96. A.Bean, G.Efstathiou, R.Ellis, B.Peterson & T.Shanks, Mon. Not. R. astron. Soc., 205:605 (1983).

97. K.Olive, D.Schramm, G.Steigman, M.Turner & J.Yang, Astrophys. J., 246:557 (1981).

98. P.J.Peebles, Astrophys. J., 284:439 (1984).

99. J.D.Barrow, R.Juszkiewicz & D.H.Sonoda, Nature, 305:397 (1983).

100. J.D.Barrow, R.Juszkiewicz & D.H.Sonoda, Mon. Not. R. astron. Soc., 213:917 (1985).

101. G.F.Smoot, M.V.Gorenstein & R.A.Muller, Phys. Rev. Lett., 39:898 (1979).

102. D.Fixsen, E.Cheng & D.Wilkinson, Phys.Rev. Lett., 50:620 (1983).

103. R.Fabbri, I.Guidi, F.Melchiorri & V.Natale, Phys. Rev. Lett., 44:1563 (1981).

104. P.M.Lubin, G.Epstein & G.F.Smoot, Phys. Rev. Lett., 50:616 (1983).

105. I.Strukov & D.P.Skulachev, Sov. Astron. Lett., 10:1 (1984).

106. S.W.Hawking, Mon. Not. R. astron. Soc., 142:129 (1969).

107. C.B.Collins & S.W.Hawking, Mon. Not. R. astron. Soc., 162:307 (1973).

108. A.Doroshkevich, V.Lukash & I.D.Novikov, Sov. Astron., 18:554 (1975).

109. K.Thorne, Astrophys. J., 148:51 (1967).

110. L.P.Grishchuk, A.Doroshkevich & I.D.Novikov, Sov. Phys., JETP 28:1210 (1969).

111. R.A.Matzner, Astrophys. J., 157:1085 (1969).

112. I.D.Novikov, Sov. Astron., 12:427 (1968).

113. S.Bajtlik, R.Juskiewicz, M.Prozynski & P.Amsterdamskii, Astrophys. J. 000:000 (1986).

114. V.Lukash & I.D.Novikov, Nature, 316:46 (1985).

115. J.D.Barrow, Mon. Not. R. astron. Soc., 175:359 (1976).

116. J.D.Barrow & J.Stein-Schabes, Phys. Lett. B 167:173 (1986)

117. P.J.Peebles, "Physical Cosmology", Princeton U.P.,NJ (1971).

118. A.Webster, Mon. Not. R. astron. Soc., 175:61 (1976)

119. R.S.Warwick, J.P.Pye & A.C.Fabian, Mon. Not. R. astron. Soc., 190:243 (1980).

120. E.Hubble, "The Realm of the Nebulae", Dover, NY (1985).

121. C.W. Misner, Nature, 214:30 (1967).

122. C.W.Misner, Astrophys. J., 151:431 (1968).

123. E.A.Milne, "Relativity, Gravitation and World Structure" O.U.P., Oxford (1935).

124. W.R.Stoeger, G.F.R.Ellis & C.Hellaby, Cape Town preprint (1986).

125. F.Hoyle & J.V.Narlikar, Proc. Roy. Soc. A., 273:1, (1963).

126. J.D.Barrow, Mon. Not. R. astron. Soc., 178:625 (1977).

127. M.J.Rees, Phys. Rev. Lett., 28:1969 (1972).

128. J.M.Stewart, Mon. Not. R. astron. Soc., 145:347 (1969).

129. A.G.Doroshkevich, Y.B.Zeldovich & I.D.Novikov, Sov. Phys. JETP, 26:408 (1968).

130. L.Parker, in "Asymptotic Structure of Space-Time", eds. F.P.Esposito & L.Witten, Plenum, NY, (1977).

131. Y.B.Zeldovich & A.Starobinskii, Sov. Phys. JETP 34: 1159 (1972).

132. C.W.Misner, Phys. Rev. Lett., 22:1071 (1969).

133. A.G.Doroshkevich & I.D. Novikov, Sov. Astron., 14:763 (1971).

134. J.D.Barrow & R.A.Matzner, Mon. Not. R.astron. Soc., 181:719 (1977).

135. C.B.Collins & S.W.Hawking, Astrophys. J., 180:317 (1973).

136. J.D.Barrow, Q. Jl. R. astron. Soc., 23:344 (1982).

137. V.N.Lukash, Sov. Phys. JETP, 40:792 (1975).

138. V.N.Lukash, Nuovo Cim. B, 35:268 (1975).

139. A.Taub, Ann. Math., 53:472 (1951).

140. G.Gibbons & S.W.Hawking, Phys. Rev. D, 15:2738 (1977).

141. S.W.Hawking & I.Moss, Phys. Lett. B, 110:35 (1982).

142. J.D.Barrow, in "The Very Early Universe", eds. G.Gibbons, S.W.Hawking & S.T.C.Siklos, C.U.P., Cambridge, (1983).

143. W.Boucher & G.Gibbons, in "The Very Early Universe" eds. G.Gibbons, S.W.Hawking & S.T.C.Siklos, C.U.P.

Cambridge, (1983).

144. W.Boucher, in "Classical General Relativity", eds. W.Bonnor, J.Islam & M.A.H.MacCallum, C.U.P., Cambridge (1984).

145. A.A.Starobinskii, Sov. Phys. JETP, Lett., 37:66 (1983).

146. R.Wald, Phys. Rev. D, 28:2118 (1983).

147. I.Moss & V.Sahni, Phys. Lett. B, in press, (1986).

148. L.Jensen & J.Stein-Schabes, Fermi Lab preprint (1986).

149. M.S.Turner & L.Widrow, Fermi-Lab preprint (1986).

150. J.D.Barrow & J.Stein-Schabes, Phys. Lett.A 103:315 (1984).

151. J.D.Barrow & O.Gron, Sussex preprint (1986).

152. F.Lucchin & S.Matarrese, Phys. Lett.B 164:282 (1985).

153. J.D.Barrow, "The Deflationary Universe: an instability of the de Sitter universe", Sussex preprint (1986).

154. J.D.Barrow, "Deflationary universes with quadratic lagrangians", Sussex preprint (1986).

155. G.Murphy, Phys. Rev.D 8:4231 (1973).

156. Y.B.Zeldovich, Sov. Phys. JETP Lett. 12:307 (1980).

157. B.L.Hu, Phys. Lett.A 90:375 (1982).

158. R.Treciokas & G.F.R.Ellis, Comm.Math.Phys. 23:1 (1971).

159. A.Starobinskii, Phys. Lett. B 115:295 (1982).

160. S.W.Hawking, Phys. Lett.B 115:295 (1982).

161. L.F.Abbott & M.B.Wise, Nucl. Phys.B 244:541 (1984).

162. V. Belinskii & I.Khalatnikov, Sov.Phys. JETP Lett. 21:99(1975).

163. V.Belinskii & I.Khalatnikov, Sov.Phys. JETP 42:205 (1976).

164. J.D. Barrow & R.A. Matzner, Phys. Rev. D 21:336 (1980).

YET ANOTHER SCENARIO FOR GALAXY FORMATION

P. J. E. Peebles
Joseph Henry Laboratories
Princeton University
Princeton, NJ 08544
USA

ABSTRACT. I consider the prospects for developing a model for the formation of galaxies and clusters of galaxies using only what we are fairly sure is present-baryons and radiation. This leads to a primeval entropy perturbation scenario in which galaxies form at $z \sim 100$ as the last generation to be substantially held up by Compton drag. The consequences for galaxy formation seem attractive; the situation for cluster formation is unclear because of ambiguities in the scenario developed so far.

1. INTRODUCTION

A theory for galaxy formation requires a choice of what are the dominant elements and then an elaboration of details in a model whose predictions can be compared to the observations. I have taken to calling the first part the scenario. Some of my friends have expressed dislike of the word, perhaps because of its militaristic connotation; I have in mind the dictionary meaning, "an outline or synopsis of a play." I would call the general set of ideas behind the cold dark matter computations described by Simon White in these Proceedings a scenario; a model would include the specific prescription for biased galaxy formation.

We have three scenarios for galaxy formation that have been discussed in some detail and have been shown to be promising. However, as outlined in the next section, each has features that seem questionable or even potentially fatal, so it seems worthwhile to continue to cast about for possibilities. Since the standard scenarios invoke hypothetical objects - exotic matter, cosmic strings, seeds for explosions - the approach adopted here, to serve as a foil, will be to use only the components we are fairly sure are present, and to extrapolate the evolution of the mass distribution as closely as physics will allow from the evolution we think we observe. I conclude that this exercise works through readily enough, that is, that known phenomenology and accepted physics do not force us to hypothetical objects. However, the price is the adoption of initial conditions

203

W. G. Unruh and G. W. Semenoff (eds.), The Early Universe, 203–214.
© *1988 by D. Reidel Publishing Company.*

ad hoc, without the theoretical motivation that is such an attractive
feature of more standard scenarios.

2. THREE SCENARIOS

2.1. Canonical cold dark matter

This scenario, described by Simon White, arguably is the simplest of
all possibilities, and the beautiful N-body model studies by the team
of Marc Davis, George Efstathiou, Carlos Frenk, and Simon White
certainly look promising. However, the model in its simplest (and
most attractive) form has two peculiar features. First, with the
power spectrum, P(k), of primeval mass density fluctuations normalized
by large-scale fluctuations in galaxy counts one finds that the mass
fluctuations on small scales are relatively small, so galaxies form
late. Indeed, the computations indicate that an object we would call
a galaxy was at redshift z = 1 an extended group of star clusters
(Frenk et al. 1985). I am not aware of any thorough discussion of
this prediction, but it seems ominous that observers have not remarked
that galaxies at high redshift look more like compact groups of
galaxies.

The second feature is the large-scale mass distribution. A
realization is shown in Figure 2 and is compared to the distribution
of rich Abell clusters on about the same scale in Figure 1 (Bahcall
and Soneira 1984). The realization in figure 2 is computed in linear
perturbation approximation as described in Peebles (1982), but with
Hubble's constant H = 50 km sec^{-1} Mpc^{-1}; background temperature
T = 2.76 K (Johnson and Wilkinson 1986); and three types of massless
neutrinos. Short wavelengths are eliminated by multiplying the power
spectrum by exp - (k r_c)^2, r_c = 5 Mpc. The realization has period
L = 600 Mpc. The normalization is fixed so the rms mass fluctuation
on the scale d ~ 40 Mpc is one third of the rms fluctuation in galaxy
counts. (This is the biasing needed to reconcile the low mass per
galaxy observed in groups and clusters with the high mass in an
Einstein-de Sitter model.) The lowest contours shown in Figure 2 are
$\delta\rho/\rho$ =1, computed in linear perturbation approximation, which means
these contours enclose regions that are now breaking away from the
general expansion. The contours in Figure 1 enclose groups of rich
Abell clusters at similar redshifts, to a depth ~ 600 Mpc, about the
same as the width of the realization (if H = 50).

In comparing the two figures bear in mind that the effect of
projection in the map in Figure 1 is to suppress the appearance of
clustering relative to what would be seen in a two-dimensional cut
like Figure 2. My impression is that the realization fails to capture
the curious large-scale clumping of peaks in an over-all rather smooth
galaxy background.

To summarize, two things I will be following as the exploration
of this scenario continues will be structure on the very largest
scales and the observational consequences of late galaxy formation.

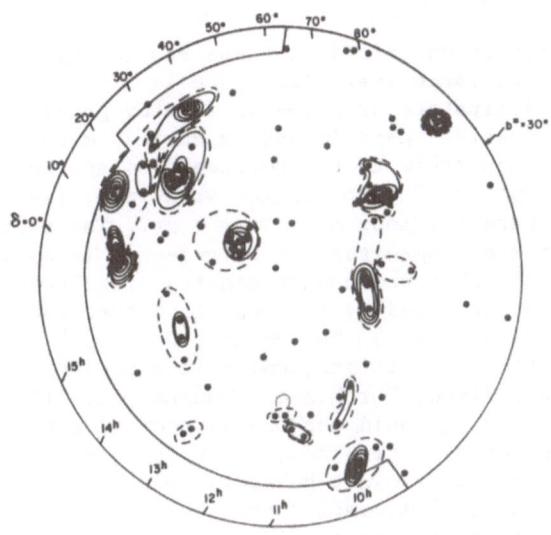

Figure 1. Map of rich clusters of galaxies (Bahcall and Soneira [1984]). The depth is comparable to the width of the realization in Figure 2. The contours enclose clusters of clusters at about the same redshift.

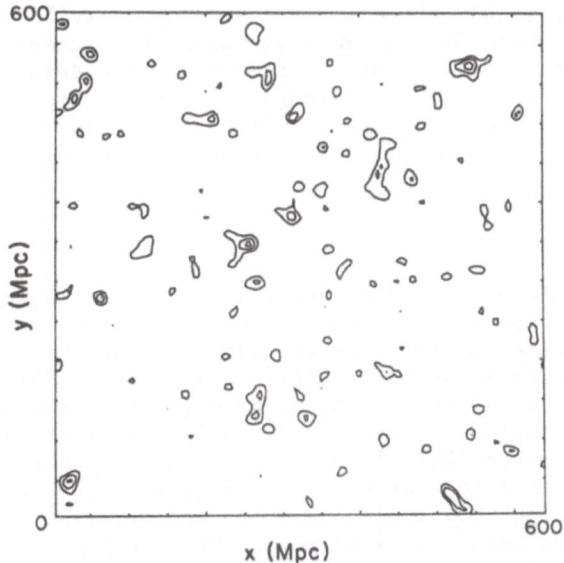

Figure 2. Realization of the mass distribution in the model. The outermost contours are at $\delta\rho/\rho = 1$ computed in linear perturbation theory; the highest contours are at $\delta\rho/\rho = 2$.

2.2. Massive cosmic strings

This picture, which is described in these Proceedings by Alex Vilenkin, deals with large-scale structure in a direct and elegant way. Galaxies and clusters of galaxies form by gravitational accretion around massive loops formed as strings move to keep their coherence length comparable to the horizon. On scales of interest the comoving number density of loops scales with loop size, r, as $n(> r) \propto r^{-3/2}$, where the loop mass is $m = \mu r$. The typical size r_g, of the loops that are progenitors of galaxies is found by setting $n(> r_g)$ equal to the observed number density of galaxies. The mass per unit length, μ, is fixed by the condition that the loop mass μr_g, accrete the mass of a galaxy. The result, $G\mu \sim 10^{-6}$, is considered reasonable by particle physicists, which is noteworthy. With the parameters r_g and μ fixed, Turok and Brandenberger (1986) showed that loops of size $r \sim 300$ r_g would produce by accretion objects with the mass and abundance of rich clusters, which is noteworthy indeed.

Two peculiar features might be mentioned. First, galaxies are produced by gravitational accretion onto objects with a scale-free frequency distribution of masses, $n(> m) \propto m^{-3/2}$. How can this account for the strikingly hard upper bound on the mass of a galaxy within a fixed radius, say r = 10 kpc (Giovanelli et al. 1986)? Why didn't the super-large loop that produced a rich cluster of galaxies produce a super-massive galaxy at the center? The second peculiar feature is that cluster formation by gravitational accretion around a large loop would pull together galaxies and mass alike, so one would expect that the mean mass per galaxy in a cluster is a fair measure, which would imply that the mean mass density is well below the critical Einstein-de Sitter density. Since the successful prediction of the abundance of rich clusters depends on the density this is a problem for the scenario. I discuss in more detail elsewhere (Peebles 1986a) these and other features I will be following with interest as models based on the cosmic string scenario are worked out in more detail.

2.3. Explosions

An attractive feature of this scenario is that it is fashioned on what is observed to happen, on a smaller scale, as stars are born and die in the interstellar medium. Energy deposited in the intergalactic medium, perhaps by supernovae (Ostriker and Cowie 1981), or quasars (Ikeuchi 1981), or by the vibration of magnetized superconducting cosmic string loops (Ostriker et al. 1986), or by the decay of unstable dark matter (Daly 1986), piles matter into ridges that can cool and fragment into star clusters or galaxies. An attractive consequence is that one would expect to see galaxies distributed along the ridges where mass was piled up in the last generation of explosions to have made galaxies, as was subsequently observed in redshift maps (de Lapparent et al. 1986; Haynes and Giovanelli 1986). On the other hand, it may be a problem that our neighborhood, extending to cosmological redshifts ~ 600 km sec^{-1}, does not give the impression of a place born in violent explosions. The nearby isolated galaxies and small groups all are drifting away from the Local Group at about the Hubble velocity. (Where reliable relative distances are available redshifts are at least within 25% of Hubble's law: Sandage 1986; Brown and Peebles 1986). Is this what we would expect from explosions energetic enough to have rearranged matter on the scale ~ 3000 km sec^{-1}? Galaxies on thin sheets can have low relative peculiar velocities, but the galaxies in our neighborhood are not distributed on a single sheet - the plane of the Local Supercluster is visible at v ~ 600 km sec^{-1} toward the Virgo Cluster but most galaxies in our neighborhood are not on this plane. The original sheet-like distribution would be obscured if sheets intersected in our neighborhood, but then we might have expected to have seen large relative peculiar velocities from the original relative motions of the sheets. If we and our neighbors formed at very high redshifts our original peculiar motions would have died out. But then one might wonder why we and our neighbors look so much like the galaxies in the "froth" that would have formed in the last generation in an environment substantially different from that of the early generations.

3. YET ANOTHER SCENARIO

The goal here is to see where we are led if we work with what we are fairly sure is present - baryons and radiation (which would include massless neutrinos) - and we extrapolate evolution from what is happening on large scales now. It seems to be a good bet that the present large-scale structure is growing by the gravitational instability of the expanding universe, so we will suppose that that is the way galaxies grew.

The computation of nucleosynthesis in the Hot Big Bang indicates that the density parameter in baryons could be no more than about 0.1 (Yang et al. 1984), and the Loh-Spillar (1986) measurement says $\Omega \gtrsim 0.4$ if $\Lambda = 0$, so as a compromise I assume

$$\Omega = 0.2.$$

It is pleasant to note that this number agrees with the dynamical mass estimates obtained in a considerable range of ambient densities, from our neighborhood to rich clusters (Peebles 1986b; Brown and Peebles 1986).

To complete the cosmological model, I assume $\Lambda = 0$ and adopt

$$H = 60 \text{ km sec}^{-1} \text{ Mpc}^{-1}.$$

This makes the age of the universe 14 billion years, which is in the range of estimates of globular cluster stellar evolution ages.

In this model linear density perturbations grow only slowly, by a factor ~ 200 from decoupling to the present, so to avoid problems with the limits on anisotropy of the microwave background let us assume primeval isocurvature entropy perturbations (Peebles 1980, § 94; Kodama and Sasaki 1986), where the mass distribution is homogeneous to begin with but the entropy is a random function of position. As entropy fluctuations appear within the horizon before decoupling they drive acoustic waves that dissipate by diffusion, leaving a nearly smooth sea of radiation and the original clumpy distribution of baryons, with density contrast $(\delta\rho/\rho)_{Bi}$. The large-scale part of the entropy fluctuation spectrum that appears on the horizon at $z < z_{eq}$ and after the universe has become transparent appears mainly in the radiation density, because the baryon mass density dominates, so the large-scale background temperature perturbation is

$$\left(\frac{\delta T}{T}\right)_0 \cong \frac{1}{3} \left(\frac{\delta\rho}{\rho}\right)_{Bi}.$$

Since $(\delta T/T)_0 \lesssim 10^{-4}$ on large angular scales (Wilkinson 1986), we want $(\delta\rho/\rho)_{Bi}$ to be small on large scales. To get galaxies to form fairly early we must assume $(\delta\rho/\rho)_{Bi}$ is large on small scales. As will be discussed, a white noise spectrum seems about right.

It will be noted that the assumption of large initial fluctuations on small scales would not be possible in the adiabatic case where baryons and photons have the same space distribution. The net density contrast, $\delta\rho/\rho$, on the horizon, cannot exceed unity on any scale (for that would trigger catastrophic production of black holes) so in the adiabatic case the baryon distribution at decoupling would have had to have been only mildly inhomogeneous prior to decoupling.

Now let us consider the expected stages of evolution of a strongly clumpy baryon distribution near the epoch of decoupling of matter and radiation (Hogan 1978; Rees 1984).

At redshifts $z > z_{dec} \sim 1000$ the radiation is hot enough to keep the bulk of the matter ionized, and Compton drag is strong enough to make mildly non-linear fluctuations in the baryon distribution expand with the radiation. Strongly non-linear concentrations of baryons can contract at $z > z_{dec}$ as Compton drag becomes weaker than gravity, to form objects with mass greater than or comparable to the linear matter Jeans mass $\sim 10^5 \, M_0$. Because the initial contraction is damped it would be close to radial, producing a very compact star cluster or even a superstar (Rees 1984). Because of the strong Compton drag on free electrons this first generation may not do much to the remaining matter prior to decoupling.

Table I shows some properties of the matter and radiation at decoupling, redshift $z = 1000$, and at a factor of 10 expansion after that, for the cosmological parameters mentioned above. The velocity dissipation time is

$$t_d = v/|dv/dt|,$$

where dv/dt is the acceleration due to Compton drag on fully ionized matter moving at speed v through the radiation. The Compton cooling and recombination times are similarly defined. The latter takes account of the inhibiting effect of the recombination Lyman α photons under the assumption that the plasma is expanding with the general

expansion of the universe. If dust were present to absorb the Lyman α
photons it would speed recombination at z ~ 1000.

TABLE I.

	z = 1000	z = 100
ρ_r/ρ_m	0.6	0.06
ρ_m (protons cm^{-3})	800	0.8
expansion time (y)	7.7×10^5	2.4×10^7
velocity dissipation time (y)	4.1×10^3	4.1×10^7
cooling time (y)	2.2	2.2×10^4
recombination time (y)	1×10^3	1×10^4

In the absence of ionizing radiation from stars the plasma at
z ~ 1000 combines to almost fully neutral hydrogen. The hydrogen
readily slips through the radiation, so wherever $(\delta\rho/\rho)_B \gtrsim 1$ the gas
collapses within an expansion time to form clouds with masses $\gtrsim$ the
Jeans mass ~ 10^5 M_Θ within which stars presumably form. Massive stars
would ionize the surroundings. Where that happens Compton drag would
do work on the clouds of plasma, lowering their binding energy,
pulling the plasma away from the stars. This is similar to the
situation observed in giant molecular clouds in our galaxy, but here
the feedback not only inhibits runaway star formation but also
inhibits the formation of stable gaseous systems at the ambient
density at decoupling, n ~ 10^3 cm^{-3}. This offers a way to understand
why galaxies did not form at decoupling at unacceptably high
densities.

At redshifts z ~ 100 the second part of the feedback becomes
ineffective, so patches of gas at density contrast $(\delta\rho/\rho)_B$ ~ 1 would
break away from the general expansion to form clouds at the ambient
density ~ 1 proton cm^{-3}. Since cooling times are short these clouds
could not support themselves by gas pressure; gas and embedded star
clusters would have to suffer violent collapse and, one trusts, rapid
star formation, so that, by redshift z ~ 30, say, the systems would be
dominated by stars with heavy concentrations of gas and dust. This is
the classic collapse picture of Eggen, Lynden-Bell and Sandage (1962),
with the slight modification that the collapsing material would have
suffered earlier sporadic bursts of star formation.

To make galaxies form at the right size we must adjust a
parameter: the amplitude of the primeval entropy fluctuations must be
such that at z = 100 $(\delta\rho/\rho)_B$ ~ 1 on the comoving scale (normalized to
the proper length at the present epoch) λ_g ~ 10 Mpc characteristic of
the present separation of galaxies.

It is an encouraging coincidence that the mass density at the
epoch at which the decay of Compton drag would have triggered the main

burst of formation is the density characteristic of the bulk of the bright matter in galaxies. Put a somewhat different way, the present typical separation of galaxies, $\lambda_g \sim 10$ Mpc, extrapolates back to ~ 100 kpc at $z \sim 100$, which is comparable to sizes of galaxies, if we allow room for observed dark halos. This means galaxies would have roughly filled space when they formed, as we might have expected.

In the more commonly discussed scenarios for biased galaxy formation in an Einstein-de Sitter model the large spacing of field galaxies is in part the result of suppression of galaxy formation in low density regions (Frenk et al. 1985). If suppressed galaxies leave massive dark remnants, as is also commonly proposed, one might wonder why we do not see the effect of the gravitational perturbation of these remnants on the local Hubble flow (Sandage 1986; Brown and Peebles 1986). In the present scenario there is no need for suppressed galaxies; the vast space between seen galaxies is the result of their early formation.

The early formation of galaxies avoids the following problem. Suppose galaxies formed at $z \sim 3$, to match quasars. This would require collapse to increase the density by a factor $\sim 10^4$ from the mean baryon density at $z = 3$. Stars achieve considerably greater collapses, but there is a striking regularity in the mean distribution of mass around galaxies that does not apply to stars. The mean value of the mass density at distance r from the center of a bright galaxy scales as $\rho(r) \sim r^{-1.8}$, from the body of the galaxy out into the general clustering of neighbors. (This just expresses the observation that velocity dispersions in and among galaxies are about the same. For the mean mass distribution within spirals see Burstein and Rubin 1985; for the statistical connection to the mean density of neighbors outside the galaxy see Davis and Peebles 1983 and Peebles 1984, fig. 6). The situation is very different for stars. The mean value of the mass density as a function of distance r from the center of an ensemble of stars such as we see in our neighborhood drops by some 20 orders of magnitude from the mean stellar surface to a mean density in the environment. This is because stars contracted a good deal more than did star clusters. If galaxies formed late, contraction would have to have increased their densities by a large factor, and the scaling law $\rho \propto r^{-\gamma}$ would require a conspiracy in the dissipative collapse of galaxies and the non-dissipative formation of groups and clusters. In the present picture we want galaxies to collapse only by the fixed factor needed for virialization. In this case the usual scaling argument says that if the primeval power spectrum of density fluctuations is $P \propto k^n$ then the mean mass distribution is

$$\rho \propto r^{-\gamma}, \quad \gamma = (9 + 3n)/(5 + n)$$

This reproduces the observations if $n \sim 0$. Galaxies would be at the end of this clustering hierarchy because they are the objects that formed by violent collapse when Compton drag first allowed it.

Two other constraints on large-scale structure might be mentioned. There are indications that the motion of the Local Group relative to the microwave background, $v \sim 600$ km sec^{-1} (Wilkinson

1986), is coherent over scales as large as ~ 100 Mpc (Burstein et al.
1986). If the large-scale peculiar velocity field, v(r), were
produced by the gravitational field of large-scale mass fluctuations,
then in linear perturbation theory the velocity autocorrelation
function would be (Peebles 1980)

$$\langle \vec{v}(0) \cdot \vec{v}(r) \rangle \cong \Omega^{1 \cdot 2} \, H^2 [\frac{1}{r} \int_0^r r^2 dr \xi(r) + \int_r^\infty r dr \xi(r)],$$

where $\xi(r)$ is the dimensionless mass autocorrelation function. Let us
model $\xi(r)$ after the galaxy function (Davis and Peebles 1983),

$$\xi(r) = (r_0/r)^\gamma, \quad r < r_x,$$

$$= 0, \quad r \cdot > r_x,$$

$$Hr_0 = 540 \text{ km sec}^{-1}, \quad \gamma = 1.77.$$

Let us evaluate the velocity autocorrelation function at $Hr = 5000$ km
sec^{-1}. With $Hr_x = 2000$ km sec^{-1}, which is a reasonable lower bound
from the galaxy distribution, and $\Omega = 0.2$, we get

$$\langle \vec{v}(0) \cdot \vec{v}(100) \rangle^{1/2} = 140 \text{ km sec}^{-1}.$$

With $Hr_x = 10,000$ Mpc, we get

$$\langle \vec{v}(0) \cdot \vec{v}(100) \rangle^{1/2} = 330 \text{ km sec}^{-1}.$$

This spans the range of what seems to be wanted. As shown by Vittorio
et al. (1986), if $\xi < 0$ on large scales (corresponding to power
spectrum $P \propto k^n$ with $n > 0$), it would make the coherence length of the
peculiar velocity field smaller than current estimates. If the $r^{-\gamma}$
power law extended to the horizon, which would correspond to power
spectrum $P \propto k^{-1.23}$ at $k \to 0$, it would make the coherence length
uncomfortably large. Thus it appears that we want something fairly
close to white noise, $P \sim$ constant on large scales, consistent with
the above scaling result.

The second constraint comes from the bounds on anisotropy of the
microwave background. We want $(\delta\rho/\rho)_{B,i} \lesssim 1$ at comoving length λ
~ 10 Mpc. If $P \propto k^n$ the large-scale anisotropy of the microwave
background at the present horizon would be

$$\frac{\delta T}{T} \sim \frac{1}{3} (\frac{10}{5000})^{(3 + n)/2}.$$

If $n = 0$ this would amount to $\delta T/T \sim 3 \times 10^{-5}$, which is well below the
observational limit. If $n = -1$, $\delta T/T \sim 10^{-4}$, which is marginally
acceptable.

On small angular scales perturbations to the microwave background
by galaxy formation could be strongly suppressed by scattering at
$z \gtrsim 30$. A serious test will be the anisotropy on the angular scale
~ 5° subtended by the horizon at $z \sim 25$, when the universe would have

become transparent. However, that will require a more detailed calculation.

The next step ought to be a discussion of large-scale features in the galaxy distribution - bubbles (de Lapparent et al. 1986), clusters and superclusters - but here we encounter an ambiguity. We have seen that the power spectrum of $(\delta\rho/\rho)_{B,i}$ ought to be roughly flat. To complete the prescription we must specify the higher moments. The standard assumption of random phases (random Gaussian process) would not do if the entropy per baryon, $s(x)$ were bounded by $s > 0$, because we want $(\delta\rho/\rho)_{B,i} > 1$ on small scales. We could allow $s < 0$ by admitting regions of antimatter; or we could assume that log $s(x)$ is Gaussian or that $s(x)$ is a non-Gaussian fractal. The simplest possibility would be that $s(x)$ approximates a Gaussian process on large scales where the fluctuations are small. I am betting that if $P \sim$ constant that will not do for the reason illustrated in Figures 1 and 2. (On the scales shown in Figure 2 the spectrum in the cold dark matter scenario is close to flat.) The scenario does allow a primeval power law power spectrum $P \propto k^n$, with $n \sim -1$, which would result in isothermal perturbations with about the same shape shortward of the matter-radiation Jeans length. (A power law $P \sim k^{-1}$ on large scales is not acceptable in the adiabatic case because the Sachs-Wolfe effect would perturb the microwave background by more than is observed.) This might make more interesting-looking structures on the scale ~ 100 Mpc. Also remaining to be seen is whether a "natural" non-Gaussian prescription might look more promising (Peebles 1983).

4. DISCUSSION

I argued in Section 3 that the more standard scenarios all have potentially serious problems, and to be fair I should list the critical difficulties here. First, the approach lacks the motivation from fundamental physics - inflation, baryosynthesis, and all that - that is such an attractive feature of standard scenarios. This is aesthetically unappealing, and, more serious, a barrier to unambiguous tests, as we see reflected in the large-scale structure problem. Second, it is easy enough to list potential problems with the physics. For example, the development and survival of galaxy rotation in the crowded and viscous environment at $z \sim 100$ arguably is problematic. Third, the scenario places all exotic phenomena in the initial conditions, and one certainly can debate the virtues of this feature. It would be nice to be able to work with a dynamical system that is insensitive to initial conditions, such as we find in fully developed turbulence, or star formation, or inflation. The assertion of the ancient right of choice of initial conditions thus is a step backward from the vision of inflation, to be justified only if we can see strong evidence for the results.

Freedom of choice of initial conditions does not mean freedom of choice of the final state. Having chosen to avoid hypothetical objects we are led to the present scheme in a fairly unambiguous way.

To my mind the main virtue of the scheme is that it captures what seems to me to be the essence of galaxies - isolated stable islands. In the scenario there are not dark massive objects between the galaxies, and there is not even all that much dark mass outside the great clusters, so we can understand why the Hubble flow is so remarkably quiet in our neighborhood (Sandage 1986). Rapid merging of the pieces of galaxies would have been accomplished in a few galaxy rotation periods, at $z \sim 30$, so we can understand why galaxy counts show nothing remarkable, and why disc galaxies seem to have lived such tranquil lives. And cluster formation would be a scaled version of galaxy formation, as the observations suggest.

On the theoretical side, the wanted roughly flat spectrum of entropy perturbations seems difficult to arrange, but we have learned not to underestimate the ingenuity of physicists of the early universe.

5. ACKNOWLEDGMENTS

This work, though still quite preliminary, was improved by lively discussions with the participants of this conference. I am grateful to Bill Unruh for his hospitality at the conference, where much of the work was done, and to Adrian Melott, Jerry Ostriker, and Chris Thompson for stimulating comments after the conference. This research also was supported in part by the U.S. National Science Foundation.

REFERENCES

Bahcall, N. A. and Soneira, R. A. 1984, Ap. J. **277**, 27.
Brown, M. E. and Peebles, P. J. E. 1986, submitted to Ap. J.
Burstein, D., Davies, R. L., Dressler, A., Faber, S. M., Lynden-Bell, D., Terlevich, R. and Wegner, G. 1986, preprint.
Burstein, D. and Rubin, V. C. 1985, Ap. J. **297**, 423.
Daly, R. A. 1986, submitted to Ap. J.
Davis, M., and Peebles, P. J. E. 1983, Ap. J. **267**, 465.
de Lapparent, V., Geller, M. J., and Huchra, J. P. 1986, Ap. J. **302**, L1.
Eggin, O. J., Lynden-Bell, D., and Sandage, A. R. 1962, Ap. J. **136**, 748.
Frenk, C. S., White, S. D. M., Efstathiou, G., and Davis, M. 1985, Nature **317**, 595.
Giovanelli, R., Haynes, M. P., Rubin, V. C., and Ford, W. K. 1986, Ap. J. **301**, L7.
Haynes, M. P., and Giovanelli, R. 1986, Ap. J. **306**, L55.
Hogan, C. 1978, M.N.R.A.S. **185**, 889.
Ikeuchi, S. 1981, Publ. Astron. Soc. Japan **33**, 211.
Johnson, D. G., and Wilkinson, D. T. 1986, submitted to Ap. J. Letters.
Kodama, H., and Sasaki, M. 1986, Intl. J. Mod. Phys. **A1**, 265.
Loh, E. D., and Spillar, E. J. 1986, Ap. J. **307**, L1.
Ostriker, J. P., Cowie, L. L. 1981, Ap. J. **243**, L127.

214

Ostriker, J. P., Thompson, C., and Witten, E. 1986, Phys. Lett. B (Nov. 13).

Peebles, P. J. E. 1980, The Large-Scale Structure of the Universe (Princeton: Princeton University Press).

Peebles, P. J. E. 1982, Ap. J. 263, L1.

Peebles, P. J. E. 1983, Ap. J. 274, 1.

Peebles, P. J. E. 1984, Science 224, 1385.

Peebles, P. J. E. 1986a, a privately circulated screed.

Peebles, P. J. E. 1986b, Nature 321, 27.

Rees, M. J. 1984, in Formation and Evolution of Galaxies and Large Structure in the Universe, NATO ASI Series, eds. J. Andouze and J. Tran Thanh Van, p. 239.

Sandage, A. R. 1986, Ap. J., 307, 1.

Turok, N. and Brandenberger, R. H. 1986, Phys. Rev. D 33, 2175.

Vittorio, N., Juszkiewicz, R., and Davis, M. 1986, Nature 323, 132.

Wilkinson, D. T. 1986, Science 232, 1517.

Yang, J., Turner, M. S., Steigman, G., Schramm, D. N., and Olive, K. A. 1984, Ap. J. 281, 493.

NON-GAUSSIAN FLUCTUATIONS

Mark B. Wise
California Institute of Physics
Pasadena, California 91125
U.S.A.

ABSTRACT.

Natural primordial mass density fluctuations are those for which the probability distribution, for the mass density fluctuations averaged over the horizon volume, is independent of time. This criterion determines the two-point correlation of the mass density fluctuations to have a Zeldovich power spectrum but allows for many types of higher correlations. If the connected higher correlations vanish the primordial fluctuations are Gaussian. In this case the probability distribution develops into a non-Gaussian one due to the non-linear time evolution. The nature of this non-Gaussian distribution and its effects on the large scale distribution of galaxies or clusters of galaxies and their large scale streaming velocities is explored. Next the possibility of natural primordial non-Gaussian fluctuations is examined. These can give rise to a very different large scale distribution of galaxies (or clusters of galaxies) than the Gaussian primordial fluctuations.

I. INTRODUCTION

The large scale structure of the universe probably arose from small fluctuations in the energy density that grew due to their gravitational instability. The main purpose of these lectures is to introduce some techniques that I believe will prove useful if the primordial fluctuations in the energy density are not Gaussian[1]. As preparation for the study of non-Gaussian fluctuations many of the features of Gaussian primordial mass density fluctuations will be explored. Throughout these lectures I assume that the Universe is at critical density ($\Omega = 1$) and that the cosmological constant vanishes. I denote the Hubble constant by H and the Robertson Walker scale factor by a. The horizon length is defined to be equal to H^{-1}. Note that this is not necessarily the size of a region that has been in casual contact. In an inflationary cosmology[2], for example, the size of a region that has been in casual contact is very much greater than the horizon length.

Since the energy density fluctuations were once small linear perturbation theory should provide an adequate description of their evolution at early times. In linear perturbation theory it is convenient to Fourier transform the energy density fluctuations.

W. G. Unruh and G. W. Semenoff (eds.), The Early Universe, 215–238.

$$\frac{\delta\rho(\vec{x},t)}{\langle\rho(t)\rangle} = \int d\vec{k}\frac{\delta\rho(\vec{k},t)}{\langle\rho\rangle}e^{i\vec{k}\cdot\vec{x}}. \tag{1}$$

In eq. (1) $\vec{x}$ is the comoving coordinate and k is the comoving wavenumber. In linear perturbation theory modes of different wavevectors $\vec{k}$ evolve independently. Note that the physical wavelength associated with a mode of wavenumber k is $a(t)/k$. Since the horizon length is increasing linearly with cosmic time t while the Robertson Walker scale factor grows like $t^{1/2}$ during the radiation dominated era, and $t^{2/3}$ during the matter dominated era, fluctuations with physical wavelengths that are less than the horizon length today had wavelengths greater than the horizon length at early times. It is this feature that makes it hard to come up with reasonable ways for generating the primordial fluctuations.

Modes with physical wavelength less than the horizon length do not grow during the radiation dominated era and grow like $t^{2/3}$ during the matter dominated era (in linear perturbation theory). The evolution of modes with wavelength greater than the horizon length is gauge dependent[3]. I shall work in a gauge where they do not grow in the radiation or matter dominated eras.

2. THE PRINCIPLE OF SCALE INVARIANCE

There are many different possibilities for the form of the primordial fluctuations in the energy density. In order to make progress it is necessary to introduce some principle that narrows down the number of possibilities. For fluctuations which cross the horizon in the matter dominated era we can write

$$\frac{\delta\rho(\vec{k},t)}{\langle\rho\rangle} = a(\vec{k})(t/t_{\text{h.c.}})^{2/3}, \tag{2}$$

where $a(\vec{k})$ is a random variable and $t_{\text{h.c.}}$ is the time that the fluctuations with comoving wavenumber k entered the horizon. This is

$$k/a(t_{\text{h.c.}}) = H(t_{\text{h.c.}}). \tag{3}$$

Using

$$a = (t/t_0)^{2/3}, \quad H = 2/3t, \tag{4}$$

the time of horizon crossing can be expressed in terms of the present time t_0 and the comoving wavenumber k (Note that I have normalized the scale factor so that today the scale factor is unity and physical and comoving wavenumbers coincide).

$$t_{\text{h.c.}}^{-1/3} = \frac{3}{2}kt_0^{2/3}. \tag{5}$$

Putting eq. (5) into eq. (2) gives

$$\frac{\delta\rho(\vec{k},t)}{\langle\rho\rangle} = \frac{9}{4}a(\vec{k})k^2t^{2/3}t_0^{4/3} \equiv \epsilon(\vec{k})t^{2/3}t_0^{4/3}. \tag{6}$$

For fluctuations which cross the horizon in the radiation dominated era this gets modified by a computable function of k that goes to unity a k goes to zero.

In the matter dominated era fluctuations in the energy density become fluctuations in the mass density. Since the Fourier transform of the fluctuations in the mass density $\frac{\delta\rho(\vec{k},t)}{\langle\rho\rangle}$ have dimensions of $(\text{length})^3$, (I shall use the particle physics convention that $c = \hbar = 1$) eq. (6) implies that ϵ has dimensions of (length). It is the probability distribution for ϵ that determines the nature of the primordial fluctuations. Equivalently one can specify the primordial fluctuations by the moments of the probability distribution (assuming of course that all the moments exist).

$$\langle \epsilon(\vec{k}_1)\ldots\epsilon(\vec{k}_n)\rangle = \int [d\epsilon]\epsilon(\vec{k}_1)\ldots\epsilon(\vec{k}_n)P[\epsilon]. \tag{7}$$

In equation (7) the measure of integration $[d\epsilon]$ means that the value of ϵ at each wavevector $\vec{k}$ is integrated over. This is called a functional integral[4].

If the length scales associated with the process which generated the primordial fluctuations are small compared to astrophysically relevant length scales, then they can be neglected. Under this circumstance dimensional analysis gives[5]

$$\langle \epsilon(\lambda\vec{k}_1)\ldots\epsilon(\lambda\vec{k}_n)\rangle = \lambda^{-n}\langle\epsilon(\vec{k}_1)\ldots\epsilon(\vec{k}_n)\rangle. \tag{8}$$

Eq. (8) is called the principle of scale invariance. I shall adhere religiously to this principle during these lectures. Equation (8) implies that the mass density fluctuations averaged over the horizon volume have a probability distribution that is independent of time.

Scale invariance plus the homogeneity and isotropy of space determines the two-point correlation of ϵ up to normalization to have the Zeldovich[6] form.

$$\langle \epsilon(\vec{k}_1)\epsilon(\vec{k}_2)\rangle \propto k_1\delta^3(\vec{k}_1 + \vec{k}_2). \tag{9}$$

In eq. (9) the δ function of wavevectors is required by the homogeneity of space. It ensures that the correlation $\langle\epsilon(\vec{x})\epsilon(\vec{y})\rangle$ depends only on the separation between the points $\vec{x}$ and $\vec{y}$. Scale invariance gives that the coefficient of the δ function must go like a single power of a wavevector. For example, the magnitude of $\vec{k}_1$ or its component along some particular axis. The only rotationally invariant choice is the magnitude of $\vec{k}_1$.

Scale invariance does not determine the constant of proportionality in eq. (9). However, positivity of the probability distribution restricts this constant to be positive (recall that since the mass density fluctuations are real $\epsilon^*(\vec{k}) = \epsilon(-\vec{k})$ so $\langle\epsilon(\vec{k})\epsilon(-\vec{k})\rangle$ is the average value of a positive quantity).

3. GAUSSIAN SCALE INVARIANT FLUCTUATIONS

The simplest probability distribution consistent with the principle of scale invariance is a Gaussian one.

$$P[\epsilon] = \frac{1}{Z}\exp-\frac{1}{2}\int d\vec{k}\epsilon(\vec{k})\epsilon(-\vec{k})f(\vec{k}). \tag{10}$$

Since $\epsilon(\vec{k})$ has dimensions of length, scale invariance demands that $f(k) \propto 1/k$ (where the constant of proportionality is dimensionless). The function f determines the two point function

$$\langle\epsilon(\vec{k}_1)\epsilon(\vec{k}_2)\rangle = (1/f(k_1))\delta^3(\vec{k}_1 + \vec{k}_2). \tag{11}$$

The probability distribution (10) is not completely correct since it assigns a nonzero probability to configurations $\epsilon(\vec{k})$ that correspond to a negative mass. Fortunately the error made is incredibly small because these configurations are highly exponentially suppressed.

One can also characterize a Gaussian probability distribution by the behavior of all its moments. For Gaussian primordial mass density fluctuations all the connected correlations of $\epsilon(\vec{k})$ vanish, except the two-point correlation function. For example, the four-point function can be written in the form (using the property that the average value of ϵ is zero).

$$\langle\epsilon(\vec{k}_1)\epsilon(\vec{k}_2)\epsilon(\vec{k}_3)\epsilon(\vec{k}_4)\rangle = \langle\epsilon(\vec{k}_1)\epsilon(\vec{k}_2)\rangle\langle\epsilon(\vec{k}_3)\epsilon(\vec{k}_4)\rangle + \langle\epsilon(\vec{k}_1)\epsilon(\vec{k}_3)\rangle\langle\epsilon(\vec{k}_2)\epsilon(\vec{k}_4)\rangle$$

$$+\langle\epsilon(\vec{k}_1)\epsilon(\vec{k}_4)\rangle\langle\epsilon(\vec{k}_2)\epsilon(\vec{k}_3)\rangle + \langle\epsilon(\vec{k}_1)\epsilon(\vec{k}_2)\epsilon(\vec{k}_3)\epsilon(\vec{k}_4)\rangle_c. \tag{12}$$

In eq. (12) the subscript c denotes the connected part of the correlation. For Gaussian fluctuations this part of the four-point correlation vanishes.

The main motivation for Gaussian fluctuations is the small value of the primordial mass density fluctuations. The mass density fluctuations averaged over the horizon volume are about equal to 10^{-5}. This suggests that it was a weakly interacting field which gave rise to the mass density fluctuations; in the standard new inflationary model[7] for how the mass density fluctuations are generated[8] this is indeed the case. The small value of primordial fluctuations restricts the inflaton field to be very weakly interacting and hence for the resulting fluctuations to be approximately Gaussian. Of course there is no observational evidence to support the standard inflationary model. In fact since the final cosmological constant must be fine tuned to zero, it seems silly to imagine that the details of this model are correct, (although the general idea may be).

4. NON-LINEAR TIME EVOLUTION

Even if the primordial mass density fluctuations are Gaussian the non-linear time evolution[9] will ensure that at late times the mass density fluctuations are highly non-Gaussian. It is important to understand the nature of the connected correlations of the mass density induced by the non-linear time evolution in order to distinguish their effects from those of the primordial non-Gaussian fluctuations. In this section the connected correlations of the mass density and the peculiar velocity are studied at wavenumbers corresponding to distances that are large compared with those that have undergone very non-linear evolution, but small compared to the horizon length. The mass is treated as a pressureless Newtonian fluid. The time evolution of the mass density fluctuation field $\delta(\vec{x}, t) = \frac{\rho(\vec{x},t) - \langle \rho \rangle}{\langle \rho \rangle}$ and the peculiar velocity field $\vec{v}(\vec{x}, t)$ is governed by the equation of continuity, Eulers equation and the Newtonian expression for the gravitational field $\vec{g}(\vec{x}, t)$.

$$\frac{\partial \delta}{\partial t} + \frac{1}{a} \vec{\nabla} \cdot (1 + \delta) \vec{v} = 0 \tag{13a}$$

$$\frac{\partial \vec{v}}{\partial t} + \frac{\dot{a}}{a} \vec{v} + \frac{1}{a} (\vec{v} \cdot \vec{\nabla}) \vec{v} = \vec{g} \tag{13b}$$

$$\vec{\nabla} \cdot \vec{g} = -4\pi G \langle \rho \rangle a \delta, \quad \vec{\nabla} \times \vec{g} = 0. \tag{13c}$$

A perturbative expansion for the mass density fluctuation field and the peculiar velocity field is developed by linearizing these equations about the background

$$\delta = 0 \quad \text{and} \quad \vec{v} = 0, \tag{14}$$

and keeping only the fastest growing mode, then linearizing about this solution and again keeping only the fastest growing mode, etc. The resulting perturbation expansions for δ and $\vec{v}$ have the form

$$\delta(\vec{x}, t) = \sum_{n=1}^{\infty} \epsilon_n(\vec{x}_1) t^{2n/3}, \tag{15}$$

$$\vec{v}(\vec{x}, t) = a \sum_{n=1}^{\infty} \vec{v}_n(\vec{x}) t^{\frac{2n}{3} - 1}, \vec{\nabla} \times \vec{v}_n = 0. \tag{16}$$

At early times (small t) the series is dominated by the first term so ϵ_1 and $\vec{v}_1$ characterize the primordial mass density fluctuations. The equation of continuity determines $\vec{v}_1$ in terms of ϵ_1. The equations of motion (13) determine ϵ_n and $\vec{v}_n$ in terms of the primordial fluctuations. In wavenumber space this relationship has the form:

$$\epsilon_n(\vec{k}) = \int \frac{d\vec{q}_1}{(2\pi)^3} \cdots \int \frac{d\vec{q}_n}{(2\pi)^3} (2\pi)^3 \delta^3(\vec{q}_1 + \ldots + \vec{q}_n - \vec{k}) P_n^{(s)}(\vec{q}_n \ldots, \vec{q}_n) \qquad (17)$$

$$\cdot \epsilon_1(\vec{q}_1) \ldots \epsilon_1(\vec{q}_n),$$

$$\vec{v}_n(\vec{k}) = \frac{i\vec{k}}{k^2} \int \frac{d\vec{q}_1}{(2\pi)^3} \cdots \int \frac{d\vec{q}_n}{(2\pi)^3} \delta^3(\vec{q}_1 + \ldots + \vec{q}_n - \vec{k})$$

$$\cdot Q_n^{(s)}(\vec{q}_1, \ldots, \vec{q}_n) \epsilon_1(\vec{q}_1) \ldots \epsilon_1(\vec{q}_n). \qquad (18)$$

Here $P_n^{(s)}$ and $Q_n^{(s)}$ are symmetric homogeneous functions of the wavevectors $\vec{q}_1, \ldots \vec{q}_n$ of degree zero. Recursion relations relate $P_n^{(s)}$ and $Q_n^{(s)}$ to $P_1^{(s)} = 1$ and $Q_1^{(s)} = 2/3$. For example

$$P_2^{(s)}(\vec{p}, \vec{q}) = \frac{5}{7} + \frac{\vec{p} \cdot \vec{q}}{2pq} \left(\frac{p}{q} + \frac{q}{p} \right) + \frac{2}{7} \left(\frac{\vec{p} \cdot \vec{q}}{pq} \right)^2 \qquad (19a)$$

$$Q_2^{(s)}(\vec{p}, \vec{q}) = \frac{2}{7} + \frac{\vec{p} \cdot \vec{q}}{3pq} \left(\frac{p}{q} + \frac{q}{p} \right) + \frac{8}{21} \left(\frac{\vec{p} \cdot \vec{q}}{pq} \right)^2. \qquad (19b)$$

For $n > 2$ $P_n^{(s)}$ and $Q_n^{(s)}$ are quite complicated. The Zeldovich approximation[10] is sometimes used to describe the non-linear time evolution. Since it is exact in one dimension it should be useful when the dominant non-linear effect is the collapse of pancakes. In appendix A it is shown that the Zeldovich approximation is equivalent to the choice[11]

$$P_n^{(s)}(\vec{q}_1, \ldots \vec{q}_n) = \frac{1}{n!} \frac{(\vec{k} \cdot \vec{q}_1)}{q_1^2} \cdots \frac{(\vec{k} \cdot \vec{q}_n)}{q_n^2} \qquad (20)$$

$$Q_n^{(s)}(\vec{q}_1, \ldots \vec{q}_n) = \frac{2}{3} P_n^{(s)}(\vec{q}_1, \ldots \vec{q}_n), \qquad (21)$$

where $\vec{k} = \vec{q}_1 + \ldots + \vec{q}_n$. From the recursion relations it can be deduced that the $P_n^{(s)}(\vec{q}_1, \ldots \vec{q}_n)$ and $Q_n^{(s)}(\vec{q}_1, \ldots \vec{q}_n)$ have the following special properties.

(a) As $\vec{k} = \vec{q}_1 + \ldots + \vec{q}_n$ goes to zero (but the individual $\vec{q}_i$ do not) $P_n^{(s)} \propto k^2$. This property is essentially a consequence of momentum conservation. In the Zeldovich approximation this property is modified to $P_n^{(s)} \propto k^n$. It appears that the Zeldovich approximation makes successive orders in the perturbative expansion for δ less important at large scales (i.e. small k) than they actually are.

(b) As some of the arguments of $P_n^{(s)}$ (or $Q_n^{(s)}$) get large but the vector sum of all the arguments of $P_n^{(s)}$ (or $Q_n^{(s)}$) stays fixed, $P_n^{(s)}$ (or $Q_n^{(s)}$) vanishes like a power of the large arguments. For example, for $p \gg q_j$,

$$P_n^{(s)}(\vec{q}_1, \ldots \vec{q}_{n-2}, \vec{p}, -\vec{p}) \propto 1/p^2 \quad Q_n^{(s)}(\vec{q}_1, \ldots \vec{q}_{n-2}, \vec{p}, -\vec{p}) \propto 1/p^2. \tag{22}$$

This is essentially a decoupling property. It limits the importance of primordial fluctuations of high wavenumber to the mass density fluctuations at small wavenumbers.

(c) If one of the arguments of $P_n^{(s)}(\vec{q}_1, \ldots \vec{q}_n)$ or $Q_n^{(s)}(\vec{q}_1, \ldots \vec{q}_n)$ goes to zero then there is an infrared divergence of the form

$$\vec{q}_i / q_i^2. \tag{23}$$

This property is of kinematical origin. These infrared singularities would be absent if the fluctuations had been expanded in terms of $\vec{v}_1$ instead of ϵ_1. There are no infrared divergences as partial sums of several wavevectors go to zero.

The effects of nonlinearities in the time evolution can be taken into account once the probability distribution for ϵ_1 is specified. In this section it is assumed that ϵ_1 is a Gaussian random variable with a Zeldovich spectrum.

$$\langle \epsilon_1(\vec{k}_1) \epsilon_1(\vec{k}_2) \rangle t^{4/3} = (2\pi)^3 A(k_1) k_1 \delta^3(\vec{k}_1 + \vec{k}_2), \tag{24}$$

where A goes to a constant as k_1 goes to zero. With cold dark matter $A(k) \propto (\ln k/k^2)^2$ for large k. A arises because fluctuations which cross the horizon in the radiation dominated era only grow logarithmically until the time of matter domination.

To compute correlations of δ and $\vec{v}$ one expands them in a "power series" in ϵ_1 and then evaluates the correlations of ϵ_1 by factorizing them into products of two-point functions (see for example eq. (12)). As a simple example consider the leading perturbative contribution to the three-point correlation of $\langle \delta(\vec{k}_1, t) \delta(\vec{k}_2, t) \delta(\vec{k}_3, t) \rangle$. It arises when one of the δ's is expanded to second order in ϵ_1

$$\langle \delta(\vec{k}_1, t)\delta(\vec{k}_2, t)\delta(\vec{k}_3, t)\rangle = t^{8/3}\left\{\langle \epsilon_2(\vec{k}_1)\epsilon_1(\vec{k}_2)\epsilon_1(\vec{k}_2)\rangle + \text{perms}\right\}$$

$$= t^{8/3}\int \frac{d\vec{q}_1}{(2\pi)^3}\int \frac{d\vec{q}_2}{(2\pi)^3}P_2^{(s)}(\vec{q}_1, \vec{q}_2)(2\pi)^3\delta^3(\vec{q}_1 + \vec{q}_2 - \vec{k}_1)$$

$$\cdot\langle \epsilon_1(\vec{q}_1)\epsilon_1(\vec{q}_2)\epsilon_1(\vec{k}_2)\epsilon_1(\vec{k}_3)\rangle + \text{perms.} \tag{25}$$

Evaluating the average value of the product of four in eq. (25) using eq. (24) gives

$$\langle \delta(\vec{k}_1, t)\delta(\vec{k}_2, t)\delta(\vec{k}_3, t)\rangle$$

$$= 2\int \frac{d\vec{q}_1}{(2\pi)^3}\int \frac{d^3 q_2}{(2\pi)^3}P_2^{(s)}(\vec{q}_1, \vec{q}_2)(2\pi)^3\delta^3(\vec{q}_1 + \vec{q}_2 - \vec{k}_1)A(k_1)k_1$$

$$\cdot(2\pi)^3\delta^3(\vec{k}_1 + \vec{q}_1)A(k_2)k_2(2\pi)^3\delta^3(\vec{k}_2 + \vec{q}_2) + \text{perms}$$

$$= (2\pi)^3\delta^3(\vec{k}_1 + \vec{k}_2 + \vec{k}_3)2\left[P_2^{(s)}(\vec{k}_1, \vec{k}_2)A(k_1)A(k_2)k_1 k_2\right.$$

$$\left. + P_2^{(s)}(\vec{k}_1, \vec{k}_3)A(k_1)A(k_3)k_1 k_3 + P_2^{(s)}(\vec{k}_2, \vec{k}_3)A(k_2)A(k_3)k_2 k_3\right]. \tag{26}$$

There is a simple diagrammatic way to visualize this computation. This is shown in Fig. (1). Each δ is denoted by a solid line and the primordial fluctuation ϵ_1

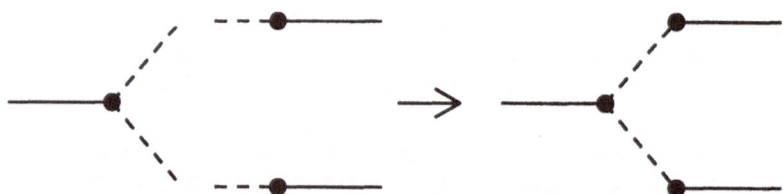

$Fig.(1)$: Diagrammatic view of the computation in eq. (25).

is denoted by a dotted line. One of the δ's split into two ϵ_1's via second order perturbation theory while the other two δ's just went into a single dotted line (linear perturbation theory). Then the four ϵ_1's (dotted lines) are sewn together two at a time as is appropriate for Gaussian primordial fluctuations. This diagrammatic approach can be generalized to all orders in the perturbative expansion and is analogous to the Feynman diagrams used to compute correlations of quantum fields. Contributions to a connected n-point correlation of δ, come from connected diagrams with n external solid lines and an arbitrary number of internal dotted lines. Each internal line is labeled by a wavevector that is integrated over. The vertices which join the lines and the factors associated with them and the internal lines are shown in Fig. (2). There are also combinatorial factors associated with the number of ways of putting the ϵ_1's into two-point correlations (24). Some of the perturbative contributions to the two-point function are shown in Fig. (3).

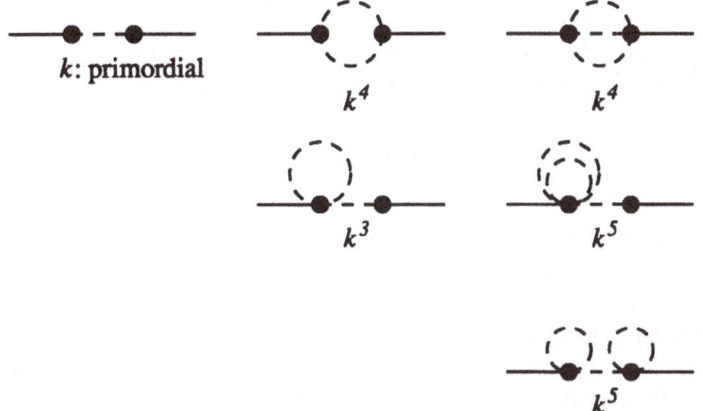

$$\frac{1}{(2\pi)^{3n}}(2\pi)^3\delta^3(\vec{q}_1+\cdots+\vec{q}_n-\vec{k})P_n^{(s)}(\vec{q}_1,\cdots,\vec{q}_n)$$

$$(2\pi)^3 A(q)q$$

$Fig.(2)$: Factors associated with vertices and dotted lines.

k: primordial

$$k^4 \qquad k^4$$

$$k^3 \qquad k^5$$

$$k^5$$

$Fig.(3)$: Some diagrams contributing to the two-point correlation of δ.

According to the diagrammatic rules the diagram at the top of the second column in Fig. (3), for example, gives the following contribution to the two-point function.

$$(2\pi)^3\delta^3(\vec{k}_1 + \vec{k}_2)2 \int \frac{d\vec{q}}{(2\pi)^3} \left(P_2^{(s)}(\vec{k}_1 - \vec{q}, \vec{q})\right)^2 A(q)A(\vec{k} - \vec{q})q|\vec{k} - \vec{q}|. \qquad (27)$$

Note that there is an integral over the internal wavevector $\vec{q}$ in eq. (27). It is straightforward to see that in general the number of integrations over internal wavenumbers in the contribution of a diagram is equal to the number of loops in the diagram. Note also that with $A(q)$ falling as $1/q^4$ for large q this integral is convergent.

To evaluate the mean square value, of the mass density fluctuations averaged over a ball of large radius R (with a fuzzy edge) $\delta(R)$, the diagrammatic contributions to the two-point correlation of δ must be examined at small external wavenumbers k. Fig. (3) gives the leading k behavior for small external wavenumber of each of the diagrams. Most of these factors follow straightforwardly from the properties of the $P_n^{(s)}$ mentioned earlier. For example the first diagram in the second column of Fig. (3) has a factor of k^4. This comes from the part of this diagram that has large loop wavenumbers compared with k. At loop wavenumbers comparable with k the diagram has a factor of k^3 associated with the measure of integration and one factor of k for each internal dotted line. There are no factors of k associated with the two vertices since they are homogeneous functions of degree zero and all their arguments are of order k. Thus the contribution from loop wavenumbers comparable with k is of order k^5. At large loop wavevectors there are no factors of k associated with the measure of integrations or the internal dotted lines. There are however, two factors of k^2 associated with the vertices. According to the decoupling property each vertex falls off like the square of the wavenumber associated with the loop integration. Since they are homogeneous functions this necessitates a compensating factor of k^2 in each vertex. (The presence of this factor could have also been deduced from property (a).)

Note that there appear to be an infinite number of diagrams that behave like k^4 (this is different from what happens in the Zeldovich approximation where successive orders of perturbation theory produce additional factors of k). Clearly the contribution of order k^4 cannot be computed. However, only the first diagram (linear perturbation theory) in Fig. (3) goes like k. Thus for large R

$$\langle \delta(R)^2 \rangle \propto 1/R^4, \tag{28}$$

with the constant of proportionality calculable from linear perturbation theory.

To get information on the behavior of the higher moments of $\delta(R)$, connected correlations with more external lines must be considered. Consider the average value, for large R, of $\delta(R)^3$. This is determined by the behavior of the three-point correlation function at small wavenumbers k (here k stands for any of the three wavenumbers that are needed to specify this function). Fig. (4) shows some low order diagrams and the power of k associated with their contribution. Only the first diagram produces a factor of k^2. This behavior persists to all orders in the perturbative expansion. Thus for large R

$$\langle \delta(R)^3 \rangle \propto 1/R^8, \tag{29}$$

with the constant of proportionality given by a computation of the first diagram in Fig. (4) (see eq. (26)).

Diagrams without loops are called trees. In general the tree graphs dominate the large R behavior of $\langle \delta(R)^n \rangle_c$. Since the trees produce a factor of k for each internal dotted line

$$\langle \delta(R)^n \rangle_c \propto \frac{1}{R^{4(n-1)}}. \tag{30}$$

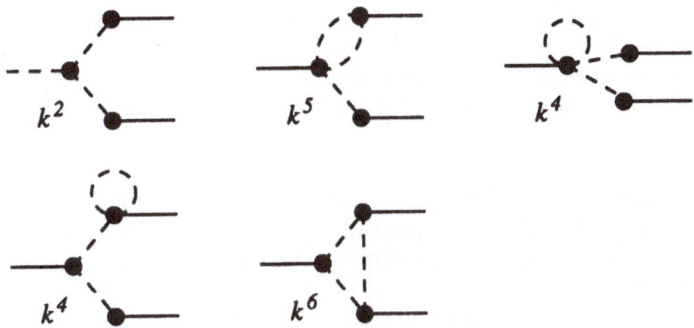

$Fig.(4)$: Some diagrams contributing to the three-point correlation of δ.

For n even the primordial fluctuations contribute to $\langle \delta(R)^n \rangle$ through it's disconnected part. They give

$$\langle \delta(R)^n \rangle_{\text{primordial}} \propto 1/R^{2n}. \tag{31}$$

This dominates over the connected contribution (30) at large R. So for large R, $\delta(R)$ becomes a Gaussian random variable with a variance determined by the primordial fluctuations. The leading corrections to this Gaussian behavior are computable through the tree graphs.

A similar analysis can be done for the peculiar velocity field. Perturbative contributions to connected correlations of $\vec{v}$ come from connected diagrams with n external wavy lines and an arbitrary number of internal dotted lines. The factors associated with the vertices are shown in Fig. (5). The behavior of moments of the velocity field averaged over a large ball (with a fuzzy edge) $\vec{v}(R)$, is determined by the behavior of the correlations of $\vec{v}$ at small external wavenumbers $\vec{k}$. In general

$$\langle \vec{v}(R)^{2n} \rangle_c \propto 1/R^{6n-4} \tag{32}$$

with the constant of proportionality determined by the tree graphs.

$$\frac{i\vec{k}}{k^2} \frac{a(t)}{t} \frac{(2\pi)^3}{(2\pi)^{3n}} \delta^3(\vec{q}_1 + \cdots + \vec{q}_n - \vec{k}) \varrho_n^{(s)}(\vec{q}_1, \cdots, \vec{q}_n)$$

$Fig.(5)$: Vertices needed for correlations of $\vec{v}$.

The perturbative expansions for δ and $\vec{v}$ will not be valid for late times t; the pressureless mass density fluid develops a pressure at late times due to orbit crossing. Clearly the present time t_0 is late enough for orbit crossing to have occurred. Nonetheless, the conclusions of this section probably remain valid since they are completely insensitive to the form of the primordial power spectrum at high wavenumbers.

5. OBJECTS

In the previous section properties of the large scale distribution of mass were explored. Unfortunately it is not possible to observe directly the distribution of mass. What is really needed are predictions for the large scale distribution of the various types of luminous objects observed (e.g., galaxies or rich clusters of galaxies). Since the details of galaxy formation are not well understood one is forced to resort to models for the relationship between the number density n of the objects observed and the underlying mass density fluctuations. The simplest possible assumption one can make is that the objects trace the mass. That is, the number density of the objects is proportional to the mass density and the peculiar velocity of the objects is equal to the peculiar velocity of the mass at the location of the objects. One way the objects could trace the mass is if all the mass ends up in the objects; another way is if the objects are a fair sample of the mass distribution. It is clear from observations that not all of the objects observed can trace the mass since different types of objects have different spatial distributions. For example, the two-point correlation for galaxies is unity at about $5h^{-1}$Mpc. While the two-point correlation for rich clusters of galaxies[12] is unity at about $25h^{-1}$ Mpc.

In order to make progress, I will assume that the number density of the various classes of luminous objects can be written as a local function of the mass density fluctuations, (and derivatives of the mass density fluctuations), filtered on the comoving scale that collapsed to the object observed.

$$n(\vec{x}) = \sum_{n=0}^{\infty} C_n \delta_f^n(\vec{x}) + \sum_{n=0}^{\infty} C_n' \vec{\nabla} \delta_f(\vec{x}) \cdot \vec{\nabla} \delta_f(x) \delta_f^n(\vec{x}) + \ldots \tag{33}$$

where

$$\delta_f(\vec{x}) = \int d\vec{y} W(\vec{x} - \vec{y}) \delta(\vec{y}), \tag{34}$$

and W is the appropriate filter. I shall also assume that the peculiar velocity of the objects is equal to that of the mass density fluid at the location of the objects.

An example of a number density of the type in eq. (33) is

$$n_>(\vec{x}) = C \exp T \delta_f(\vec{x}). \tag{34}$$

For large T (i.e. $T\delta_f(0) >> 1$) this corresponds to the objects forming preferentially where the mass density fluctuations are unusually large[13]. Eq. (34) has the advantage that it is easy to display how the two point correlation of such objects, $\xi_>$, defined by

$$1 + \xi_>(|\vec{x} - \vec{y}|) = \frac{\langle n_>(\vec{x}) n_>(\vec{y}) \rangle}{\langle n_> \rangle^2} \tag{35}$$

depends on the correlations of the underlying mass density fluctuations. For a source J it can easily be shown that the generating functional

$$Z[J] = \int [d\,\delta] P[\delta] e^{\int d\vec{x} J(x) \delta(x)} \tag{36}$$

can be expressed in terms of the connected correlations of δ in the following fashion

$$Z[J] = \exp\left\{ \sum_{n=2}^{\infty} \frac{1}{n!} \int d\vec{x}_1 \ldots d\vec{x}_n \langle \delta(\vec{x}_1) \ldots \delta(\vec{x}_n) \rangle_c J(\vec{x}_1) \ldots J(\vec{x}_n) \right\}. \tag{37}$$

To verify this, just expand the exponentials in eqs. (36) and (37) in a power series and equate powers of the source J. The two-point function for the objects is essentially the generating functional evaluated at a particular source.

$$1 + \xi_>(|\vec{x} - \vec{y}|) = \frac{Z[J(\vec{z}) = W(\vec{x} - \vec{z}) + W(\vec{y} - \vec{z})]}{Z[J(\vec{z}) = W(\vec{z})]^2}. \tag{38}$$

Using eq. (37) this becomes

$$1 + \xi_>(|\vec{x}_1 - \vec{x}_2|) = \exp\left\{ \sum_{n=2}^{\infty} \frac{T^n}{n!} \int d\vec{y}_1 \ldots d\vec{y}_n \sum_{m=1}^{n-1} \binom{n}{m} \langle \delta(\vec{y}_1) \ldots \delta(\vec{y}_n) \rangle_c \right.$$

$$\left. \cdot [W(\vec{x}_1 - \vec{y}_1) \ldots W(\vec{x}_1 - \vec{y}_m)][W(\vec{x}_2 - \vec{y}_{m+1}) \ldots W(\vec{x}_2 - \vec{y}_n)] \right\}. \tag{39}$$

The two-point correlation of the objects depends on all the connected correlations of the mass density fluctuations. If the mass density fluctuations are approximated as Gaussian so that only the connected two-point correlation of δ is non-zero the above becomes[13,14]

$$\xi_>(|\vec{x}_1 - \vec{x}_2|) = \exp T^2 \xi_f(|\vec{x}_1 - \vec{x}_2|) - 1. \tag{40}$$

where

$$\xi_f(r) = \int d\vec{k}\, e^{i\vec{k}\cdot\vec{r}} k A(k) \tilde{W}(k)^2. \tag{41}$$

For T large the two-point correlation of the objects is enhanced compared with that of the mass and the enhancement increases with T. This seems to be what is needed to explain the enhanced rich cluster correlations mentioned earlier. There is, however, one problem with this. Since the two-point correlation of the filtered mass distribution ξ_f has a Fourier transform that vanishes at $k = 0$, the two-point correlation must integrate to zero. That means that the filtered two-point correlation of the mass must cross zero somewhere (the zero crossing of $\xi_>$ coincides with that of ξ_f, see eq. (40)). The location of the zero crossing depends (with cold dark matter) on the time of matter domination and the filtering scale. For rich clusters it is at $17h^{-2}$ Mpc. Using a more realistic model where the objects form at peaks (local maxima) of the mass filtered mass density fluctuations[15] only makes matters worse. The peak condition introduces an anticorrelation (peaks do not occur right next to each other) which reduces the size of the enhancement and moves the zero crossing in[16]. If the mass density fluctuations are approximately Gaussian it does not seem likely that this model can accommodate significant (positive) rich-cluster two-point correlations at a distance of $30h^{-1}$ Mpc.

Even if the primordial fluctuations are Gaussian there are connected correlations of δ induced by the non-linear time evolution. It is easy to characterize how the connected correlations can effect the behavior of $\xi_>(r)$ at large r[17]. An n-point correlation of δ depends on the locations of n-points $\vec{x}_1, \ldots, \vec{x}_n$. Suppose m of those points are kept close together but separated a large distance r from the remaining $n - m$ points (which are also close to each other). If a connected correlation falls like r^{-p} in this limit then from eq. (39) it is evident that typically this will give rise to a two-point correlation $\xi_>(r)$ that also falls as r^{-p} for large r. This condition can also be given in wavenumber space. The Fourier transform of a connected n-point correlation of the mass density is a function of n wavevectors $\vec{k}_1, \ldots, \vec{k}_n$. If it diverges like k_T^{-p} as a partial sum $\vec{k}_T$ of those wavevectors (e.g., $\vec{k}_T = \vec{k}_i$, $\vec{k}_T = \vec{k}_i + \vec{k}_j$ with $i \neq j$, etc.,) goes to zero then at small k the power spectrum $P_>(k)$ typically also diverges like k^{-p}. Fig. (6a) shows a contribution to the connected four-point correlation of δ that behaves like k_T^0 as the partial sum $\vec{k}_T = \vec{k}_1 + \vec{k}_2$ goes to zero (but the individual wavevectors k_1 and k_2 do not). This gives a contribution to $P_>(k)$ that behaves like k^0 as k gets small. Clearly there are an infinite number of diagrams that behave in this fashion. Fig. (6b) shows a tree graph that contributes to the six-point correlation of δ. Since each of the vertices diverges linearly as $\vec{k}_T/k_T^2$, where $\vec{k}_T = \vec{k}_1 + \vec{k}_2 + \vec{k}_3$, this diagram seems like it would give a contribution to $P_>(k)$ that diverges like k^{-1} as k goes to zero. This is not the case, however. It is easy to see that rotational invariance causes the most divergent piece to cancel when the integrals over $\vec{k}_1$, $\vec{k}_2$, and $\vec{k}_3$ are done.

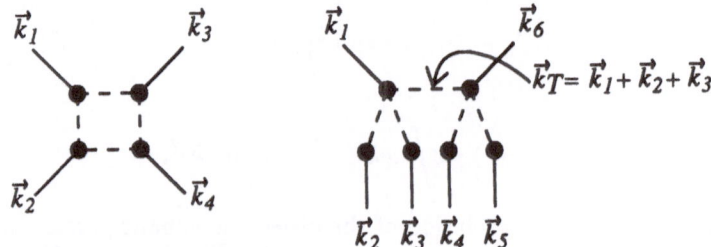

$Fig.$(6) : Two diagrams contributing to correlations of δ.

This behavior persists to all orders in the perturbative expansion. In general $P_>(k)$ goes to a constant as k goes to zero. Unless the threshold T is so large that linear perturbation theory is valid on the comoving scale that collapsed to the objects, the behavior of $P_>(k)$, at small k, is not computable since an infinite number of diagrams (like Fig.(6a)) contribute significantly. These diagrams are sensitive to physics on short distance scales (i.e., distances of order the comoving scale that collapsed to form the objects). Clearly numerical simulations of the non-linear evolution must correctly handle physics at these short distances if they are to draw believable conclusions about the large scale distribution of objects that arise wherever the mass density fluctuations are unusually large. Similar conclusions hold for any biasing (of the type in eq. (33)) in the definition of the objects.

It is possible to observe not only the positions of objects but also their peculiar velocities[18]. Imagine that the peculiar velocities of some type of object are averaged over a large ball of radius R (with fuzzy edges). The resulting average velocity is $\vec{v}(R)$ where,

$$\vec{v}(R) \simeq \frac{1}{\bar{n}} \int d\vec{r} \, W_R(\vec{r}) n(\vec{r}) \vec{v}(\vec{r}). \tag{42}$$

In eq. (42) n is the number density of the objects observed, $\bar{n}$ its average value and W_R is a smooth function that integrates to unity and becomes small outside the ball. Note that if n is set to unity in eq. (42) then the results of Section 4 imply that, for large R, $\vec{v}(R)$ becomes a Gaussian random variable with a variance determined by linear perturbation theory.

In general the mean square value of $v(R)$ is given by

$$\langle \vec{v}(R)^2 \rangle = \int d\vec{r} \, W_R(\vec{r}) \int d\vec{r}' \, W_R(\vec{r}') \cdot \left(\frac{1}{\bar{n}^2}\right) \langle n(\vec{r}) \vec{v}(\vec{r}) n(\vec{r}') \vec{v}(\vec{r}') \rangle. \tag{43}$$

Let

$$n(\vec{r}) = \bar{n}(1 + \lambda(\vec{r})), \tag{44}$$

then the relevant correlation in the expression for $\vec{v}(R)$ breaks into four pieces

$$(1/\bar{n}^2)\langle n(\vec{r}) \vec{v}(\vec{r}) n(\vec{r}') \vec{v}(\vec{r}') \rangle = \langle \vec{v}(\vec{r}) v(\vec{r}') \rangle + \langle \lambda(\vec{r}) \vec{v}(\vec{r}) \vec{v}(\vec{r}') \rangle$$

$$+ \langle \vec{v}(\vec{r}) \lambda(\vec{r}') \vec{v}(\vec{r}') \rangle + \langle \lambda(\vec{r}) \vec{v}(\vec{r}) \lambda(\vec{r}') \vec{v}(\vec{r}') \rangle. \tag{45}$$

I have already argued that (if $\vec{v}$ is the peculiar velocity of the mass) the first of the four terms in eq. (45) is dominated by the primordial fluctuations (after the integrations over $\vec{r}$ and $\vec{r}'$, are performed). It is easy to see that the other two terms are negligible if λ is uncorrelated with $\vec{v}$ or if the only connected correlations that λ has with $\vec{v}$ is a two-point correlation.

If the objects being observed trace the mass

$$\lambda(\vec{r}) \propto \delta(\vec{r}). \tag{46}$$

It appears naively that in this case the non-linear time evolution will cause the last three terms in eq. (45) to contribute as significantly as the first (even after integrating $\vec{r}$ and $\vec{r}'$, over the large balls). Writing

$$\langle \delta(\vec{r})\vec{v}(\vec{r}) \cdot \vec{v}(\vec{r}') \rangle = \int \frac{d\vec{q}}{(2\pi)^3} \int \frac{d\vec{k}}{(2\pi)^3} \int \frac{d\vec{k}'}{(2\pi)^3} e^{i\vec{k}\cdot\vec{r}} e^{i\vec{k}'\cdot\vec{r}'} \cdot \langle \delta(\vec{q})\vec{v}(\vec{k}-\vec{q})\vec{v}(\vec{k}')\rangle, \tag{47}$$

the diagrams which diverge $1/k$ as k goes to zero and contribute to $\langle \delta(\vec{q})\vec{v}(\vec{k}-\vec{q})\vec{v}(\vec{k}')\rangle$ in the lowest non-trivial order of gravitational perturbation theory are shown in Fig. (7). (Note homogeneity implies $\vec{k}' = -\vec{k}$.) Each of these graphs diverges like $1/k$ as k goes to zero, however, when they are summed the most divergent piece cancels. This cancellation is a consequence of the equation of continuity.

$$\frac{\partial \delta}{\partial t} + \frac{1}{a}\vec{\nabla}(1+\delta)\vec{v} = 0. \tag{48}$$

$Fig.(7)$: Contributions to $\langle \delta(\vec{q})\vec{v}(\vec{k}-\vec{q})v(\vec{k}')\rangle$ that appear to diverge like $1/k$.

It is straightforward to show that at small k, $[(1+\delta)\vec{v}](k)$ is curl free (i.e. $(1+\delta)\vec{v}\times\vec{k} = 0$). This implies that

$$[(1+\delta)v](\vec{k}) = \frac{ia\vec{k}}{k^2}\frac{\partial\delta(\vec{k})}{\partial t}. \tag{49}$$

It has already been shown that at small k, δ is dominated by linear perturbation theory and eq. (49) implies that the same is true of $(1+\delta)\vec{v}$.

Although for objects which trace the mass, $\vec{v}(\vec{R})$ is dominated by linear perturbation theory that is not true is general. As a simple example imagine a biased model in which

$$\lambda(\vec{r}) = \delta_f(\vec{r}) + b\delta_f^2(\vec{r}) - b\langle\delta_f^2(\vec{r})\rangle \tag{50}$$

but with the peculiar velocities of the objects the same as that of the mass density fluid at the location of the objects. Now the second term in eq. (45) will be dominated by

$$b\langle \delta_f^2(\vec{r})\vec{v}(\vec{r})\vec{v}(\vec{r}')\rangle_c = b \int \frac{d\vec{q}}{(2\pi)^3} \int \frac{d\vec{p}}{(2\pi)^3} \int \frac{d\vec{k}'}{(2\pi)^3} \int \frac{d\vec{k}}{(2\pi)^3} \tilde{W}(\vec{p})\tilde{W}(\vec{q})$$

$$\cdot e^{i\vec{k}\cdot\vec{r}} e^{i\vec{k}'\cdot\vec{r}'} \langle \delta(\vec{p})\delta(\vec{q})\vec{v}(\vec{k}-\vec{p}-\vec{q})\vec{v}(\vec{k}')\rangle_c. \tag{51}$$

In the lowest non-trivial order of perturbation theory the graphs which give a contribution to (51) that goes like $1/k$ at small k are shown in Fig. (8). Explicit computation of these graphs reveals that the $1/k$ piece does not cancel. In addition they are sensitive to physics a short distances since large wavevectors $\vec{q}$ and $\vec{p}$ are important.

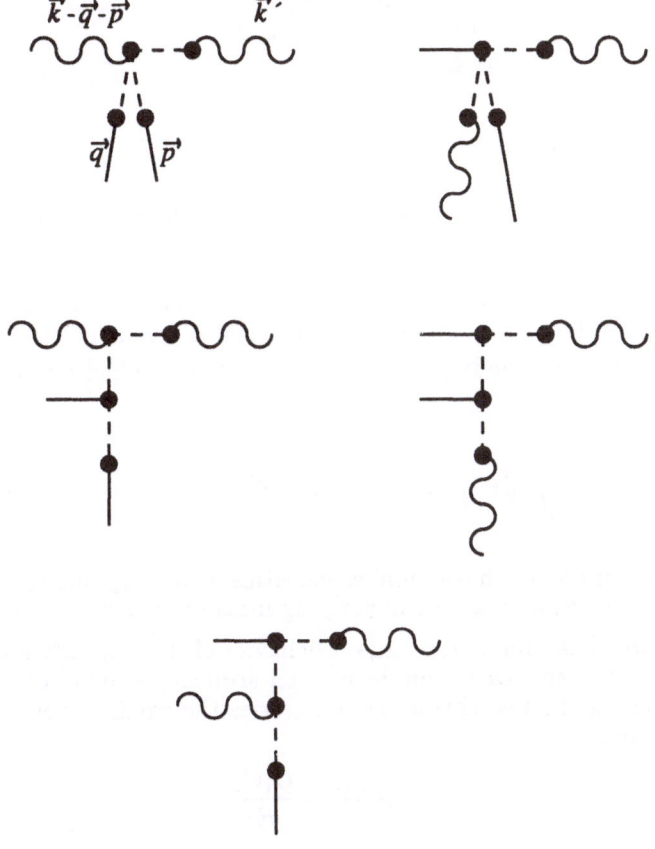

$Fig.(8)$: Contributions to $\langle \delta(\vec{p})\delta(\vec{q})\vec{v}(\vec{k}-\vec{p}-\vec{q})\vec{v}(\vec{k}')\rangle$ that diverge like $1/k$.

232

At higher orders in gravitational perturbation theory there are also contributions of order $1/k$ (e.g. Fig. (9)). Not every diagram has a $1/k$ piece however. The diagram in Fig. (10), for example, goes like k^4 as k goes to zero.

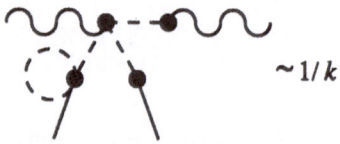

$\sim 1/k$

$Fig.$(9) : Higher order contribution to $\langle \delta(\vec{p})\delta(\vec{q})\vec{v}(\vec{k}-\vec{p}-\vec{q})\vec{v}(\vec{k}')\rangle$ that diverges like k^{-1} (for small k).

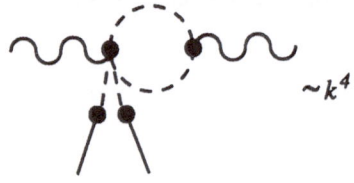

$\sim k^4$

$Fig.$(10) : Contribution to $\langle \delta(\vec{p})\delta(\vec{q})\vec{v}(\vec{k}-\vec{p}-\vec{q})\vec{v}(\vec{k}')\rangle$ that doesn't diverge as k goes to zero.

. Clearly $\vec{v}(R)$ is sensitive to the definition of the objects and to non-linear dynamics even for arbitrarily large R: different objects yield different $\vec{v}(R)$. The relative orientations of $\vec{v}(\vec{R})$'s corresponding to different objects is studied by examining the two-point correlation.

$$\langle \vec{v}_1(R) \cdot \vec{v}_2(R)\rangle = \frac{1}{\bar{n}_1 \bar{n}_2} \int d\vec{r} \int d\vec{r}' \, W_R(\vec{r}) W_R(\vec{r}') \, \langle n_1(\vec{r})\vec{v}(\vec{r}) \cdot n_2(\vec{r}')\vec{v}(\vec{r}')\rangle, \quad (52)$$

where the two types of objects have number densities n_1 and n_2 but are assumed to have the same peculiar velocity as the underlying mass density fluid.

The diagrams which dominate the large R behavior of this quantity have the form shown in Fig. (11). The shaded region denotes an arbitrary number of dotted lines. The dotted line joining the two shaded circles carries the small wavevector k. This implies that for large R

$$\langle \vec{v}_1(R) \cdot \vec{v}_2(R)\rangle = \frac{C_1 C_2}{R^2} \quad (53)$$

with C_1 and C_2 constants independent of R that depend on the definitions of objects one and two respectively.

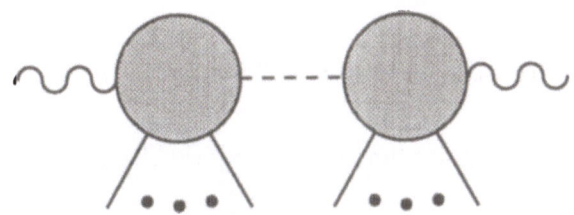

$Fig.(11)$: Generic diagram contributing to the two-point correlation of $(n\vec{v})(\vec{k})$ that diverges like $1/k$ for small k.

Therefore

$$\langle (C_2\vec{v}_1(R) - C_1\vec{v}_2(R))^2 \rangle = 0 \tag{54}$$

which means that

$$C_2\vec{v}_1(R) = C_1\vec{v}_2(R). \tag{55}$$

All the large scale streaming velocities $\vec{v}(R)$ are proportional to that of the mass. The constant of proportionality, however, gets contributions from non-linear gravitational dynamics. For objects that arise where mass density fluctuations are very large one expects the discrepancy between the large scale streaming velocity of the objects and that of the mass to be small if linear perturbation theory is valid on the comoving scale that collapsed to form the objects.

6. PRIMORDIAL NON-GAUSSIAN FLUCTUATIONS

Previously, in a few sections, I have explored the consequences of natural primordial Gaussian fluctuations. In this section the possibility of primordial non-Gaussian fluctuations will be examined. It will be shown that natural primordial non-Gaussian fluctuations can give rise to very different expectations for the large scale distribution of the observed luminous objects (e.g., galaxies or rich clusters of galaxies).

The strongest motivation for Gaussian primordial mass density fluctuations is the small value of the fluctuations (when averaged over the horizon volume). In the new inflationary cosmology this restricts the field that is doing the inflating to be very weakly coupled so that fluctuations in that field (which are equivalent to energy density fluctuations) are approximately Gaussian. However, there can be other contributions to the energy density fluctuations. Suppose that there is another field a that has a negligible mass in the inflationary era but in a later era develops a mass for dynamical reasons. At late times the field a contributes to the mass density but does not dominate it. Even if the field a contributes a small amount to the mass density quantum fluctuations in the field a, generated in the inflationary era, could dominate the fluctuations in the mass density if they are not small. Thus the field a is not restricted to be weakly coupled in the inflationary era and the fluctuations in the mass density that it gives rise to can be highly non-Gaussian.

For scale invariant fluctuations (as shown in eq. (8)),

$$\langle \epsilon(\lambda \vec{k}_1) \ldots \epsilon(\lambda \vec{k}_n) \rangle = \lambda^{-n} \langle \langle \epsilon(\vec{k}_1) \ldots \epsilon(\vec{k}_n) \rangle \rangle. \tag{8}$$

This is determined the two-point function, up to normalization (as shown in eq. (9)),

$$\langle \epsilon(\vec{k}_1) \epsilon(\vec{k}_2) \rangle \propto k_1 \delta^3(\vec{k}_1 + \vec{k}_2) \tag{9}$$

However the higher correlations are not uniquely determined. Writing

$$\langle \epsilon(\vec{k}_1) \ldots \epsilon(\vec{k}_n) \rangle_c = \xi_n(\vec{k}_1, \ldots \vec{k}_n)(2\pi)^3 \delta^3(\vec{k}_1 + \ldots + \vec{k}_n), \tag{56}$$

the function $\xi_n(\vec{k}_1, \ldots \vec{k}_n)$ is restricted by eq.(8) to be homogeneous of degree $3 - n$ (of course it must also be a symmetric rotationally invariant function of the wavevectors $\vec{k}_1, \ldots, \vec{k}_n$). It is easy to generalize the diagrammatic rules introduced in section (4) to the case when the primordial fluctuations are non-Gaussian. One must add new vertices that take into account the fact that the ϵ's can now be sewn together with the correlations in eq. (56). The new vertices and the factors associated with them are shown in Fig. (12). All the dotted lines that are joined to these new vertices must originate from a solid (or wavy) line. The diagrammatic rules forbid there to be dotted lines joining the new vertices to each other. Unfortunately, it is difficult to make any general statements about the effect of non-linear gravitational evolution on primordial non-Gaussian fluctuations, without a particular model for these fluctuations.

$$\frac{(2\pi)^{3(1-n)} \delta^3(\vec{q}_1 + \cdots + \vec{q}_n)}{n! \, q_1 \sqrt{A(q_1)} \cdots q_n \sqrt{A(q_n)}} \xi_n(\vec{q}_1, \cdots, \vec{q}_n)$$

$Fig.(12)$: New vertices for primordial non-Gaussian fluctuations.

Not every set of correlations (56) that are consistent with the principle of scale invariance (8) and the homogeneity and isotropy of space is permitted. One must be careful to make sure that the $\xi_n(\vec{k}_1, \ldots \vec{k}_n)$ are consistent with the positivity of the probability distribution for the primordial mass density fluctuations. For example, positivity of the probability distribution forbids any of the $\xi_n(\vec{k}_1, \ldots \vec{k}_n)$ to diverge like k_i^{-p}, $p > 1$, as a single wavevector $\vec{k}_i$ gets small. (Strictly speaking I assume this divergence does not vanish when integrated against a function of k_i, therefore my remarks don't apply to a divergence of the form $\vec{k}_i/k_i^{p+1}$.) To see this, note that positivity of the probability distribution for the primordial fluctuations means that

$$\langle f(\epsilon)g(\epsilon)\rangle^2 \leq \langle\{f(\epsilon)\}^2\rangle\langle\{g(\epsilon)\}^2\rangle \tag{57}$$

for any two functions f and g. Suppose now that a connected m-point function diverges as k_i^{-p} as one of its arguments gets small. Choose

$$f(\epsilon) = \int d\vec{k}\epsilon^{-k^2 R^2} \epsilon(\vec{k}) \tag{58}$$

and

$$g(\epsilon) = \left\{\int d\vec{q}\epsilon^{-q^2 r^2}\epsilon(\vec{q})\right\}^{m-1}. \tag{59}$$

Putting this into eq. (57) it is easy to see that for R large (but r fixed) the divergence of the connected m-point function as one of its arguments gets small, causes the left-hand side to scale as R^{2p-6}. The R behavior of the right-hand side however, is just determined by the primordial two-point function. It scales like R^{-4} for large R. Therefore the inequality in eq. (57) will not be satisfied for very large R if $p > 1$.

Primordial non-Gaussian could have a significant effect on the power spectrum for the two-point correlation of biased objects[17,19]. Recall that in section (5) it was shown that if one of the correlations $\xi_n(\vec{k}_1, \ldots \vec{k}_n)$ diverges as a partial sum of its wavevectors $\vec{k}_T = \vec{k}_1 + \ldots + \vec{k}_i$, $i < n$, gets small like k_T^{-p}, then the power spectrum for the objects (typically) diverges like k^{-p} for small k. The case $p = 1$ is very interesting since in coordinate space this corresponds to a two-point function that falls like r^{-2} for large r. This could explain the significant correlations of rich clusters of galaxies at large distances. In ref. (17) it was shown that it is possible to construct a natural positive probability distribution that gives rise to this behavior. This model was quite awkward and it required a fine tuning of counter terms to preserve the naturalness of the probability distribution (in the sense of eq. (8)). It is not clear how general this difficulty is. It is worth noting however, that models for the probability distribution of the primordial fluctuations that do not require a fine tuning of counter terms, can easily be constructed to give a primordial power spectrum for objects that goes to a constant at small wavenumbers. This may be enough to make the zero crossing of the two-point correlation for rich clusters of Galaxies occur at distances significantly larger than $17h^{-2}$ Mpc in a model where they occur at peaks of the filtered mass density field that are unusually high.

APPENDIX[11]

The Zeldovich approximation[10] takes the Lagrangian coordinates of the mass to be

$$\vec{r}(\vec{s},t) = a(t)\vec{s} + b(t)\vec{\alpha}(\vec{s}), \qquad (A.1)$$

where the mass is evenly distributed in $\vec{s}$-space. The comoving coordinates are obtained by dividing by the Robertson Walker scale factor $a(t)$. In eq (A.1) $b(t) = t^{2/3}a(t)$. Writing the peculiar velocity as

$$\vec{v} = a(t)\frac{d}{dt}(\vec{r}/a(t)) = \frac{2}{3}\left(\frac{a(t)}{t}\right)\vec{p}. \qquad (A.2)$$

Eq.(A.1) implies that in the Zeldovich approximation $\vec{p} = \vec{\alpha}(\vec{s})$. To show that eqs. (20) and (21) are the Eulerian equivalent of the Zeldovich approximation (A.1), it is convenient to introduce a phase space distribution; $f(\vec{x},\vec{p}:t)d\vec{x}d\vec{p}$ is the number of particles (which I assume to have mass m) in the six-dimensional phase space volume $d\vec{x}d\vec{p}$ at the time t. In the Zeldovich approximation

$$f(\vec{x},\vec{p}:t) = \frac{\langle\rho\rangle}{m}\delta^3(\vec{p} - \vec{\alpha}(\vec{s})) \qquad (A.3)$$

with

$$\vec{s} = \vec{x} - (b(t)/a(t))\,\vec{p}. \qquad (A.4)$$

Before there is orbit crossing one can expand the Dirac δ function yielding

$$f(\vec{x},\vec{p}:t) = \frac{\langle\rho\rangle}{m}\sum_{n=0}^{\infty}\frac{(-1)^n}{n!}\alpha_{i_1}\ldots\alpha_{i_n}\frac{\partial}{\partial p_{i_1}}\ldots\frac{\partial}{\partial p_{i_n}}\delta^3(\vec{p}). \qquad (A.5)$$

Putting this into the equation for the mass density

$$\rho(\vec{x},t) = m\int f(\vec{x},\vec{p}:t)d\vec{p}, \qquad (A.6)$$

integrating by parts to remove the derivatives from the delta functions and using eq. (A.4) gives

$$\frac{\rho(\vec{x},t)}{\langle\rho\rangle} = \sum_{n=0}^{\infty}\frac{(-1)^n}{n!}\left(\frac{b(t)}{a(t)}\right)^n\frac{\partial}{\partial x_{i_1}},\ldots\frac{\partial}{\partial x_{i_n}}(\alpha_{i_1}(\vec{x})\ldots\alpha_{i_n}(\vec{x})). \qquad (A.7)$$

The $n = 1$ term in the sum corresponds to linear perturbation theory and it implies that $\epsilon_1(\vec{x}) = -\vec{\nabla}\cdot\vec{\alpha}$. Eqs. (20) and (21) follow (when $\vec{\alpha}$ is curl free) from Fourier transforming eq. (A.7).

For a one dimensional perturbation

$$x = a(t)s_1 + b(t)\alpha_1(s_1)$$
$$y = a(t)s_2 \tag{A.8}$$
$$z = a(t)s_3$$

where $\vec{r} = (x, y, z)$. In this case the Zeldovich approximation in Eulerian coordinates is given by eqs. (20) and (21) where the wavevectors are replaced by scalars and vector dot products by ordinary multiplication, (one only Fourier transforms the coordinate x since the density perturbation is independent of the y and z coordinates). The Zeldovich approximation is actually the exact solution in the case of one-dimensional perturbations. The equations of motion that the exact solution must satisfy are

$$\nabla^2\phi = -4\pi G\rho, \quad \frac{\partial^2\vec{r}}{\partial t^2} = \vec{\nabla}\phi \tag{A.9}$$

where the gradient is derivatives with respect to x, y, and z. Using the fact that $\ddot{b}/b = -2\ddot{a}/a$ and eq. (A.8) it is straightforward to show that

$$\vec{\nabla} \cdot \frac{\partial^2\vec{r}}{\partial t^2} = \left(\frac{3\ddot{a}}{a}\right)\frac{1}{1 + (b/a)(d\alpha_1/ds_1)}. \tag{A.10}$$

Since the particles are evenly distributed in $\vec{s}$-space the mass density is proportional to the inverse of the Jacobian for the transformation from x, y, z coordinates to s_1, s_2, s_3 coordinates. Therefore

$$\rho = \langle\rho\rangle\frac{1}{1 + (b/a)(d\alpha_1/ds_1)}. \tag{A.11}$$

Since $4\pi G\langle\rho\rangle = -3\ddot{a}/a$ eq. (A.10) implies that the Zeldovich approximation solves the equations of motion (A.9).

REFERENCES

1. The suggestion that primordial non-Gaussian fluctuations might have an interesting effect on the large scale distribution of objects was originally given in: Peebles, P.J.E., *Ap.J.* **274**, 1 (1983).

2. Guth, A. *Phys. Rev.* **D23**, 347, (1981).

3. For an exposition on a gauge invariant formulation for density perturbations see: Bardeen, J.M., *Phys. Rev.* **D22**, 1882, (1980).

4. For a review of functional integration and a discussion of its application to probability theory see: Feynman, R.P. and Hibbs, A.R., *Quantum Mechanics and Path Integrals*, McGraw Hill, New York (1965).

5. Otto, S., Preskill, J., Politzer, H.D. and Wise, M.B., *Ap.J.* **304**, 62, (1986).

6. Harrison, E., *Phys. Rev.*, **D1**, 2726 (1970); Peebles, P.J.E. and Yu J., *Ap. J.* **162**, 815 (1970); Zeldovich, Ya.B., *M.N.R.A.S.*, **160**, 1P, (1972).

7. Linde, A.D. *Phys. Lett.*, **108B**, 289 (1982); Albrecht, A. and Steinhardt, P.J., *Phys. Rev. Lett.* **48**, 1220 (1982).

8. Guth, A. and Pi, S.Y.,*Phys. Rev. Lett.* **49**, 1110 (1982); Bardeen, J.M., Steinhardt, P.J. and Turner, M., *Phys. Rev.*, **D28**, 679 (1983); Starobinaskii, A., *Phys. Lett.*, **117B**, 175 (1982); Hawking, S., Phys. Lett., **115B**, 295, (1982).

9. The formalism discussed in this section was developed in; Goroff, M., Grinstein, B., Rey, S-Y. and Wise, M.B., *Ap. J.* **311**, 6, (1986). See also Fry, J., *Ap. J.* **279**, 499 (1984).

10. Zeldovich, Ya. B. *Astron and Astrophys*, **5**, 84 (1970).

11. Grinstein, B., Kamien, R., and Wise, M.B. (1986), unpublished.

12. Hauser, M. and Peebles, P.J.E., *Ap. J.*, **185**, 757 (1973); Kopolov, A. and Klypin, A., *Soviet. Astr. Lett.*, **9**, 41 (1983); Bachall, N. and Soniera, R., *Ap. J.*, **276**, 20 (1983).

13. Kaiser, N., *Ap. J. Lett.*, **284**, L9 (1984).

14. Politzer, H.D. and Wise, M.B., *Ap. J. Lett.*, **285**, L1, (1984).

15. Bardeen, J.M., Bond, J.R., Kaiser, N. and Szalay, A.S., *Ap. J.*, **304**, 15 (1986).

16. Otto, S., Politzer, H.D., Wise, M.B., *Phys. Rev. Lett.*, **56**, 1878 (1986).

17. Grinstein, B. and Wise, M.B., *Ap. J.*, **310**, 19 (1986).

18. This discussion follows from the paper: Grinstein, B., Politzer, H.D., Rey, S-J. and Wise, M.B., to appear in *Ap.J.*, **314** (1987).

19. Matarrese, S., Lucchin, F., Bonometto, S.A., *Ap.J. Lett.*, **310**, L21 (1986).

20. Zeldovich, Ya. B., 'Fluid Dynamics Transactions'; Volume **8**, *Polish Scientific Publishers*, Warsaw Poland (1976).

N-BODY METHODS AND THE FORMATION OF LARGE-SCALE STRUCTURE

Simon D.M. White
Steward Observatory
University of Arizona
Tucson, AZ 85721
U.S.A.

ABSTRACT. I review the N-body techniques that have been used to study the evolution of large-scale structure in the Universe. After a brief summary of the initial conditions expected in universes dominated by weakly interacting massive particles, I discuss the nonlinear structure found in neutrino-dominated and cold dark matter dominated models. The cold dark matter model currently appears the most attractive possibility and is able to reproduce observed structures from galaxy halos up to rich galaxy clusters. The distribution of galaxy formation sites in this model may be biased in the manner required to make a flat universe consistent with dynamical constraints from galaxy groups and clusters and from the Local Supercluster.

1. N-BODY METHODS

In recent years gravitational N-body simulations have become the standard tool for comparing the observed structure of the universe with the predictions of various theories for its early evolution. Such theories usually characterise structure by decomposing the density fluctuations into a superposition of small amplitude plane waves, and then specifying the relative amplitude of waves of different scale. While observations of the microwave background can provide, at least in principle, a direct test of such theoretical predictions, any confrontation with the wealth of data on the large-scale galaxy distribution requires some treatment of nonlinear evolution. N-body methods are an attractive way to do this for two reasons. They allow nonlinear gravitational effects to be handled in complete generality. In addition, each model can be set up so that its initial conditions are a random realisation of the statistical process predicted to result from earlier evolution; its final state can then be considered to represent an appropriately sized random volume which can be studied in exactly the same way as observers study nearby volumes of real space.

The attractive features of N-body methods are offset by both technical and physical limitations. Technical limitations stem primarily from the limited resolution of calculations that can be attempted on present computers; they can often be avoided by carefully designed experiments. More fundamental physical limitations cannot be avoided, however, because it is clear that gravitational processes are not solely responsible for the structure we see. Hydrodynamic shocks and radiative processes play a crucial role in the formation of individual stars and are very

W. G. Unruh and G. W. Semenoff (eds.), The Early Universe, 239–260.

likely important in determining the structure of galaxies. They thus determine how the matter distribution is lit up for our appreciation. In addition, several mechanisms have recently been suggested by which such processes could produce all the observed large-scale structure (Ostriker and Cowie 1981; Ikeuchi 1981; Hogan and Kaiser 1983; Hogan and White 1986). If any of these mechanisms proves viable, the present galaxy distribution may contain no information about the early universe. Even if they fail, the connection between what we see and the true large- scale mass distribution remains the major uncertainty in using simulations to confront theory with observation.

An N-body model for the evolution of cosmic structure represents the mass distribution in some region of the universe by a finite number, N, of particles which are assumed to move under the influence of their mutual gravitational interactions. The value of N is limited by the scheme used to solve for the gravitational forces and by the size of the computer available. It is the primary limitation on the resolution available. For example a cubic region of our universe 6000 km/s on a side has about the volume of the CfA redshift survey and contains about $10^{17} M_\odot$ (for $\Omega h^2 = 0.25$). Thus even with $N = 10^6$ the mass of each particle is about equal to that contained in our Galaxy within the solar radius. This radius corresponds to a Hubble velocity of only 0.5 km/s; as a result a simulation of the large region would need to model structures 10^4 times smaller than the side of the cube in order to make full use of the available mass resolution. Such a large dynamic range cannot be attained for such large N by any current hardware/software combination.

For relatively small N ($\lesssim 10^4$) it is feasible to calculate forces by a direct sum over all particle pairs. There is then no limit, in principle, to the accuracy with which the forces can be obtained. However, close interactions between pairs of particles have no equivalent in the physical system being simulated and can cause substantial evolution of clumps made up of tens or hundreds of particles. It is clearly desirable to suppress such unwanted two-body relaxation effects. This goal can be partially achieved by an artificial reduction or "softening" of the short-range gravitational force. A frequently used softening is based on the pairwise interaction potential

$$\phi_{ij} = Gm_i m_j (r^2 + \epsilon^2)^{-1/2} \tag{1}$$

The softening length, ϵ, limits the force resolution on small scales, and is normally chosen to be of order the mean interparticle separation in the densest multiparticle clumps expected in a calculation. Such softening leads to a modest reduction in the rate of diffusive relaxation by distant encounters; however, it eliminates a number of more dramatic unwanted effects associated with close encounters of pairs and triplets of particles. For example, bound pairs of particles are no longer a significant sink for the binding energy of clumps and are unable to have strong superelastic encounters with "field" particles. In addition, the elimination of strong and rapidly changing forces in close encounters greatly enhances the numerical stability of the integration of the equations of motion. Aarseth (1984) gives a detailed discussion of the numerical methods available for carrying out this kind of simulation. Recent cosmological work using this technique is presented by Dekel, West and Aarseth (1984) and Quinn, Salmon and Zurek (1986); I will not discuss it further here.

For larger values of N, forces must be obtained by a finite difference solution of Poisson's equation on a density grid derived from the particle positions. For the cosmological problem, equal resolution is required *a priori* at every point in space, and so a cubic mesh with periodic boundary conditions provides a simple and appropriate way to represent the density distribution. In addition, fast algorithms are available for the solution of the field equations on such a mesh. A straightforward way to solve for the grid potential is to carry out a Fast Fourier Transform (Cooley and Tukey 1965) on the density grid, followed by multiplication by a Green's function and an inverse transform. As discussed in detail by Hockney and Eastwood (1981), more efficient algorithms are available for this problem. However, they are more difficult to program, they vectorise less readily on current supercomputers, they do not allow the freedom to adjust the Green's function as desired, and they do not yield the power spectrum of the density distribution as an intermediate result. All published cosmological simulations which use grids have been based on Fourier Poisson solvers.

There are three stages to obtaining the forces by grid methods. First the particle positions are used to assign a density to each grid point. Then a grid potential is obtained using the Poisson solver. Finally this potential is differenced and interpolated to the particle positions in order to get the forces. The resulting force between any pair of particles depends not only on their separation, but also on their location and orientation relative to the grid. On scales comparable to the grid spacing, these anisotropies can become quite large. In addition the typical force between two particles falls below that expected for point particles, implying an effective softening of the interaction law. A variety of choices are possible for the density assignment, differencing and interpolation schemes, as well as for the Green's function used by the Poisson solver. Typically the smoother the density assignment or the potential differencing, the smaller the force anisotropies, the greater the effective softening of the force, and the more computer time needed to calculate forces. Further, a proper matching of density assignment scheme and differencing and interpolation scheme can produce interparticle forces which satisfy Newton's Third Law, and can ensure that particles do not exert forces on themselves. As a consequence, such a choice results in particle motions which conserve linear momentum (for sufficiently accurate time integration). An alternative choice can produce a scheme in which energy is conserved. Unfortunately no simple choice has both characteristics, although high order schemes can approach this ideal at the expense of a very soft force and considerable computer time. These questions are discussed by Hockney and Eastwood (1981) in general, and by Efstathiou et al. (1985; hereafter EDFW) in the specific context of cosmological simulations. The latter authors conclude that a momentum-conserving, cloud-in-cell scheme offers the best compromise between resolution, force anisotropy and expense, but a survey of the literature shows that this choice has been far from uniformly adopted.

The force resolution of the particle-mesh (PM) methods can be improved by including force corrections obtained by direct summation over all particle pairs for which the mesh force is significantly softened. Such schemes were first developed by Eastwood and Hockney (see Hockney and Eastwood 1981) and are known as P^{3}M methods. Their adaptation to cosmological problems was discussed by Efstathiou and Eastwood (1981) and EDFW. The calculation of the force corrections is time-consuming, particularly for highly clustered distributions, and so considerably reduces the number of particles that can be followed. However, provided the underlying PM scheme is high enough order for force anisotropies to be small, P^{3}M Poisson solvers can give accurate inverse square law interparticle forces to arbitrarily small scale. In practice a softened potential with characteristic scale a small fraction of the grid spacing is usually adopted; as discussed above, this avoids some of the relaxation and integration difficulties associated with small N subsystems and hard potentials. With presently available supercomputers (e.g. a CRAY-XMP with at least 4 Megawords of main memory) a PM simulation with a million particles on a 128 grid can be carried out in a few hours. A P³M simulation with arbitrarily high force resolution but five times fewer particles can be carried out for about the same cost per timestep; however, shorter timesteps are likely to be needed to maintain adequate accuracy.

For large simulations, computer memory limitations usually require a time-centred leapfrog or other similar low order scheme for integrating particle trajectories. Integration efficiency can be improved by an appropriate choice of time variable, although the results are not very sensitive to this choice (EDFW). The overall accuracy of a simulation can be gauged using the Layzer-Irvine cosmic energy equation (Layzer 1963, Irvine 1965) in one of the integral forms:

$$a^4T + aU - \int U da = \text{const.}$$

or

$$a^3T + U + \int a^2 T da = \text{const.}$$

(2)

In these equations T is the mean cosmic kinetic energy density, U is the mean gravitational potential energy density in excess of that for a uniform distribution, and a is the expansion factor of the universe. Precise definitions and derivations can be found in Peebles (1980). These equations remain valid for softened force laws if the softening scale is taken to expand with the universe, but some care is needed to determine the appropriate interaction potential when calculating U for PM schemes (EDFW). With a small enough time step and an appropriate definition of U, a momentum-conserving CIC scheme typically satisfies these conservation equations to about 0.5% of the potential energy term. For P³M schemes this number can be arbitrarily small, but for the parameters suggested by Efstathiou et al. it is about 0.1%. Note that this is rather a crude measure of numerical accuracy. Efstathiou et al. show that errors in the individual terms of equations (2) have a strong tendency to cancel and can be up to 10 times the size of the error in their right-hand sides. Furthermore, integration errors may cause certain scientifically interesting regions of a simulation (for example, the cores of dense clusters) to be poorly modelled without noticeably affecting the total energy.

Three major criteria determine the utility of cosmological simulations. The initial conditions should correspond as closely as possible to those predicted by the theoretical framework to be investigated. The code should follow evolution from these initial conditions without undue distortion from artificial effects associated with softening of the forces, the discreteness of particles, or the finite size of the model. Finally some well defined procedure is required to relate the final distribution to observables such as the galaxy distribution.

For some theoretical models specification of the inital conditions is a major obstacle to carrying out realistic simulations. For example, in most versions of the explosive theories first proposed by Ostriker and Cowie (1981) and Ikeuchi (1981), structure on large scales forms by gravitational amplification of the irregularities produced by blast waves associated with galaxy formation. Although this final phase of evolution is, in principle, amenable to direct simulation, the preceding explosive phase is too poorly understood to provide general initial conditions. In this case N-body methods may be useful to investigate various idealised problems (for example the stability of an expanding shell or of the interface between two colliding shells) but they cannot provide a general final distribution for direct comparison with observation. Similarly, because galaxy formation is directly related to the explosions and the nonlinear instability of shells, it is very uncertain how the observable galaxy distribution should be related to the mass distribution in such models.

The situation is simpler in more traditional models, where galaxies and large-scale structure grow directly from linear perturbations. It is particularly simple if weakly interacting, massive particles dominate the mass, since gravitational effects alone can structure their distribution. Although hydrodynamic and radiative processes may affect the baryonic gas it seems likely that the associated potential fluctuations will be too small to affect the dark matter. To set up initial conditions for a simulation of such a model, particle positions and velocities must be chosen so that the deviation from a uniform, uniformly expanding distribution is a good representation of linear, growing mode fluctuations with random phase and the desired power spectrum. This can be accomplished by distributing particles on a comoving lattice and then assigning them displacements, $\mathbf{x}$, and peculiar velocities, $\dot{\mathbf{x}}$, according to Zel'dovich's (1970) general Lagrangian formulation of linear theory:

$$\mathbf{x}(\mathbf{q}) = -b\nabla U; \quad \dot{\mathbf{x}}(\mathbf{q}) = -\dot{b}\nabla U. \tag{3}$$

Here $b(t)$ is the amplitude of the growing mode and $U(\mathbf{q})$ is the potential at Lagrangian coordinate, $\mathbf{q}$, obtained from a particular random phase realisation of the linear density field (EDFW). The longest wavelength mode that can be imposed corresponds, of course, to the side, L, of the computational volume. The shortest mode corresponds to the Nyquist frequency of the *particle* grid, $2N^{-1/3}L$. For 10^6 particles this gives a dynamic range in length of 50 in the initial conditions, corresponding to a dynamic range of 10^5 in mass. As a simulation evolves, the reliable dynamic range in mass cannot increase. However, clustering increases the reliable dynamic range in length, because bound systems collapse and cease to expand with the computational volume. As a result the major resolution limit in evolved simulations usually comes from softening of the forces rather than from discreteness effects. For this reason P^{3}M codes often give better results than PM codes even though they cannot handle as many particles.

Figure 1 shows some tests of the limitations of N-body methods. Figure 1a compares the result of evolving the same 32768 particle distribution with a cloud-in-cell PM code on a 32 grid and with a P^{3}M code with 8 times better linear resolution. The initial condition had no power imposed for waves with k outside a cube of side twice the fundamental frequency. At the final time the distribution contains large pancakes, filaments and clusters (see Fig. 5 of EDFW) which correspond in the power spectra to a nonlinear transfer of power to high frequencies. Very strong phase correlations are present. There is a substantial power deficit at high frequencies in the PM model, resulting from its poor force resolution. Such a model is clearly only suitable for studying structure on large scales.

Figure 1b shows the result of setting up the same realisation of a near power law power spectrum using $N = 32768$ and $N = 8000$. Apart from the normalisation difference, the distributions have the same spectrum at frequencies below the Nyquist frequency ($k = 16$ and $k = 10$ in the two cases); at higher frequencies the power spectrum approaches the white noise level. The later output times in these two P^{3}M model differ by the amount required for linear theory to predict they should have the same clustering amplitude. This expectation is extremely well fulfilled, showing that the discreteness of the particles has had almost no effect on the power spectrum. The initial difference in high frequency power has been swamped by power generated from lower frequencies by nonlinear effects. Notice that each individual structure is represented by 4 times fewer particles in the smaller N model; as a result its 2-body relaxation time is 4 times shorter compared to the overall evolution time. Despite this, 2-body effects lead to no obvious difference between the power spectra or other statistical properties of the models.

Figure 1c shows a test of the influence of the finite size of the computational volume. It compares the power spectra of two ensembles of P^3M models designed to study universes filled with cold dark matter. In one set the side of the computational volume is twice that of the other set. The later times were chosen so that linear theory would require the spectra to coincide where they overlap. At high frequency a systematic difference is visible which is due to the manner in which the power spectra were calculated. However, at lower frequencies the agreement is good. Thus in this case edge effects (i.e. the absence of longer wavelength modes) have not influenced the evolution of the Fourier components accessible to the smaller calculations. Similar tests for models in which the low frequency modes reach larger amplitudes do show substantial power deficits which result from the lack of coupling to longer wavelength modes.

Empirical tests like those in Figure 1 are necessary to delineate the range of validity of any set of cosmological simulations. It is clear that reliable results can be obtained over a moderate range of scales, and for certain types of physical model. N-body models can do little to support or refute theories which do not provide well specified initial conditions, or in which nongravitational effects play a determining role.

2. LINEAR FLUCTUATIONS

The initial conditions for a nonlinear simulation of structure formation must be taken from some theory for the origin of structure together with calculations of the evolution prior to the relatively late epochs that can be treated by N-body methods. The linear density fluctuation field is usually specified in terms of $\delta(r) = \rho(r)/\overline{\rho} - 1$, or of its Fourier transform δ_k. The fluctuation field at relatively late times (i.e. after recombination) may be thought of as

$$\delta_k(t) = T(k,t)(\delta_k)_p, \tag{4}$$

where $(\delta_k)_p$ is the field generated initially, and $T(k,t)$ is a transfer function describing its subsequent evolution. Because it seems unnatural that the generation process should have a characteristic scale in the range of interest here, the primordial power spectrum is usually taken to be a power law,

$$|\delta_k|^2 \propto k^n. \tag{5}$$

If the Fourier components of the fluctuation field are assumed to have random phase, its statistical properties are completely specified by Eq. (3). This random phase assumption seems natural and is almost always employed, but it can be rigorously justified only for specific models of the origin and evolution of fluctuations. Fluctuations generated by quantum effects during inflation do have random phase, and their power spectrum is of the form (3) with the index taking the value $n = 1$. This is the Harrison-Zeldovich "constant curvature spectrum" and has the property that

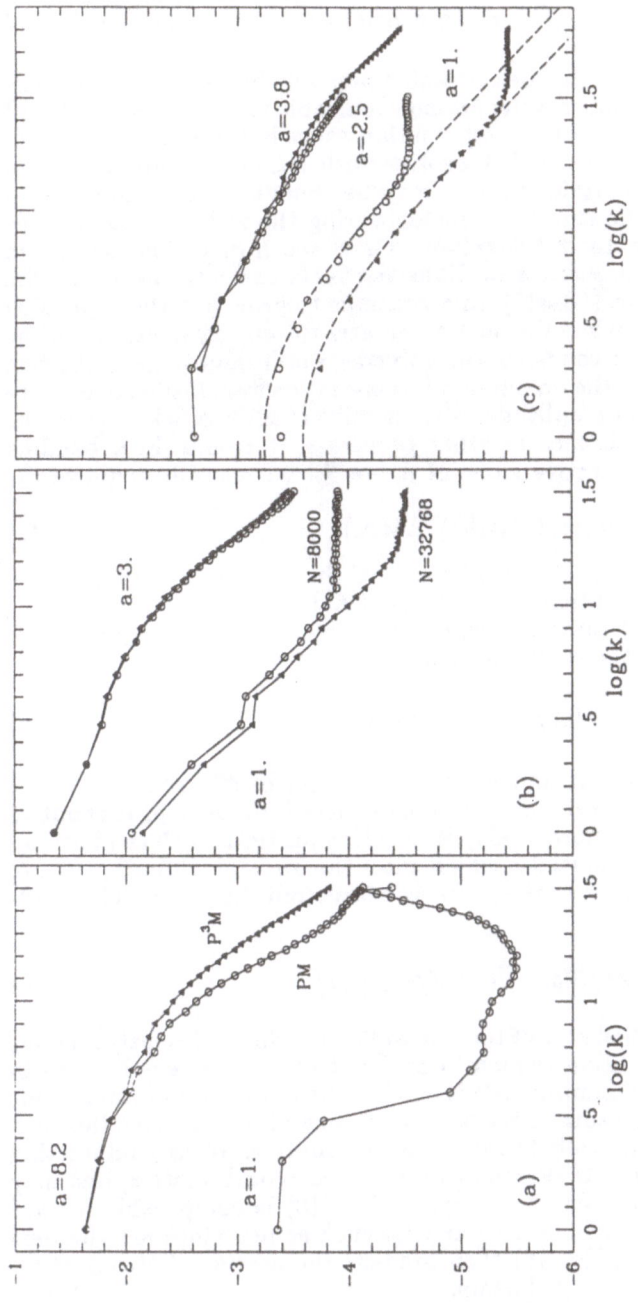

Fig. 1 Power spectra illustrating the tests of N-body methods discussed in the text. In (a) the same initial conditions were evolved using codes with high and low resolution force calculations. Excessive force softening results in inadequate modeling of small-scale (high frequency) structures. In (b) the same initial fluctuation distribution was modeled using different numbers of particles. The expansion factor at the later time is 2.7 rather than 3 for the 8000-particle model in order to compensate for its larger initial fluctuation amplitude. In (c) two ensembles of simulations are compared which model the theoretical spectrum shown as a dashed curve over regimes differing by a factor of two in scale. The difference in the later times compensates for the different initial fluctuation amplitudes.

the rms gravitational potential depth of fluctuations is independent of their scale. Thus in its simple form the inflationary model provides a complete specification of $(\delta_k)_p$.

The transfer function $T(k,t)$ can be calculated precisely for any chosen particle mix for such fluctuations. Reliable results are now available for all cases of interest here. When the wavelength of a given perturbation exceeds the particle horizon, causal effects cannot act across it, and it evolves with a constant fluctuation in curvature. The associated density fluctuation may be thought of as amplifying in proportion to the square of the expansion factor during the radiation-dominated era, and in direct proportion to it thereafter. Once the horizon has grown to encompass the perturbation, pressure and diffusive effects can alter its evolution. Perturbations in a pressure-free ("dust") fluid continue to grow like the expansion factor if they enter the horizon when the dust dominates the energy density. On the other hand, if they enter the horizon when the universe is still radiation-dominated, their growth is arrested because the dominant photon-baryon fluid begins to execute acoustic oscillations at constant amplitude. This slowing of growth is known as the Meszaros effect, and, in the absence of other processes, it results in a bending of the power spectrum to an effective slope of $n-4$ for wavenumbers above the characteristic value

$$k_{eq} = 2\pi/(13(\Omega h^2)^{-1}Mpc), \qquad (6)$$

corresponding to the present size of the diameter of the horizon at the epoch of equal density of matter and radiation ($z_{eq} = 2.5.10^4 \Omega h^2$).

A collisionless particle of mass m, which was once in thermal equilibrium, becomes nonrelativistic when the radiation temperature is,

$$T_c = (g_*/3.9)^{1/3}mc^2/k_B, \qquad (7)$$

where k_B is Boltzmann's constant and g_* is the number of effective degrees of freedom of the relativistic species present when the particle decoupled; g_* is about 11 for neutrinos, and about 100 for more weakly interacting particles such as photinos. Fluctuations in the collisionless particles which enter the horizon while $T > T_c$ are wiped out by relativistic streaming of the particles away from the peaks. This leads to a characteristic damping scale,

$$k_d = 2\pi/(6g_*^{-1/3}(30eV/m)Mpc), \qquad (8)$$

corresponding to the present diameter of the horizon at T_c. For sufficiently massive particles, or for particles like axions which formed out of thermal equilibrium in states of low momentum, this damping scale is smaller than any scale of interest for the formation of large-scale structure. Because of their small thermal velocities such dark matter candidates are generically termed cold dark matter. Weakly interacting particles with $g_* \sim 100$ and $m \sim 1keV$ might provide warm dark matter, but they are currently out of fashion; the damping scale of Eq. (8) is comparable to that of a bright galaxy in this case. Abundant particles such as neutrinos are the best candidates for hot dark matter; Eq. (8) then predicts the erasing of structure on all scales below those of large galaxy clusters.

Figure 2 shows power spectra at late times for adiabatic, constant curvature initial fluctuations imposed on a universe now dominated by collisionless particles. I have plotted $k^3|\delta_k|^2$ in this diagram, because this quantity is closely related to the amplitude of density fluctuations averaged over regions of size k^{-1}, and thus gives an

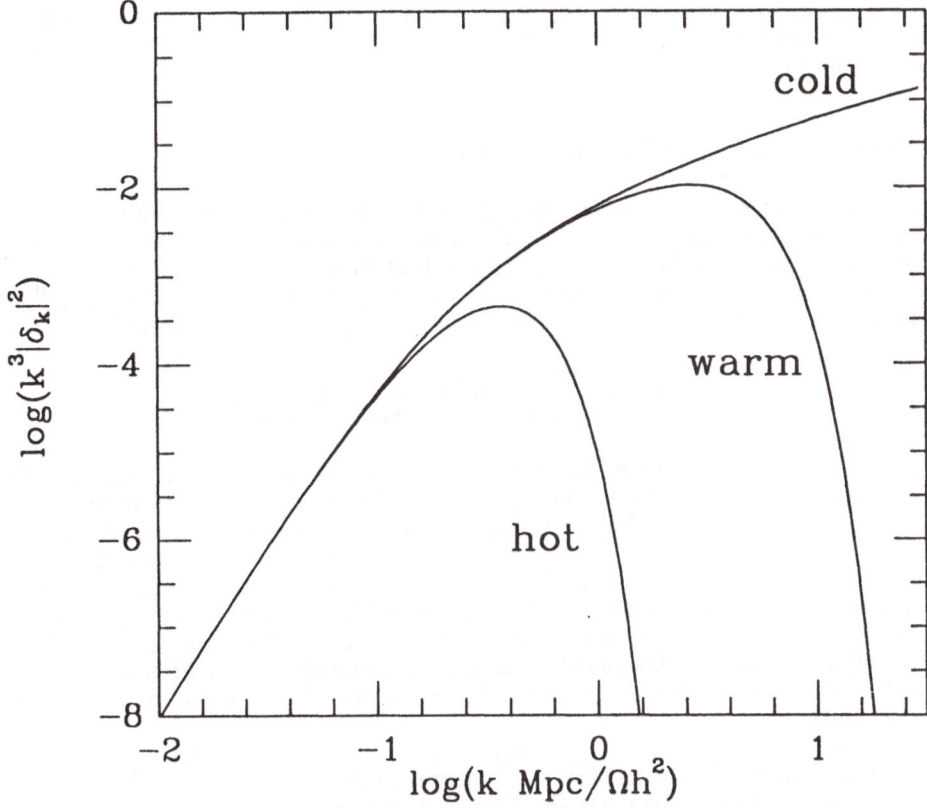

Fig. 2. The power per decade as a function of spatial frequency for the density fluctuations expected in a universe dominated by collisionless elementary particles. These are the linear power spectra at late times in a universe which initially had the adiabatic, constant curvature fluctuations predicted by inflationary models. The three cases shown are differentiated by the magnitude of the random velocities of the particles. These results are taken from work by Bond and Szalay (1983) and Bond and Efstathiou (1984).

indication of the order in which objects of different scale fragment out of the general expansion. For cold dark matter the curve has no peak; the only scale corresponds to k_{eq} and characterises the gradual transition between the asymptotic regimes $n = 1$ and $n = -3$. For warm dark matter the spectrum is truncated exponentially above k_d, while for hot dark matter the truncation scale has lengthened until it almost coincides with k_{eq}. Together with the random phase assumption, these power spectra determine the statistical properties of the mass distribution until the quite recent epochs when nonlinear structures first form. For hot dark matter these first structures are expected to be large "pancakes" which must fragment to form galaxies, whereas for warm dark matter the pancakes themselves will be galaxy sized. For cold dark matter a whole hierarchy of structures is expected to grow rapidly from star cluster scales up to those of galaxy clusters. These patterns are seen clearly in simulations of neutrino and CDM dominated universes. I now discuss each of these in turn.

3. NEUTRINO-DOMINATED UNIVERSES

Massive neutrinos were the first weakly interacting particle candidate for the dark matter, and they are still *a priori* the most plausible, since we do at least know that neutrinos exist. The most important feature of the fluctuation spectrum in a neutrino-dominated universe is the free streaming cutoff which eliminates all small-scale structure. The characteristic scale usually quoted for this cutoff is

$$k_c = \frac{2\pi}{13Mpc} \, \Omega h^2 = \frac{2\pi}{41Mpc} \left(\frac{m_\nu}{30eV} \right) \tag{9}$$

Fig. 2 shows there is no significant power at shorter wavelengths than $\lambda_c = 2\pi/k_c$. As pointed out long ago by Zel'dovich and his colleagues (Zel'dovich 1970; Doroshkevich *et al.* 1980), the initial nonlinear phases of the evolution of a distribution of this kind will involve the formation of caustics (or "pancakes") with a characteristic scale given by Eq. (9). A cell-like structure of such caustics is expected to form and then evolve into a network of filaments which eventually collapse into dense clumps. This general picture has been confirmed through a series of one-, two- and three-dimensional simulations of evolution from distributions with a large coherence length (Klypin and Shandarin 1983; Centrella and Melott 1983; White, Frenk and Davis 1983).

Figure 3 shows a series of projections of one of the models of White, Frenk and Davis (1983). These models used initial conditions based directly on the linear calculations of Fig. 2; they set the side of the calculation volume equal to $5\lambda_c$ and the rms initial relative density fluctuation at 22%. The three left-hand panels of the figure show projections of all the points after the model has expanded by factors of 4.0, 6.1 and 10; the general expansion has been taken out by rescaling. Over this period the model goes from a relatively uniform state to one in which much of the mass is concentrated in condensed clusters. It is important to realise, however, that these pictures represent the *neutrino* distribution. In this kind of universe galaxies can only form from baryonic material in regions which have undergone local collapse; the spaces between the collapsed objects in the left-hand panels are thus expected to be devoid of galaxies. We have tried to identify those particles which *could* represent galaxies by tagging particles when the local region surrounding them collapses. One percent of the particles are tagged as "galaxies" by an expansion factor of 2.9; this seems a reasonable time to identify with the onset of significant galaxy formation. Later times can then be identified by their expansion factor, $1 + z_{GF}$, since this epoch; the panels of Fig. 3 correspond to $z_{GF} = 0.4, 1.1$ and 2.5.

Fig. 3. Projected particle positions in a simulation of a neutrino-dominated, Einstein-de Sitter universe. The three left-hand panels show all the particles in the simulations, while those on the right show only those particles tagged as "galaxies." Time increases from top to bottom.

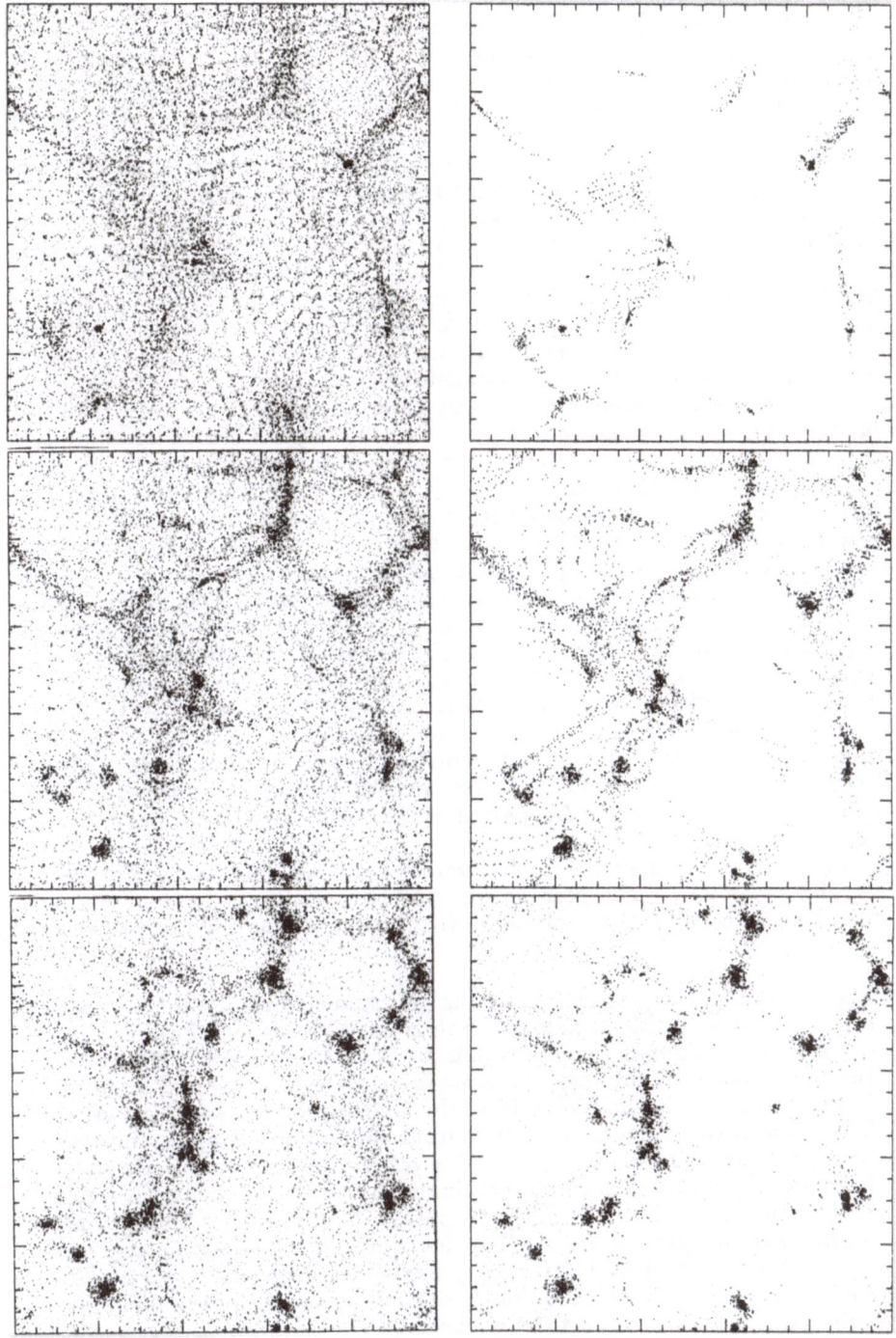

The right-hand panels show the evolution of the projected distribution of "galaxies", and contain 5, 23, and 52% of the particles at the three times shown. The formation of sheets, filaments and clusters is much clearer in the "galaxy" pictures than in the "neutrino" pictures. Notice that clusters begin to form very soon after the first collapse of structure, and that any cell-like structure is a transient state that lasts only for a short time.

In the absence of a cosmological constant, $h < 0.54$ is required to get the age of a flat universe to exceed 12 Gyr. This appears to be a very conservative lower bound on the ages of globular star clusters. The side of the box in Fig. 3 must thus exceed 200 Mpc, and it is clear that the model predicts strong inhomogeneity on very large scales. In fact the large-scale lumpiness of the "galaxy" distribution is always much greater than that of real galaxies. This discrepancy can be illustrated graphically by imagining oneself to be an observer situated at some point in a properly scaled neutrino model, and choosing particles in the same way as observers have chosen galaxies for inclusion in complete redshift surveys of nearby space. The observations can then be compared directly with the simulated catalog. The upper left-hand picture in Fig. 4 shows an equal area projection of the distribution on the sky of part of the Center for Astrophysics redshift survey. Fig. 4 compares it to artificial catalogs over the same area of sky and with the same mean distribution in depth; three of these were made from neutrino-dominated models with $\Omega = 1, z_{GF} = 2.5$ and $h = 0.54$, and two were made from the cold dark matter models discussed below. In the neutrino models "galaxies" are shown as open triangles, whereas other points are shown as dots. It is important to remember that the triangles mark those points which could, in principle, represent real galaxies; in a real neutrino-dominated universe only a biased subset of them might succeed in making visible objects. On the other hand, none of the dots can represent a real galaxy, so it is significant that there are much larger empty areas in the neutrino models than in the real data. Very large empty regions are also seen in simulations of 3D surveys (cf. Fig. 3) and it is worth emphasizing that these regions are predicted to be entirely devoid of galaxies. This picture is very different from that presented by the surveys of de Lapparent et al. (1986) and Haynes and Giovanelli (1986).

Galaxy formation is a sufficiently complex and poorly understood process that Figs. 3 and 4 cannot be accepted as conclusively rejecting a neutrino-dominated universe. It is conceivable that various feedback mechanisms might suppress galaxy

Fig. 4. Equal area projections of the galaxy distribution on the northern sky and in artificial galaxy catalogs made from simulations. The outer circular boundary corresponds to Galactic latitude +40°, while the inner dashed boundary corresponds to declination 0°. The upper left-hand picture is the real data of the CfA survey (Davis et al. 1982). The bottom two pictures are made from simulations of open universes dominated by cold dark matter. The other three pictures were made from simulations of a neutrino-dominated universe; the triangles represent "galaxies," while the dots are points in whose neighborhood the matter has not yet collapsed. The neutrino models assume a flat universe with $h = 0.54$ in which galaxy formation began at $z = 2.5$. The cold dark matter models assume $h = 1.1$ in order to match their two-point correlation function to that observed for galaxies; however, other h values could give equally good results.

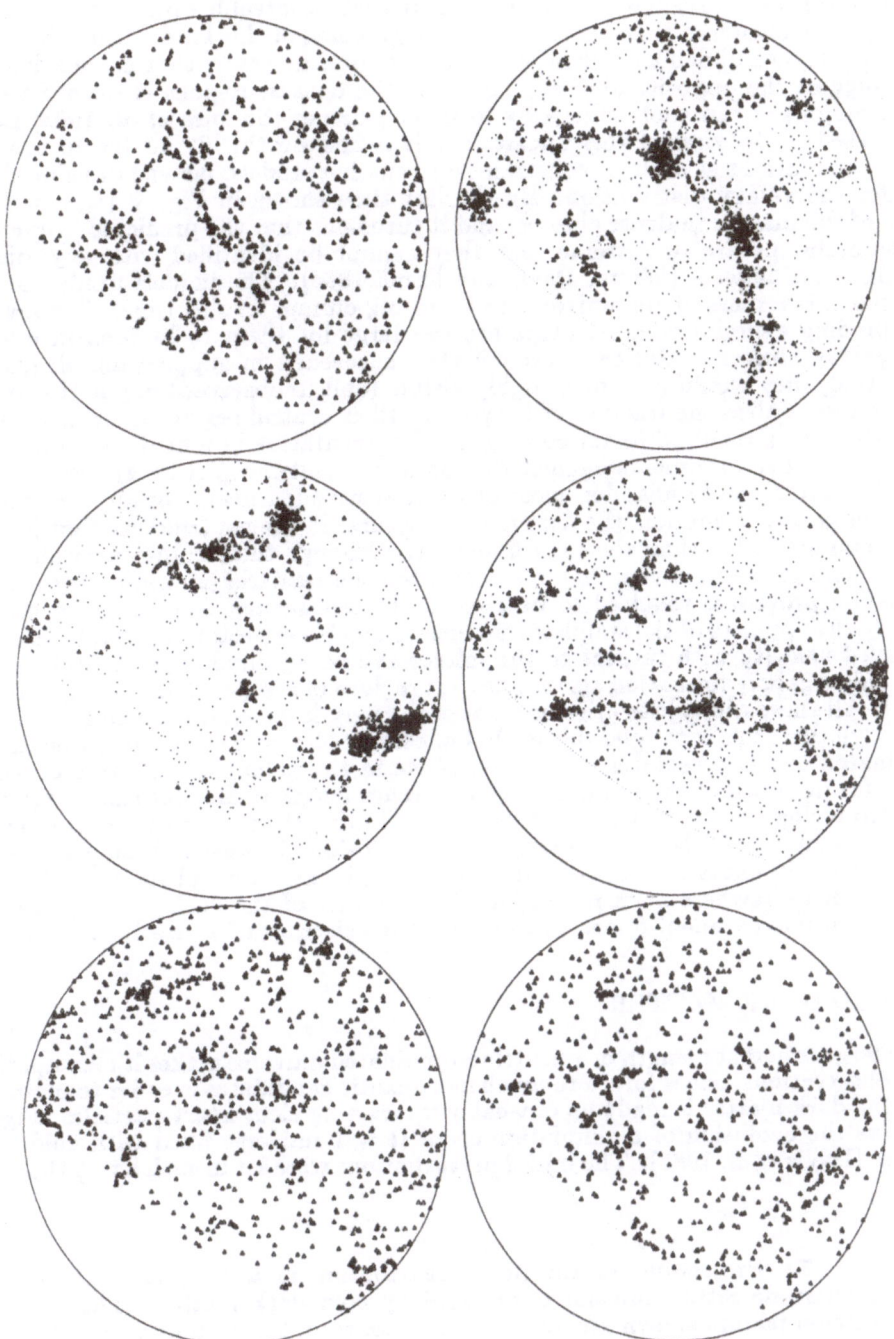

formation in dense regions to such an extent that observable objects are actually more weakly clustered than the neutrinos. Arguments of this kind, although rather implausible, are difficult to dismiss entirely. (Note, however, that calculations to date suggest the opposite difficulty. Even in the densest regions it is hard to see how enough gas can cool sufficiently to make galaxies (Shapiro et $al.$ 1983; Bond et $al.$ 1984).) In view of this uncertainty it seems worthwhile to try to use the mass distributions as a direct constraint on neutrino models, instead of using them to infer the galaxy distribution. By the last time shown in Fig. 3 ($z_{GF} = 2.5$), much of the mass is in dense clumps, and it turns out that the predicted properties of these clumps are so extreme that they cannot be identified with any known population of objects (White, Davis and Frenk 1984). The models predict about half the entire mass of the universe to be in big clumps at this time. The masses and binding energies of the clumps are too large for them to be identified with real galaxy clusters. However, even if they succeeded in suppressing all galaxy formation, they would provide a highly visible (and unobserved) population of X-ray sources, unless the fraction of baryons in their central regions is much smaller than the $\sim$10% required to get cooling and fragmentation in a neutrino pancake.

The above arguments against the standard neutrino picture appear strong, and they force one to abandon most of its attractive features. For any acceptable epoch of galaxy formation, the dominant structures in the neutrino distribution are dense clumps rather than sheets or filaments. Although these clumps contain most of the mass of the universe, they must somehow be rendered invisible. The galaxy formation process is required to be strongly biased against dense regions; this is exactly the opposite of the simplest and most plausible expectation. In addition our observed velocity with respect to the microwave background appears implausibly low, unless galaxy formation also singles out regions of low peculiar velocity (Kaiser 1983). All these difficulties arise because of the very large scale implied by Eq. (9). Although this scale is comparable to that of certain observed structures, a neutrino-dominated universe requires density contrasts in the galaxy distribution on these scales far in excess of observation. The coherence length of Eq. (9) can be reduced by introducing a cosmological constant, or by allowing the universe to be dominated by an unstable particle at some stage of its evolution. However, these possibilities are sufficiently unattractive, and introduce enough extra difficulties, that it seems best to look elsewhere for a resolution of the nature of the dark matter; at least, that is, until experiment forces us to accept the existence of a massive neutrino.

4. COLD DARK MATTER

If the large coherence length of a neutrino-dominated universe makes it incompatible with observations, one is forced to much more exotic candidates in order to maintain that the dark matter is made up of weakly interacting elementary particles. Figure 5 shows the evolution of a simulation of an $\Omega = 1$ universe filled with cold dark matter (Davis et $al.$ 1985). The initial perturbations were set in such a way that the

Fig. 5. Projections of the point distribution in a simulation of an Einstein-de Sitter universe dominated by cold dark matter. The expansion factors shown are 1.0, 1.4, and 1.8, from top to bottom on the left side, and 2.4 and 3.0 at the top on the right. The plot at bottom right shows the "galaxy" distribution at an expansion factor of 1.8 (see text).

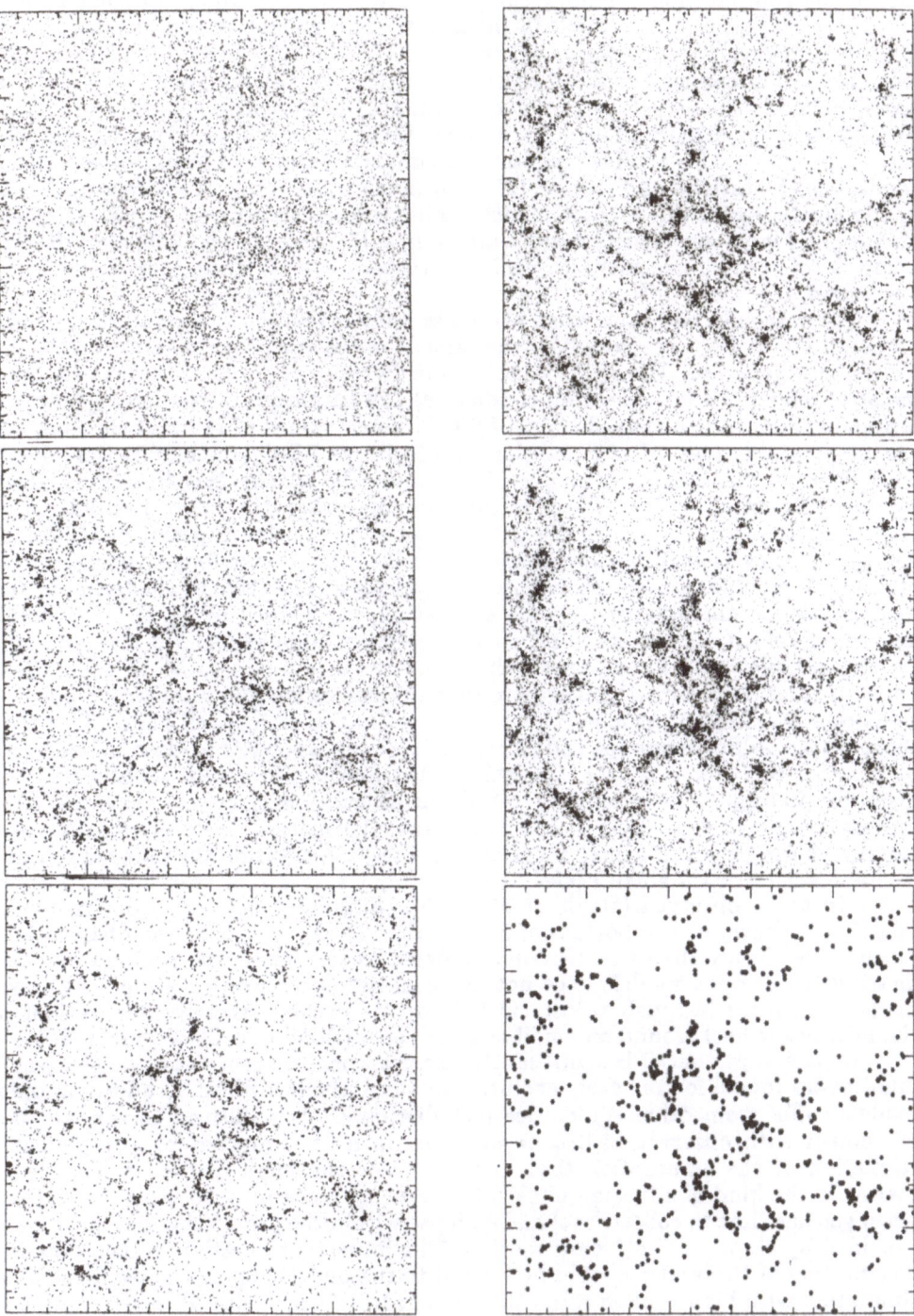

side of the computational volume corresponds to 32.5 h^{-2} Mpc, implying a mass of $3.10^{11}h^{-4}M_\odot$ for each particle in the model. This should be compared with the mass per observed bright galaxy which is $\sim 3.10^{13}\ h^{-1}M_\odot$ if our Universe is flat. The simulation cannot therefore represent aggregates smaller than the halo of a bright galaxy. The structure in this simulation clearly evolves rapidly from small to large scale in a more or less hierarchical fashion. It is interesting to compare these point plots with similar diagrams for simulations evolved from Poisson initial conditions (Efstathiou and Barnes 1983). A much wider range of scales is apparent in the clustering in the present models. Filamentary structures, superclusters of clumps, and large low density regions are all in evidence in these plots; quantitative study is, of course, necessary to decide if these are truly analogous to observed structures.

There is only one time when the mass autocorrelation function, $\xi(r)$, of the model in Fig. 5 has approximately the same slope as the observed autocorrelation function of galaxies. However, at this time (which occurs after expansion by a factor of only 1.8 from the initial conditions) the amplitude of the model function is smaller than that observed unless $h \lesssim 0.22$. Such values of the Hubble constant are unacceptably low. After expansion by a factor of 3.5 the model function agrees with observation in the region $\xi \sim 1$ for $h = 0.5$, but by this time it is much too steep on small scales. There is thus no time when the *mass* distribution of these Einstein-de Sitter models is a good match to the observed *galaxy* distribution. An even more serious difficulty with such an identification is found when particle motions are compared with observations of galaxies. The scale of clustering in the models may be characterised by the separation, r_o, defined by $\xi(r_o) = 1$. If the expansion velocity across r_o is set to its observed value of 500 km s^{-1}, the rms relative peculiar velocity of close pairs of particles is found to be of order 1800 km s^{-1}, whereas observation suggests that the corresponding number for pairs of galaxies is 500 km s^{-1} or less. This is a manifestation of the well known fact that the mass per galaxy in groups and clusters of galaxies is too low to close the universe by a factor of 5–10; if $\Omega = 1$, the galaxy distribution cannot be identical to the mass distribution. A possible way out of this difficulty is clearly to consider CDM universes with $\Omega < 1$. If $\Omega = 0.2$ at present, the mass autocorrelation function can match the observed galaxy function fairly well; in addition the inferred relative peculiar velocities on small scales drop by about a factor of two and so become closer to those observed (Davis *et al.* 1985). Simulated catalogs from two such models are shown at the bottom of Fig. 4; they clearly resemble the observations much more closely than the neutrino models. Nevertheless, the peculiar velocity distribution in open models does not agree in detail with observation, and those models which are in rough agreement turn out to predict excessive fluctuations in the microwave background on small angular scale (Bond and Efstathiou 1984).

If we accept that Ω is significantly less than one, we cast doubt on the whole inflationary model for the early evolution of the Universe. Since, in addition, open models suffer from other difficulties just discussed, it is natural to see whether agreement with observation can be obtained in a flat universe for any plausible relaxation of the assumption that galaxies trace the mass distribution on large scales. In the kind of situation under discussion, galaxies may be expected to form when gas has time to collect, cool and collapse within the preexisting potential wells provided by clumps of cold dark matter. The important question is then how the distribution of these sites should be related to the underlying mass distribution. A simulation with high enough resolution could identify such sites directly. However, present techniques do not permit such resolution when studying the large scale galaxy distribution. A plausible *model* for the sites of galaxy formation is obtained by identifying them with high peaks in a suitably smoothed version of the linear fluctuation distribution. This hypothesis leads to a galaxy distribution which is more strongly clustered than the mass distribution. It has become known as the biased galaxy formation model (Davis *et al.* 1985; Bardeen *et al.* 1986). As shown in Fig. 6 there is actually a rather good correspondence between such high peaks and the centers of the deeper potential wells which form in high resolution CDM simulations. The high peak model may therefore give a good representation of the galaxy distribution which arises naturally in CDM universes.

The predicted distribution of galaxies for a high peak formation model of the kind just described is shown in the bottom right-hand frame of Fig. 5. This distribution corresponds to the mass distribution at bottom left. The same structures are apparent in both pictures but they have much higher contrast in the galaxy distribution. Thus the mass per galaxy is much lower than the mean in the high

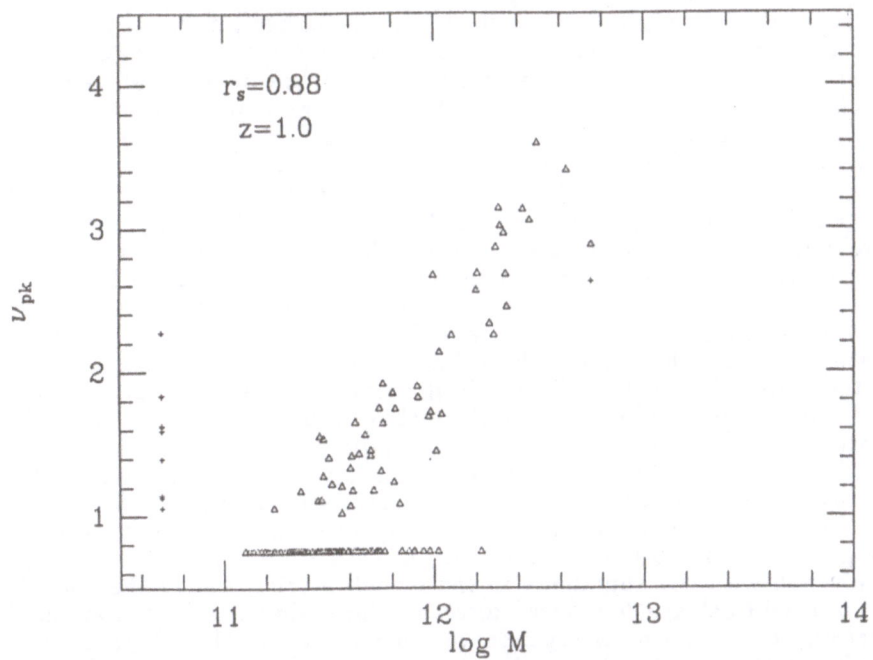

Fig. 6. The correspondence between peaks of the initial linear density field and the halos which are present at $z = 1$ in a set of 3 flat CDM models of a 14 Mpc cubic region (assuming $h = 0.5$). The initial conditions were smoothed with a gaussian, $\exp(-r^2/2r_s^2)$ with $r_s = 0.88$ Mpc before the peaks were located. ν_{pk} is the height of the peak in units of the rms density fluctuation. All peaks higher than one sigma and all halos heavier than $1.2 \times 10^{11} M_\odot$ are plotted. Triangles at the bottom are halos that do not contain an initial peak; crosses at the left are peaks without a halo. All the more massive halos contain a peak and all the higher peaks correspond to a halo. The single most massive halo is the only one to contain more than one peak.

density clusters and groups, and much higher than the mean in the low density "void" regions. This bias is just what is needed to reconcile a flat universe with the observed dynamics of groups and clusters of galaxies. In the present model it is simply a consequence of the CDM spectrum of Fig. 2 and of the statistical properties of high peaks (2.5σ) in a Gaussian noise field (Bardeen *et al.* 1986). The biasing model has two free parameters which can be thought of as specifying the abundance of galaxies and the strength of the bias. In our first investigation of the model we found values of these parameters which were plausible *a priori* and which led to a galaxy distribution in good agreement with observation on scales between 0.5 and $10h^{-1}$ Mpc where our simulations were reliable (Davis *et al.* 1985). Agreement required $h \approx 0.5$. In fact these biased models with $\Omega = 1$ actually gave a rather better fit to observed statistics on the positions and velocities of galaxies than did our open models in which the mass and galaxy distributions were assumed identical. This conclusion has since been reinforced by a more detailed comparison of the structure and dynamics of observed and simulated groups and clusters (Nolthenius and White 1987).

If the particular model just described is indeed correct, it should also explain the structure, abundance and characteristic parameters of individual galaxy halos. In addition it should reproduce the very large-scale filaments, voids and superclusters which have been the focus of many recent observational studies of the galaxy distribution. In investigating the behavior of the model on smaller and larger scales in this way, no further free parameters are available. As a result, such comparisons provide quite a stringent test of its performance. Work has now been completed on extensions in both directions (White *et al.* 1987; Frenk *et al.* in preparation).

Preliminary results for small scales show that the clumps of dark matter do indeed form with the requisite properties to be identified with galaxy halos (Frenk *et al.* 1985). Figure 7 shows the mass distribution within the ten most massive objects which formed in a simulation of a region 14 Mpc on a side. In this diagram the velocity of a particle on a circular orbit within the clump potential is plotted against the radius of the orbit. The resulting curve can thus be compared with the observed rotation curves of material in the disks of spiral galaxies. It is remarkable that most "rotation curves" in the model are quite flat, in agreement with observation. Note, however, that the model does not have enough resolution to follow the curves in to the radii where real rotation curves are measured (2–50 kpc). In a volume of the size simulated about two spiral galaxies are expected with rotation velocity greater than 250 km/s and about ten with $V_c > 125$ km/s. This agrees reasonably well with the abundance in the model. Notice that in this model the location of bright galaxies is inferred directly. Three similar models were used to construct Fig. 6. The good correspondence demonstrated by that diagram suggests that biasing may turn out to be purely a consequence of the properties of nonlinear gravitational clustering.

Studies of simulations of much larger volumes of space show that many of the morphological features of the large-scale galaxy distribution can also be matched by this same flat CDM model (White *et al.* 1987). In particular, if rich galaxy clusters are identified in the simulations in the same way that Abell (1958) identified real rich clusters, the abundance of systems he found is matched almost exactly. In addition the mass per galaxy in the model clusters is found to be about a fifth of that in the simulations as a whole. This agrees well with dynamical measurements of real systems which find about one fifth the mass per galaxy required to close the Universe. A similar result is that in a galaxy system with the size and population of

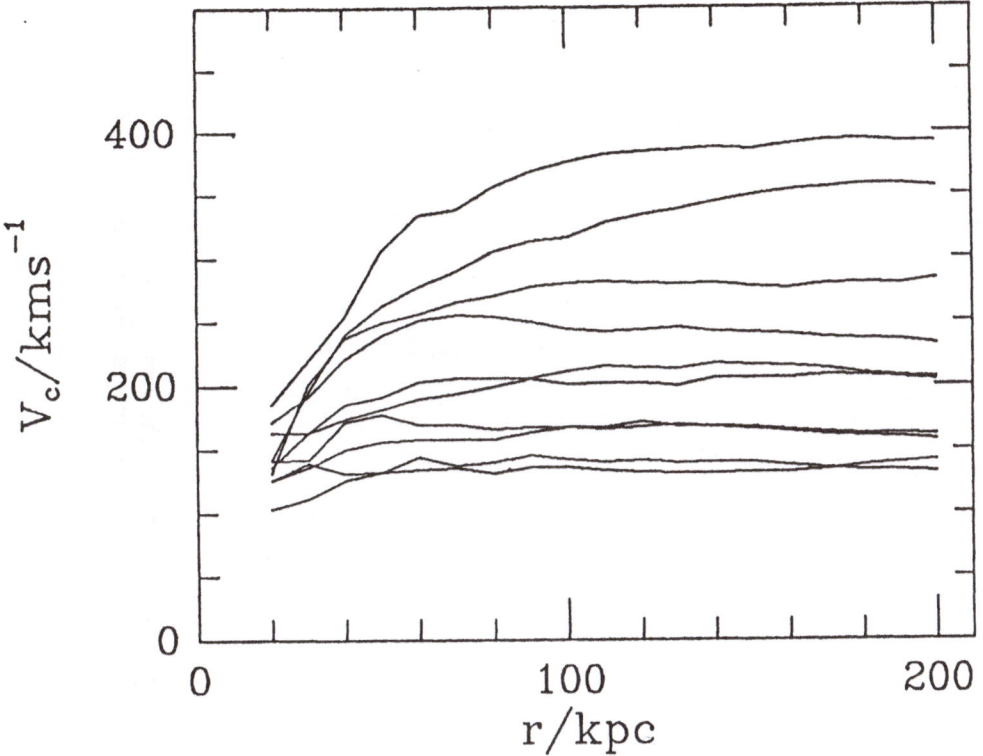

Fig. 7 "Rotation curves" for the ten most massive clumps present at the end of a simulation of a cubic region of a flat CDM universe. For each object circular velocities were calculated as a function of radius using $V_c^2 = GM(r)/r$ where $M(r)$ is the mass contained in a sphere radius r centred on the densest region of that clump.

the Local Supercluster, the mass per galaxy implied by the observed perturbation of the Hubble flow is about that expected in the biased model. A direct identification of the mass and galaxy distributions, on the other hand, implies $\Omega \approx 0.3$ (Davis and Peebles 1983). In addition to producing the correct abundance of rich clusters, the models also contain voids almost as large as the famous void in Bootes. This is a spherical region with with a diameter of 6,200 km/s which contains no galaxies in the survey of Kirshner *et al.* (1983, 1987). When we conducted an analogous survey of our simulations, several voids with diameters in excess of 5,600 km/s turned up. These results show that it is very important to duplicate observational procedures as closely as possible when using simulations to test theoretical models.

Figure 8 shows three plots of simulation data constructed in a similar manner to a recent survey of all the brighter galaxies in a 6° strip of the northern sky (de Lapparent *et al.* 1986). In each diagram recession velocity and the angular coordinate along the strip are used as radial and angular coordinates. The upper plot contains a rich cluster (which is not, however, as rich as the Coma cluster in the real data) and the lower plots correspond to the two immediately adjacent 6° strips. The radial extent of these plots is 15,000 km/s. They contain several

258

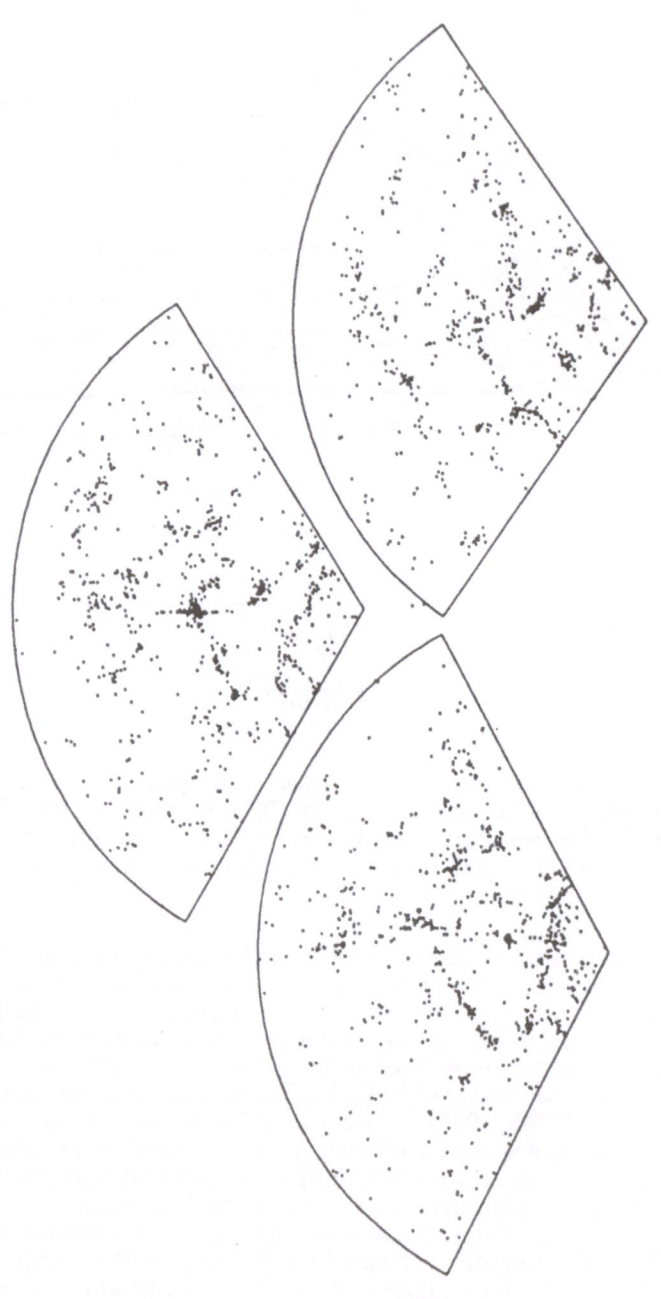

Fig. 8. Distributions of "galaxies" for three simulated redshift surveys with the geometry chosen by de Lapparent et al. (1986). The simulation surveyed is a flat biased CDM model of a cubic region 360 Mpc on a side (assuming h = 0.5).

filamentary structures about 5,000 km/s in length in addition to a number of low density regions with a diameter of this order. Some of the filaments can be traced from one strip to the next, suggesting that they may be sections through sheets. The data give an impression of a bubble-like structure which is simiiar enough to that of the real survey to suggest that it would be premature to conclude that the observations require anything other than purely gravitational processes and random phase initial conditions. Quantitative comparison of the topology of structures in the models and in the real Universe is required to reach stronger conclusions (Gott *et al.* 1986). Preliminary results show rather good agreement with the model discussed here (Weinberg *et al.* 1987).

The most serious discrepancies between this biased CDM model and observation concern two measurements of structure on very large scales. Rich clusters of galaxies are themselves weakly grouped into very large, loose aggregates called superclusters. One measure of this tendency is the spatial autocorrelation function of the clusters, which is estimated to fall to unity on a scale 4–5 times larger than the corresponding scale for galaxy correlations (Bahcall and Soneira 1983; Klypin and Kopylov 1983). In our CDM model the cluster correlation scale is indeed larger than the galaxy scale but only by about a factor of two. Uncertainties in these measurements are difficult to assess because of possible systematic problems both in the real data and in the simulations. However, the discrepancy appears to be significant at about the 90 percent level (White *et al.* 1987; Ling *et al.* 1986). In addition the model predicts that large regions of the universe should have quite small peculiar velocities with respect to the microwave background. This is in agreement with one recent observational study (Aaronson *et al.* 1986), but in conflict with three others (Bahcall and Soneira 1983; Collins, Joseph and Robertson 1986; Burstein *et al.* 1986). Some time will be required for the observational situation to clarify, but such data could ultimately prove fatal for the model in the simple form which has so far been investigated. Nevertheless, it is impressive that a single model with $\Omega = 1$ can reproduce such a wealth of structural information on objects ranging from the halos of individual galaxies to the largest known clusters, filaments and voids.

This work was supported by NSF Presidential Young Investigator Award AST-83352062 and by NATO travel grant 689/84.

REFERENCES

Aaronson, M., Bothun, G., Mould, J., Huchra, J. P., Schommer, R. A., and Cornell, M. E. (1986), *Ap. J.*, **302**, 306.

Aarseth, S. J (1984), in *Methods of Computational Physics*, ed. J. U. Brackbill and B. I. Cohen (New York: Academic), p.1.

Abell, G. O. (1958), *Ap. J. Suppl.*, **3**, 211.

Bahcall, N. A., and Soneira, R. (1983), *Ap. J.*, **270**, 20.

Bardeen, J. M., Bond, J. R., Kaiser, N., and Szalay, A. S. (1986), *Ap. J.*, **304**, 15.

Bond, J. R., Centrella, J., Szalay, A. S., and Wilson, J. R. (1984), *M.N.R.A.S.*, **275**, 413.

Bond, J. R., and Szalay, A. S. (1983), *Ap. J.*, **174**, 443.

Bond, J. R., and Efstathiou, G. (1984), *Ap. J. (Letters)*, **285**, L45.

Burstein, D., Davies, R. L., Dressler, A., Faber, S. M., Lynden-Bell, D., Terlevich, R., and Wegner, G. (1986), preprint.

Centrella, J., and Melott, A. (1983), *Nature*, **305**, 196.

Collins, A., Joseph, R. D., and Robertson, N. A. (1986), *Nature*, in press.
Davis, M., Efstathiou, G., Frenk, C. S., and White, S. D. M. (1985), *Ap. J.*, **292**, 371.
Davis, M., Huchra, J., Latham, D., and Tonry, J. (1982), *Ap. J.*, **253**, 423.
Davis, M., and Peebles, P. J. E. (1983), *Ann. Rev. Astr. Ap.*, **21**, 109.
Dekel, A., West, M. J., and Aarseth, S. J. (1984), *Ap. J.*, **279**, 1.
De Lapparent, V., Geller, M. J., and Huchra, J. P. (1986), *Ap. J. (Letters)*, **302**, L1.
Doroshkevich, A. G., Khlopov, M. Yu., Sunyaev, R. A., Szalay, A. S., and Zel'dovich, Ya. B. (1980), *Ann. N. Y. Acad. Sci.*, **375**, 32.
Efstathiou, G., and Barnes, J. (1983), *Formation and Evolution of Large Structures in the Universe*, ed. J. Audouze and J. Tran Thanh Van (Dordrech: Reidel), p. 316.
Efstathiou, G., Davis, M., Frenk, C. S., and White, S. D. M. (Ap. J. Suppl.), **57**, **241**, .
Efstathiou, G., and Eastwood, J. W. (1981), *M.N.R.A.S.*, **194**, 503.
Frenk, C. S., White, S. D. M., Efstathiou, G., and Davis, M. (1985), *Nature*, **317**, 595.
Gott, J. R., Melott, A. L., and Dickinson, M. (1986), *Ap. J.*, **306**, 341.
Haynes, M. P., and Giovanelli, R. (1986), *Ap. J. (Letters)*, **306**, L55.
Hockney, R. W., and Eastwood, J. W. (1981), *Computer Simulation Using Particles*, (New York: McGraw Hill).
Hogan, C. J., and Kaiser, N. (1983), *Ap. J.*, **274**, 7.
Hogan, C. J., and White, S. D. M. (1986), *Nature*, **321**, 575.
Ikeuchi, S. (1981), *Pub. Astr. Soc. Japan*, **33**, 211.
Kaiser, N. (1983), *Ap. J. (Letters)*, **273**, L17.
Kirshner, R. P., Oemler, A., Schechter, P. L., and Shectman, S. A. (1983), *Early Evolution of the Universe and its Present Structure*, ed. G. O. Abell and G. Chincarini (Dordrecht: Reidel), p. 197.
Kirshner, R. P., Oemler, A., Schechter, P. L., and Shectman, S. A. (1987), *Ap. J.*, in press.
Klypin, A. A., and Kopylov, A. A. (1983), *Sov. Astr. Lett.*, **9**, 41.
Klypin, A. A., and Shandarin, S. F. (1983), *M.N.R.A.S.*, **202**, 593.
Ling, E. N., Frenk, C. S., and Barrow, J. (1986), *M.N.R.A.S.*, **223**, 21P.
Nolthenius, R., and White, S. D. M. (1987), *M.N.R.A.S.*, in press.
Ostriker, J. P., and Cowie, L. L. (1981), *Ap. J. (Letters)*, **243**, L127.
Peebles, P. J. E. (1980), *The Large Scale Structure of the Universe*, (Princeton: Princeton University Press).
Quinn, P. J., Salmon, J., and Zurek, W. (1986), *Nature*, **322**, 329.
Shapiro, P. R., Struck-Marcell, C., and Melott, A. (1983), *Ap. J.*, **275**, 413.
Weinberg, D. H., Gott, J. R., and Melott, A. L. (1987), *Ap. J.*, in press.
White, S. D. M., Davis, M., and Frenk, C. S. (1984), *M.N.R.A.S.*, **209**, 27P.
White, S. D. M., Frenk, C. S., and Davis, M. (1983), *Ap. J. (Letters)*, **274**, L1.
White, S. D. M., Frenk, C. S., Davis, M., and Efstathiou, G. (1987), *Ap. J.*, in press.
Zel'dovich, Ya. B. (1970), *Astr. Ap.*, **5**, 84.

NUMERICAL RELATIVITY AND COSMOLOGY

Tsvi Piran
The Racah Institute for Physics,
The Hebrew University, Jerusalem, ISRAEL
and The Institute for Advanced Study, Princeton, NJ, USA

ABSTRACT

We discuss two schemes for general relativistic numerical cosmological solutions. The scheme are based on the 3+1 ADM formalism and on the 3+1 Regge Calculus and they are designed for a full three dimensional time dependent solution of the Einstein equations.

1. Introduction

Computers and numerical experiments play an essential role in Cosmology. As it is impossible to perform Cosmological experiments the computer replaces the laboratory and with its aid we test and compare different cosmological hypothesis. The central objective of most of the current research in numerical Cosmology is to model the evolution of the large scale structure in the Universe by studying the growth of perturbation, and formation of groups of galaxies, clusters and super clusters. The approach is based on N body simulations employing Newtonian gravitational forces in an expanding Friedmann background. An excellent review of this work was given by White (this school) and I will not discuss these topics here.

The homogeneity of the microwave background (and some theoretical bias) suggests that our attention should be focused on small perturbations from homogeneous Cosmological models. For these the above techniques provide an excellent approximation and there is no reason to go beyond Newtonian calculations on an homogeneously expanding background. However, to study Cosmological questions, like the onset of inflation under inhomogeneous initial conditions (Piran and Williams. 1985;1986; Kurki-Suonjo et al., 1987), Nucleosynthesis in an inhomogeneous medium (Centrella et al, 1986) or bubble formation during the quark confinement phase transition (Pantano and Miller, 1986) it is essential to turn to the full set of Einstein equations and to Numerical Relativity.

In this review I describe two different approaches to Numerical Relativity, the ADM formalism and Regge Calculus. I focus on specific problems that occur in Cosmological solutions and I formulate two schemes for general, i.e. three dimensional time dependent Cosmological solutions.

W. G. Unruh and G. W. Semenoff (eds.), The Early Universe, 261–282.

2. The 3+1 Formalism and Cosmology

2.1 The 3+1 Equations

The 3+1 formalism (Arnowitt et al, 1962) is the common approach to Numerical Relativity. There are some excellent expositions of this formalism Misner, Thorne and Wheeler, 1973; Choquet-Bruhat and York, 1979), and some detailed discussion of the application of the 3+1 formalism to Numerical Relativity (York, 1979; York, 1983; York and Piran 1982). I refer the reader who is not familiar with this formalism to these works for details. In this review I will discuss aspects that are unique to Cosmological applications.

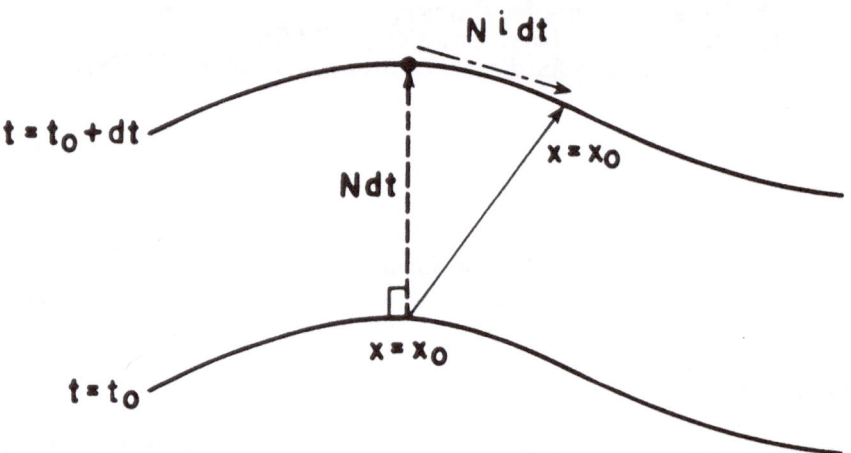

Fig. 1: *The lapse function, N, and the shift vector, N^i, describe the perpendicular distance between two infinitesimal close hypersurfaces and the motion of the coordinates.*

In the 3+1 formalism we break the space-time symmetry of general relativity and foliate the four dimensional space time to three dimensional spacelike hypersurfaces. The four metric $g_{\mu\nu}^{(4)}$ is written in terms of the three metric $g_{ij}^{(3)}$ $(i, j, = 1..3)$, the lapse function, N, and the shift vector, N^i:

$$ds^2 = -N^2 dt^2 + g_{ij}(dx^i + N^i dt)(dx^j + N^j dt) \quad , \tag{2.1}$$

where we have eliminated the superscript (3) from the three metric. Fig. 1 , displays graphically the roles of the lapse and the shift (see York, 1979, for more details). The three metric determines the intrinsic geometry of a hypersurface. the extrinsic curvature K_{ij} $(K_{ij} = \Gamma^{(4)0}_{ij})$, describes how the hypersurface is embedded in the four dimensional space time. Together g_{ij} and K_{ij} specify the geometry of a hypersurface, with an identity relating the time derivative of the metric to K_{ij}:

$$\frac{\partial g_{ij}}{\partial t} = -2NK_{ij} + N^k g_{ij,k} + g_{ik}N^k_{,j} + g_{jk}N^k_{,i} \quad . \tag{2.2}$$

To complete the initial data on a hypersurface we add variables like the density, pressure and velocity that describe the matter.

The four $G_{0\mu}$ components of Einstein's tensor and the corresponding equations do not contain second time derivatives of the metric. These equations, the constraint equations in the 3+1 terminology, limit the initial data, which cannot be specified arbitrarily. The Hamiltonian constraint is the 00 Einstein equation:

$$\mathcal{R} + (trK)^2 - K_{ij}K^{ij} = 16\pi G\bar{\rho} \quad , \tag{2.3}$$

where $\bar{\rho} = N^2 T^{00}$ and $\mathcal{R}$ is the three dimensional Ricci scalar. The 0i equations are the momentum constraints:

$$K^j_{\ i;j} - trK,i = 8\pi GJ^i \quad , \tag{2.4}$$

where $J_i = NT^0_i$ and where ; denotes a covariant three derivative. In a homogeneous and isotropic Cosmology the Hamiltonian constraint becomes the Friedmann equation, while the momentum constraints vanish identically.

Once the initial data is specified, its future evolution is obtained from the ij Einstein's equations which in the 3+1 notations are:

$$\frac{\partial K_{ij}}{\partial t} - N^k K_{ij,k} - K_{ik}N^k_{\ ,j} - K_{jk}N^k_{\ ,i} =$$

$$-N_{;ij} + N\left(\mathcal{R}_{ij} - 2K_{ik}K^k_{\ j} + trKK_{ij} - 8\pi G[S_{ij} - \frac{1}{2}g_{ij}(trS - \bar{\rho})]\right) \quad , \tag{2.5}$$

where

$$S_{ij} = T_{ij} \quad . \tag{2.6}$$

2.2 The Initial Value Problem and the York Procedure

The initial data on a three dimensional spatial hypersurface must satisfy the constraint equations. To obtain such data we must solve, in the 3+1 formalism Eqs. 2.4 and 2.5. in some special cases (e.g. Wilson, 1979; Bardeen and Piran, 1983) one can solve these equations for particular components of the metric or the extrinsic curvature. For a Cosmological solution it is better to follow the general York procedure (for detailed reviews see York 1979; Choquet-Bruhat and York, 1979; York and Piran, 1982). To solve the Hamiltonian constraint we (following Lichnerowicz 1944) isolate a conformal metric, γ_{ij} and a conformal factor ϕ such that:

$$g_{ij} = \phi^4\gamma_{ij} \quad . \tag{2.7}$$

The Hamiltonian constraint becomes:

$$-8\phi^{-5}\Delta_\gamma\phi + \phi^4 \mathcal{R}(\gamma) + (trK)^2 \cdot K_{ij}K^{ij} = 16\pi G\bar{\rho} \quad . \tag{2.8}$$

It seems that for a given γ_{ij} we can solve Eq. 2.8 for ϕ. However in its present form Eq. 2.8 is ill posed. A nonlinear elliptic equation:

$$\Delta\psi = F(x)\phi^n \quad , \tag{2.9a}$$

has a solution if

$$nF > 0 \quad . \tag{2.9b}$$

The terms $K_{ij}K^{ij}$ and $16\pi G\bar{\rho}$ in Eq. 2.8 do not satisfy this condition. To overcome this problem we first split K_{ij} to various components and latter we rescale these components and $\bar{\rho}$. For any tensor, and in particular for K_{ij}, we define $\tilde{K}_{ij}$ as the as the traceless part of this tensor, i.e.:

$$\tilde{K}_{ij} = K_{ij} - \frac{1}{3}\gamma_{ij}trK \quad . \tag{2.10a}$$

We rescale $\tilde{K}_{ij}$ as:

$$\tilde{K}_{ij} = \phi^{-10}\hat{K}_{ij} \quad , \tag{2.10b}$$

and

$$\bar{\rho} = \phi^{-8}\hat{\rho} \tag{2.11}$$

to obtain:

$$8\Delta_\gamma\phi = \mathcal{R}(\gamma)\phi + \frac{2}{3}(trK)^2\phi^5 - \hat{K}_{ij}\hat{K}^{ij}\phi^{-7} - 16\pi G\hat{\rho}\phi^{-3} \quad . \tag{2.12}$$

The last three terms on the r.h.s. of Eq. 2.12 satisfy conditions 2.9. However the sign of the first term may vary. Cantor (1979) has obtained necessary and sufficient conditions, limiting $\mathcal{R}$, so that Eq. 2.12 has a unique positive solution. The essence of these conditions (Wheeler, 1964) is that $\mathcal{R}$ should not be too negative over too large region. If these conditions are not satisfied ϕ will vanish (on a topologically spherical region) and the resulting metric, g_{ij}, will have a "bag of gold" singularity (Wheeler, 1964, Cantor and Piran, 1983).

The momentum constraints are solved for a vector W^i such that

$$\hat{K}^{ij} = \hat{K}_*^{ij} + (\hat{l}W)^{ij} \quad , \tag{2.13a}$$

where:

$$(\hat{l}W)^{ij} = W^{i;j} + W^{j;i} - \frac{2}{3}\gamma^{ij}W^k_{;k} \quad , \tag{2.13b}$$

$$\hat{K}_{*\ ;j}^{ij} = tr\hat{K}_* = 0 \quad , \tag{2.13c}$$

i.e. $\hat{K}_*^{ij}$ is the transverse, traceless rescaled part of the K^{ij}. Substitution of Eqs. 2.10a,b and 2.13a,b,c to Eq. 2.4 yield an elliptic equation for W^i:

$$(\hat{l}W)^{ij}_{\ ;j} - \frac{2}{3}\phi^6 trK^{,i} = 8\pi G\hat{J}^i \quad , \tag{2.14}$$

where the momentum density has been rescaled as:

$$\bar{J}^i = \phi^{-10}\hat{J}^i \quad . \tag{2.15}$$

2.3 A 3+1 Cosmological Scheme - Geometry

So far I have described the general 3+1 formalism. Some modifications of the general formalism are useful when we approach a Cosmological solution. We would like to factor out the overall Cosmological Expansion. This will enable us to isolate the expansion which takes place in the background from the evolution of the deviation from homogeneity and isotropy. This should reduce the numerical errors and ease the interpretation of the results. Along the same lines we would like that these deviations from homogeneity and isotropy will be manifestly clear. To do so we should employ variables that vanish in the homogeneous and isotropic limit. This will enable us to identify easily when the solution approaches the Friedmann Solution, and when perturbations are growing. Finally we would like to cover all of the spacetime and avoid local singularities, which might arise due to gravitational collapse of a local region.

To achieve these goals we have to turn to specific coordinate conditions. The temporal coordinate conditions depends on the lapse function, i.e. on the slicing condition. The constant mean curvature slicing conditions (York, 1972; Piran, 1980; Centrella 1980) is by far the most suitable for Cosmological solution. York (1972) even calls trK Cosmic time. If we are on a $trK = Constant$ hypersurface, this amounts to:

$$\frac{\partial trK}{\partial t} = F(t) \quad , \tag{2.16}$$

where $F(t)$ is an arbitrary function, e.g. a constant and $K(t) = \int_{t_0}^{t} F(t')dt' + K(t_0)$. Contraction of Eq. 2.5:

$$\frac{\partial trK}{\partial t} = -N_{;i}^{\ i} + N[\mathcal{R} + (trK)^2 + 4\pi G(trS - 3\bar{\rho})] + N^k trK_{,k} \quad , \tag{2.17}$$

together with Eq. 2.16 yields an linear elliptic equation for the lapse function, N.

An immediate consequence of the constant mean curvature condition is that it enables us to identify instantaneously whether we are in, or are we approaching, a homogeneous solution . The only constant mean curvature slices of a homogeneous solution are its natural homogeneous constant time slices. With this condition we avoid the potential gauge ambiguity of inhomogeneous slicing of an actually homogeneous solution (see Fig. 2). An additional feature of this condition is its singularity avoidance properties (Smarr and York, 1978b, Eardley and Smarr 1979). If a local singularity tends to form, this condition will slow the evolution exponentially near it so that the constant mean curvature slices do not reach the singularity and the evolution of the rest of the universe is not disturbed by it. Finally, with this condition York's initial value equations, which we have discussed earlier, decouple.

Since we are interested in the onset of inflation it is worthwhile to discuss what are the implication of this slicing conditions in an inflationary Universe. A DeSitter Universe is homogeneous and time invariant. Therefore, all the (standard) $t = constant$ hypersurfaces of a DeSitter Universe are $trK = Const$ hypersurfaces. However all these hypersurfaces have the same trK, which is independent of t. Clearly, we can use the $trK = Const$ condition in a DeSitter Universe only with $F = 0$. A physical Universe undergoing inflation is, however, only approximately in a DeSitter Phase. The scalar field is slowly changing and with it the expansion rate

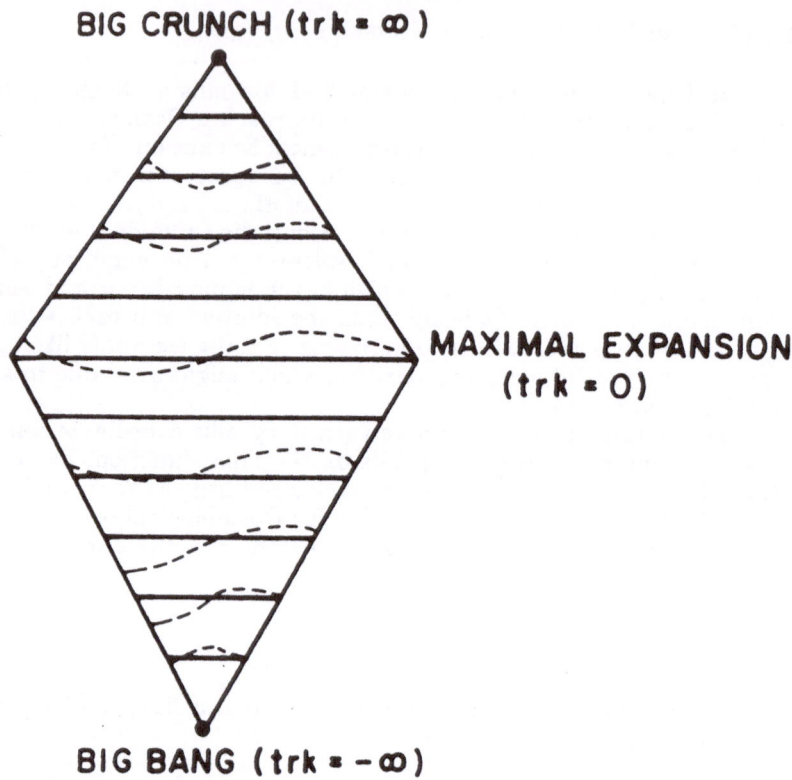

BIG CRUNCH (trk = ∞)

MAXIMAL EXPANSION (trk = 0)

BIG BANG (trk = -∞)

Fig. 2: *Two foliations of the Friedmann universe. The solid lines represent homogeneous spacelike hypersurfaces while the dashed lines describe inhomogeneous hypersurfaces.*

and trK. trK varies, by a very small amount, but however monotonically, from the beginning to the end of the inflationary phase. If the Universe is inhomogeneous and only parts of it undergo inflation we can distinguish, using $trK = Const$, easily between the regions that undergo inflation and the ones that do not. In both regions trK will change from, say C_1 to C_2. However the proper time spend during this change will be long in the inflationary region and short elsewhere. In other words if N is of order unity in the inflationary region it will be exponentially small elsewhere. We find therefore another advantage of this slicing condition. It allows us to identify easily when inflation occurs and where.

The advantage of the constant mean curvature slicing condition for a cosmological solution are obvious. Still I feel that it is essential to mention two other slicing conditions, which are potential pitfalls. One should not use them. These are the constant lapse condition:

$$N = 1 \; , \tag{2.18}$$

and the comoving slicing condition for zero vorticity flow:

$$Nu^t = -1 \; . \tag{2.19}$$

GENERAL SHIFT MINIMAL DISTORTION SHIFT

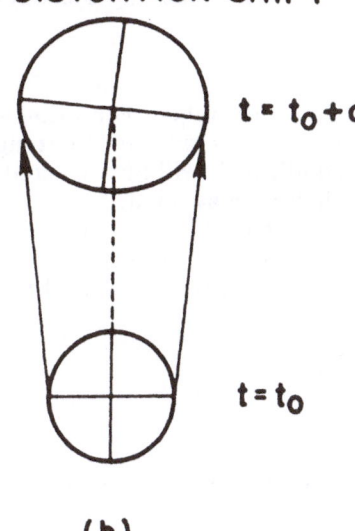

$t = t_0 + dt$

$t = t_0$

(a) **(b)**

Fig. 3: *The distortion of a circle during the evolution (a) is minimized when we use the minimal distortion gauge (b).*

Both conditions look simple and therefore tempting. However both types of slicing converge rather then diverge towards singularities. When local singularities form, with such conditions the evolution accelerates towards the singularity and the computation ceases prematurely!

We need also spatial coordinate conditions. For a Cosmological solution we will generally be interested in configurations for which there are no special points to which we could assign the role of a center. In this case there are three natural coordinate choices: Normal coordinates, Minimal distortion coordinates and Comoving coordinates. For Normal coordinates the shift simply vanishes:

$$N^i = 0 . \tag{2.20}$$

With this simple condition the spatial coordinates flow is orthogonal to the three dimensional spacelike hypersurfaces. This condition has two important features: it will not lead to a coordinate singularity as long as the slicing condition does not lead to singular slice and it simplifies significantly the calculations as all the shift vector terms vanish. I think that this is the best coordinate condition for a Cosmological work.

Smarr and York, (1978a) have suggested the minimal distortion condition. In general a sphere on a given slice is distorted as it is mapped along the coordinate flow into the future (see Fig. 3). The minimal distortion shift vector minimizes this distortion. The minimal distortion shift vector satisfies a vector elliptic equation:

$$(lN)^{ij}{}_{;j} = 2[N \check{K}^{ij}]_{;j} , \tag{2.21}$$

in which the vector elliptic operator is the one appearing in York's momentum constraint. It is not clear whether this condition will lead to an entangled coordinate flow and to numerical problems (although this was not seen in an actual solution, Piran 1980).

The comoving coordinates follow the matter:

$$N^i = V^i = u^i/u^t \quad . \tag{2.22}$$

This Lagrangian condition simplifies the hydrodynamics calculations, however it leads to an extremely entangled coordinate flow and just like the comoving slicing condition it will be attracted toward local singularities. I believe that this condition should be avoided.

To achieve our goal of separating the global expansion from the rest of the evolution we factorize a local expansion factor $R = R(x^i, t)$ from the spatial metric. We choose the conformal metric γ_{ij} such that

$$g_{ij} = R^2(x^i, t)\gamma_{ij} \quad , \tag{2.23a}$$

with

$$\gamma = det\gamma_{ij} = 1 \quad . \tag{2.23b}$$

Rewriting Eq. 2.2 as an equation for γ_{ij} and an equation for R we obtain:

$$\frac{1}{R}\frac{\partial R}{\partial t} = -\frac{1}{3}NtrK + \frac{1}{R}N^k R_{,k} + \frac{1}{3}N^k_{,k} \quad , \tag{2.24a}$$

where

$$\frac{\partial\gamma_{ij}}{\partial t} = -2N\gamma_{ik}\tilde{K}^i_j + N^k\gamma_{ij,k} + \gamma_{ik}N^k_{,j} + \gamma_{jk}N^k_{,i} - \frac{2}{3}\gamma_{ij}N^k_{,k} \quad . \tag{2.24b}$$

Eq. 2.24a reveals the nature of trK, for a homogeneous Universe it is just the expansion rate, H, in a disguise. For an inhomogenous Universe we still define a local expansion rate, $H = H(x^i, t)$ as:

$$H \equiv \frac{1}{R}\frac{\partial R}{\partial t} \quad . \tag{2.24c}$$

Eq. 2.24b suggests that we use $\tilde{K}^i_j$ as the conjugate momentum. Substitution of Eq. 2.17 into Eq. 2.5 provides an equation for $\tilde{K}^i_j$:

$$\frac{\partial\tilde{K}^i_j}{\partial t} - N^k\tilde{K}^i_{j,k} - N^k_{,j}\tilde{K}^i_k + N^i_{,k}\tilde{K}^k_j =$$

$$-\tilde{N}^i_{;j} + N[\tilde{\mathcal{R}}^i_j + \tilde{K}^i_j trK - 8\pi G\tilde{S}^i_j] \quad . \tag{2.25}$$

To obtain $\mathcal{R}$ from γ rather then from g we use:

$$\tilde{\mathcal{R}}^i_j = \frac{1}{R^2}\left(\tilde{\mathcal{R}}^i_j(\gamma_{ij}) - \frac{2}{\sqrt{R}}[(\sqrt{R})^{;i}_{;j} - \frac{1}{3}\delta^i_j(\sqrt{R})^{,k}_{,k}] + \frac{3}{2R^2}[R^{;i}R_{;j} - \frac{1}{3}\delta^i_j R^{;k}R_{;k}]\right)$$

$$= \frac{1}{R^2}\left(\tilde{\mathcal{R}}^i_j(\gamma_{ij}) - \frac{2}{\sqrt{R}}[(\sqrt{R})^{,i}_{,j} - \frac{1}{3}\delta^i_j(\sqrt{R})^{,k}_{,k}] - \frac{R_{,k}}{R}[\gamma^{il}\Gamma^k_{lj}(\gamma_{ij}) - \frac{1}{3}\delta^i_j\Gamma^k_{kj}(\gamma_{ij})]\right)$$

$$+\frac{3}{2R^2}[R^{,i}R_{,j}-\frac{1}{3}\delta^i{}_j R^{,k}R_{,k}]\Bigg) \quad , \tag{2.26a}$$

and

$$\mathcal{R}=\frac{1}{R^2}\left(\mathcal{R}(\gamma)-\frac{8}{\sqrt{R}}(\sqrt{R})_{;}{}^i{}_{;j}\right)=\frac{1}{R^2}\left(\mathcal{R}(\gamma)-\frac{4}{R}(\gamma^{ij}R_{,i})_{,j}+\frac{2\gamma^{ij}R_{,i}R_{,j}}{R^2}\right) \quad , \tag{2.26b}$$

where ; in Eqs. 2.26a,b, denotes a covariant derivatives with respect to γ_{ij}.

We substitute Eqs. 2.10a and 2.24c in the Hamiltonian constraint, Eq. 2.3 to obtain a generalization of the Friedmann equation (for $N^i=0$:

$$\left(\frac{H}{N}\right)^2+\frac{1}{6R^2}\left(\mathcal{R}(\gamma)-\frac{4}{R}(\gamma^{ij}R_{,i})_{,j}+\frac{2\gamma^{ij}R_{,i}R_{,j}}{R^2}\right)=\frac{8\pi G}{3}\bar{\rho}+\frac{1}{6}\tilde{K}^j_i\tilde{K}^i_j \quad . \tag{2.27}$$

The momentum constraint:

$$\tilde{K}^j{}_{i;j}=8\pi G\bar{J}_i \quad , \tag{2.28}$$

is also slightly simpler when expressed using $\tilde{K}^i{}_j$. In the homogenous case it simply vanishes identically.

2.4 A 3+1 Cosmological Scheme - Matter Fields

The evolution of matter sources is determined by the conservation equations:

$$T^\mu{}_{\nu;\mu}=0 \quad . \tag{2.29}$$

We are interested in three types of matter sources: a radiation field, a perfect fluid (or dust) and a scalar field.

A perfect fluid is described by, n, the particle number density, the total energy density, $\rho=mn+\epsilon_m$, the internal energy density, ϵ_m, the pressure, p_m and the flow velocity, u^μ_m. Since particle number is conserved we introduce a number conservation equation:

$$(nu^\mu)_{;\mu}=0 \quad . \tag{2.30}$$

In the following we assume that the matter obeys a simple equation of state of the form:

$$p_m=(\Gamma-1)\epsilon_m \quad , \tag{2.31}$$

where Γ is an adiabatic index. The matter equations of motion are best described using normalized variables:

$$D=R^3 nNu^t_m \quad , \tag{2.32a}$$

$$E=\epsilon_m(R^3 Nu^t_m)^\Gamma \quad , \tag{2.32b}$$

$$S_i=R^3 Nu^t_m(\rho+p)u_{mi} \quad , \tag{2.32c}$$

$$V^i_m=\frac{u^i_m}{u^t_m} \quad , \tag{2.32d}$$

and obtain the equations of motion:

$$\frac{\partial D}{\partial t} + \frac{\partial}{\partial x^i}\left(DV_m^i\right) = 0 \quad , \tag{2.33a}$$

$$\frac{\partial E}{\partial t} + \frac{1}{(V_m^i)^{\Gamma-1}}\frac{\partial}{\partial x^i}\left(E(V_m^i)^{\Gamma}\right) = 0 \quad , \tag{2.33b}$$

and

$$\frac{\partial S_i}{\partial t} + \frac{\partial}{\partial x^i}\left(S_i V_m^i\right) = -(\Gamma - 1)\frac{\partial}{\partial x^i}\left(\frac{E}{(u^t N R^3)^{\Gamma}}\right)$$

$$+\frac{S^t}{R^3}\frac{\partial N}{\partial x^i} - \frac{S_j}{NR^3}\frac{\partial N^j}{\partial x^i} + \frac{1}{2NR^5}\frac{S_i S_j}{S^0}\frac{\partial \gamma^{lj}}{\partial x^i} - \frac{1}{NR^6}\frac{\gamma^{lj} S_l S_j}{S^0}\frac{\partial R}{\partial x^i} \quad , \tag{2.33c}$$

The radiation field is characterized by an energy density $\epsilon_{(rad)}$ and velocity field $u_{(rad)}$. The corresponding equations of motion for the radiation field are similar to those of the perfect fluid with the substitutions $n_{(rad)} = 0$ and $\Gamma_{(rad)} = 4/3$.

The above equations are valid at late times when the matter and the radiation field are not coupled. When there is a coupling a particle moving relative to the radiation background feels a drag and a corresponding term should be added to the equations of motion.

The energy momentum tensor of the scalar field ψ is:

$$T^{\mu}_{(\psi)\ \nu} = \psi^{,\mu}\psi_{,\nu} - \frac{1}{2}\delta^{\mu}_{\nu}[\psi^{,\sigma}\psi_{,\sigma} + V(\psi)] \quad . \tag{2.34}$$

where V is the scalar potential. The resulting equation of motion, for $N^i = 0$ is:

$$-\frac{1}{R^3 N}\frac{\partial}{\partial t}\left(\frac{R^3}{N}\frac{\partial \psi}{\partial t}\right) + \frac{1}{R^3 N\sqrt{\gamma}}\frac{\partial}{\partial x^i}\left(NR\gamma^{ij}\frac{\partial \psi}{\partial x^j}\right) + \frac{dV(\psi)}{d\psi} = 0 \quad . \tag{2.35}$$

ψ has different R dependence in different regimes (see Piran, 1986), therefore, unlike the matter variables, it is not useful to factor out R dependence from ψ.

2.5 Boundary Conditions

A mathematical set of differential equations and a numerical scheme to solve them are never complete without specification of boundary conditions. The boundary conditions in a cosmological solution will depend on whether we model an open or a closed Universe. A closed Universe do not have a boundary and for it we do not need boundary conditions. Our numerical grid will still have boundaries but the data on them will be determined by regularity conditions which are usually simple to derive even thought they might be difficult to implement.

The situation is more difficult for an open Universe. It is possible to embed a perturbed inhomogeneous section in a Friedmann background with open outgoing waves boundary conditions. However , small perturbations may not have enough time to grow in such a case and the meaning of the solution is not clear. An essential part of a Cosmological scenario is the continuous introduction external perturbations which does not exist with this boundary condition.

Alternatively it is possible to employ periodic boundary conditions. In this case we have the opposite situations, a given perturbation does not leave the system any more and it returns again and again to the same location. In fact we are not modeling an open Universe any longer, with a periodic boundary conditions we model a three torous and again the meaning of the results is not clear. In view of these difficulties my inclination will be to consider, at this stage, only closed Universes with a topology of a three sphere.

2.6 Examples

For a spatially homogeneous and isotropic universe the Ricci scalar is: $\mathcal{R}(\gamma) = -6, \ 0, \ 6$ for a closed flat or open universe, and $\tilde{R}^i_{\ j} = 0$. Since the matter is not moving relative to the background, $\bar{J}^i = 0$ and therefore, $\tilde{K}^i_{\ j} = 0$, by virtue of Eq. 2.28. As the matter is not moving $\bar{\rho} = \rho$ i.e. the energy density in the matter frame is the same as the energy density in the coordinate frame. Eqs. 2.17 and 2.27 suggest that N is a constant, say $N = 1$, and that Eq. 2.27 becomes the Friedmann equation. The r.h.s. of the evolution equations, Eqs. 2.24b and 2.25 vanish and therefore γ_{ij} is independent of t, and $\tilde{K}^i_{\ j} = 0$.

Gravitational radiation in a Friedmann universe appear as a perturbation to the metric:

$$\gamma_{ij} = \gamma_{ij}^{(F)} + h_{ij} \quad . \tag{2.36}$$

To first order

$$\mathcal{R}^i_j(\gamma) = \mathcal{R}^i_j(\gamma^{(F)}) + D^2(h)_{ij} \quad , \tag{2.37}$$

where D^2 is a second order operator (see Weinberg, 1972). Eq. 2.25 becomes:

$$\frac{\partial \tilde{K}^i_{\ j}}{\partial t} \approx \frac{\tilde{D}^2(h)^i_j}{R^2} \quad , \tag{2.38}$$

and Eq. 2.24b:

$$\frac{\partial h_{ij}}{\partial t} \approx \frac{\tilde{D}^2(h)^i_j}{R^2} \quad . \tag{2.39}$$

Recalling that $\mathcal{R}^i_j(g) = \mathcal{R}^i_j(\gamma)$, one can immediately realize that

$$h \propto \frac{1}{R^2} \quad ; \quad \tilde{K} \propto \frac{1}{R^2} \quad . \tag{2.40}$$

The gravitational radiation gives rise to two terms in the perturbed Friedmann equation, Eq. 2.27. These terms appear as an effective density of the gravitational radiation and as radiation field they falls of like R^{-2}.

The scheme was devised to deal effectively with Friedmann like solutions, since we are interested in a realistic solution and we know that the observed universe resembles the Friedmann model to a large extend. Still it is useful to consider applying it to universes that are very different from Friedmann. One such universe in the Kasner universe, which is a vacuum universe with:

$$ds^2 = -dt^2 + t^{2p_1} dx^{1\ 2} + t^{2p_2} dx^{2\ 2} + t^{2p_3} dx^{3\ 2} \quad , \tag{2.41}$$

where p_1, p_2, p_3 satisfy:

$$p_1 + p_2 + p_3 = 1 \quad ; \quad p_1^2 + p_2^2 + p_3^2 = 1 \quad . \tag{2.42}$$

This universe is expanding with an overall expansion rate $R = t$, however it is not expanding isotropically, it is expanding in two directions and contracting in the third. The $t = Constant$ slices of this universe are homogeneous and flat. They are, therefore, slices of $trK = Constant$, in fact $trK = -3/Nt$. N is also constant. If we want to use Kasner's t as our time coordinate we must set $N = 1/3$, to satisfy Eq. 2.17. We obtain γ from Eq. 2.41 $\gamma_{ij} = g_{ij}/R^2 = g_{ij}/t^2$, hence $\gamma_{ii} = t^{2(p_i-1)}$. From Eq. 2.24b, it follows that $\tilde{K}^i_{\ i} = -(3/2)(2p_i - 1)/t$ and we can verify, using Eq. 2.42 that $tr\tilde{K} = 0$. Final consistency is demonstrated when we substitute $\tilde{K}^i_{\ j}, trK$ and N to the evolution equation, Eq. 2.25, and obtain an identity.

Examination of this solution reveals that factorization of R here does not necessarily improve the features of the numerical solution. Two of the metric functions, the expanding directions with $p_i > 0$, grow slower (compared to an unfactorized solution). However the third, the one with $p_i < 0$ decays faster by an additional power of t. Without numerical experiments it is difficult to judge how this scheme will work for a Kasner universe or for its Mixmaster generalization. It is clear however that for a Friedmann like solution this scheme is advantageous.

3. 3+1 Regge calculus

3.1 Four Dimensional Regge Calculus

Regge calculus is a finite elements approach to Numerical Relativity. In the original version of this calculus (Regge, 1961) the four dimensional spacetime is divided to flat elements (see Fig. 4). These elements are put together with a delta function curvature on the two dimensional area elements joining them. Hilbert's action's,

$$I_H = \int \sqrt{-g}(\mathcal{R} + L_{matter})d^4x) \quad , \tag{3.1a}$$

(with L_{matter} the action of the matter fields) becomes:

$$I_R = \sum_i A_i \epsilon_i + \sum_A L_{matter\ A} V_A \quad , \tag{3.1b}$$

where the first summation is over all two dimensional elements in the skeleton and the second summation is over all four dimensional elements. A_i is the area of a two dimensional element (labeled i) and ϵ_i is its deficit angle.

$$\epsilon_i = 2\pi - \sum_{\alpha \ni \epsilon} \theta_{\alpha i} \quad , \tag{3.2}$$

the summation here is over all tetrahedron (labeled with α) that include the area element i, and $\theta_{\alpha i}$ is the dihedral angle of this are in the tetrahedron. V_A is the four dimensional volume of the four dimensional element (labeled A). The analogue of Einstein equation is obtained by varying the action with respect to the squared edge lengths l_a^2:

$$\sum_{i \ni a} \frac{\partial A_i}{\partial l_a^2} \epsilon_i = 0 \ . \tag{3.3}$$

The contribution to this equation from variation of the deficit angle with respect to the edge lengths vanished identically by virtue of the Bianchi identities (Regge, 1961). This equations seem very attractive, they are the gravitational analogue of lattice gauge calculations in which one eliminates the gauge dependence by using only gauge independent quantities. The edge lengths, which are the proper distances between events in space-time, are gauge independent quantities. This motivated Hamber and Williams (1984) to perform gravitational lattice calculations (with a R^2 action) using this approach. From a cosmological point of view, in which we are interested in evolution of random perturbation fields, Regge calculus is attractive since, unlike in the common 3+1 approach discussed earlier, we do not introduce a computational grid, which may influence or bias our calculations.

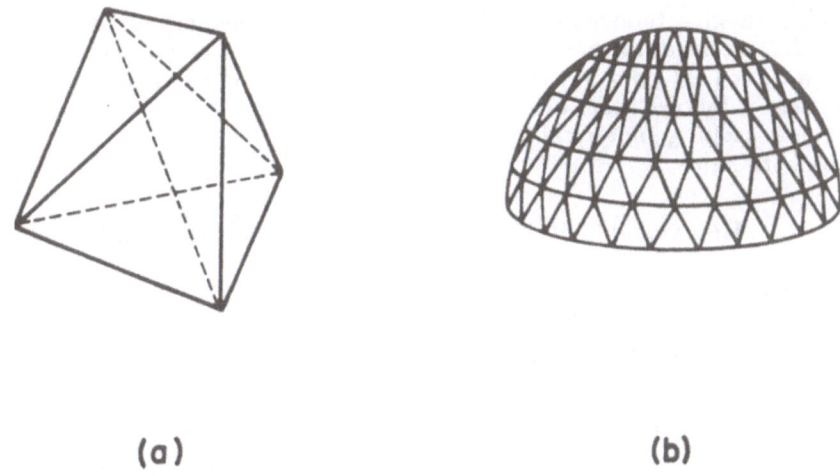

(a)   (b)

Fig. 4: (a) The five simplex is the building block of the four dimensional Regge calculus. (b) A geodesic dome made of triangles is a two dimensional analogue of the four dimensional Regge calculus.

To obtain a complete set of equations we have to add matter terms to the action and to include its variation with respect to the edge length in Eq. 3.3. to obtain a complete set of equations. Collins and Williams (1972, 1973, 1974) and latter Brewin (1986) have solved these equations for the analogue of the Friedmann universe, i.e. for a homogeneous cosmological solution.

While Eq. 3.3 is a relatively simple set of algebraic equations it is not clear yet how to convert these set of equations into a computational scheme. To solve these equations in an evolutionary manner we have to order the equations in a consistent temporary manner so that initial data could be formulated on some edges and we could solve systematically the rest of the equations for edge length that are to the future of this initial data (see however Sorkin, 1975, Dubal 1986, Miller, 1986).

In view of these difficulties we turn, in the rest of this work, to a 3+1 approach based on an original idea of Lund and Regge (1974) that was modified recently by Piran and Williams (1986) and by Friedman and Jack (1986). Like in the ADM (Arnowitt et al 1962.) formalism we cover the four dimensional spacetime by a foliation of three dimensional spacelike hypersurfaces. We then approximate the continuous three dimensional hypersurfaces by a three dimensional lattice and use Regge calculus to calculate the curvature of these slices. Time is kept continuous in this formalism. We obtain a set of ordinary differential equations for the edge lengths, (or in fact the edge length squared) $g_a = l_a^2$ and their conjugate momentum: p^a.

3.2 Geometric Formulae:

Our three dimensional hypersurfaces are approximated by a three dimensional lattice whose building blocks are tetrahedra. Lower dimensional elements in the lattice are vertices, edges and triangles. We label the vertices with $i, j, k...$, the edges with $a, b, c...$ and the tetrahedra with $\alpha, \beta, \gamma...$ (we will not need a label for the triangles).

We arrange a tetrahedron with a canonical arrangement of vertices with $i < j < k < l$. The edges are related to the vertices that they connect with: $a = i - j$; $b = i - k$; $c = i - l$; $d = j - k$; $e = j - l$; $f = k - l$ and from the arrangement of the vertices it follows that: $a < b < c < d < e < f$.

A few useful geometrical formulae are:
Area of a triangle:

$$A(g_a, g_b, g_c) = \sqrt{(2(g_a g_b + g_b g_c + g_c g_a) - g_a^2 - g_b^2 - g_c^2)/4} \qquad (3.4)$$

Volume of a tetrahedron:

$$V^2(g_a, g_b, g_c, g_d, g_e, g_f) = \frac{1}{144}(g_a g_f(g_d + g_b + g_c + g_e - g_a - g_f) +$$

$$g_d g_c(g_a + g_b + g_f + g_e - g_d - g_c) + g_b g_e(g_a + g_d + g_f + g_c - g_b - g_e) - \qquad (3.5)$$

$$g_a g_b g_d - g_a g_c g_e - g_d g_e g_f - g_b g_c g_f)$$

Dihedral angle on edge a :

$$\cos\theta(g_a, g_b, g_c, g_d, g_e, g_f) = \frac{g_a \Sigma(g_c, g_b, g_f) - \Sigma(g_a, g_b, g_d)\Sigma(g_a, g_c, g_e)}{4A(g_a, g_b, g_d)A(g_a, g_c, g_e)} = \qquad (3.6a)$$

$$\frac{2g_a(g_c + g_b - g_f) - (g_a + g_b - g_d)(g_a + g_c - g_e)}{\sqrt{(2(g_a g_b + g_b g_d + g_d g_a) - g_a^2 - g_b^2 - g_d^2)}\sqrt{(2(g_a g_c + g_c g_e + g_e g_a) - g_a^2 - g_c^2 - g_e^2)}}$$

or

$$\sin\theta(g_a, g_b, g_c, g_d, g_e, g_f) = \frac{3\sqrt{g_a}V(g_a, g_b, g_c, g_d, g_e, g_f)}{2A(g_a, g_b, g_d)A(g_a, g_c, g_e)} \qquad (3.6b)$$

The essence of the Regge calculus is that the spacelike hypersurfaces are piecewise linear. In other words each hypersurface is composed of flat elements that are connected in a curved way. Each tetrahedron is internally flat and we define a

metric in terms of its edge lengths. In fact we have a few metrics which are related
The first one is the trace metric:

$$g^{(\alpha)a} = -\frac{1}{V_{(\alpha)}^2} \frac{\partial V_{(\alpha)}^2}{\partial g_a} \quad , \tag{3.7a}$$

which is defined such that

$$(TrT)_{(\alpha)} = g^{(\alpha)a}T_a = g^{(\alpha)a}T_a = g^{(\alpha)a}l_a^\mu l_a^\nu T_{\mu\nu} \quad , \tag{3.7b}$$

and where summation over a repeated index (in this case a) is implicit. The index
a does not only label edges, it also replaces the spatial indices μ, ν (in this section
Greek indices, μ, ν range from 1 to 3) with a single index replacing two indices. As
a spatial index it can be raised and lowered. To emphasize the special role of the
spatial indices, $a, b, \ldots$ we put all other indices in brackets. We will also put a label
like a, b in brackets when they just label an edge. Comparison of

$$g_{\mu\nu}g^{\mu\nu} = 3 \quad , \tag{3.8a}$$

and

$$g_a g^{(\alpha)a} = 3 \quad , \tag{3.8b}$$

reveals that the inverse of the trace metric is just the edge length squared:

$$g_a = l_a^2 \quad . \tag{3.9}$$

While g^a depends on the tetrahedron, and therefore we label it with (α), the lower
metric g_a is independent of the tetrahedron. The trace metric is used to calculate
the trace of a symmetric two tensor (note that a single index like a stands for two
symmetric spatial indices). We can raise the indices of a symmetric tensor with the
raising metric:

$$G^{(\alpha)ab} = -\frac{\partial^2 \ln V_{(\alpha)}^2}{\partial g_a \partial g_b} \quad , \tag{3.10a}$$

$$T^{(\alpha)a} = G^{(\alpha)ab}T_b \quad . \tag{3.10b}$$

The inverse metric, the lowering metric is simply:

$$G_{ab}^{(\alpha)} = (G^{(\alpha)ab})^{-1} \quad . \tag{3.11}$$

These metrics are sufficient for our formalism. However, the 3+1 formalism
becomes somewhat simpler if we define the DeWitt's super metric, which provides
a measure for the distance between two symmetric two tensors:

$$\mathcal{G}^{\mu\nu\sigma\delta} = g^{\mu(\sigma}g^{\delta)\nu} - g^{\mu\nu}g^{\sigma\delta} \tag{3.12a}$$

$$\mathcal{G}^{(\alpha)ab} = -\frac{1}{V_{(\alpha)}^2} \frac{\partial^2 V_{(\alpha)}^2}{\partial g_a \partial g_b} \quad . \tag{3.12b}$$

The inverse of the metric is defined simply as:

$$\mathcal{G}_{ab}^{(\alpha)} = (\mathcal{G}^{(\alpha)ab})^{-1} \; . \tag{3.13}$$

A density for the DeWitt metric is obtained simply by multiplication

$$\hat{\mathcal{G}}^{(\alpha)ab} = V_{(\alpha)}\mathcal{G}^{(\alpha)ab} \; , \tag{3.14a}$$

or division

$$\hat{\mathcal{G}}_{ab}^{(\alpha)} = \frac{1}{V_{(\alpha)}}\mathcal{G}_{ab}^{(\alpha)} \; , \tag{3.14b}$$

by the volume element (which corresponds to $\sqrt{g}$).

3.3 The Einstein equations.

So far we discussed the geometry inside a tetrahedron. To proceed we must define the four dimensional geometry. We introduce, just like in the ADM formalism, the lapse function, $N_{(\alpha)}$, which is defined on each tetrahedron and measures the proper time along the world line of this tetrahedron. We should introduce a shift vector but, for simplicity, we ignore it in this discussion (a discussion of the shift vector in this formalism can be found in this formalism can be found in Friedman and Jack 1986). We introduce now p^a, the conjugate momentum to g_a. The 3+1 Hamiltonian is:

$$\mathcal{H} = \int dt \left[\sum_a p^a \dot{g}_a - \sum_\alpha N_{(\alpha)} \left(\frac{1}{4} \sum_{a,b\epsilon\alpha} \frac{\hat{\mathcal{G}}_{ab}^{(\alpha)}}{c_{(a)}c_{(b)}} p^a p^b + 2 \sum_{a\epsilon\alpha} g_{(a)}^{1/2}\theta_{(\alpha a)} \right) \right] , \tag{3.15}$$

where:

$$\theta_{(\alpha a)} = \frac{2\pi}{c_{(a)}} - \delta_{(\alpha a)} \; , \tag{3.16}$$

$\delta_{(\alpha a)}$ is the dihedral angle of tetrahedron α on edge a and $c_{(a)}$ is the number of tetrahedra hanging on edge a. Following Friedman and jack (1986) we have used the following definition :

$$\int_{V(\alpha)} NR\sqrt{g}d^3x = 2\sum_{a\epsilon\alpha} N_{(a)}g_{(a)}^{1/2}\theta_{(a)} = 2N_{(\alpha)}\sum_{a\epsilon\alpha} g_{(a)}^{1/2}\theta_{(\alpha a)} \; , \tag{3.17}$$

where $N_{(a)}$ is defined as:

$$N_{(a)}\theta_{(a)} = \sum_{\alpha \ni a} N_{(\alpha)}\theta_{(\alpha a)} \quad . \tag{3.18}$$

Note that there is a potential problem when $\theta_{(a)} = 0$.

We proceed to determine the Einstein equations from this Hamiltonain. First we vary Eq. 3.15 with respect to $N_{(\alpha)}$ and obtain the Hamiltonian constraint:

$$\frac{\delta \mathcal{H}}{\delta N_{(\alpha)}} = -\frac{1}{4} \sum_{a,b\epsilon\alpha} \frac{\hat{\mathcal{G}}_{ab}^{(\alpha)}}{c_{(a)}c_{(b)}} p^a p^b + 2 \sum_{a\epsilon\alpha} g_{(a)}^{1/2} \theta_{(\alpha a)} = 0 \ . \tag{3.19}$$

The g_a evolution equation is:

$$\frac{\delta \mathcal{H}}{\delta p^a} = \dot{g}_a = -\sum_{\alpha \ni a} \frac{1}{2} N_{(\alpha)} \sum_{b\epsilon\alpha} \frac{\hat{\mathcal{G}}_{ab}^{(\alpha)}}{c_{(a)}c_{(b)}} p^b \ , \tag{3.20}$$

and the p^a evolution equation is:

$$\frac{\delta \mathcal{H}}{\delta g_a} = -\dot{p}^a = -\sum_{\alpha \ni a} N_{(\alpha)} \left(\frac{1}{4} \sum_{b,c\epsilon\alpha} \frac{1}{c_{(b)}c_{(c)}} \frac{\partial \hat{\mathcal{G}}_{bc}^{(\alpha)}}{\partial g_a} p^b p^c + g_{(a)}^{-1/2} \theta_{(\alpha a)} \right) =$$

$$\sum_{\alpha \ni a} N_{(\alpha)} \left(\frac{1}{4} \sum_{b,c\epsilon\alpha} \frac{1}{c_{(b)}c_{(c)}} \frac{\partial \hat{\mathcal{G}}^{(\alpha)bc}}{\partial g_a} \hat{\mathcal{G}}_{bd}^{(\alpha)} \hat{\mathcal{G}}_{ce}^{(\alpha)} p^d p^e - g_{(a)}^{-1/2} \theta_{(\alpha a)} \right) = \tag{3.21}$$

$$\sum_{\alpha \ni a} N_{(\alpha)} \left(\frac{1}{4} \sum_{b,c\epsilon\alpha} \frac{1}{c_{(b)}c_{(c)}} \frac{\partial \hat{\mathcal{G}}^{(\alpha)bc}}{\partial g_a} \frac{p_b p_c}{V_{(\alpha)}^2} - g_{(a)}^{-1/2} \theta_{(\alpha a)} \right)$$

Just like in the continuous case we have to define coordinate conditions. For a cosmological scheme we use the constant mean curvature slicing. The condition $tr K = Const$ is replaced here by:

$$\frac{d \ln V_{(\alpha)}^2}{dt} = Const \ , \tag{3.22}$$

which yields:

$$\frac{d \ln V_{(\alpha)}^2}{dt} = \sum_{a\epsilon\alpha} \frac{1}{V_{(\alpha)}^2} \frac{\partial V_{(\alpha)}^2}{\partial g_a} \dot{g}_a = -\sum_{a\epsilon\alpha} g^{(\alpha)a} \dot{g}_a = \tag{3.23}$$

$$\sum_{a\epsilon\alpha} g^{(\alpha)a} \sum_{\beta \ni a} 2 N_{(\beta)} \sum_{b\epsilon\beta} \frac{\hat{\mathcal{G}}_{ab}^{(\beta)}}{c_{(a)}c_{(b)}} p^b = Const \ .$$

This condition has to be imposed on the initial data. The condition on the lapse function: $d tr K/dt = F(t)$ becomes:

$$\frac{d^2 \ln V_{(\alpha)}^2}{d^2 t} = F(t) \ , \tag{3.24}$$

which yields, after substitution of Eqs. 3.20 and 3.21 a set of linear algebraic equations for $N_{(\alpha)}$.

3.4 Matter Terms

It is not clear yet how to introduce a general matter terms to Regge calculus. In particular it seems that to include a perfect fluid one has to introduce velocity potentials (Shutz, 1970; Porter, 1986) which I will not discuss here. Luckily inclusion of a scalar field is fairly simple. We consider a scalar field with a general potential $\mathcal{V}(\phi)$. One can introduce a scalar field, $\phi_{(i)}$, located on the vertices (Friedberg and Lee., 1984; Piran and Williams, 1985, 1986). The matter Lagrangian becomes:

$$\mathcal{L}_\phi = \sum_i [-\frac{1}{2} \frac{\dot{\phi}_{(i)}^2}{N_{(i)}} + N_{(i)} \mathcal{V}(\phi_{(i)})]V_{(i)}^* + \sum_{j>i} \frac{1}{2} \frac{(\phi_{(i)} - \phi_{(j)})^2}{g_a} N_{(a)} V_{(a)}^* \quad , \quad (3.25a)$$

where in the last sum a is the edge connecting vertex j and vertex j, $V_{(i)}^*$, $V_{(a)}^*$ are the dual volumes to vertex i and edge a respectively (see e.g., Hamber and Williams 1984) and $N_{(i)}$ and $N_{(a)}$ are suitable averaged lapse function on the vertices and edges respectively. The dual volume elements are volume elements that can be associated with a vertex or an edge. Rather then give a formal definition of this elements we explain these concepts using a two dimensional analogue. Fig. 5 displays the dual area elements associated with a vertex and an edge in a triangle. The generalization to three dimension is obvious.

In section 3.3 we have defined the lapse function at the center of the tetrahedra. Here we need the lapse on the vertices and we have to use some averaging procedure. It might be advantageous therefore, to define the scalar field on the tetrahedra as well. We present an alternative Lagrangian with this choice:

$$\mathcal{L}_\phi = \sum_\alpha [-\frac{1}{2} \frac{\dot{\phi}_{(\alpha)}^2}{N_{(\alpha)}} + N_{(\alpha)} \mathcal{V}(\phi_{(\alpha)})]V_{(\alpha)} + \sum_{\beta>\alpha} \frac{1}{2} \frac{(\phi_{(\alpha)} - \phi_{(\beta)})^2}{l_{\alpha-\beta}^2} \frac{N_{(\alpha)} + N_{(\beta)}}{2} V_{(\alpha-\beta)}^* \quad ,$$

$$(3.25b)$$

where $l_{\alpha-\beta}^2$ is the distance between the centers of the tetrahedra α and β and $V_{(\alpha-\beta)}^*$ is the dual volume of the triangle which belongs to the tetrahedra α and β. The evolution equations for the scalar field and its contribution to the Hamilton-ain constraint and to the geometric evolution equations are obtained simply from variation of Eqs. 3.25a or 3.25b. If we wish to use a first order formalism we define a conjugate momentum $\psi_{(i)} = N_{(i)}\dot{\phi}_{(i)}$ or $\psi_{(\alpha)} = N_{(\alpha)}\dot{\phi}_{(\alpha)}$ and proceed to define a corresponding Hamiltonian.

3.5 Examples

To demonstrate the features of the 3+1 Regge calculus we consider a simple example of a homogeneous Universe. It is simplest to describe such a Universe by a tessellation in which all vertices, all edges, all triangles and all tetrahedra are equivalent. The simplest such tessalation is α_4 a skeleton with 5 vertices and 5 tetrahedra (see Fig. 4a).

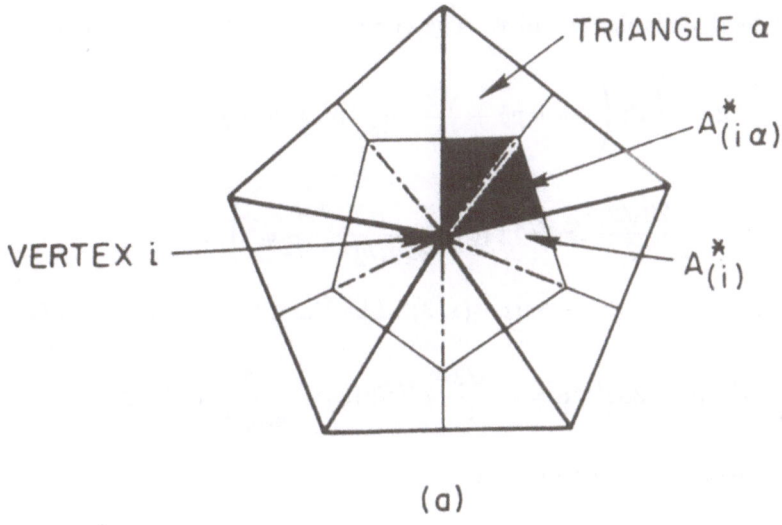

(a)

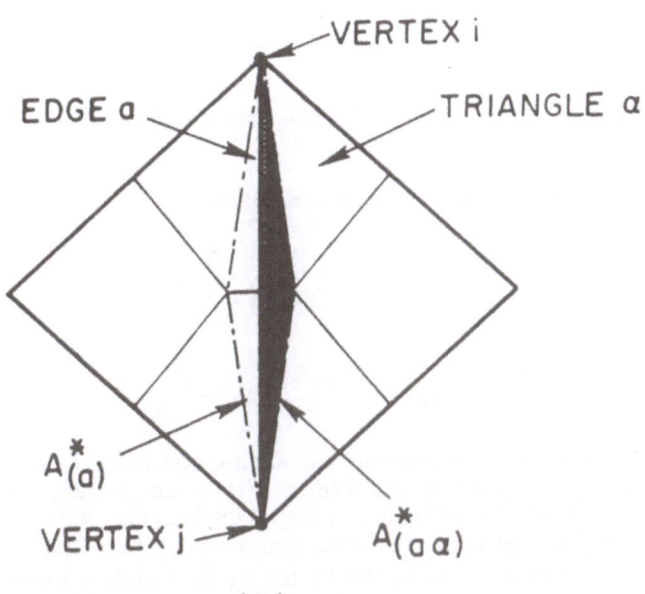

(b)

fig 5: *(a) The shaded area is $A^*_{(i)}$ the dual area to the vertex i. The very dark area is $A^*_{(i\alpha)}$ the part of the dual area to i which is in the triangle α. (b) The shaded area is $A^*_{(a)}$ the dual area to the edge a. The very dark area is $A^*_{(a\alpha)}$ the part of the dual area to a which is in the triangle α.*

Since all edges are equal we have only three geometric variables: g, p and N and two scalar field variables ϕ and ψ. The complete Hamiltonian becomes:

$$\mathcal{H} = \int dt \left(\psi\dot{\phi} + p\dot{g} + \frac{\sqrt{2}}{5} N g^{1/2} p^2 + 20 N g^{1/2} \delta \right.$$

$$\left. - \frac{20\sqrt{2}\pi}{3} N g^{3/2} \mathcal{V}(\phi) - \frac{3}{40\sqrt{2}\pi} \frac{N}{g^{3/2}} \psi^2 \right) , \tag{3.26}$$

where the deficit angle $\delta = 2\pi - 3arcos(1/3)$. The Hamiltonian constraint is:

$$\frac{\sqrt{2}}{5} g^{1/2} p^2 - 20 g^{1/2} \delta = \frac{20\sqrt{2}\pi}{3} g^{3/2} \mathcal{V}(\phi) + \frac{3}{40\sqrt{2}\pi} g^{-3/2} \psi^2 \tag{3.27}$$

The evolution equations for g and p are:

$$\dot{g} = -\frac{2\sqrt{2}}{5} N g^{1/2} p \tag{3.28}$$

and

$$\dot{p} = \frac{\sqrt{2}}{10} N g^{-1/2} p^2 + 10 N g^{-1/2} \delta - 10\sqrt{2}\pi N g^{1/2} \mathcal{V}(\phi) + \frac{9}{80\sqrt{2}\pi} g^{-5/2} \psi^2 \tag{3.29}$$

The evolution equations for the scalar field are:

$$\dot{\phi} = \frac{3}{20\sqrt{2}\pi} \frac{N}{g^{3/2}} \psi \tag{3.30}$$

and

$$\dot{\psi} = -\frac{20\sqrt{2}\pi}{3} N g^{3/2} \frac{d\mathcal{V}(\phi)}{d\phi} \tag{3.31}$$

Since the Universe is homogeneous the slicing condition $tr K = Const$ is satisfied trivially. The lapse equation: $dtr K/dt = F(t)$ becomes just an algebraic equation relating N and $F(t)$. Substitution of Eq. 3.28 into Eq. 3.27 yields the Friedmann equation for a homogeneous Universe (note that $\dot{g}/g = 2H$) coupled to a scalar field. The only difference is in the factor $4\sqrt{2}\delta$ which is unity in the continuous case. The origin of this term is the scalar curvature of the spatial slice. It is not surprising therefore, that this approximation for the geometry of the three-dimensional hypersurface does not agree exactly with the continuous calculation. This factor approaches unity if we work with finer tesselations.

I would like to thank Dalia Goldwirth for helpful discussions. This research was partially supported by NSF grant number PHY-8217352 to the IAS.

REFERENCES

Arnowitt, R. Deser, S. and Misner, C. W. (1962) in *Gravitation: Introduction to Current Research*, L. Witted, ed. Willey, New York.

Bardeen, J. M, (1983) in *Gravitational Radiation* Eds. Deruelle, N. and Piran, T. North Holland, Amsterdam.

Bardeen, J., M. and Piran, T., (1983) Physics Reports,96,20.

Belinski, V. et al. (1985) Phys. Lett B. **55**, 977, 1985.

Brewin, L., (1986) Paper presented at the GRG 11 meting, Stockholm.

Cantor, M., (1983) J. Math. Phys. **20** 1741.

Cantor, M. and Piran, T., (1983) General Relativity and Gravitation, **15**, 1077-1082.

Centrella, J. (1980) Ap. J. **241**, 875.

Centrella, J. and Matzner, R. A. (1983) Ap. J. **273**, 428.

Centrella, J., and Wilson, J. R. (1983) Ap. J. **273**, 428.

Centrella, J., Matzner, R. A., Rothman, T. and Wilson, J., (1986) Nucl. Phys. **B266**, 171.

Choquet-Bruhat and York, J. M. (1980)in *General Relativity and Gravitation* ed. A. Held, Plenum Press, New York.

Collins, P. A. and Williams, R. M., (1972) Phys. Rev. **D5**, 1908.

Collins, P. A. and Williams, R. M., (1973) Phys. Rev. **D7**, 965.

Collins, P. A. and Williams, R. M., (1974) Phys. Rev. **D11**, 3537.

Eardley D. and Smarr L. L. (1979) Phys. Rev. **D19**, 2239.

Friedberg, R. and Lee, T. D., (1984) Nucl. Phys. **B252**, 145.

Friedman J. and Jack, I. (1986) J. Math. Phys., in press.

Hamber, H. W. and Williams, R. M. (1984) Nucl. Phys. **B248**, 392.

Kurki Suonio, H., Centrella, J., Matzner, R. A., and Wilson, J., (1987) Phys. Rev. **D35**, 435.

Lichnerowicz, A. (1944) J. Math,. Pures er Appl., **23**, 37.

Lund H. and Regge, T., (1974) unpublished work.

Miller, W. (1986) private communication.

Misner C. W., Thorne, K. S. and Wheeler, J. A. (1973) *Gravitation* Freeman, San Francisco.

Pantano O. and Miller, J. (1986) preprint.

Penrose R. (1979) in *General Relativity an Einstein Centenary Volume* eds. Hawking, S. W. and Israel, W., Cambridge University Press.

Piran, T. (1980) J. Comp. Phys. **35**, 254.

Piran, T. (1983) in *Gravitational Radiation* Eds. Deruelle, N. and Piran, T. North Holland, Amsterdam.

Piran, T. and Williams, R. M. (1985) Phys. Lett **163B**, 331.

Piran, T. and Williams, R. M. (1986) Phys. Rev. **D33**, 1622.

Porter, R. (1986) preprint.

Regge, T. (1961) Nuovo Cimento **19** 558.

Schutz, B. F., (1970) Phys. Rev. **D2**, 2762.

Smarr L. L. and York, J. W. (1978a) Phys. Rev. **D17**, 1945.

Smarr L. L. and York, J. W. (1978b) Phys. Rev. **D17**, 2529.

Smarr, L. L., (1979) in *Sources of Gravitational Radiation* Ed. Smarr, L. Cambridge Univ. Press, Cambridge, England.

Sorkin, R. (1975) Phys. Rev. **D12** 385.

Weinberg, S. (1972) *Gravitation and Cosmology* , Willey, New York.

Wheeler J. A. (1964) in *Relativity Groups and Topology* ed. C. DeWitt and B. DeWitt (Gordon & Breach, N.Y.

Williams, R. M., (1985) Gen. Rel. and Grav., **17**, 559.

Williams R, M, and Piran, T. (1985) Contributed Talk by Williams at the IV Marcel Grossmann meeting , to be published in the proceeding.

Wilson, J. (1972) Ap. J. 173, 431.

Wilson, J. (1979) in *Sources of Gravitational Radiation* ed Smarr, L. Cambridge Univ. Press, Cambridge, England.

York, J. M. (1972) Phys. Rev. Lett. **28** 1982.

York, J. M. (1979) in *Sources of Gravitational Radiation* ed Smarr, L. Cambridge Univ. Press, Cambridge, England.

York, J. M. (1980) in *Essays in General Relativity* ed. Tipler F. , Academic Press, New York.

York, J. M. (1983) in *Gravitational Radiation* Eds. Deruelle, N. and Piran, T. North Holland, Amsterdam.

York, J.W. Jr. and Piran, T. (1982) in *Space-time and Geometry*, Shapely L. and Matzner R. Eds., Texas University Press, Austin, TX.

DISTORTIONS AND ANISOTROPIES

OF THE COSMIC BACKGROUND RADIATION

J. Richard Bond
Canadian Institute for Theoretical Astrophysics
University of Toronto
Toronto, Ontario
M5S 1A1, Canada

ABSTRACT: The general theory of spectral distortions and angular anisotropies in the cosmic background radiation is reviewed. Constraints on the amount of energy injection allowed in the early universe are discussed. Predictions of fluctuation levels are given for small and large angle anisotropy experiments for theories of structure formation in which the fluctuations are initially Gaussian and scale-invariant, as expected in inflationary models. The texture of maps of the CMB fluctuations is discussed using the theory of two-dimensional Gaussian random fields. The implications of the recently reported large scale streaming velocities for CMB fluctuations on scales of a few degrees are considered. Using a simple model for the fuzziness of recombination, simple equations are derived which give a physical overview of the various mechanisms which lead to anisotropy in the microwave background. Secondary anisotropies that result from Compton upscattering of the CMB photons off inhomogeneous hot gas and redshifted dust emission from primeval galaxies are estimated.

1. INTRODUCTION

As experimental upper limits on the anisotropy and spectral distortions in the microwave background improve, it is becoming increasingly important for our theoretical predictions in our various candidate models of structure formation to be quantitatively detailed. In this paper, the theory of the microwave background is reviewed, with emphasis on computation of the statistical predictions for models in which structure arises from scale invariant Gaussian initial fluctuations. These lectures are based on a review article that I am writing with George Efstathiou, and most of the results quoted here are part of our collaborative work.

The current status of CMB observations is reviewed by Wall in these proceedings. The main result is a distortionless and isotropic microwave sky. However, armed with the remarkably tight upper limits to the distortions and anisotropies that currently exist, we can slay many of the candidate models of the universe, while others are a hair's breadth away from being under the knife. When anisotropies or distortions are finally discovered, a probe of great power will be unleashed to reveal fatal diseases in all models but also, presumably, that one healthy survivor.

W. G. Unruh and G. W. Semenoff (eds.), The Early Universe, 283–334.
© *1988 by D. Reidel Publishing Company.*

2. GENERAL RELATIVISTIC TRANSPORT IN RANDOM MEDIA

The theoretical framework used to calculate the anisotropies and distortions of the CMB is general relativistic kinetic theory. In this section, we review the general aspects of this field required for the microwave background problem. The important equations for the treatment of distortions and anisotropies arising from this section are given in §2.4, so the reader may wish to skip §2.1-2.3 in this rather formal section. Kinetic theory in general relativity was actively developed in the late sixties and early seventies. (See Ehlers 1971 and Stewart 1971 for reviews.) The general relativistic version of the Liouville equation leading to the BBGKY hierarchy has difficulties in a general relativistic setting due to the necessity of following a large number of particles, each with their own proper time. Fortunately, we do not require such a theory to adequately treat the cosmological transport problem. Only a set of Boltzmann transport equations for single particle distribution functions is required of the kinetic theory; and this subject has been well developed.

2.1 The Distribution Function

The distribution function $f(q^i, x^\alpha)$ is the mean occupation number of the state of momentum q^i in the neighbourhood of the spacetime point x^α. (Spin is discussed below.) f is a general relativistic scalar under coordinate transformations of position and momentum. The number current vector J^a and stress-energy tensor T^{ab} are related to f by

$$J^a = \sum_q q^a f/q_0, \quad = \sum_q q^a q^b f/q_0, \quad \text{where} \quad \sum_q \cdots \equiv \int (-g)^{1/2} d^3q/(2\pi)^3 \cdots \quad (2.1)$$

denotes momentum or wavenumber space integrations. Here, q_0 is the covariant time component of the momentum, and g is the determinant of the metric tensor. In this paper, we (almost) always take $\hbar = c = 1$.

Due to the nonlocalizability of position and momentum, one must be more precise in defining f. One way is to use the Wigner distribution function approach, whereby f would be defined in terms of a two-particle equal-time propagator (Greens function):

$$f(q, x, t) = \sum_k e^{ik \cdot x} \langle a^\dagger_{q-k/2}(t) a_{q+k/2}(t) \rangle$$

$$= \int d^3x' \, e^{iq \cdot x'} \langle \psi^\dagger(x + x'/2, t) \psi(x - x'/2, t) \rangle. \quad (2.3)$$

Here, $a_{q+k/2}$ annihilates a particle of momentum $q+k/2$, $a^\dagger_{q-k/2}$ creates a particle of momentum $q-k/2$, and $\langle\,\rangle$ denotes an ensemble average; the ensemble will generally be a nonequilibrium one. Baym and Kadanoff (1962) follow this development in the nonrelativistic case. f defined this way is not a positive definite quantity and so the interpretation of f as phase space density is invalid. Further, the use of Fourier transforms is of limited utility in a general relativistic setting. Its advantage, however, is that much of many body theory is cast in the language of propagators, and the equation for the evolution of f is just a special case.

Another type of definition involves choosing a complete othonormal basis labelled by splitting phase space (x, q) into boxes of volume $(2\pi\hbar)^3$ and labelling the centers by (X, Q). Wave functions $\langle x|XQ\rangle$ which are zero outside of the spatial part of the box and which are box-normalized plane waves, $\langle x|XQ\rangle = exp(iQ.x)$, inside, form a complete orthonormal set of wavefunctions, each with an associated annihilation and creation operator. The occupation indices of each such fundamental phase space cell,

$$f(Q, X, t) = \langle a^\dagger_{XQ}(t)a_{XQ}(t)\rangle, \tag{2.4}$$

define the distribution function. Osborn and Yip (1973) have developed the theory of (nonrelativistic) neutron transport using this definition. The relative degree to which the boxes are spatially elongated is at our disposal provided the quantum volume constraint is maintained. The volumes we deal with are certainly macroscopically small, so gravitational field variations are generally extremely weak over the boxes. This prescription thus has some formal advantages over the Wigner distribution function approach. A conceptual advantage of this approach is that it emphasizes the fundamental graininess imposed by quantum mechanics on phase space. Coarse graining of phase space only involves making the boxes of much larger volume than that required by the Heisenberg uncertainty principle. A disadvantage of this approach is that boundary terms involving transport from one box to another are complicated.

Either of the above two approaches give exact quantum evolution equations which reduce to the usual form of the Boltzmann transport equation (2.8 below) in the limit when the spatial inhomogeneities of the distribution function are of long wavelength compared with the typical de Broglie wavelength of the particles, q^{-1}. We also require that the gravitational field curvature be small over the distance q^{-1}. Coherent effects - such as the modification of the photon propagator by collective plasma effects - must be taken into account by appropriately defined quasiparticles which have these collective interactions included. In this example, the transport would be of transverse plasmons rather than of bare free photons.

2.2 Polarization and Stokes Parameters

If the particles have spin (or polarization) labelled by s, then a state is specified by the the ket $|XQs\rangle$. The distribution function then becomes a matrix in spin space,

$$f_{ss'}(Q, X, t) = \langle a^\dagger_{XQs'}(t)a_{XQs}(t)\rangle, \tag{2.5}$$

Not only are there the diagonal components f_{ss} giving the occupation number of the state with spin s, but also there are off-diagonal components $f_{ss'}$, $s \neq s'$, which contain phase information, describing the probability amplitude for propagation from a state of spin s to a state of spin s'.

For photons, there are two polarizations, hence a 2×2 'polarization matrix' transverse to $\hat{q} \otimes \hat{q}$ is required for photons in the direction $\hat{q}$. Consider linear polarization. If we expand the polarization distribution function in terms of the basis consisting of the Pauli matrices σ_i, $i = 1, 2, 3$ and the identity $\sigma_0 = diag(1, 1)$,

$$\mathbf{f} = (f_{ss'}) = \frac{1}{2}s^\mu\sigma_\mu, \tag{2.6}$$

then the 4 real distribution functions s^μ are just conventional Stokes parameters:

$$s^0 = I, \ s^1 = U, \ s^2 = V, \ s^3 = Q. \tag{2.7}$$

The polarization matrix approach was developed by Fano (1954) and others in optics.

For Thomson scattering, the mechanism responsible for the generation of polarization in the microwave background, $V=0$. Further, only Q and of course I are needed for the computation of polarization in a plane parallel atmosphere (Chandrasekhar 1960). If Fourier plane wave decomposition is valid, as in the anisotropy calculations described in §5, only Q is required. There we call s^0 f_T and s^3 f_P. Upon reconstruction of the distribution function in position space by inverse Fourier transform, U appears as well, though V remains zero.

For neutrinos, there is chirality conservation, so only one distribution function is required for 'left-handed' chirality. If right-handed neutrinos exist, then a separate distribution function would be required for them. If there were a Dirac-type mass term which couples left and right chiralities, then in general a 2×2 matrix would be required. If neutrinos can oscillate between one flavor and another, and there are three neutrino flavors, then the distribution function would in that case become a 3×3 matrix in flavor space. Majorana neutrinos require just two distribution functions, one for the left-handed component, the neutrino, and one for the right-handed component, the antineutrino.

The general rule is that if phase information is irrelevant, so only probabilities are important, only the diagonal components associated with a given set of quantum variables need be considered. However, if transformations among the quantum variables are important as in the rotation to a new set of polarization states, then the off-diagonal terms must in general be included. The general propagation equation equation for $f(X, Q, t)$ involves the off diagonal amplitudes $\langle a^\dagger_{X'Q} a_{XQ} \rangle$ which we have ignored in adopting the Boltzmann transport equation.

If the transport of nonrelativistic electrons is to be described by a Boltzmann equation, then a 2×2 matrix in spin space would generally be required. The timescales of equilibration due to Coulomb interactions are so fast that electrons can be treated as Maxwell Boltzmann distributed with a bulk velocity, temperature and chemical potential, in which the off-diagonal components, $f_{1/2,-1/2}$ and $f_{-1/2,1/2}$, vanish. That is, we need only follow the transport of gas particles in the hydrodynamical limit.

2.3 Boltzmann Transport Equation

Our transport model considers the particles propagating along geodesics in spacetime. The particles may undergo absorptions or emissions or scatterings at single points. For such a description of collisions to be valid it is also necessary that the interaction regime be small in spatial and temporal extent compared with the scale of inhomogeneity in f. Also. in order for the equations to be closed off at the single-particle distribution level, the only correlations allowed to be explicitly included are those due to the particle statistics, Bose-Einstein or Fermi-Dirac. In this section, we discuss the transfer equation following the development given in Bond and Szalay (1983).

In general relativity, the one-particle phase space is a seven-dimensional submanifold of the eight-dimensional tangent bundle in the spacetime manifold. That is, particles of momentum p at the spacetime point x^α are constrained to lie on the mass shell $g_{ab}(x)p^a p^b = m^2$ in the momentum space located at the point x. A picturesque way to see this is to consider the tangent space at each point x^α radiating out normally from the point as a fiber, one of many in a fiber bundle. The trajectory of a particle then consists of motion through the fibres. A 7-dimensional coordinate system for the phase space can be arbitrary. It is usual to take the spatial coordinate system x^α and the 3 coordinate components $p^\alpha = \langle dx^\alpha, q \rangle$, where dx^α denotes the coordinate basis, and $\langle \omega, v \rangle = \omega(v)$ denotes the action of the 1-form ω on the vector v. (See Misner, Thorne and Wheeler 1973 for basic differential geometry notation.) In such a coordinate system, the equation for the evolution of the distribution function is

$$q^\alpha \frac{\partial f}{\partial x^\alpha}\bigg)_q - \Gamma^i_{\alpha\beta} q^\alpha q^\beta \frac{\partial f}{\partial q^i}\bigg)_x = q^0 S[f], \tag{2.8}$$

where $S[f]$ is the source function which include annihilations, creations and scatterings. $q^0 S[f]$ transforms as a general relativistic scalar, which conveniently allows transformation from one gauge to another. We discuss specific contributions to this term in §3. Equation (2.3) expresses conservation of the mean occupation number (i.e., of phase space density) along the particle trajectory. Though it is not manifestly covariant due to the summation in the second term running over spatial indices only, it is, in fact, covariant. This is a consequence of being constrained to lie on the mass shell. Any other choice of the 3 coordinates parameterizing the mass shell will do; for example, the energy and the angles defining momentum direction give a spherical coordinate system.

We choose to use the spatial momentum components relative to an orthogonal tetrad $e^a_\mu(x^\alpha)$ which is not generally derivable from a coordinate basis (i.e. it is a non-holonomic basis). Relative to an arbitrary tetrad, the transfer equation takes the form in which we shall use it:

$$q^a e_a(f) - \Gamma^{\hat i}_{ab} q^a q^b \frac{\partial f}{\partial q^i}\bigg)_x = q^{\hat 0} S[f], \quad e_a(f) \equiv e^\mu_a \frac{\partial f}{\partial x^\mu}\bigg)_q. \tag{2.9}$$

The connection coefficients relative to the tetrad e_a are $\Gamma^a_{bc} = \langle e^a, \nabla_{e_b} e_c \rangle$.

For problems involving an expanding geometry, it is convenient to adopt an orthogonal tetrad obeying $e^\mu_a g_{\mu\nu} e^\nu_b = a^{-2} \eta_{ab}$, where $a = (1+z)^{-1}$ is the expansion factor of the universe in terms of the redshift and $\eta_{ab} = (1, -1, -1, -1)$ for the sign convention adopted here. Thus ae_b is orthonormal. The momentum components relative to e_b, $q^b = \langle e^b, q \rangle$, are *comoving momenta* which remain constant as the universe expands in the absence of gravitational perturbations. If the particles have nonzero mass, the comoving energy $q^{\hat 0} = (q^2 + m^2 a^2)^{1/2}$ does change as the universe expands.

In perturbation theory, the tetrad

$$e_a = a^{-2}(\delta^\nu_a - \frac{1}{2} h^\nu_a) \frac{\partial}{\partial x^\nu}, \quad e^a = a^2(\delta^a_\nu + \frac{1}{2} h^a_\nu) dx^\nu \tag{2.10}$$

is orthogonal to linear order in the perturbed gravitational field, $h_{\mu\nu} = a^{-2} g_{\mu\nu} - \eta_{\mu\nu}$, and is a convenient one to adopt. Raising and lowering indices is done with respect to η, so $h^a_\mu = \eta^{a\nu} h_{\mu\nu}$. The connection coefficients relative to this tetrad are

$$\Gamma^a_{bc} = a^{-2}(h^a_{b,c} - h^{;a}_{bc})/2 + \delta^a_b e_c(\ln a) - \delta^a_c e_b(\ln a) - \eta^{ad} e_d(\ln a)\eta_{bc}. \qquad (2.11)$$

(Partial derivatives are denoted by ,.) The transfer equation appropriate to conformal linearized Einstein theory is then

$$q^a \frac{\partial f}{\partial x^a}\bigg)_{q^i} + \frac{1}{2}q^a q^b(h^i_{b,c} - h^{;i}_{bc})\frac{\partial f}{\partial q^i}\bigg)_x + \frac{1}{2}\left[q^a h^b_a \frac{\partial f}{\partial x^b}\bigg)_{q^i}\right] = q^{\hat{0}}S[f]. \qquad (2.12)$$

The only assumption made is that the comoving gravitational field be linearized. This equation is valid for arbitrary gauge choices. Note that the change of momentum coordinates on the mass shell is itself part of a gauge transformation.

The equation describing the transfer of $f_{ss'}$ with polarization is identical, except that the source function becomes $S_{ss'}[\mathbf{f}]$, a matrix in polarization space.

2.4 Background and Perturbed Boltzmann Equations

In perturbation theory, we expand the distribution function $f = \bar{f} + \delta f$ in terms of a background distribution function $\bar{f}$ and a first order perturbation δf. In first order perturbation theory, the term in square brackets in eq. (2.12) is dropped, f is replaced by $\bar{f}$ in the momentum derivative term, and only first order terms δS in the source function are kept, involving, for example, the perturbed baryon density as well as δf.

In §5, we adopt the synchronous gauge in which $h_{0\mu} = 0$, so only the spatial part remains. The unperturbed zeroth order transfer equation and the first order transfer equation in this gauge are then simply

$$\partial \bar{f}/\partial \tau\big)_q = \bar{S}[\bar{f}],$$
$$\frac{\partial f}{\partial \tau}\bigg)_q + \frac{q}{q^0}\hat{q}^i \frac{\partial f}{\partial x^i}\bigg)_q + \frac{1}{2}\dot{h}_{ij}\,\hat{q}^i\hat{q}^j q \frac{\partial \bar{f}}{\partial q} = \delta S[\bar{f}, \delta f]. \qquad (2.13)$$

where q denotes the magnitude of q^i and $\hat{q}$ denotes a unit vector (which is identical to the $\vec{\gamma}$ of Peebles and Yu 1970). Thus, in evaluating the background equation, which is the appropriate one for the study of *distortions*, the unperturbed source term should be expressed as a function of comoving frequencies with the redshift taken out rather than of physical frequencies to solve for deviations from Planckian $f_P = [\exp(q/(aT_\gamma)) - 1]^{-1}$. Note that f_P satisfies the background equation, being τ independent in both the tight coupling and free streaming regimes provided there is no energy injection so aT_γ is constant. The explicit form of the perturbed equations for photon transport for the study of *anisotropy*, including polarization ($f_{ss'}$ replacing f in 2.13), is given in §5.

3. SOURCE FUNCTIONS FOR THE TRANSPORT EQUATIONS

3.1 Processes

Provided the temperature of the universe is well below $m_e c^2$ where $e^+ e^-$ pairs recombine, only a small number of processes have to be included to adequately describe the photon transport. In order of importance, these are:

3.1.1 Compton scattering, $\gamma e \to \gamma e$.

For the nonrelativistic electrons appropriate to the period after pair recombination, Compton scattering is primarily Thomson scattering, a conservative scattering process in which the outgoing photon energy ω' equals that of the incoming one ω, so momentum but not energy is transferred. The associated source function can describe the development of anisotropy, but will give rise to no spectral distortion. Its form is given in Table 1 for the cases of unpolarized and partially polarized light. For this source function to vanish, it is necessary that the radiation field be isotropic in the comoving frame of the electrons.

Thomson scattering is the zeroth order term in an expansion in T_e/m_e, where T_e is the electron temperature. Small energy transfers $\Delta\omega = \omega - \omega'$ do occur in the next order:

$$\langle \frac{\Delta\omega}{\omega} \rangle = 4\frac{T_e}{m_e} - \frac{\omega}{m_e}, \quad \langle (\frac{\Delta\omega}{\omega})^2 \rangle = 2\frac{T_e}{m_e}. \tag{3.1}$$

The averages here are taken with respect to the Compton scattering kernel which, in this case, is related to the Klein-Nishina cross section averaged over the thermal electron distribution, $n_e \langle d\sigma_{KN}/[(2\pi)^3 d^3 q] \rangle$. These small transfers are of great importance in spectral modification, and are included as the first order Fokker-Planck expansion of Compton scattering, describing the drift and diffusion in energy space following energy injection. The source function S_K for an isotropic radiation field is given in Table 1. This equation was derived by Kompaneets (1957). It vanishes when the distribution function is of Bose-Einstein form,

$$f_{BE} = [\exp(\omega/T_e + \alpha) - 1]^{-1}; \quad \alpha = -\mu_\gamma/T_e \tag{3.2}$$

relates α to the photon chemical potential μ_γ. A chemical potential enters because photon number is a conserved quantity in Compton scattering. Such a 'kinetic equilibrium' distribution causes the full Compton scattering source function to vanish.

We can ignore terms in the source function which describe the small $\mathcal{O}(T_e/m_e)$ corrections to the nonzero Thomson scattering terms. Thus, the source function can be expanded as $S_{\gamma e} = S_K + \delta S_K$. The Kompaneets term describes anisotropies as well as distortions, since the Kompaneets rate, $\Gamma_K = 4n_e\sigma_T T_e/m_e$, can have spatial fluctuations as a result of pressure perturbations in the (ionized) gas. We delay the discussion of this, the Sunyaev-Zeldovich effect (Zeldovich and Sunyaev 1969), which is of importance primarily at late times once the universe has gone large scale nonlinearity, until §7.

3.1.2 Bremsstrahlung, $ep \to ep\gamma$, $\gamma ep \to ep$.

The source function is given in Table 1. Since free-free processes are characterized by the logarithmic divergence, $dn_\gamma \sim d\omega/\omega$, in the photon number, bremsstrahlung is very efficient at filling in an equilibrium Planck distribution (with zero chemical potential) at low energies, but is very inefficient at high energies; nonetheless, given enough time, that is what the distribution will drive towards.

3.1.3 Double Compton scattering, $ee \to ee\gamma$, $\gamma ee \to ee$,

in which the Compton scattered electron shakes off a soft photon. This is like the bremsstrahlung process

Table 1: Transport Source Functions (§3.1)

1. COMPTON SCATTERING

● THOMSON SCATTERING (anisotropy):

Rate $= \Gamma_s = n_e\sigma_T = (1400\,y)^{-1}Y_e\omega_B z_3^3$

unpolarized:

comoving gauge $\mathcal{S}[f] = -n_e\sigma_T(f - f_0 - P_2 f_2/2)$

synchronous gauge $\mathcal{S}[f] = -n_e\sigma_T(1 - \hat{q}\cdot\vec{v})(f - f_0 - P_2 f_2/2 + \hat{q}\cdot\vec{v}\omega\partial f_0/\partial\omega)$

partially linearly polarized:

comoving gauge $\mathcal{S}[f_T, f_P] = -n_e\sigma_T(f_T - f_{T0} - P_2 f_{T2}/2 + P_2(f_{P0} - f_{P2})/2)$

synchronous gauge $\mathcal{S}[f_T, f_P] = -n_e\sigma_T(f_T - f_{T0} - P_2 f_{T2}/2 + P_2(f_{P0} - f_{P2})/2 + \hat{q}\cdot\vec{v}\omega\frac{\partial f_0}{\partial\omega})$

Momentum Transfer per Baryon (Compton Drag):

$'\dot{v}'_B = (T_\gamma s_\gamma \Gamma_s/m_N)(v_\gamma - v_B) = (440\,y)^{-1}Y_e z_3^{-1}(v_\gamma - v_B)$

● KOMPANEETS SCATTERING (distortion and late-time anisotropy):

Rate $= \Gamma_K = 4n_e\sigma_T(T_e/m_e) = (7.7\times10^8\,y)^{-1}Y_e\omega_B z_3^3(T_e/T_\gamma)$

Source Function: $\mathcal{S}[f] = \Gamma_K \frac{1}{4T_e\omega^2}\frac{\partial}{\partial\omega}\left[\omega^4\left(T_e\frac{\partial f}{\partial\omega} + f(1+f)\right)\right]$

Energy Transfer per Baryon (Compton Cooling):

$'\dot{\epsilon}'_B = \frac{3}{4}s_\gamma\Gamma_K\frac{T_\gamma}{T_e}(T_\gamma - T_B) = (0.8\,y)^{-1}Y_e z_3^4(T_\gamma - T_B)$

$'\dot{u}_\gamma/u'_\gamma = \Gamma_K(\frac{T_\gamma}{T_e} - 1)$

2. BREMSSTRAHLUNG

Absorption Rate: $\Gamma'_B = \Gamma_B(1 - e^{-x})$

$\Gamma_B = \frac{(2\pi)^{3/2}}{3^{1/2}}\frac{n_B}{T_e^3}\alpha n_e\sigma_T\left(\frac{m_e}{T_e}\right)^{1/2}\frac{g(x)}{x^3}$

$\quad = (2.1\times10^{10}\,y)^{-1}\omega_B^2 Y_e z_3^{5/2}\left(\frac{T_\gamma}{T_e}\right)^{7/2}\frac{g(x)}{x^3}$

$\quad g(x) \approx 1, \quad x > 1, \quad \frac{\sqrt{3}}{\pi}ln[2.25/x], \quad x < 1$

Source Function: $\mathcal{S}[f] = -\Gamma'_B(f - f_{eq})$

3. DOUBLE COMPTON SCATTERING

Absorption Rate: $\Gamma'_{DC} = \Gamma_{DC}(1 - e^{-x}), \quad x = \omega/T_e, \quad x' = \omega/T_\gamma$

$\Gamma_{DC} = \frac{16\pi^3}{45}\alpha n_e\sigma_T\left(\frac{T_e}{m_e}\right)^2\frac{g_{DC}(x)}{x^3} = (1.2\times10^{11}\,y)^{-1}\omega_B Y_e z_3^3 T_e^2\left(\frac{T_\gamma}{T_e}\right)^5\frac{g_{DC}(x')}{x^3}$

$g_{DC}(x') = \frac{15}{4\pi^4}\int_{2x'}^\infty f(y)(1 + f(y - x'))y^4 dy[wF(w)/2]$

$[wF(w)/2] = \frac{1}{2}(1 - w)[(1 + (1 - w)^2 + w^2(1 + w^2)/(1 - w)^2 + w^4 + w^2(1 - w)^2],$
$\quad (\to 1 \text{ for } small\ w),$

$g_{DC}(x) \approx \frac{32}{33}e^{-x}\left[1 + 2x + 2x^2 + 4x^3/3 + 2x^4/3 + \frac{1}{32}e^{-x}[1 + 4x + 8x^2 + 32x^3/3 + 32x^4/3]\right].$

Source Function: $\mathcal{S}[f] = -\Gamma'_{DC}(f - f_{eq})$

4. GRAIN ABSORPTION

Absorption Rate: $\Gamma'_d = \Gamma_d(1 - e^{-x}), \; x = \omega/T_d$

$\Gamma_d = n_d\sigma_d(1 + (\omega_d/\omega)^\alpha)^{-1}, \; \omega_d = C_d/r_d, \; C_d \sim 1, \; \alpha \sim 1 - 2$

Source Function: $\mathcal{S}[f] = -\Gamma'_d(f - f_{eq})$

in that there is a logarithic divergence in number of low energy photons emitted. However. the rate of low energy emission is dominated by bremsstrahlung for typical early universe conditions. The source functions for Double Compton scattering have not been previously been adequately treated in the literature. In Table 1, we give the detailed rates, using the cross sections derived by Gould (1984). The Double Compton process has a different dependence on photon energy than bremsstahlung at high energies. The net effect is that it never dominates in the cosmologically interesting regime, as can be seen from Table 1.

3.1.4 Grain absorption and emission. If metals created by an early generation of stars (of Population III as discussed in §6) can condense into grains, then ultraviolet and visible radiation from pregalactic objects can be absorbed. Re-emission by the dust at CMB wavelengths can result in CMB spectral distortions, and also in anisotropies in the distortions. Dust associated with protogalaxies may re-emit a significant infrared cosmic background radiation. A typical source function is given in Table 1. The uncertain parameters describing the dust absorption rate, $\Gamma_d \approx n_d \sigma_d (1 + (\omega/\omega_d)^\alpha)^{-1}$, are (a) the UV absorption rate, which is generally not far from the rate associated with the geometrical cross section for grains of (spherical) size r_d, $\sigma_d = \pi r_d^2$; (b) the energy at which the absorption rate begins

Notes to Table 1: In these expressions we have taken $\hbar = c = k_B = 1$. Temperatures are in energy units, keV in numerical expressions. The energy rates are those appropriate to near equilibrium transfer from photons to plasma. All processes but the special case of Thomson scattering are written in the comoving-electron gauge. Thus ω should be interpreted as $\langle U, q \rangle$, where U is the bulk 4-velocity of the gas. T_e and T_γ are the electron and photon temperatures, $x = \omega/T_e$, $\omega_B = \Omega_B h^2/0.1$, $Y_e = n_e/n_B$, $z_3 = (1+z)/1000$, $s_\gamma = 4u_\gamma/(3n_B T_\gamma) = 1.28 \times 10^9 \omega_B^{-1}$, $u_\gamma = aT_\gamma^4 = 0.25(1+z)^4$ ev cm^{-3}, m_N = nucleon mass, m_e = electron mass, $\sigma_T = (8\pi/3)\alpha^2/m_e^2$ is the Thomson cross section and α is the fine structure constant. $f_{T\ell} = \int d\mu/2\, f_T P_\ell(\mu)$ and $f_{P\ell} = \int d\mu/2\, f_P P_\ell(\mu)$ are multipole moments with respect to Legendre polynomials P_ℓ of the total distribution function $f \equiv f_T$ and of the polarization distribution function f_P (as defined in §2.2 and §5.1) for plane parallel situations, where $\mu = \hat{q} \cdot \hat{z}$ is the cosine of the angle between the photon direction $\hat{q}$ and the z-direction normal to the plane. For our perturbative radiative transfer case, $\hat{z} = \hat{k}$ for each plane wave with wavevector $\vec{k}$ in a Fourier decomposition of the perturbed distribution functions. The unpertubed f_T, $\bar{f}$, is a Planckian and the unperturbed f_P, $\bar{f}_P$, vanishes. We define the perturbed temperature distortions and anisotropies by $\Delta T/T \equiv \Delta_T/4 \equiv (f_T - \bar{f})[\bar{f}(1+\bar{f})\omega/T]^{-1}$, with a similar definition for the perturbed polarization $\Delta_P/4$ (§5.1).

an approximately power law decline, $\omega_d = C_d/r_d$, where $C_d \sim 1$; and (c) the power law index α which is frequency dependent, typically ranging from ≈ 1 at energies below ω_d to ≈ 2 at very low energies, though resonances can substatially modify the general power law decline in specific wavelength regions. The major uncertainty is the grain size, which may plausibly vary from $\sim 10^{-7}$ cm to $\sim 10^{-4}$ cm, and the overall grain abundance n_d which many theories of galaxy formation would have zero until redshift ten or so. The work on grain-generated spectral distortions is reviewed in §6.

3.1.5 Rayleigh scattering, $\gamma H \rightarrow \gamma H$. This process has an identical source function to that for Thomson scattering except that the neutral hydrogen and helium abundances replace that of the electrons. As such, the process could only be of importance after recombination. The ratio of Rayleigh scattering to Thomson obeys the fourth power law, $\Gamma_R/\Gamma_T \sim (Y_H/Y_e)(\omega/\omega_\alpha)^4$, where $\omega_\alpha = 10.2$ eV is the Lyman α transition energy. For typical photon energies at $z = 1000$, this is $2 \times 10^{-5} Y_H/Y_e$, much smaller than unity. The decline at lower temperatures is precipitous, and Rayleigh scattering never enters a regime where it can exceed the Thomson scattering associated with the residual ionization ($Y_e \sim 10^{-5} - 10^{-3}$, §4). Here,

$$Y_j \equiv n_j/n_B \qquad (3.3)$$

denotes the number of particles of a given type per baryon.

3.1.6 Line Radiation. Lines formed during the recombination of helium at T in the $10^{4-5} K$ range and of hydrogen at $T \sim 1/4$ eV are either too weak to be easily observable, or are buried in the background associated with interstellar dust emission (Lyubarski and Sunyaev 1983). Such processes play a very important role in the recombination process itself of course, and this is discussed in §4.

3.2 Cosmic Photosphere and Perturbative Spectral Distortions

At redshifts higher than that of recombination, photons diffuse very slowly from region to region due to the efficiency of Thomson scattering. Accompanying each random walk will be many Compton upscatters or downscatters; which depends primarily on whether $\omega < 4T_e$ or $\omega > 4T_e$. There may also be a rare free-free or Double Compton absorption, especially if the photon energy is low. The photon number will be enhanced at the low ω end by bremsstrahlung and Double Compton emission, aiding in the drive toward electron-photon 'chemical' equilibrium ($\mu_e + \mu_p \rightarrow \mu_e + \mu_p + \mu_\gamma$ hence $\mu_\gamma = 0$), which will occur for high energies on a much longer timescale than the drive toward kinetic equilibrium powered by Kompaneets scattering (($\mu_\gamma + \mu_e \rightarrow \mu_e + \mu_\gamma + \mu_\gamma$ hence μ_γ is indeterminate). The photon distribution function thus can rapidly approach a situation in which the frequency-dependent chemical potential is zero for low frequencies, frequency-independent but time-dependent for high frequencies, with a small transition region in between.

3.2.1 Bose Einstein Distortions: An explicit solution of the transfer problem describing this behavior based on Zeldovich and Sunyaev (1969) is now described. We assume the distribution function has the form $f = [\exp(x + \alpha(x,t)) - 1]^{-1}$ and linearize the transport equation in α, which can be expressed in terms of a frequency dependent distortion in the thermodynamic temperature: $(\Delta T/T)(\omega, t) = -\alpha/x$.

If we require that $S_K + S_B$ vanishes, we obtain, for small $x = \omega/T_e$, a solution of form $\alpha = \alpha_0(t)\exp(-x_0/x)$. A similar form holds, though with different values of x_0, if Double Compton scattering is included as the number source, or if both are included:

$$x_{0B} = (4x^3\Gamma_B(x)/\Gamma_K)^{1/2},$$
$$x_{0DC} = (4x^3\Gamma_{DC}(x)/\Gamma_K)^{1/2}, \quad x_{0t} = (4x^3(\Gamma_B + \Gamma_{DC})/\Gamma_K)^{1/2}. \tag{3.4}$$

The approximate constancy of $\Gamma_B(x)x^3$ has been exploited in obtaining this formula. For $x < x_{0t}$, the distribution is Planckian, and above it is of the Bose-Einstein form. The time dependent coefficient $\alpha_0(t)$ is obtained by again linearizing the photon number and energy densities in α_0, $n_\gamma = n_\gamma^{(0)}(1 - 1.3685\alpha_0)$ and $u_\gamma = u_\gamma^{(0)}(1 - 1.11058\alpha_0)$, and obtaining evolution equations for α_0 and the photon temperature in terms of the photon number and energy injection rates. The primary contribution to the photon number will come from bremsstrahlung and/or Double Compton scattering. If we assume that there are $\dot{Y}_\gamma$ photons per baryon being injected with average energy E_γ at time t, then we find α_0 growth driven by the injection source term competing with a damping term driving it toward zero (hence a Planck distribution) on a timescale τ_D:

$$\frac{d\alpha_0}{dt} = -\frac{\alpha_0}{\tau_D} + (\frac{E_\gamma}{3.6T_\gamma} - 1)1.87\frac{\dot{Y}_\gamma}{Y_{\gamma 0}},$$
$$\tau_D = 1.29x_{0t}[(\Gamma_B + \Gamma_{DC})x^3)^{-1} = 1.29(\Gamma_K/4)^{-1}x_{0t}^{-1}. \tag{3.5}$$

When the damping time is shorter than the expansion rate of the universe, any injected energy input would be rethermalized into a Planckian in equilibrium with the electrons within one Hubble time. The condition $H\tau_D = 1$ defines a redshift $z_P \sim 10^{6.4}$ for a CDM-dominated universe with $\Omega = 1$, $\Omega_B = 0.1$ and $h = 0.5$, as indicated in Figure 3.1. Between z_P and z_{BE}, injected photon energy would result in the formation of a Bose-Einstein distribution, which is Planckian only for very low frequencies, with a characteristic distortion signature $\Delta T/T \propto -\omega^{-1}$.

The current constraint on $\alpha < 0.01$ (Smoot et al. 1985) allows us to infer that between z_P and z_{BE}, the fractional energy input relative to the primeval radiation must have been below a percent.

3.2.2 y-distortions: After z_{BE}, defined by $\Gamma_K = H$, energy injection will lead to a distribution whose perturbation is linearly proportional to

$$y \equiv \int_t^{t_0} \Gamma_K dt/4. \tag{3.6}$$

The spectral signature of this y-distortion is uniquely characteristic, $-2y$ on the Rayleigh Jeans side, xy on the far Wein side, passing through zero at $x = 3.83$:

$$\Delta T/T \equiv \delta f(T\frac{\partial f_P}{\partial T})^{-1} = -2y\psi_K(x), \quad \psi_K(x) \equiv [2-(x/2)(e^x+1)(e^x-1)^{-1}]. \tag{3.7}$$

The current constraint on y of a few percent is not yet as strong as that on α since it relies on measurements on the Wein side. The rate $\dot{u}_\gamma/u_\gamma$ in Table 1 gives

$\Delta u/u = 4y$ for the time integrated fractional input below z_{BE}, constraining the energy addition to be below the 10% level.

Limits on α and y constrain the properties allowed for massive radiatively decaying relics of the Big Bang (e.g., Silk and Stebbins 1983) and constrain such currently unpopular energy injectors as pre-decoupling dissipation of turbulence, exploding primordial black holes, and antimatter. As structure evolves and matter is heated by the release of gravitational or nuclear energy, a late time y-distortion can result. Indeed, the primary gas coolant prior to $z \sim 10$ is Compton cooling, so the CMB is where much of the released energy would finally reside, whether the object is a cooling HII region or shock-heated gas. In many cases, such as the explosive galaxy formation picture of Ostriker, Cowie and Ikeuchi, CMB anisotropies due to the inhomogeneities in y lead to more powerful constraints than this one using the angular-averaged distortion.

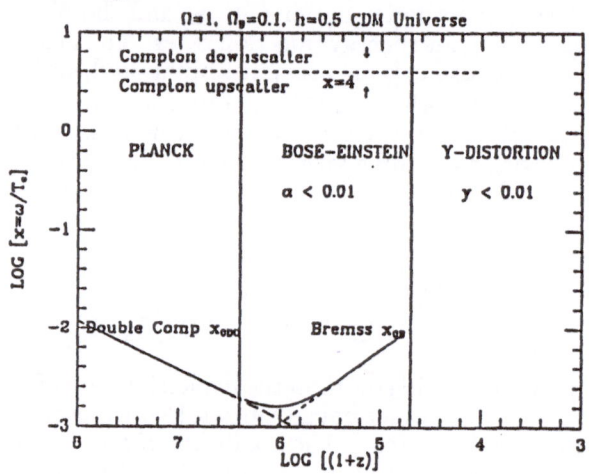

Figure 3.1 The regions in which energy injection leads to distinct spectral distortion signatures are shown for the universe indicated. We may take the line $z_P = 10^{6.4}$ above which the damping time is shorter than the Hubble time to define the *cosmic photosphere*. The boundary line prior to which kinetic equilibration should occur is $z_{BE} \sim 10^{4.7}$ in this case. The flow in energy space from low and high energies is toward the energy $4T_e$, and the higher the energy of the radiation, the longer will be the migration time down to the $4T_e$ region; in addition, there is diffusion spreading the injected energy throughout energy space. For radiation below x_{0B}, bremsstrahlung fills in the photon occupation up to the Planck level, and there is a slow bleed from this equilibrated region upward, as well as the downward processing. The equivalent curves for Double Compton scattering x_{0DC} and the two combined x_{0t} are also shown.

3.2.3 Primeval Dust Another source of distortion is primeval dust, created by a first generation of stars, either pregalactic or as members of primeval galaxies (see Carr, Bond and Arnett 1983 and Bond, Carr and Hogan 1986, BCH, for references). In BCH, we showed that if primeval galaxies formed by redshift 5-10, they would cover the sky (Ostriker and Heisler 1984), and if endowed with a significant dust content *all* radiation above the optical coming from the early universe would be absorbed and re-emitted in the infrared. We found that the ratio of the characteristic temperature of the dust to the CMB temperature T_c would be insensitive to the redshift,

$$T_d/T_c \approx 2\left[\frac{10\Delta u}{u}\frac{.01\mu}{r_d}(10a_e)\right]^{1/5},\tag{3.8}$$

where $\Delta u/u$ is the ratio of injected high frequency energy to that in the CMB, $(1+z_e) = a_e^{-1}$ is the emission redshift of the dust, and the dust cross section of Table 1 with $\alpha=1$ was assumed. Values of $aT_d \sim$ 5-7K are typical for fractional energy releases of a few percent, leading to a potentially large far infrared background. This BCH result allows us to solve for the spectral distortion on the Wein side,

$$\frac{\Delta T}{T}(q) \equiv (\delta f)_{dust}\left[T_c\partial f_{cmb}/\partial T_c\right]^{-1} = \zeta_d(q = aT_c)\psi_d(x) = \frac{T_c a}{\omega_d}\bar{N}_d\sigma_d 0.6(1 + z_d)\psi_d(x)$$

$$\psi_d(x) \equiv \exp[x(1 - T_c/T_d)]\left(1 - e^{-x}\right)^2, \quad x \equiv q/(aT_c).\tag{3.9}$$

Here $\zeta_d(q)$ is the average optical depth for radiation of comoving frequency q due to dust, N_d is the total column density of dust between us and the last scattering surface, and z_d is the redshift of dust creation. (More precisely, $0.6(1 + z_d)N_d \equiv \int n_d d\tau$.) $\bar{N}_d\sigma_d$ is the optical depth to high frequency radiation (which should be of order one for this model to be interesting). However, the multiplier is only 10^{-4} for $.01\mu$ dust generated at $z = 10$, so $\Delta T/T$ does not reach unity until $\sim 300\mu$. Since $\Delta T/T$ grows faster than a y-distortion at short wavelengths, the two should be distinguishable. More precise predictions for the two types of distortions are made in §7.

4. RECOMBINATION

4.1 The Physics of Recombination

Though the basic process of recombination in the standard hot Big Bang model simply involves the transition of a hydrogen ion into a neutral atom, how this occurs in the early universe is rather novel. As we saw in the last section, the perturbations to Planckian must be small, hence recombination should proceed as in the standard hot Big Bang model. This will not be true if a considerable number of UV photons have been injected after z_{BE}, as may occur, for example, through decaying particles or an early generation of stars.

The theory of recombination was developed by Peebles (1968) and Zeldovich, Kurt and Sunyaev (1968). We let $Y_{n\ell}$ denote the abundance (per baryon) of hydrogen atoms in the atomic state with principal and angular quantum numbers n, ℓ. Recombination develops according to a network of ODEs, with equations describing, respectively: (a) Equilbrium of the state $\{n,\ell\}$ with the $2s$ state. The availability of photons per baryon in the background radiation illustrates that there are not enough photons above $\omega_\alpha=10.2$ ev to guarantee equilbrium of the $1s$ state with those above it, though there are plenty below the Balmer continuum:

	1s-c	1s-2p	2s-3p	2s-c
	> 13.6	10.2	1.9	> 3.4
$Y_\gamma(Z_3 = 1)$	1.2(-15)	2.7(-9)	8.4(6)	1.3(4)

Here $Z_3 = (1 + z)/10^3$. Above the Lyman continuum, Y_γ is still only 6×10^{-3} at $Z_3=2$ and 100 at $Z_3=3$. That the rates for the transitions $\{n,\ell\} \rightleftharpoons \{n',\ell'\}$ are indeed fast enough to ensure equilibration is illustrated by the absorption and production timescales for the $2s \rightarrow 3p$ transitions, $0.1Z_3^{-3}$ s and $60Z_3^{-3}$ s, respectively. (b) Baryon conservation. (c) The loss of free electrons through recombination. (d) The production of hydrogen in the $1s$ state through, primarily, $2s \rightarrow 1s + \gamma\gamma$ (lifetime $\tau_{2\gamma}=0.12$ s) with some contribution from the small out-of-equilibrium imbalance in Lyα radiation arising from $2p \rightarrow 2s$ transitions in the redward wings of the line which can be redshifted completely out of the line before reabsorption:

$$Y_{n\ell} = \frac{g_{n\ell}}{4} Y_{2s} e^{-(B_n - B_2)/T}, \tag{4.1a}$$

$$Y_p + Y_{1s} + Y_{2s}\mathcal{Z}(T) = 1, \quad \mathcal{Z}(T) = \sum_{n \geq 2, \, \ell} \frac{g_{n\ell}}{4} e^{-(B_n - B_2)/T}, \tag{4.1b}$$

$$\dot{Y}_e = -\alpha_c n_B Y_e Y_p (1 - \theta_{2s}), \quad \theta_{2s} = \frac{e^{-B_2/T} Y_{2s}}{n_B Y_p Y_e} \left(\frac{m_e T}{2\pi}\right)^{3/2}, \tag{4.1c}$$

$$\dot{Y}_{1s} = -\dot{Y}_{1s} = \frac{Y_{2s}}{\tau_{2\gamma}} - \frac{Y_{1s}}{\tau_{2\gamma}} e^{-\omega_\alpha/T} + \frac{\omega_\alpha^3}{n_B \pi^2} \frac{\dot{a}}{a} \left(\frac{Y_{2s}}{Y_{2s}} - e^{-\omega_\alpha/T}\right). \tag{4.1d}$$

Here, *dot* is with respect to cosmic time t and B_n is the binding energy of the state n. Saha equilibrium, with $\dot{Y}_{1s}=0$, hence $Y_{2s} = Y_{1s} e^{-\omega_\alpha/T}$, and $\dot{Y}_e=0$, hence $\theta_{2s}=1$, gives recombination which is much too sharp, with no residual ionization predicted. Equations (4.1) should be coupled to the Friedman equation for the expansion factor $a(t)$ and to an equation for the evolution of the electron temperature T_e as it breaks equality with the photon temperature T_γ to follow the $(1 + z)^2$ redshift evolution of a nonrelativistic ideal fluid:

$$\frac{1}{a^2} \frac{d}{dt} T_e a^2 = -\frac{8\sigma_T u_\gamma}{3 m_e c} Y_e (T_e - T_\gamma). \tag{4.2}$$

The large value of the photon energy density u_γ ensures that this *Compton heating* keeps T_e and T_γ nearly equal until a redshift below about 400.

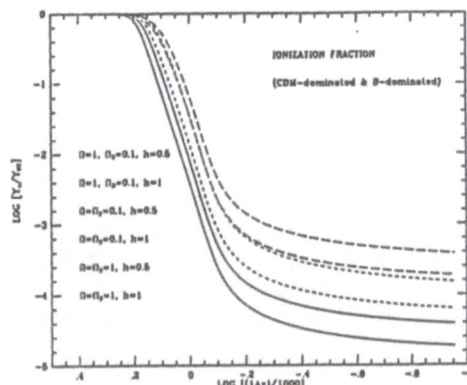

Figure 4.1 Ionization Histories

These equations must be integrated numerically. Solutions for Y_e for a variety of model universes are given in Figure 4.1. The final values of the residual ionization are significant since those few free electrons present catalyze the formation of molecular hydrogen, through the reactions H+e$\rightarrow$H$^-\gamma$, H$^-$+H$\rightarrow$H$_2$+e. H$_2$ is an important coollant in the first generation of objects forming in the early universe if present. If one is just interested in the development of anisotropies, then the critical region is not around the redshift ~ 1500 when the universe passes from 95% to 10% ionized, but rather a redshift interval from about 1200 to 900 when the radiation passes from being tightly coupled to freely streaming. We discuss this emerging visibility in the context of a simple model for the behaviour of the optical depth which we use extensively in our discussion of the physics of anisotropy generation.

4.2 The Visibility of Decoupling

For the purposes of radiative transport in the expanding universe, the redshift of recombination should be defined as the point where the differential visibility $de^{-\zeta}$ has its peak. Here $\zeta(\tau) = \int_\tau^{\tau_0} n_B Y_e \sigma_T a c d\tau$ is the optical depth from the current (conformal) time τ_0 back to time τ. Figure 4.2 illustrates that the differential visibility function $de^{-\zeta}/d\ln a$ for a variety of CDM-dominated universes is only weakly dependent on cosmological parameters, and that, though the distribution is somewhat skew, a Gaussian fit is not a bad approximation.

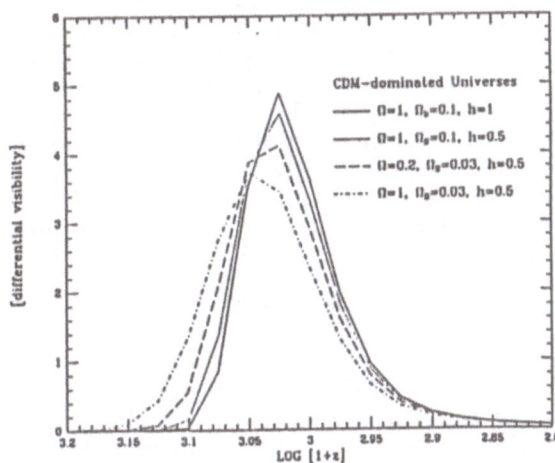

Figure 4.2 Visibility Function

Thus we write

$$de^{-\zeta} \approx \frac{\exp[-(\ln a - \ln a_r)^2/2\sigma_a^2]}{(2\pi\sigma_a^2)^{1/2}} d\ln a,$$

$$\approx \frac{\exp[-(\tau - \tau_r)^2/2\sigma_r^2]}{(2\pi\sigma_r^2)^{1/2}} d\tau,$$

$$\tag{4.3}$$

$$\tau = 190 h^{-1} \Omega_{nr}^{-1/2} (10^3 a)^{1/2} [(1+u)^{1/2} - u^{1/2}], \quad u \equiv a_{eq}/a,$$

$$(n_e \sigma_T c a)_r = 2(2+p)\tau_r^{-1} [1 + u - (u+u^2)^{1/2}],$$

$$\sigma_r = (H(a)a)^{-1}\sigma_a = 9.5(10\sigma_a)\Omega_{nr}^{-1/2} h^{-1} Mpc [(10^3 a)^{1/2}(1+u)^{-1/2}],$$

$$\sigma_a = (2+p)^{-1}\Big[1 - \frac{(1+2u)}{2(2+p)(1+u)} + (2+p)^{-2}\frac{dp}{d\ln a}\Big]^{-1/2},$$

$$a_{eq} = \rho_{er}/\rho_{nr} = [25000\Omega_{nr}h^2], \quad p \equiv -d\ln Y_e/d\ln a.$$

A reasonable way of estimating σ_a is to determine the full width at half maximum from Figure 1, then multiply by 0.425. Values of $p \sim 10$ and σ_a in the range 0.04-0.08 are typical for the CDM dominated universes of Figure 4.2. The last scattering surface is therefore quite thin.

4.3 Reionization of the Universe

If there is no recombination $p = 0$, hence the redshift of decoupling becomes $Z_r = 53(\Omega_B h)^{-2/3}\Omega_{nr}^{1/3}$, while the fuzziness is large, $\sigma_a = 3^{-1/2}$. Small-scale anisotropies can therefore be effectively erased on scales below $\sim \tau_r/4$. For $\Omega_B = 0.1$ baryon-dominated universes, which might undergo early reionization in isocurvature models, $Z_r = 180$ and $\sigma_r = 410 \ h^{-1}Mpc$. For typical CDM parameters, $Z_r = 390$ and $\sigma_r = 88 \ h^{-1}Mpc$. However, erasure is significant only if re-ionization occurs earlier than the minimum redshift required to make the optical depth to us unity, $z_{reh} \approx 50(\Omega_B h_{50}/.07)^{-2/3}$, and this seems unlikely in CDM dominated models as we now show.

The Gunn-Peterson test shows that the cumulative optical depth to Lyα radiation back to the most distant quasars at $z \sim 3$ is less than 0.1 implying the universe is extremely highly ionized with neutral hydrogen fraction $Y_H < 10^{-6}$. Did this reionization occur early enough to affect small angle anisotropies of the microwave background? An early population of massive stars is one of the few ways to reionize at such early epochs, by the overlapping of their HII regions. Carr, Bond and Arnett (1984) have estimated the fraction of the closure density in massive stars of various types required for re-ionization to occur via the overlapping of their HII regions. Even in the optimal case of Population III Very Massive Objects, we found that to reionize by z_{reh} requires a cosmic abundance in VMOs

$$\Omega_* = K10^{-6}\left[\left(\frac{\Omega_B h_{50}}{.07}\right)^{.8}(1 + \bar{\delta}_g)^{1.5}\right], \quad K \approx 2. \qquad (4.4)$$

This is lowest if the gas is unclumped. One expects the gas in the neighbourhood of the stars to be overdense by a clumpiness factor $1 + \delta_g$ in excess of one. The HII region would first have to break out of this gas before entering into the $\delta_g \approx 0$ background medium. It is unclear what to take for the average $\bar{\delta}_g$ entering eq.(4.4). For Population III stars of mass $\sim 30 \ M_\odot$, $K \approx 50$ and for VMOs of Population II, $K \approx 8$.

To assess whether it is plausible that such relatively large fractions of the universe can have gone into massive stars by z_{reh}, we estimate the amount of gas that would have gone nonlinear by this epoch. The redshift at which the rms density fluctuations in the gas reach unity, z_{nl}, depends upon the normalization of the spectrum of fluctuations, but the values are typically below 30 for CDM models (Bond and Efstathiou 1984). Assuming the fluctuations are Gaussian, the fraction of the gas which has $\delta\rho/\rho > 1$ at z_{reh} is given by $\Omega_{Bnl}/\Omega_B = \frac{1}{2}\mathrm{erfc}\left(\sigma_{\rho B}^{-1}(z_{reh})/\sqrt{2}\right)$, where $\sigma_{\rho B}$ denotes the rms level of the gas density fluctuations at z_{reh}. If $\Omega_B = 0.03$ and mass traces light, then $\Omega_{Bnl}(z_{reh})$ is 10^{-5} for an open universe with $\Omega = 0.2$ and $h = 0.5$. $\Omega=1$ CDM models typically have Ω_{Bnl} smaller by an order of magnitude or more. Thus extremely high efficiencies ($\sim 100\%$) of massive star formation from nonlinear gas would be required. This illustrates that only exotic models of pregalactic star formation which amplify the amount of nonlinear gas through, for example, explosions could lead to early enough reheating to damp anisotropies in CDM models. On the other hand, in the isocurvature baryon models with white noise initial conditions popular in the late seventies, and advocated by Peebles in this volume, the first objects collapse at $z \sim 300$, making reionization easy, and, indeed, expected.

Another speculative reionizing source is a massive decaying relic of the Big Bang which has a sizable branching ratio to decay into photons. The decay redshift must be effectively tuned to be between z_P and z_{reh} in order to have a major impact on damping small scale microwave fluctuations. Particle theory offers us no compelling candidate relics for such a task.

5. SYNCHRONOUS GAUGE CALCULATIONS OF ANISOTROPY

5.1 Perturbation Equations and Numerical Methods

In this section we summarize the main features of the Bond and Efstathiou (1984, 1986, 1987) calculations. The basic steps that must be made in setting up such perturbation computations are: (1) Choose a gauge. (2) Choose a set of spacelike hypersurfaces upon which to measure the perturbations. (3) Derive the background equations. (4) Derive the perturbed Einstein equations and the perturbed transport equations for all of the fluids presents (photons, baryons, cold dark matter (X), massless and massive neutrinos). (5) Expand the perturbation equations in a complete orthonormal set of spatial eigenfunctions (of $^{(3)}\nabla^2$, the Laplacian operator for the background 3-space, just plane waves for $k = 0$ cosmologies). This leads to a decomposition into scalar, vector and tensor modes, as well as into the independent waves. (6) Build a catalogue of solutions as a function of wavenumber and mode to the independent linear evolution equations. (7) Compute correlation functions $etc.$ to compare with observations.

Almost all detailed computations of anisotropy have been performed in the synchronous gauge (e.g., Peebles and Yu 1970, Doroshkevich, Zeldovich and Sunyaev 1978, Wilson and Silk 1981, Vittorio and Silk 1984, Bonometto $et\ al.$ 1984) which we also have adopted. For the spacelike hypersurfaces, we choose the comoving hypersurfaces for cold matter. The gauge is then completely fixed up to changes of the spatial coordinates on the CDM hypersurfaces. In the synchronous gauge, the metric $ds^2 = a^2(\tau) dr^\mu dr^\nu (\eta_{\mu\nu} + h_{\mu\nu})$, where $r^0 = \tau$ is conformal time and $\eta_{\mu\nu} = (1, -1, -1, -1)$ is the background metric, has $h_{0\mu} = 0$. For scalar perturbations, h_{ij} can be written in terms of two scalars, $h_{ij} = A\delta_{ij} + \nabla_i \nabla_j B$. The flat metric η_{ij} gives an accurate description of the background in open models at early times provided one considers temperature anisotropies on angular scales $\theta < (\Omega/2)(1 - \Omega)^{-1/2}$. On larger scales, curvature of the spacelike hypersurfaces cannot be neglected (e.g., Wilson 1983).

Adopting the flat background metric implies a rather transparent mode decomposition of a generic perturbation variable $D_i(x, \tau)$:

$$D_i(x, \tau) = \frac{1}{2} \sum_{\vec{k}s} u_{\vec{k}s}^{(i)}(\tau) e^{i\vec{k}\cdot\vec{x}} a_{\vec{k}s} + \text{complex conjugate.}$$

Here s denotes the different modes (growing or decaying, isocurvature scalar, adiabatic scalar, vector or tensor) and $a_{\vec{k}s}$ is a random variable which specifies the initial mode amplitude. We take it to be Gaussian with a power spectrum $\langle a_{\vec{k}_s}^* a_{\vec{k}_s} \rangle$ specified at some very early time. Such a form is predicted in inflationary models. The $u_{\vec{k}_s}^{(i)}(\tau)$ are mode functions which describe the evolution. The Gaussian nature of the statistics is not modified until mode-mode coupling occurs in the nonlinear regime.

We have made a concerted effort to exactly solve the full transport equations for the $u_{\vec{k}_s}^{(i)}(\tau)$ all the way from well before recombination to the present. We believe these results can serve as a benchmark for comparison of many of the approximations previously used in the literature, some of which are quite useful. For a detailed discussion of these technical points see Bond and Efstathiou (1987). We now describe the perturbation equations for the mode functions in the synchronous gauge and the numerical techniques we use to solve them.

5.1.1 Radiation We follow the transfer of the linear polarization of the photons as well as of total radiation intensity since the two are coupled by Thomson scattering (Table 1). The evolution of the radiation anisotropy is described by the Boltzmann transport equation for the perturbation to the photon distribution function, $\delta f_T(q, k, \tau) = (T \partial \bar{f}/\partial T) \Delta_T/4$, where $\bar{f}$ is the unperturbed photon distribution function, T is the photon temperature, and q is the comoving momentum. For Thomson scattering and plane waves, there is only one other nonzero Stokes parameter (Chandrasekhar 1960, §2.2), the net linear polarization, $\delta f_P = (T \partial f/\partial T) \Delta_P/4$. Both Δ_P and $\Delta T/T = \Delta_T(\hat{q}, k, \tau)/4$ are independent of photon energy, depending only on the direction of the photon momentum $\hat{q}$. The Boltzmann equation for Δ_T and Δ_P follow directly from eq.(2.13) and the source terms for Thomson scattering in Table 1:

$$\dot{\Delta}_T + ik\mu\Delta_T = \dot{h}(1 - \mu^2) + \dot{h}_{33}(3\mu^2 - 1)$$
$$- n_e\sigma_T a\{\Delta_T - \Delta_{T0} - 4\mu v_{BS} - \frac{1}{2}P_2(\mu)(\Delta_{T2} + \Delta_{P2} - \Delta_{P0})\},$$

$$\dot{\Delta}_P + ik\mu\Delta_P = -n_e\sigma_T a\{\Delta_P - \frac{1}{2}(1 - P_2(\mu))(\Delta_{P_0} - \Delta_{T_2} - \Delta_{P_2})\}.$$

$$(5.1)$$

Here, *dots* now denote differentiation with respect to τ, $h = -h_i^i$, $\mu = \hat{k} \cdot \hat{q}$ with $\hat{k}$ in the z-direction, v_{BS} is the Fourier transform of the baryon velocity measured in the synchronous gauge and σ_T is the Thomson cross section. The sources of anisotropy are therefore metric perturbations (Sachs Wolfe effect), baryon velocities, the photon density perturbation $\delta_\gamma = \Delta_{T0}$ and a quadrupole term arising from the angular anisotropy of Thomson scattering. There is also a sink for anisotropy with a damping time $(n_e\sigma_T)^{-1}$. The physics of these contributions is discussed in §6.

We solve Eqs.(5.1) by expanding Δ_T and Δ_P in angular moments, $\Delta_{T\ell}$ and $\Delta_{P\ell}$, with respect to Legendre polynomials $\Delta(\hat{q}, \vec{k}, \tau) = \sum(2\ell + 1)\Delta_\ell(k, \tau)P_\ell(\mu)$. Until one approaches recombination, it is satisfactory to solve eq.(5.1) in a tight coupling limit which truncates the expansion at $\ell = 2$. The full equations are solved from a k-dependent time when this approach is no longer accurate down to a time τ_s, taken to be at a redshift ~ 300 for short wavelengths, and as large as τ_0 for the longest waves. The expansions are truncated at a sufficiently large ℓ to ensure high accuracy (*e.g.* $\ell \sim 300$ for $k \sim 1$ Mpc^{-1}). To take our solutions forward to the present for short waves, we utilize a series solution to the free-streaming equations (when the Thomson coupling terms are negligible) which takes us from τ_s to τ_0

in one step. To accurately compute anisotropies on *all* angular scales we need to evaluate $\Delta_{T\ell}(k, \tau_0)$ for ℓ up to ~ 5000, a nontrivial numerical task. As we shall see, ℓ (in inverse radians) is analogous to the Fourier transform variable conjugate to angle for large ℓ, so this value of ℓ corresponds to about an arcminute. Fortunately, the natural fuzziness of the last scattering surface damps $\Delta_{T\ell}$ for higher ℓ's, so we need not determine them.

5.1.2 Baryons The baryons obey number and momentum conservation laws:

$$\dot{\delta}_B + ikv_{BS} = \dot{h}/2$$

$$a^{-1}\frac{d}{dt}av_{BS} = -\frac{4}{3}\frac{\rho_\gamma}{\rho_B}n_e\sigma_T a[v_{BS} - \frac{3}{4}\Delta_{T1}]. \tag{5.2}$$

The latter term is the familiar Thomson drag which is quite efficient in stopping gas motion down to $z \sim 300$ if the universe remains ionized, but lets up at z below ~ 1000 for normal recombination.

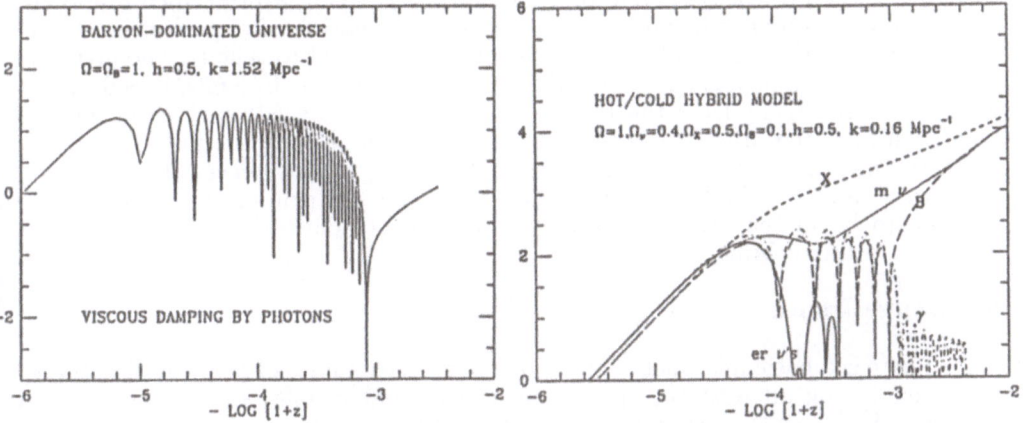

Figure 5.1 The evolution of the perturbation mode functions for the specific wavevectors shown for a $\Omega_B = 1$ adiabatic model illustrating severe Silk damping of δ_B (a) and for a hybrid hot/cold universe which demonstrates how all of the main scales discussed in §5.3 interplay (b).

5.1.3 Neutrinos and Cold Dark Matter We include massless neutrinos and, when appropriate, massive neutrinos. The Boltzmann equations for both massless and massive neutrinos are solved by direct integration, following the methods described by Bond and Szalay (1983). For low wave numbers ($k < 0.15 \ Mpc^{-1}$), we expand the massless neutrino distribution function in Legendre polynomials. For cold dark matter, the density perturbation obeys

$$\dot{\delta}_X = \dot{h}/2. \tag{5.3}$$

5.1.4 Perturbed Einstein Equations Only 2 equations are required, one a dynamical one, for $\dot{h}$ and the other a constraint equation determining h_{33} in terms of

$\dot{h}$ and velocities:

$$\delta R_0^0 = \frac{1}{2a^2}(\ddot{h} + \frac{\dot{a}}{a}\dot{h}) = 4\pi G(\delta\rho + 3\delta P),$$

$$\delta R_3^0 = -\delta R_0^3 = \frac{ik}{2a^2}(\dot{h} - \dot{h}_{33}) = -8\pi G(\rho_\gamma \Delta_{T1} + \rho_B v_B + \text{neutrino terms}).$$

(5.4)

Here, $\delta\rho = \rho_\gamma\delta_\gamma + \rho_B\delta_B + \rho_X\delta_X + \delta\rho_\nu$ is the total energy density perturbation as seen by CDM observers and $\delta P = \rho_\gamma\delta_\gamma/3 + \delta P_\nu$ is the total pressure perturbation. The explicit forms for the massless and massive neutrino contributions $\delta\rho_\nu$ and δP_ν are given in Bond and Szalay (1983).

5.1.5 Initial Conditions For scalar perturbations, there are two main modes: **adiabatic**, for which $\delta_\gamma = \delta_\nu = 4\delta_B/3 = 4\delta_X/3$ initially, with $\delta_\gamma \propto k^{n/2}$ having $n = 1$ for scale invariant fluctuations; **isocurvature** for which the total energy density fluctuation vanishes initially, $\delta\rho = 0$, with either the cold dark matter (for isocurvature axion perturbations) or the baryons (for isocurvature baryon perturbations, the old classic isothermal fluctuations) having a prescribed δ and the radiation and neutrinos having a compensating fluctuation of opposite sign to ensure $\delta\rho=0$. For the isocurvature axion model it is natural to take a scale invariant spectrum, $n = -3$ in this case (Efstathiou and Bond 1986). In a CDM seed model, with structure arising from the response to uncorrelated seeds such as primordial black holes created in the early universe, $n = 0$ would specify the initial condition down to the scale of the horizon when seed formation took place. The standard isothermal model of the late seventies had δ_B having the primary fluctuation, with n taken to be in the range -1 to 0 to fit the galaxy clustering data, as Peebles discusses in this volume.

5.2 Gauge Issues

There is a basic misconception among some aficionados of perturbation theory in the expandng universe that gauge invariant variables offer some magical elixer to ward off the evils of spurious gauge modes lurking in the computer ready to attack the unwary. In particular, the venerable synchronous gauge introduced so long ago by our perturbative pioneer, Lifshitz (1946), is often considered not only old fashioned, but also diseased. However, after specifying the hypersurfaces upon which to measure perturbations (as one must do) to be those comoving with CDM, the *synchronous comoving-X* gauge is fully specified up to spatial coordinate changes upon the hypersurfaces which change the Fourier coefficients $h_{ij}(k,\tau)$ by a time independent constant, while leaving $\dot{h}_{ij}$ intact. Braving the risk of appearing overly pedantic in order to illustrate that practically any variables can be considered gauge invariant, we define the following gauge invariant variables

$$\Psi_h \equiv \frac{\dot{h}}{6} - \frac{\dot{a}}{a}\frac{h_{00}}{2} + \left[(\frac{\dot{a}}{a})^2 - \frac{d}{d\tau}\frac{\dot{a}}{a}\right]\frac{v_X - h_{03}k}{ik} + \frac{kv_X}{3i}$$

$$\Psi_{\delta_\gamma} \equiv \delta_\gamma - 3(1 + P_\gamma/\rho_\gamma)\frac{\dot{a}}{a}\frac{v_X - h_{03}k}{ik}$$

which reduce to $\dot{h}$ and δ_γ in the synchronous gauge with $v_X=0$. A similar combination $\Psi_{h_{33}}$ reducing to $\dot{h}_{33}$ in the synchronous gauge can also be defined. Since

$v_{\gamma S} \equiv 3\Delta_{T1}/4 = v_\gamma - v_X$ and $v_{BS} \equiv v_B - v_X$ and $\Delta_{T\ell}$, $\ell \geq 2$ are also gauge invariant, *all of our equations can be considered as gauge invariant ones*. It is also a natural gauge to use if cold dark matter is present, since the world from its view is surely more interesting than that from, for example, the comoving frame where the total momentum current of all of the species present vanishes. If there is no CDM present at any level, then v_X is the velocity of the 'hypothetical freely falling observers' which, for example, Press and Vishniac (1980) find so distasteful.

One should take a lesson from the community of general relativity numericists who would adapt the lapse and shift functions defining the propagation of the spacelike hypersurfaces to best solve the numerical problem at hand. This we think was the main lesson of the famous and authoratative Bardeen (1980) paper on the gauge problem in perturbation theory which emphasized which hypersurfaces suffered from large warpings during evolution. There was nothing sacrosanct about the specific variables that were introduced in that paper. Nonetheless, we do find certain combinations physically or numerically appealing:

$$\Phi \equiv -\Phi_H = 4\pi G a^2 \nabla^{-2}(\delta\rho)_{com} = -\frac{1}{4}(h - h_{33}) + k^{-2}\frac{1}{4}\frac{\dot{a}}{a}(\dot{h} - 3\dot{h}_{33}),$$

$$(\delta\rho)_{com} = \sum_i \frac{\rho_i}{\rho_{tot}}[\delta_i - 3(1 + P_i/\rho_i)\frac{\dot{a}}{a}\frac{v_i}{ik}],$$

$$\zeta \equiv \frac{3}{4}(h - h_{33}) + \sum_i \frac{\rho_i}{\rho_{tot} + P_{tot}}\delta_i.$$

Φ reduces to an analogue of the 'Newtonian' gravitational potential, satisfying the Poisson-Newton equation indicated with $(\delta\rho)_{com}$ taken to be in the comoving gauge for the total stress energy. Φ remains constant outside the horizon in the radiation dominated phase. The curvature of the comoving X-hypersurface is $a^2\mathcal{R} \equiv 4k^2\phi_{Xc}$, where $\phi_{Xc} = (h - h_{33})/4$. Once the universe becomes dominated by cold nonrelativistic matter, $\phi_{Xc} = -\Phi/3$. Provided one is far outside the horizon, ζ, introduced by Bardeen, Steinhardt and Turner (1983), has the amazing property that it remains constant in passing from radiation to matter dominated, and even from vacuum dominated to radiation dominated throughout the reheating phase in inflationary models (see also Bardeen *et al.* 1986, hereafter BBKS, Appendix G and Bardeen 1987).

When confronted with the multicomponent fluid that we have here, certain gauges have advantages for certain tasks. For example, the perturbed source function (Table 1) was derived in the comoving gauge of the baryons, then transformed to the synchronous gauge. Also, for long wavelength isocurvature CDM perturbations, we found the manifestly gauge invariant photon entropy per axion, $S_\gamma = 3\delta_\gamma/4 - \delta_X$ and baryon to axion ratio $S_B = \delta_B - \delta_X$ have advantages in accuracy over δ_γ and δ_B, though the latter were better at short wavelengths.

What becomes of the synchronous comoving-X gauge as the universe evolves into the nonlinear regime? Once the universe has become dominated by nonrelativistic matter, $h_{33} \to h$, and $h = -(k\tau)^2\Phi/3 = 2\delta_X$, where δ_X is the magnitude of CDM fluctuations. Therefore, in the matter dominated (cold) regime, the synchronous metric perturbation is $h_{ij} = \frac{4}{3}\Phi_{,ij}/(H(a)a)^2 = 2\epsilon_{ij}$ in terms of the tidal field $\Phi_{,ij}$ and the strain tensor that enters in the Zeldovich approximation $\epsilon^i_j = \partial x^i(r,\tau)/\partial r^j - \delta^j_i$. The triad $e^b_j = a^2(\delta^b_j + \epsilon^b_j)$, the deformation tensor apart from the a^2 factor, is just the spatial part of the tetrad eq.(2.10) introduced as our canonical basis for momentum measurements. The coordinates $x^i(r, \tau)$ are Eulerian ones appropriate to the longitudinal gauge in which only the 'Newtonian' potential Φ is required to specify the perturbation; the r^i coordinates are Lagrangian ones labelling the cold matter particles. Our gauge breaks down only once e^b_j becomes singular, i.e., caustics form as shell crossing occurs in the first pancakes and clusters to collapse. In our gauge, collapse is viewed as a motionless distortion of the geometry.

5.3 Density Transfer Functions, Characteristic Scales and Normalization

An immediate outcome of our calculations is the transfer function for the density fluctuations, which maps the initial density fluctuation spectrum in the very early universe into the final post-recombination one. From this, fluctuation spectra appropriate to the linear regime for the density, velocity and gravitational potential can be constructed. Density spectra for a variety of model universes are displayed in Figure 5.2.

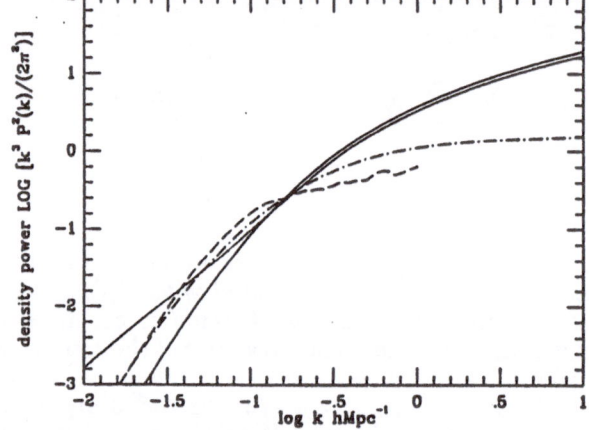

Figure 5.2 The fluctuation spectra are plotted for the following model universes: a standard biased adiabatic (solid) and isocurvature (dot-dash) CDM model with $\Omega = 1$ and $h = 0.5$, a phenomenological CDM model with a mountain of extra power (an extended $n = -1$ tail) introduced by Bardeen, Bond and Efstathiou (1987) to reproduce the observations of the cluster-cluster and cluster-galaxy correlation functions and the large 600 km s^{-1} streaming motions found by Dressler *et al.* (1986) (dot). The spectrum of the hybrid hot/cold model of Fig. 5.1 is also shown (dash).

Various (comoving) wavenumber scales determined by the transport of the many species of particles present in the universe characteristize these spectra:

$$k^{-1}_{Heq} = 10 \ \theta^{1/2} (\Omega_{nr} h_{50})^{-1} \ h^{-1} Mpc = \tau_{eq}/\pi, \quad \theta \equiv \rho_{er}/(1.69\rho_\gamma)$$
$$k^{-1}_{\nu damp} = 6 \ (\Omega_{nr}\Omega_\nu h^2_{50})^{-1/2}(g_{m\nu}/2)^{1/2} \ h^{-1}Mpc$$
$$k^{-1}_{Hrec} = 41 \ (\Omega_{nr})^{-1/2} \ h^{-1}Mpc = \tau_{rec}(z = 1000)/\pi$$

$$k_{Silk}^{-1} = 1.3 \; (\Omega_B^{-1/2} \Omega_{nr}^{-1/4})^{10/9} h_{50}^{-2/3} \; h^{-1} Mpc$$
$$k_{JBrec}^{-1} = 0.0016 \; (\Omega_{nr})^{-1/2} \; h^{-1} Mpc \qquad\qquad (5.5)$$
$$k_{curv}^{-1} = 3000 \; (1 - \Omega_{nr})^{-1/2} \; h^{-1} Mpc$$

These are respectively (1) the horizon at the epoch when the energy densities in relativistic and nonrelativistic matter are equal, the collisionless damping scale for hot dark matter (massive neutrinos), (2) the horizon scale at recombination, the Silk damping scale, (3) the baryon Jeans length at recombination and (4) the curvature scale for open universes (in which case k is not exactly wavenumber). For comparison, the scale corresponding to the masses of rich clusters is $k_{cl}^{-1} \approx 5 \; h^{-1} Mpc$. For example, the cold dark matter (CDM) spectra with $\Omega_B \ll \Omega_X$ form a one parameter family of curves characterized by k_{Heq}^{-1}. The spectrum can be shifted to larger scales by lowering h, by raising θ ($= 1$ with the canonical three massless neutrino species present), or by lowering Ω_{nr}, the energy density in nonrelativistic matter present at the epoch of matter-radiation equality. Baryons in the CDM model also have a transfer function which is effectively filtered below (5) the baryon Jeans length at recombination k_{JBrec}^{-1}. In addition there are scales characterizing the (effective) transfer function for CMB photons, k_{curv}^{-1}, k_{Hrec}^{-1} above which causal processes cannot occur at the recombination epoch, k_{Silk}^{-1} and (6) $k_{LS}^{-1} \approx = 5\Omega_{nr}^{-1/2} \; h^{-1} Mpc$, the fuzziness of the last scattering surface below which destructive interference damps CMB anisotropies. Associated with these physical scales are angular scales $\theta_{LS} \approx 3'\Omega_{nr}^{1/2}$ and $\theta_{Hrec} \approx 2°\Omega_{nr}^{1/2}$, evaluated using the angle-distance relation $\theta(d) = 0.95°\Omega_{nr} \; d/100 \; h^{-1} Mpc$ appropriate for an $\Omega = \Omega_{nr} = 1$ universe and for an $\Omega = \Omega_{nr} \ll 1$ universe. In §6 we show how these scales arise to determine the nature of CMB fluctuations.

The overall amplitude of the density spectra is still at our disposal. Normalization of models is conveniently parameterized in terms of a *biasing factor* b which is one if mass traces light and greater than one if galaxies are more clustered than the mass distribution. Technically, the spectrum is normalized by matching $J_{3\rho}(x_0) = \int_0^{x_0} \xi_{\rho\rho}(x) x^2 \, dx$ to $J_{3g}(x_0)/b^2$, where J_{3g} is estimated from the the CfA redshift survey (Davis and Peebles 1983) at $x_0 = 10h^{-1} \; Mpc$. Here, $\xi_{\rho\rho}$ is the two-point correlation function of the mass distribution at the present time as determined by linear theory which can be calculated directly from our density fluctuation spectra. J_3 is insensitive to the details of nonlinear clustering provided $\xi_{\rho\rho}(x_0) \ll 1$ (Peebles and Groth 1976, Efstathiou 1979). x_0 has been chosen so that observed fluctuations on this scale are close to linear. One would like x_0 to be much larger to ensure linearity, but unfortunately the error bars are too large.

The mass distribution in low Ω CDM models provides a reasonable fit to the galaxy correlations, so b is taken to be one in these models. It is not possible to have a consistent picture of structure formation in $\Omega = 1$ CDM models unless there is biasing with a value of $b \sim 2$ (see White, this volume, for a discussion). For neutrino-dominated universes (Bond *et al.* 1984) we would require antibiasing, $b < 1$, in order that the nonlinear redshift z_{nl}, where $\xi_{\rho\rho}(0, z_{nl}) = 1$, is high enough; we then have $b \approx (1 + z_{nl})^{-1}$. Using the statistical distributions given by Doroshkevich and Shandarin (1978), we find that 11% of the pancakes have begun to form caustics and 1% of the gas has shocked by $1 + z = 1.44(1 + z_{nl})$; by z_{nl},

50% of the pancakes have begun to form caustics and 10% of the gas has shocked. These estimates agree with the N-body results of White *et al.* (1983).

5.4 Radiation Correlation Function and Angular Power Spectrum

The correlation function for the temperature fluctuations at the present epoch τ_0 at our position x is

$$C(\theta, \tau_0) = \langle \Delta_T(\hat{q}, x, \tau_0) \Delta_T(\hat{q}', x, \tau_0) \rangle / 16, \qquad \cos \theta = \hat{q} \cdot \hat{q}',$$

$$= \frac{1}{4\pi} \sum_{\ell > 2} (2\ell + 1) C_\ell P_\ell(\cos \theta), \tag{5.6}$$

$$C_\ell = \int \frac{k^2 dk}{8\pi} \langle |\Delta_{T\ell}(k, \tau_0)|^2 \rangle.$$

The unobservable monopole component of $\Delta T/T$ at our location has been subtracted as has the large gauge dependent dipole term. Since $\Delta_{T\ell}$ is gauge invariant for $\ell \geq 2$, so is C_ℓ and this $C(\theta)$. These quantities are plotted in Figure 5.3 for a standard adiabatic CDM model with $\Omega = 1$, $\Omega_B = .03$, $h = 0.75$, $b = 1$ and scale invariant initial conditions.

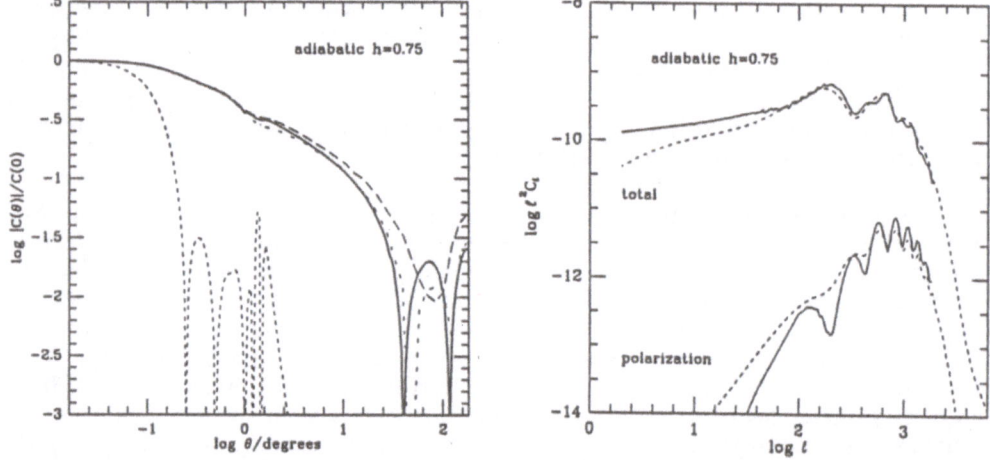

Figure 5.3 $C(\theta)$ and C_ℓ (solid lines) for a standard adiabatic CDM model with $\Omega = 1$, $\Omega_B = .03$, $h = 0.75$, $b = 1$ and scale invariant initial conditions. Notice the slow falloff of $C(\theta)$ indicative of much large scale texture in the microwave fluctuations. The polarization correlation function $C_P(\theta)$ (short dash) and power spectrum $C_{P\ell}$ (solid) demonstrate that significant polarization arises only for scales smaller than the horizon at recombination. (The dashed curves are fits described in Bond and Efstathiou 1986).

C_ℓ is the *angular power spectrum* for the radiation fluctuations. For example. $C_2^{1/2}$ is the *rms* for the quadrupole amplitude, often quoted as a_2. C_ℓ naturally comes out of our calculational procedure and is of fundamental importance in its own right. If we expand the radiation pattern in spherical harmonics,

Table 2: Primary Anisotropies of the CMB (§5.4)

Notes to Table 2: Sample small angle (θ=4.5′, θ_s=0.6′) and large angle (θ=6°, θ_s=2°) anisotropies (in units of 10^{-6}) are shown for a variety of models of structure formation from initially Gaussian fluctuations. The open, vacuum dominated and baryon dominated models are normalized assuming mass traces light, $b = 1$; the rest were normalized assuming the biasing factor $b = 1.7$. Models have $h = 0.5$ unless otherwise stated. The baryon-dominated adiabatic results have been estimated from Wilson and Silk (1981). The rest of the results were obtained in collaboration with Efstathiou. ISOC refers to isocurvature models. The other models have adiabatic initial conditions.

The classic baryon dominated isocurvature models that were very popular in the 1970s with initial spectral index $n = 0$ (Poisson seed model) or $n = -1$ (phenomenological) have been computed by Peebles (1987) and Efstathiou and Bond (1987); the quoted results were taken from the latter work. The models are labelled B-dom ISOC n if there is normal recombination and by ION B-dom ISOC n if the universe remains fully ionized as might be expected in these models since collapse on the smallest scales can even occur before recombination.

MODEL	$(\Delta T/T)(4.5')$	$(\Delta T/T)(6°)$
OBSERVATIONS	< 25	< 48
B-dom $\Omega = \Omega_B = 0.1$	1000	-
B-dom $\Omega = \Omega_B = 1$	50	-
B-dom ISOC 0 $\Omega = 1\ \Omega_B = 1$	36	7
B-dom ISOC -1 $\Omega = 1\ \Omega_B = 1$	28	14
OPEN B-dom ISOC 0 $\Omega = 0.2\ \Omega_B = 0.2$	61	3
OPEN B-dom ISOC -1 $\Omega = 0.2\ \Omega_B = 0.2$	84	16
OPEN ION B-dom ISOC 0 $\Omega = 0.2\ \Omega_B = 0.2$	11	4
OPEN ION B-dom ISOC -1 $\Omega = 0.2\ \Omega_B = 0.2$	14	31
CDM-dom $\Omega = 1\ \Omega_B = 0.03$	3	7
CDM-dom $\Omega = 1\ \Omega_B = 0.1$	5	7
CDM-dom $\Omega = 1\ \Omega_B = 0.2$	6	8
CDM+B hybrid $\Omega = 1\ \Omega_B = 0.5$	8	15
CDM-dom ISOC $\Omega = 1\ \Omega_B \ll \Omega$	-	60
HOT (m_ν=24 ev) $\Omega = 1\ \Omega = 0.1, b = .53$	20	20
HOT/COLD hybrid $\Omega_\nu = 0.4\ \Omega_X = 0.5\ \Omega_B = 0.1$	-	8
OPEN/CDM-dom $\Omega = .2\ \Omega_X = .17\ \Omega_B = .03$	70	40
OPEN/CDM/B $\Omega = .2\ \Omega_X = .1\ \Omega_B = .1\ h = .75$	80	50
VAC/CDM hybrid $\Omega = 1\ \Omega_{vac} = .8\ \Omega_X = .17\ \Omega_B = .03$	20	20
VAC/CDM/B $\Omega = 1\ \Omega_{vac} = .8\ \Omega_X = .1\ \Omega_B = .1\ h = .75$	20	25
CDM-dom + Extra Power Mountain $\Omega = 1\ \Omega_B \ll \Omega$	-	40
CDM-dom + Extra Power Plateau $\Omega = 1\ \Omega_B \ll \Omega$	-	60

$T(\hat{q}) = T_0 \sum a_{\ell m} Y_{\ell m}$, then the coefficients $a_{\ell m}$ are statistically independent and Gaussian distributed with variance C_ℓ. Construction of a map realizing the theoretical appearance of the microwave sky therefore requires the power spectrum C_ℓ. We have shown that the pattern on small ($< 10°$) square patches of the sky can be obtained by performing a standard FFT on a two dimensional Gaussian random field with power spectrum C_ℓ.

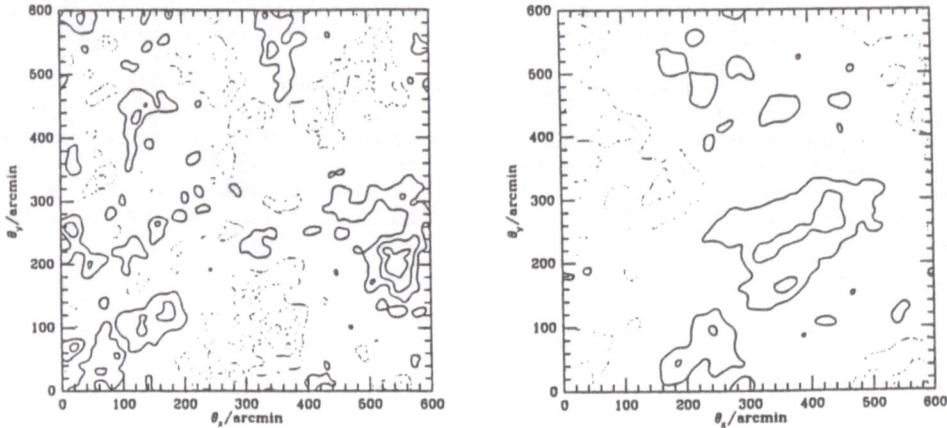

Figure 5.4 Maps for the adiabatic CDM case of Fig. 5.3 and for an isocurvature CDM model with identical cosmological parameters. The *isocurvature effect* discussed in §6 is responsible for the preponderance of large scale power which clearly differentiates this model from the adiabatic one.

Sample predictions for a variety of models of a fiducial large angle anisotropy experiment by Melchiorri *et al.* (1981) and a fiducial small angle experiment by Uson and Wilkinson (1984a,b) are given in Table 2. The theoretical prediction for the results of a typical observation, $(\Delta T/T)_{\mathcal{R}} = \int \mathcal{R}(\hat{q}) d\Omega_{\hat{q}} \, \Delta_T(\hat{q}, x, \tau_0)/4$, where $\mathcal{R}(\hat{q})$ is the angular response of the experiment, is a Gaussian probability distribution of $(\Delta T/T)_{\mathcal{R}}$ with variance

$$\langle (\Delta T/T)^2_{\mathcal{R}} \rangle = \int \mathcal{R}(\hat{q}) d\Omega_{\hat{q}} \mathcal{R}(\hat{q}') d\Omega_{\hat{q}'} \, C(\hat{q} \cdot \hat{q}', \tau_0). \tag{5.7}$$

A beam-switching experiment with beam pattern F has $\mathcal{R} = F(\hat{q} - \hat{q}_1) - F(\hat{q} - \hat{q}_2)$, hence $\langle (\Delta T/T)^2_{\mathcal{R}} \rangle = 2(C_F(0) - C_F(\theta))$, where $\cos\theta = \hat{q}_1 \cdot \hat{q}_2$ and C_F is defined by Eq. (5.7). The Melchiorri *et al.* (1981) experiment with switching angle 6° and a beam pattern with a *fwhm* of 5° is such a two beam experiment. The Uson and Wilkinson (1984a,b) experiment is a three beam one. They measure the temperature difference between the direction $\hat{q}_1$ and the average of the temperatures at $\hat{q}_2$ and $\hat{q}_3$, at the small separation 4.5′ on either side of $\hat{q}_1$: $\mathcal{R} = F(\hat{q} - \hat{q}_1) - (F(\hat{q} - \hat{q}_2) + F(\hat{q} - \hat{q}_3))/2$. Hence, we must compare their experimental results with our predictions of a different combination of correlation functions,

$$\langle (\Delta T/T)^2_{\mathcal{R}} \rangle = 2(C_F(0) - C_F(\theta)) - \frac{1}{2}(C_F(0) - C_F(2\theta)).$$

The angular power profiles of the telescopes are approximated by Gaussian profiles with standard deviation $\theta_s = 0.425\theta_{fwhm}$ in terms of the full width. The beam smeared correlation function C_F is then denoted by $C(\theta; \theta_s)$.

Bond and Efstathiou (1984) and Vittorio and Silk (1984) found that models with $\Omega \leq 0.2h^{-4/3}$ for $\Omega_B = 0.03$ can be ruled out at the 95% confidence level using the small angle constraint. The situation is worse for higher Ω_B in low Ω models. Assuming a large vacuum energy to bring Ω up to unity saves these models. The isocurvature baryon models have excessive small angle anisotropies unless early reionization occurred. The other models listed in Table 2 pass on small angles. The isocurvature CDM model fails at large angles. The phenomenological CDM model with extra power on large scales either in the form of a mountain or a plateau (§6) also has large angle anisotropies near to the observational limit.

5.5 The Texture of the Radiation Pattern

In Bond and Efstathiou (1987), we analyzed the statistics of hot and cold spots using the theory of two dimensional Gaussian random fields and we summarize the main conclusions here. The number density of hot spots (maxima in $\Delta T/T$) above a threshold $\nu_t = (\Delta T/T)/\sigma_r$, $n_{pk}(> \nu_t)$, depends upon 2 parameters which characterize the shape of the angular power spectrum. Here $\sigma_r(\theta_s) \equiv C(0; \theta_s)^{1/2}$ is the *rms* level of the radiation fluctuations smoothed on the scale θ_s. If there was no noise in the data, the number density of hot and cold spots could then be used to determine $C(0; \theta_s)$, $C''(0; \theta_s)$ and $C^{iv}(0; \theta_s)$. In particular, one could determine the overall amplitude σ_r and the coherence angle θ_c defined by $\theta_c^{-2} \equiv -C''(0)/C(0)$ which is proportional to the fuzziness of the last scattering surface surface for $\theta_s = 0$. For the adiabatic map of Figure 5.4 (which has been smoothed on a scale $\theta_s = 5'$, if we include power only up to the patch scale, then $\sigma_r = 1.3 \times 10^{-5}/b$, where b is the biasing factor, and $\theta_c = 17'$. A comparison of the average number of hot spots expected for a region of this size with the number counted gives agreement within the expected map to map fluctuations:

ν_t	1	2	3
$n_{pk}(> \nu_t)(10°)^2$	106	27	2.7
spot count	95	19	3

In principle, such counts could be used to help verify that the fluctuations are Gaussian. We have also evaluated the mean area of a hot spot ($\propto \nu_t^{-2}$ for high ν_t), the average distance between hot and cold spots, the shapes of hot spots (eccentricities and the cross correlation $C_{spot,\Delta T/T}(\theta)$) and even the hot spot - hot spot correlation function $C_{spot,spot}(\theta)$. Though such results provide valuable insight to the texture of the fluctuations if they are Gaussian, we doubt whether they will be useful in the first stages of discovery of the anisotropies. Nonetheless, we can look forward to the day when hot and cold spots on the microwave sky will have their own '6C' catalogue numbers.

6. MECHANISMS OF PRIMARY ANISOTROPIES

In this section, we utilize the simple approximation to the fuzziness of the last scattering surface of §4.2 to develop a model for CMB fluctuations which illustrate how the physical sources of anisotropy interplay to determine $C(\theta; \theta_s)$ and provide a better feel for the results of our numerical computations.

6.1 Anisotropy Sources for Scalar Modes

Since the anisotropies are very small, the transport theory is best solved by decomposing the temperature fluctuations into spatial scalar, vector and tensor eigenmodes, each evolving independently in linear perturbation theory. For flat Universes, this is equivalent to Fourier decomposition. We concentrate on scalar modes. The anisotropy that develops from fluctuations of comoving wavenumber k for photons arriving from direction $-\hat{q}$ is approximately

$$\frac{\Delta T}{T}(k, \hat{q}, \tau_0) \approx e^{-ik \cdot \hat{q} R_r} e^{-(k \cdot \hat{q})^2 \sigma_r^2 / 2} \left[\Phi(k, \tau_r)/3 + \delta_{\gamma eff}(k, \tau_r)/4 + \hat{q} \cdot \vec{v}_{Beff}(k, \tau_r) \right],$$
(6.1)

$$R_r = \tau_0 - \tau_r \quad \text{if } \Omega = 1, \quad R_r = R_c \sinh(\tau_0/R_c) - \tau_r, \quad \text{if } \Omega < 1,$$

$$R_c = H_0^{-1} \Omega_c^{-1/2} \quad \Omega_c \equiv 1 - \Omega, \quad \tau_0 = f_r 2 H_0^{-1} \Omega_{nr}^{-1/2},$$

$$f_r = 1 \ (\Omega_{nr} = 1),$$

$$f_r = \left(\frac{\Omega_{nr}}{\Omega_c}\right)^{1/2} [\ln(\Omega_c^{1/2} + 1) - \ln(\Omega_{nr})^{1/2}] \ (\Omega = \Omega_{nr} < 1).$$

Eq.(6.1) follows directly from the synchronous gauge equation, 5.1, provided we assume (1) a flat dark matter dominated Universe with $a_{eq} < a_r$, so the dark matter fluctuations grow $\propto a$ through recombination and the gravitational potential fluctuations $\Phi(k, \tau)$ are constant, (2) the visibility law of eq.(4.3), (3) the baryon velocity $\vec{v}_{Beff}$ and the fractional photon density perturbation $\delta_{\gamma eff}$ are suitably averaged across the narrow shell of recombination, using some model for their behaviour, and (4) the other source terms proportional to the quadrupole and to polarization terms can be ignored (which is true in practice). The factor f_r must be computed by a simple numerical integration for conditions other than those stated (e.g. universes with sizable vacuum energies). This expression does not describe anisotropy in open universes accurately since the spatial eigenfunctions one expands in are not plane waves. However if we treat equation (6.1) as a phenomenological expression, then the correct results for anisotropies on angular scales $\theta \ll (\Omega/2)(1 - \Omega)^{-1/2}$ will be obtained.

6.1.1 Last Scattering Surface Damping: The term $F_f = \exp(-(k \cdot \hat{q})^2 \sigma_r^2 / 2)$ describes the damping due to the fuzziness of the last scattering surface. The effect is largest if one is looking at photons that arise along the perturbation wavevector. The recombination surface is perpendicular to the photon path to us, so the spatial oscillations are across the last scattering surface, giving destructive interference from *both* peaks and troughs for waves with $k\sigma_r > \pi$. (The velocity is purely longitudinal for scalar modes.) There is no destructive interference if the photons are only received from *either* peaks or troughs, but not both, the case if oscillations are along the last scattering surface, or if the wavenumbers are small.

6.1.2 The Sachs Wolfe Source: The Sachs Wolfe contribution is simply $\Phi/3$ if Φ is constant. For large wavelengths, Φ has the same shape as the initial Φ fluctuations ($\sim k^{-3/2}$ for scale invariant adiabatic fluctuations), but on smaller scales the initial spectrum should be multiplied by the transfer function for density fluctuations, giving a high frequency filtering beyond k_{Heq}^{-1}.

6.1.3 The Photon Density Source:

The source term $\delta_{\gamma eff}$ leads to the famous but misleading '$\Delta T/T = \frac{1}{3}\delta\rho_B/\rho_B$' for adiabatic fluctuations, since $\delta_\gamma = \frac{4}{3}\delta_B$ prior to recombination if the photon entropy per baryon has no fluctuations in the initial conditions. To parameterize the effective photon density at recombination we define a *form factor* $F_\Phi(k, \tau_r)$ by $\delta_{\gamma eff}/4 = \Phi/3(F_\Phi - 1)$. For adiabatic fluctuations in the limit in which photons and baryons are tightly coupled and $\rho_B \ll \rho_\gamma$, $F_\Phi - 1 \approx e^{-(\sigma_D k\tau_r)^2/2} (\cos(ks\tau) - 1)$, where $\sigma_D \sim 0.03$ is a parameter describing Silk damping and $s = c/\sqrt{3}$ is the sound speed. This form is valid for waves both outside and inside the horizon, and is obtained by a WKB solution to the tight coupling equations. The parameter $\sigma_D^{-2} = (2.5 + p)n_e\sigma_T ca\tau \propto \tau^{3+2p}$ becomes large by τ_r ($\sigma_D(\tau_r) = [(2 + p)(5 + 2p)]^{-1/2} \sim 0.05$) when Y_e is at best a few percent. At this stage, the approximations in the derivation cease to be accurate, so we treat σ_D as adjustable. Matching to numerical results for Silk damping in baryon dominated models gives the asymptotic value $\sigma_D \sim 0.02\Omega_{nr}^{1/4} (\Omega_B/0.1)^{-1/2} h_{50}^{-2/3}$, so values of a few percent are indeed reasonable choices.

Instead of using δ_γ evaluated at recombination to give F_Φ, it will be slightly more accurate at high k to include the influence of multiple oscillations through the last scattering surface,

$$
F_\Phi - 1 \approx - \exp(-\sigma_D^2 k^2 \tau_r^2/2)\Big[1 - \big[\cos sk\tau_r \cosh(s\mu k^2\sigma_r^2)
$$
$$
- i \sin sk\tau_r \sinh(s\mu k^2\sigma_r^2)\big] \exp(-s^2 k^2\sigma_r^2/2)\Big]. \tag{6.2}
$$

This form was presented to illustrate the expected magnitude of $\Im(F_\Phi)$ terms. We now ignore such terms to simplify the presentation. We emphasize that to be accurately determined, F_Φ must be calculated numerically.

6.1.4 The Isocurvature Effect:

If the perturbation mode is isocurvature rather than adiabatic, the fluctuations are initially perturbations in the entropy (per CDM particle for isocurvature axion perturbations or per baryon) without accompanying curvature perturbations. The isocurvature effect can be understood just by considering δ_γ in the long wavelength limit of the conservation equations for the total energy density and entropy. Consider a system consisting only of CDM (X) and photons. Then $\delta_\gamma = \frac{4}{3}\delta_s(1 + \frac{4}{3}\rho_\gamma/\rho_X)^{-1} = -(\rho_X/\rho_\gamma)\delta_X$ as $k \to 0$. Since $\rho_X/\rho_\gamma = a/a_{eq}$ grows from zero to $\sim 10^4$, the entropy fluctuation transfers from one primarily in the X-density to one of opposite sign in the radiation. At horizon crossing ($k\tau \sim 1$), $a/a_{eq} \sim (k\tau_{eq})^{-2}$, hence $\rho_X/\rho_\gamma \propto (k\tau_{eq})^{-2}$ for $k\tau_{eq} < 1$. The correct asymptotic behaviour for the δ_X transfer function is then: 1 at high k, turning over at $k \sim \tau_{eq}^{-1}$ to a term $\propto k^2$ at low k. When a given wave enters the horizon, δ_X ceases declining, and begins to grow via the usual Jeans instability; δ_γ enters with an amplitude $\propto k^2\delta_s$ at high k and $\propto \delta_s$ at low k. We can therefore model $\delta_\gamma(\tau_r)$ as a quantity proportional to $-\delta_X(\tau_r)/k^2 \propto \Phi$, a multiple of the Sachs Wolfe contribution. Efstathiou and Bond (1986) found $F_\Phi = 6$ for long wavelengths, independent of Ω_{nr}. Oscillations in F_Φ due to sub-τ_r scale phenomena are unimportant for scale invariant initial spectra. Isocurvature baryon fluctuations will have a similar shape to isocurvature CDM fluctuations for $k\tau_r < 1$. However, if steep initial spectral indices are chosen ($n_i \approx 0$ for seed models compared with $n_i = -3$ for scale invariant models), then sub-τ_r scale modifications will be significant.

6.1.5 The Electron Velocity Source: The baryon velocity is parameterized by $v_{Beff} = v_X(k, \tau_r)F_v(k, \tau_r)$, in terms of the the CDM velocity $v_X = -i\vec{k}\tau\Phi/3$ at the time of recombination τ_r and a *form factor* F_v which is $\approx e^{-(\sigma_D k\tau_r)^2/2}$ $\sin(ks\tau)/(ks\tau)$ in the tight coupling limit, and is 1 in the free streaming limit after the baryon perturbations have *caught up* to the CDM perturbations. Before catch-up is complete, there is another effect which should be included in F_v. As the baryons pass through recombination and begin to decouple, if the velocities are of opposite sign to that of the dark matter, then the baryons must first slow down, then reverse their flow to catchup to the dark matter. This can build up large velocities hence a bigger influence on $\Delta T/T$. Thus, as for F_Φ, F_v must be determined numerically for accuracy. However, the tight coupling limit does illustrate that oscillations depending upon the phase of the waves as they hit the narrow recombination band should be expected, with characteristic separation in k-space

If anisotropies are computed in the synchronous gauge, the velocity contribution actually arises partly from the baryon velocity as measured in the synchronous gauge, $v_{BS} = v_B - v_X$, and partly from a Sachs-Wolfe-like term deriving from the metric, $ih/(2k)$, which is just the velocity of cold matter $v_X = -(\tau/3)\nabla\Phi$ as measured in the longitudinal gauge. Thus some care must be taken to include all effects when computing numerically small angle anisotropies in the synchronous gauge.

6.1.6 Radiation Correlation Function: $C(\theta; \theta_s)$ can be approximated by an integration over a $\Delta T/T$ fluctuation spectrum, $W^2(k)$:

$$C(\theta; \theta_s) = \int \frac{k^2 dk}{2\pi^2} \frac{\sin kR_r\varpi}{kR_r\varpi} W^2(k)F_m^2(k), \quad \varpi \equiv 2\sin(\theta/2),$$

$$W^2(k) \equiv W_\Phi^2(k) + W_v^2(k),$$

$$W_\Phi^2(k) \approx \frac{1}{9}|\Phi(k)|^2 \frac{1}{2}\int_{-1}^{1} d\mu\ F_s^2(k)F_f^2(k)F_\Phi^2(k), \tag{6.3}$$

$$W_v^2(k) \approx \frac{1}{3}|v_X(k, \tau_r)|^2 \int \frac{d\mu}{2}\ F_s^2 F_f^2 F_\Phi^2 3\mu^2,$$

$$F_s^2 \equiv e^{-(kR_r\theta_s)^2(1-\mu^2)}, \quad F_f^2 \equiv e^{-k^2\sigma_r^2\mu^2}, \quad F_m^2(k) = 1 - j_0(kR_r)^2.$$

The form is similar to that for the 3-dimensional correlation functions for the gravitational potential ($\xi_\Phi/9$) and the CDM velocity at recombination ($\xi_v/3$), except for the monopole (high pass) filter F_m, the beam smearing (low pass) filter F_s, the fuzziness filter F_f and the form factors F_Φ and F_v. Imaginary parts of F_Φ and F_v are neglected in this expression, a reasonable approximation as eq.(6.2) illustrates. To obtain this expression, only the most rapidly varying part is included in the integral over the solid angle in $\vec{k}$-space, with the slowly varying parts replaced by the $\mu = \hat{k} \cdot \hat{q}$ average as indicated. The angular dependence of $C(\theta)$ then comes from the $j_0(kR_r\varpi)$ term only, which primarily probes the $k < \pi(R_r\varpi)^{-1}$ part of W^2.

6.1.7 Filters: Gaussian beam smearing can be taken into account by multiplying the unsmeared μ integrand by the exponential factor $\exp(-(kR_r\theta_s)^2(1-\mu^2))$ as indicated. (In this case, it is probably better to replace the $j_0(kR_r\varpi)$ μ-independent term by $J_0[kR_r\varpi(1-\mu^2)]$ inside the μ integral. This is the Doroshkevich, Zeldovich and Sunyaev (1978) small angle approximation as generalized by Bond and Efstathiou (1987) to include large angle anisotropy effects.) Simple Gaussian approximations, $F_s^2 = e^{-(kR_s)^2}$ and $F_f^2 = e^{-(kR_f)^2}$, to the beam smearing and fuzziness filters illustrate the dominant effects these smoothings have on $C(\theta)$ provided we take $R_s = (2/3)^{1/2}R_r\theta_s$ and $R_f = (1/3)^{1/2}\sigma_r$ in W_Φ^2, and $R_s = (2/5)^{1/2}R_r\theta_s$ and $R_f = (3/5)^{1/2}\sigma_r$ in W_v^2.

The effects of subtraction of the unobservable monopole component of the fluctuations is embodied in F_m^2, which is $\approx (kR_r)^2/3$ for waves larger than the current horizon size and ≈ 1 for shorter waves. This long wave filtering is essential for the $\Phi/3$ Sachs Wolfe term to converge for scale invariant spectra for which ξ_Φ has a logarithmic infrared divergence. (The largest monopole and dipole contributions to $\Delta T/T$ have already been subtracted in eq.(6.1).)

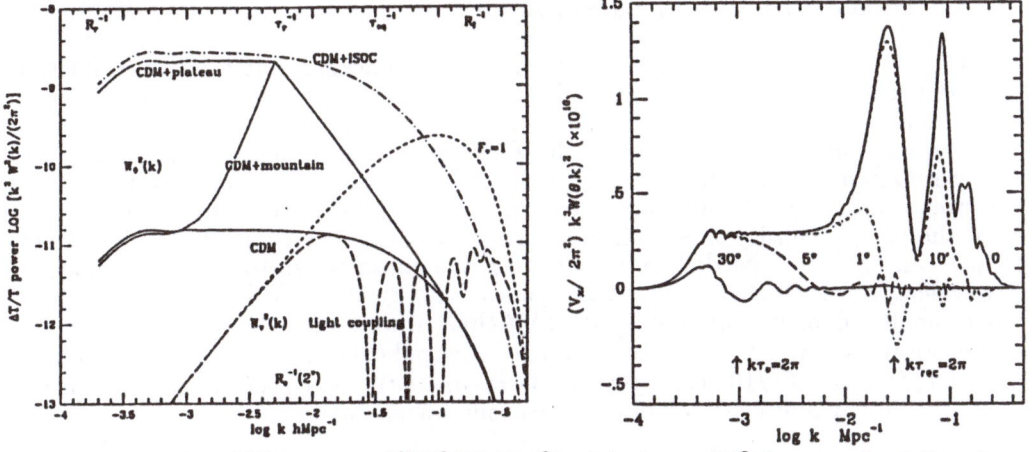

Figure 6.1a,b: $C(\theta)$ spectra $k^3 W^2(k)/(2\pi^2)$. (a) shows W_Φ^2 for standard $\Omega = 1$, $h = 0.5$, $\Omega_B \ll 1$ CDM dominated universes with initially scale invariant adiabatic (solid, assuming $F_\Phi = 1$) and isocurvature (dash, assuming $F_\Phi = 6$) fluctuations with a biasing normalization $b = 1.7$ describing the extent to which the galaxies are more clustered than the mass. For the adiabatic case, W_v^2 with $F_v = 1$ and the tight coupling value of §6.1.5 are also shown. Accurate calculations (Bond and Efstathiou 1987) for F_Φ and F_v give W^2 as in Fig. 6.1b for an $\Omega = 1$ $h = 0.75$, $\Omega_B = 0.03$ CDM universe with adiabatic fluctuations and $b = 1$. Fig. 6.1a shows how adding the mountain or the plateau of extra power to the standard CDM density fluctuation spectrum in order to satisfy the large scale streaming velocity and rich cluster correlation function observations (Bardeen, Bond and Efstathiou 1987) must lead to the large intermediate angle anisotropies indicated in Table 2.

6.2 Tensor Modes

The vector modes decay and are unlikely to be large through recombination. The tensor modes have fluctuations driven by gravitational waves. Their amplitudes

h_{ij} remain constant during inflationary and radiation dominated epochs, only declining once matter domination sets in and the waves enter the horizon (Starobinskii 1985, Veryaskin, Rubakov and Sazhin 1983). Gravitational waves will be generated by quantum fluctuations in the de Sitter vacuum during an inflationary epoch with amplitudes $\propto (\rho_{vac}/m_P^4)^{1/2}$ and with equal power per decade of wavelength, where ρ_{vac} is the energy density of the false vacuum which drives inflation and m_P is the Planck mass. The CMB anisotropy due to these waves, which is analogous to the familiar Sachs Wolfe effect for scalar perturbations discussed above, is proportional to the wave amplitude. Just as for the Sachs Wolfe effect, this mechanism is effective only for large to intermediate angles. However, as emphasized by Starobinskii (1985), current limits on the amplitude of the large angle anisotropy have allowed significant constraints to be placed upon how close the inflationary epoch can be to the Planck era: $\rho_{vac}^{1/4} < 5 \times 10^{-3}$, corresponding to a temperature below 10^{17} GeV.

7. SECONDARY ANISOTROPIES

7.1 Secondary Radiation Backgrounds

7.1.1 Observational Constraints: Secondary backgrounds arise once fluctuations become nonlinear, turnaround and collapse. We expect such backgrounds in the far infrared from dust emission, in the near infrared, optical and UV from stellar and quasar emission, and in the X-ray from quasar emission and cluster collapse. Only for X-rays has a background definitely been observed, and the energy source producing it is still unknown. In the other wavebands, only upper limits on the energy density exist. (Observations that have been reported in the literature can only be regarded as suggestive.) We characterize the energy density at wavelength λ (=1.24μ/ω, where the photon energy $\omega = q/a$ is in eV), by the dimensionless density $\Omega_R(\lambda) \equiv \rho_{cr}^{-1} du_\gamma(\lambda)/d\ln\lambda$, where $\rho_{cr} = 10.6\ h^{-2}$ keV cm^{-3} is the closure density. In terms of the angle-averaged (background) distribution function $\bar{f}$, $\Omega_R(\lambda) = (\lambda/1000\mu)^{-4}\bar{f}(q,t)/338 = (q/mev)^4\bar{f}(q,t)/800$, where mev means millivolts. Current constraints in the various wavebands are:

Waveband	$\lambda(\mu)$	$\Omega_R(\lambda)h^2/10^{-6}$	Group
X-ray	$10^{-3} - 10^{-5}$	$\sim.005$	Shafer and Fabian (1983)
UV	0.14	$<.08$	Paresce and Jacobsen (1980)
UV	0.13	.11	Weller (1983)
UV	0.15	.07	Feldman, Brune and Henry (1981)
Optical	0.44	$<.49$	Toller (1983)
Optical	0.51	$<.64$	Dube, Wick and Wilkinson (1979)
near IR	1.6-4.7	12-30	Matsumoto, Akiba and Murakami (198?
far IR	12,25,60	$<11,10,9$	Hauser et al. (1984)
far IR	100	<9	Hauser et al. (1984)
far IR	100	4	Rowan-Robinson (1986)
sub-mm	500-1000	~10	Gush (1981)
sub-mm	600	$< (4-7)$	de Bernardis et al. (1985)
CMB peak	1400	$=19$	
IRAS	100	2	90" FOV
SIRTF	100	.005	30" FOV

COBE	500-10^4	~.03	few degs FOV

Satellite sensitivities are also indicated to show how deeply we may expect to probe the far IR. SIRTF has the advantage of much smaller fields of view (FOVs) than IRAS as well as longer wavelengths. COBE has poor angular resolution, but will be able to give a tight constraint on distortions.

7.1.2 Theoretical Sources: Characteristic values for the total radiation density expected from various sources are all plausibly as large as 10^{-6}:

Sources	$\Omega_{RT}/10^{-6}$
CMB	$26h^{-2}$
stars	$4\,(\Omega_*/.01)[(1+z_*)/10]^{-1}[\epsilon/.004]$
black hole accretion	$1\,(\Omega_{B,acc}/10^{-6}[(1+z_*)/10]^{-1}[\epsilon/0.1]$
decaying particles	$5\,B_X\Omega_{X,i}[(1+z_{dec})/10^5]^{-1}$
grav collapse	$6\,\langle\Omega_{B,coll}f_{cool}\,v_{T3}^2/(1+z_c)\rangle$

Here, Ω_{Xi} is the initial density in cold particles X which are destined to decay. Ω_{Xi} may easily be in excess of unity; e.g., for keV neutrinos it is 40. B_X is the branching ratio for decay into radiative channels and z_{dec} is the decay redshift, when the lifetime equals the Hubble time. In these expressions, Ω_*, $\Omega_{B,acc}$ and $\Omega_{B,coll}$ are the densities of baryons in stars, accreted onto black holes and belonging to collapsed structures in the universe. f_{cool} is the fraction of the collapsed baryons which have actually cooled significantly.

In all wavebands there is a contribution from the gravitational energy released during the collapse of various structures in the universe. The efficiency ϵ per baryon for gravitational binding energy release (as a fraction of rest mass energy) from cosmic structures such as galaxies and clusters, $0.5(v_T/c)^2 = 6\times10^{-6}v_{T3}^2$, where v_T is the 3-dimensional velocity dispersion and $v_{T3} \equiv v_T/1000$ km s^{-1}, is much smaller than efficiencies per baryon for nuclear ($\epsilon \sim .004$ for massive stars) or accretion ($\epsilon \sim 0.1$) energy sources, though, presumably, a significantly larger fraction of the universe has collapsed than has accreted or gone into stars. In the Ω_{RT} expression for gravitational collapse, the average indicated, to be taken over all collapses, is at most a percent in CDM universes, leading to relatively small backgrounds from this contribution, even if all structures were to have completely cooled.

If Compton upscattering were the main coollant, which is typically true for energy releases occurring earlier than $z \sim 10$, then the extra energy dumped into the CMB, a fraction $4y$ of the total, is $\Delta\Omega_{RT}h^2 \approx 10^{-6}y/0.01$. If the energy sources listed above are released in the Compton cooling era, y-distortions at the percent level would then be expected, except for those driven by gravitational collapse which would be considerably smaller.

Even if reheating occurs predominantly after the Compton epoch, a sizable y-distortion can still result if the universe is heated to a high enough temperature. Indeed, the most straightforward interpretation of the hard X-ray background is that it is bremsstrahlung from the ambient IGM with a temperature T (redshifted to the present) $= 40$ kev and $\Omega_B = 0.25$. To illustrate the interplay between such energy releases and y-distortions, assume that the average pressure of the IGM depends upon the average baryon density as $p \sim n_B^\gamma \sim a^{-3\gamma}$. If energy is injected as a burst at an epoch z_x, then the 'adiabatic index' γ is 5/3 if there is no cooling; in hierarchical models, γ can be negative, reflecting the pressure increase as larger scale structures collapse. In terms of γ and the epoch z_x at which the pressure source turns on, the average distortion is $\bar{y} \approx 10^{-3}(T_0/10 \text{ kev})\Omega_B h$ $[(1+z_x)^{(3\gamma-1.5)}-1]/(3\gamma-1.5)$, where T_0 is the current temperature, and provided Compton cooling hasn't significantly lowered the temperature. For example, if the heating required to give the X-ray background occurred at $z_x = 3$, $\bar{y}$ would be as large as 0.01.

7.2 Sunyaev-Zeldovich and Primeval Dust Anisotropies

7.2.1 Overview: In this section, results from Bond (1987) on CMB anisotropies from Compton upscattering off inhomogeneously distributed hot gas and from red-shifted infrared emission from inhomogeneously distributed dust are summarized. We pay special attention to the emissions from the hierarchy of structures predicted to arise in the adiabatic cold dark matter picture of galaxy formation, which is calculationally tractable if we adopt a simple model for the collapsed structures that form. Results are compared with previous estimates of Sunyaev-Zeldovich fluctuations for the hot (Szalay, Bond and Silk 1983), explosive (Hogan 1984, Vishniac and Ostriker 1985), and cold (Ostriker and Vishniac 1986) theories of structure formation, and of dust emission fluctuations from primeval galaxies (Bond, Carr and Hogan 1986).

To illustrate the techniques used, consider Sunyaev-Zeldovich fluctuations for which the contribution to the anisotropy along a given line of sight is directly proportional to the time integral of the electron pressure. Under many circumstances, a shot noise model for the pressure is appropriate, namely that the pressure is a random field consisting of local pressure profiles centered on objects such as galaxies, clusters or pancakes which have undergone collapse, hence shock heating. The distribution of the objects defines a point process characterized statistically by a hierarchy of reduced N-point functions. The local pressure profiles may themselves be drawn from a statistical distribution.

Within this framework there are two conceptually different origins for the fluctuations. One is $\sqrt{N}$ Poisson fluctuations in the number of collapsed objects along the line of sight and the other is due to the intrinsic correlations of the scatterers, characterized by continuous N-point correlation functions. Both must be embodied in a self-consistent treatment which we now sketch.

7.2.2 General Treatment of Anisotropies: The distortions and anisotropies in the microwave background induced by the Sunyaev-Zeldovich effect and primeval dust can be treated using the same theoretical formalism, involving line of sight integrals of a random function $\mathcal{G}(x, \tau)$. The distortions are given by

$$\langle \frac{\Delta T}{T}(q, x=0, \tau_0) \rangle_{angle} = \int \bar{\mathcal{G}} d\tau \tag{7.1}$$

and the anisotropies by

$$\frac{\Delta T}{T}(q, \hat{q}, x=0, \tau_0) = \int d\chi \int \frac{d^3 k}{(2\pi)^3} \delta\mathcal{G}(q, k, \tau) e^{-ik \cdot \hat{q}\chi}, \tag{7.2}$$

where $\chi = \tau_0 - \tau$ for the flat universes we assume in this section, and with $\mathcal{G}$ given by:

Process	Form of $\mathcal{G}$

Sunyaev-Zeldovich $\mathcal{G} = -2\sigma_T m_e^{-1} a[\psi_K(q/(aT_e))P_e]$

Primeval Dust $\mathcal{G} = (aT_c/\omega_d)\pi r_d^2(1 - e^{-x})^2[e^{x(1-T_c/T_d)}n_d], \quad x = q/(aT_c)$

nonlinear flows $\mathcal{G} = \sigma_T a\hat{q} \cdot [n_e \vec{v}_e]$

The terms in square brackets are random fields whose statistical distributions will determine the statistics of the anisotropies. For comparison the term representing anisotropy development due to the asymmetry in Thomson scattering resulting from the flow of electrons is shown. As we have seen, such fluctuations are important for small angle primordial anisotropies. The motion of collapsed structures consisting of ionized gas adds a nonlinear addition to these fluctuations which can be of much smaller scale than that set by the fuzziness of recombination. We do not discuss this effect further here, but it can be calculated within the context of the model used for the other anisotropies. (See Vishniac 1987, Efstathiou and Bond 1987).

The correlation function $C(\varpi)$ then involves the correlation of two line of sight integrals, over χ_1 along a direction $\hat{q}_1$ defined by the incoming photon momentum and over χ_2 along $\hat{q}_2$. Here we calculate the power spectrum $C_{\vec{\ell}}$, where $\vec{\ell}$ is the angular wavenumber vector conjugate to $\vec{\varpi} = \hat{q}_1 - \hat{q}_2$. We assume that $\delta\mathcal{G}$ is a homogeneous random field, so that the wavevectors entering the line of sight integrals in $C(\varpi)$ have $\vec{k}_1 = -\vec{k}_2$. We split the integral over $\vec{k}$ into one over $\vec{k}_\perp \equiv \hat{\varpi} \cdot \vec{k}$ and one over $k_\parallel \approx \hat{q}_1 \cdot \vec{k}$, an adequate decomposition in the small angle approximation provided ℓ is large. The $k_\parallel$ integration approximately gives a delta function $\delta(\chi_1 - \chi_2)$ if we ignore the $k_\parallel$ dependence of $\delta\mathcal{G}$, thereby reducing the double time integral to a single one. The $k_\perp$ integration gives a delta function $(2\pi)^2\delta^{(2)}(\vec{\ell} - \vec{k}_\perp \chi)$, thereby ridding us of the $k_\perp$ integration. The power spectrum is then approximately

$$C_\ell \approx \int \frac{d\chi}{\chi^2}\langle|\delta\mathcal{G}[k_\perp = \ell/\chi, k_\parallel = 0; \chi]|^2\rangle. \tag{7.3}$$

Here we concentrate on determining the *rms* fluctuations as a function of beam smearing angle,

$$\sigma_r^2(\theta_s) = \int \frac{\ell d\ell}{(2\pi)} C_\ell e^{-\ell^2 \theta_s^2}. \tag{7.4}$$

7.2.3 Shot Noise Model:

A shot noise model is adopted for the distribution of the random variable $\mathcal{G}$: this consists of (1) a class of objects defined by parameters $\mathcal{C}$ whose positions are specified by a random point process $n_{\mathcal{C}*}(r) = \sum_j \delta^{(3)}(r - r_j)$ with the sum over points j satisfying the specifications $\mathcal{C}$; and (2) profiles for the $\mathcal{G}$ variable centered at each point, $g(r|\mathcal{C}, \tau)$. The r are taken as comoving positions, so $n_{\mathcal{C}*}$ is a comoving density. The expression for $\mathcal{G}$ is then a convolution:

$$\delta\mathcal{G}(r, \tau) = \sum_{\mathcal{C}} \int d^3r' g(\vec{r} - \vec{r}'|\mathcal{C}, \tau)\delta n_{\mathcal{C}*}(r', \tau), \tag{7.5a}$$

$$\delta\mathcal{G}(k, \tau) = \sum_{\mathcal{C}} g(k|\mathcal{C}, \tau)\delta n_{\mathcal{C}*}(k, \tau), \tag{7.5b}$$

$$g(k|\mathcal{C}, \tau) \equiv g_c(\tau|\mathcal{C})V_{\mathcal{C}}\mathcal{F}(k|\mathcal{C}), \tag{7.5c}$$

$$\bar{\mathcal{G}}(x, \tau) = \sum_{\mathcal{C}} g(k = 0|\mathcal{C}, \tau)\bar{n}_{\mathcal{C}*}. \tag{7.5d}$$

We could treat g as a random field as well as n_*, although it would not be homogeneous. The points C could define galaxies, clusters, pancakes, or even the centers of explosions (so g might define a shell), while the profiles g may be asymmetrical in some cases (e.g. pancakes). However, we do assume that n_{C*} and g_C are statistically independent. We have parameterized $g(k)$ in eq.(7.5c) by g_c, the value of g evaluated at the central point $r = 0$, the volume of the region $V_c \equiv \int g(r)/g_c d^3r$, and a form factor $\mathcal{F}(k)$ which is dimensionless and equal to unity at $k = 0$ by construction.

7.2.4 Clustering of the Shots: The form of the correlation functions $\xi^{tot}_{C_1 C_2}(r)$ for the point process consists of two terms, a Poisson contribution ($\propto \delta^{(3)}(r)$) describing the self correlation of the discrete objects and a continuous correlation $\xi_{C_1 C_2}(r)$ describing the clustering of the objects:

$$\langle \delta n_{C_1 *}(r) \delta n_{C_2 *}(0) \rangle \equiv \bar{n}_{C_1 *} \bar{n}_{C_2 *} \xi^{tot}_{C_1 C_2}(r) = \bar{n}_{C_1 *} \delta_{C_1 C_2} \delta^{(3)}(r) + \bar{n}_{C_1 *} \bar{n}_{C_2 *} \xi_{C_1 C_2}(r). \quad (7.6)$$

If the objects have a characteristic size R_C, then $1 + \xi_{C_1 C_2}(r)$ should be zero at least over that scale, reflecting the fact that the objects cannot get closer together than R_C. This volume exclusion, which is proportional to the filling factor of the objects, $n_{C*} V_C$, is usually unimportant for collapsed structures such as galaxies or clusters, but is important for shells which would form from explosions (§7.2.9).

For this paper, we adopt profiles and correlations which are analytically tractable yet still contain the expected features of more detailed treatments. For the collapsed structures, we shall only use spherically symmetric profiles of the Gaussian form: $\mathcal{F}_C(k, \tau) = \exp(-(R_C/a)^2 k^2/2)$; the volume is then $V_C = (2\pi)^{3/2} (R_C/a)^3$.

For the correlations of collapsed structures, we ignore the volume exclusion effects (valid for filling factors $\ll 1$), and adopt a form motivated by the correlations of peaks of Gaussian fields for a spectrum which gives r^{-2} correlation functions:

$$\xi_{C_1 C_2}(k) = f_{C_1}(k) f_{C_2}(k), \quad (7.7a)$$
$$f_{C_1}(k) = W_1(t) \exp[-k^2 r_{c1}^2/2]/k^{1/2}, \quad (7.7b)$$
$$W_1(t) \equiv (\langle \tilde{\nu} \rangle_1 + \sigma_{\rho_1}(t)) \sqrt{2} \pi r_{c1}, \quad r_{c1} \equiv \sqrt{6}/\langle k^2 \rangle_1^{1/2}, \quad (7.7c)$$
$$b_1 = \langle \tilde{\nu} \rangle_1 / \sigma_{\rho_1}(t) + 1. \quad (7.7d)$$

In this expression, the peaks of class C are assumed to be Gaussian filtered with a smoothing scale R_{f1}, $\langle k^2 \rangle_1$ is the average of k^2 over the density spectrum once it is Gaussian filtered, $\sigma_{\rho1}(t)$ is the *rms* level of the density fluctuations smoothed over the scale R_{f1}, and $\langle \tilde{\nu} \rangle_1$ is a time independent quantity defined in BBKS which gives the degree of statistical clustering in the initial conditions of peaks which satisfy the conditions C. For peaks of height $\nu \gg 3$, it is $\sim \nu$, but it may even be negative for low ν. It is related to the biasing factor b_1 describing the degree to which the objects are more or less clustered by the relation (7.7d). The correlation function is then $\xi_{C_1 C_2}(r, t) = b_1(t) b_2(t) \xi_{\rho\rho}(r, t)$ for $r \gg r_{c1}, r_{c2}$. Quantitatively, for the biased CDM model, this expression gives a correlation length (where $\xi = 1$) for bright galaxies of $5.7 \, h^{-1} Mpc$ and for clusters of $31 \, h^{-1} Mpc$ at the current epoch, compatible with observations; the statistical correlations which would even be present at high redshift give $2.1 \, h^{-1} Mpc$ for bright galaxies and $26 \, h^{-1} Mpc$ for clusters.

7.2.5 rms Fluctuations in a Single Beam: With these approximations, we have

$$\sigma_r^2(\theta_s) = \sigma_{rp}^2(\theta_s) + \sigma_{rc}^2(\theta_s), \tag{7.8a}$$

$$\sigma_{rp}^2 = \int d\chi \sum_C g_c^2(\tau|C) V_C^2 \bar{n}_{C*}(\tau)[4\pi]^{-1}[\theta_s^2\chi^2 + R_C^2/a^2]^{-1}, \tag{7.8b}$$

$$\sigma_{rc}^2 = \int d\chi \sum_{C_1 C_2} g_{c1}(\tau) V_{C1} W_1 \bar{n}_{C_1*}(\tau) g_{c2}(\tau) V_{C2} W_2 \bar{n}_{C_2*}(\tau)[4\sqrt{\pi}]^{-1}$$
$$[\theta_s^2\chi^2 + (R_{C_1}^2/a^2 + R_{C_2}^2/a^2)/2 + (r_{c1}^2 + r_{c2}^2)/2]^{-1/2}. \tag{7.8c}$$

Notice that the Poisson contribution scales with smoothing angle as $\sigma_{rp} \propto \theta_s^{-1}$, while the correlation contribution scales as $\sigma_{rc} \propto \theta_s^{-(\gamma-1)/2}$, with $\gamma = 2$ in our case. In the calculations which follow we do not include the cross correlations among objects of different scale.

7.2.6 Model of Virialized Structure: We now discuss the models for the collapsed objects which allow us to estimate such quantities as g_c and the virialized radii R_C and relate them to the initial conditions for the hierarchy of CDM model perturbations.

For convenience we assume that the collapsing structures in a CDM model are spherical. If the pressures of the gas and dark matter are negligible prior to collapse, then conservation of energy can be used to relate the initial amplitudes for spherical perturbations to the final average velocity and temperature. Assume $\Omega = 1$, so that unperturbed regions have zero energy. In the linear regime, the energy per mass enclosed within comoving radius r is

$$E/M(< r) = -0.5(H_0 r)^2 \nu_c \sigma_\rho(t_0) I(r), \tag{7.9}$$

$$I(r) = r^{-5} \int_0^r 15 r_1^5 [Ds(r_1)/(\nu_c \sigma_\rho r_1)] dr_1/r_1,$$

where the position of a shell initially at radius r at time t, $x(r,t) = r - D(t)s(r)$, defines $Ds(r)$. (If $\Omega_{nr} = 1$, $D = a$). The density perturbation is related to Ds in the linear regime by $\delta\rho/\rho = r^{-2}\partial r^2 Ds/\partial r$. We parameterize the amplitude of the perturbation at the center, $\delta\rho(r = 0, t)/\rho \equiv \nu_c \sigma_\rho(t)$, in terms of a time independent height ν and the rms level of the fluctuations at time t, $\sigma_\rho(t)$. (We extrapolate using the linear growth law $\sigma_\rho(t) = D(t)\sigma_\rho(t_0)$ to get the value at the current epoch t_0, $\sigma_\rho(t_0)$.) To treat the formation of an object of mass M, the density field is smoothed on a Gaussian filtering scale $R_f \approx 10(M/1.1 \times 10^{15} h_{50} M_\odot)^{1/3}$ Mpc, hence $\sigma_\rho(R_f, t)$ is also a function of R_f. For a collapsing peak, the height ν is positive; for a minimum which will evolve into a void, ν is negative.

We define a '3 dimensional virial velocity' v_T for a spherical region by $E/M \equiv -v_T^2/2$, if $\nu > 0$. If the region ultimately achieves virial equilibrium, then $-v_T^2$ is the gravitational potential energy per unit mass and $v_T^2/2$ is the specific binding energy (and also the specific kinetic energy) of the region. The average temperature T of the region is determined by equating the specific thermal energy $(3/2)Y_T T/m_N$ to the specific kinetic energy. We therefore have

$$v_T = H_0 r(\nu\sigma_\rho(t_0))^{1/2} I^{1/2} = 100 \frac{r}{h^{-1}Mpc}(\nu\sigma_\rho(t_0))^{1/2} I^{1/2} \text{ km s}^{-1}, \tag{7.10a}$$

$$T_V = \frac{m_N v_T^2}{3Y_T} = 4.0 \times 10^5 \; {}^\circ K \; Y_T^{-1} \Big(\frac{r}{h^{-1}Mpc}\Big)^2 \nu\sigma_\rho(t_0)I. \tag{7.10b}$$

Recall that m_N is the nucleon mass, Y_T and Y_e are the number of particles and electrons per baryon in the gas (1.6875 and 0.875 if it is fully ionized).

To determine other average quantities associated with the collapsed object, we need to rely on the virial equilibrium relation, $2K + 3\bar{P}V - 3P_sV = |\Omega_G|$, where the gravitational energy stored within radius R, $\Omega_G/M = -0.6KGM(< R)/R$, is parameterized by $K(R)$. Neglecting surface pressure P_s, this is balanced by v_T^2, yielding expressions for the radius R and the average overdensity and baryon density in terms of initial conditions:

$$R = \frac{3}{10}(K/I) \; r(\nu\sigma_\rho(t_0))^{-1}, \tag{7.11a}$$

$$1 + \bar{\delta} = 37\Omega h^2 [\nu\sigma_\rho(t_0)]^3 (I/K)^3, \tag{7.11b}$$

$$\bar{n}_B = 4.2 \times 10^{-4} \Omega_B h^2 [\nu\sigma_\rho(t_0)]^3 (I/K)^3 \; cm^{-3}, \tag{7.11c}$$

$$v_{circ}(R) = 1.29 v_T/K^{1/2}. \tag{7.11d}$$

In addition, the circular speed at radius R is given by eq.(7.11d).

These expressions can in turn be used to estimate the average total pressure $\bar{P}_T$ and electron pressure $\bar{P}_e$:

$$\bar{P}_T = \bar{n}_B Y_T T = 0.0146\Omega_B h^2 (hr)^2 [\nu\sigma_\rho(t_0)]^4 I^4/K^3 \; ev \; cm^{-3}, \tag{7.12a}$$

$$\bar{P}_e = (Y_e/Y_T)\bar{p}_T \approx 0.52\bar{P}_T. \tag{7.12b}$$

Eq.(7.12a) holds for isothermal regions for which there has been no gas cooling. To evaluate average pressures in the general case requires detailed knowledge of the temperature and density distributions.

The parameters I and K depend upon the initial and final collapse profiles respectively. For **top hat models** and for small r, they are both 1. In this model, the entire region collapses at the same redshift z_c given by $1 + z_c = \nu_c\sigma_\rho(R_f, t_0)/f_c$, where the collapse factor $f_c = 1.69$, and r and R in eqs.(7.6-7.8) are respectively the initial comoving top hat radius r_{th} and the final virialized top hat radius R_{vth}. The average overdensity is then $179(1 + z_c)^3$. The relation between r_{th} and R_f is as yet unspecified. For a Gaussian profile of radius R_C to have the same volume V_C as a top hat of radius R_{vth}, we require that $R_{vth} = f_{th}\sqrt{2}R_C$, with $f_{th} = 1.1$. We require the same relation to hold between r_{th} and $r_s(R_f, \nu)$, where r_s is a Gaussian length scale appropriate to peaks of height ν of the density field smoothed on the Gaussian scale R_f. The best choice for $r_s(R_f, \nu)$ is unclear, except that it is $\propto R_f$, and will differ from it by less than a factor of two. The straightforward choice $r_s = f_s(\nu)R_f$, with $f_s = 1$, tends to give velocity dispersions which are somewhat too small. If we equate r_s to the coherence length of the peak-density correlation function for peaks of height ν in the smoothed density field, we typically obtain f_s about 1.4 and 1 for high ν and for galaxy and rich cluster scales respectively, with f_s dropping somewhat as ν decreases. The far field profile of a peak follows the shape of the smoothed density density correlation function, for which the coherence length is $r_c/\sqrt{2}$, where r_c is given by eq.(7.7c). For an $n = -1$ density spectrum, this choice would give $f_s = \sqrt{3}$. A value somewhere in between might be appropriate. We

Table 3: Sunyaev-Zeldovich Effect in $\Omega = 1$ CDM Models for $\theta_s = 10''$ (§7.2.7)

	dwarf gals	bright gals	gals	small gps	large gps	poor cls	rich cls	extra rich cls	tot
$R_f(h^{-1}Mpc)$	.15	.35	.35	1	2.5	4	5	8	
std biased CDM									
$b = 1.4, f_c = 1.4$									
σ_{rp}	0	.16	.25	1.7	3.0	2.5	1.7	.08	6.1
σ_{rc}	0	.05	.17	.62	1.2	.86	.61	.003	2.5
σ_{rt}	0	.17	.30	1.8	3.3	2.7	1.8	.08	6.6
$-\langle \Delta T/T \rangle_{obj}$	0	.68	.28	1.2	3.5	6.8	10	29	
$\bar{y}$	0	.19	.70	1.5	1.4	.58	.2	.0002	11
$d(h^{-1}Mpc)$	1	4.6	2.3	6.1	16	32	53	670	
$\Omega_{cov}/4\pi$	10	.6	2.0	2.5	.68	0.14	.03	10^{-5}	50
$R_{vth}(h^{-1}kpc)$	13	30	44	220	850	1600	2200	3900	
$v_{rms}(\text{km s}^{-1})$	105	245	205	440	860	1200	1500	2300	
$P_e(ev\ cm^{-3})$	.29	1.1	.38	.32	.20	.18	.18	.29	
$T_v(kev)$	.02	.12	.09	.40	1.5	3.1	4.4	10	
$n_{Bv}(10^{-3}cm^{-3})$	11	8.8	3.6	.67	.13	.068	.044	.03	
$\delta_v/100$	370	310	130	24	4.4	2.1	1.6	1.1	
$\langle z_v \rangle$	5.4	5.4	3.5	1.6	.54	.24	.14	.03	
$b = 1.7, f_c = 1.2$									
σ_{rt}	0	.13	.22	1.2	2.2	1.7	1.1	.02	4.3
$b = 1.0, f_c = 1.69$									
σ_{rt}	0	.31	.56	3.6	7.4	6.8	5.1	.69	15
$\langle z_v \rangle$	6.4	6.4	4.2	2	.7	.31	.17	.07	
Hot DM $z_{nl}=1$									
$b = 0.5, f_c = 1.69$									
σ_{rt}	0	1.8	3.3	22	53	58	55	35	120
$-\langle \Delta T/T \rangle_{obj}$	0	15	6.1	26	57	67	73	110	
$\bar{y}$	0	2	7.4	18	25	22	18	6	180
$\langle z_v \rangle$	14	14	9	5	2.1	1	.7	.3	
Hot DM $z_{nl}=3$									
$b = 0.25, f_c = 1.69$									
σ_{rt}	0	11	21	140	360	410	410	320	890
$-\langle \Delta T/T \rangle_{obj}$	0	120	50	210	460	540	540	500	
$\bar{y}$	0	11	43	100	160	160	140	92	1200
$\langle z_v \rangle$	29	29	20	11	5.2	3.1	2.3	1	
$d(h^{-1}Mpc)$	x	x	x	x	x	x	16	41	
$\Omega_{cov}/4\pi$	x	x	x	x	x	x	.55	.33	
$T_v(kev)$	x	x	x	x	x	x	16	25	

treat f_s as a free parameter fixed to give reasonable velocity dispersions for bright galaxies and rich clusters. Values about 1.5 give reasonable results with $I = 1$ in eq.(7.10a) as required in the top hat model. This choice for f_s can be motivated by the following argument which also addresses the question of late infall.

Consider an $n = -1$ fluctuation spectrum filtered on a scale R_f. The density-density correlation function is very nearly $\sigma_\rho^2(1 + r^2/r_c^2)^{-1}$. In the far field of an average peak of height ν_c, the statistically averaged profile is $(\nu_c/\sigma_\rho)\xi_{\rho\rho}$ as is shown in BBKS. For such a profile, $s(r)$ and hence $I(r)$ can be computed. For example, $I(r_c) \approx 0.7$. For this profile, collapse does not all occur at once. Rather, the region interior to $\sim r_c$ collapses almost simultaneously ($Ds/r \approx (\nu\sigma_\rho/3)[1 - 0.6r^2/r_c^2]$), while the outer regions accrete.

Other peaks of the same scale that also have collapsed will be drawn in as turnaround occurs at larger and larger r, leading to groups and clusters with mergers of the peaks. This will be especially important for the dark matter and hot gas distributed over the entire virial scale rather than concentrated in the center as stars and dust might be.

We describe this hierarchy by a sequence of increasing filtering radii, and ignore the merging of smaller scale peaks in larger ones, and choose r_c as the collapse scale. In this paper, we take $I = 1$, $K = 1$ and $f_s = (3 \times 0.7)^{1/2} = 1.5$. Though we have argued that this is a reasonable choice, the ultimate justification is phenomenological. The bright galaxies and rich clusters are somewhat anemic for smaller values and too robust for higher values in the CDM model. We should be aware however that anemic clusters may in fact be a flaw in the biased CDM model. However, one must be careful not to overinterpret these simple analytical spherical models since collapses are almost always asymmetrical.

The density of peaks $\mathcal{N}_{pk}(\nu, R_f)d\nu$ of height within $d\nu$ of ν for a given filtering scale can be determined using the statistical theory of 3-dimensional Gaussian random fields. We can use the top hat relations which are functions of ν and R_f to estimate the contributions of the collapsed peaks of that scale to the rms anisotropies. In order to get an estimate of the total anisotropies expected from all scales, we would need a density which was differential in R_f as well and also took into account that there are always small scale peaks within bigger scale ones, the cloud-in-cloud problem, and that mergers destroy smaller scale peaks as the universe evolves. No satisfactory result for $dn_{pk}/d\nu dR_f$ exists. Here, we adopt a spectrum of filtering radii R_{fi} and perform the integral over scale by $\sum_i \mathcal{N}_{pk}(\nu, R_{fi})$ $d\nu\,(3/2)d\ln(R_{fi+1}/R_{fi-1})$, motivated by $\mathcal{N}_{pk} \propto R_f^{-3}$. The total anisotropies should therefore only be regarded as reasonable estimates.

7.2.7 SZ Anisotropies in Cold and Hot DM Models:

In Table 3, the contribution of different structures to the rms Sunyaev-Zeldovich fluctuations σ_{rt} (in units of 10^{-6}) expected in a CDM model with $\Omega = 1$, $\Omega_B = 0.1$ and $h = 0.5$ at $\theta_s = 10''$ ($\theta_{fwhm} = 24''$) for wavelengths in the Rayleigh Jeans region are given. The results should be multiplied by ψ_K (eq.3.7) for wavelengths on the Wein side. For the biased model, the 'galaxies' include all of the peaks smoothed on the Gaussian filtering scale $R_f = 0.35\ h^{-1}Mpc$ which would have collapsed by the present (satisfying $\nu\sigma_\rho > f_c$ where f_c is the collapse factor). The biasing factor $b = 1.4$ is suggested by the biasing model introduced by Bond (1986). The value of f_c has been adjusted so that the abundance of rich clusters (with the smoothing length indicated) comes out right. This procedure was also followed for the $b = 1.7$ model used by BBKS and Bardeen, Bond and Efstathiou (1987). For the mass traces light

($b = 1$) model, the clusters are slightly too abundant with $f_c = 1.69$.

The $b = 0.5$ and $b = 0.25$ models are included to demonstrate that larger effects are obtained by having cluster scale objects collapsing at high redshift. In this case, the normalization corresponds to a massive neutrino (hot dark matter) dominated universe with nonlinear redshift $z_{nl} = 1$ and $z_{nl} = 3$, respectively. However, only the $R_f = 5\ h^{-1} Mpc$ objects would be relevant in that case since smaller scales suffer collisionless damping (§5.3), and the neutrino clusters which would form (and presumably contain hot gas) are a factor 8 more abundant than Abell clusters.

These results demonstrate that hot dark matter models with nonlinear red-shifts of order 3 can be ruled out by virtue of their SZ anisotropies. The covering factor of the ν clusters is sufficiently large that it is unlikely such hot spots in the CMB would have been missed. White *et al.* (1984) pointed out that these neu-trino clusters would also be copious X-ray emitters (as their virial temperatures clearly indicate) and would have been observed by the Einstein observatory if they existed. Szalay, Bond and Silk (1983) adopted a simple model of pancakes with random orientations Poissonian-distributed in space to model the hot gas distribu-tion in neutrino-dominated models. The distribution and evolution of the electron pressure in these objects was determined using the numerical and analytic cooling pancake models constructed by Bond, Centrella, Szalay and Wilson (1984). The SZ anisotropies expected in this model from widely separated lines of sight were com-puted and shown to rule out models with pancaking occurring at $z > 5$. The cluster results presented here give stronger constraints, in part because the pressures gen-erated in 3-dimensional collapse are higher than in one-dimensional collapse. These neutrino cluster results give a better model of the evolution of a hot dark matter dominated universe with a high nonlinear redshift than that used by Szalay, Bond and Silk, since pancakes are transitory, lasting only for about one Hubble time at the epoch of collapse, while the clusters, representing the deepest potential wells, continue to accumulate hot gas, much of it draining from the pancakes. It may be more difficult for hot gas to find its way into the extra rich cluster cores than the dark matter due to opposing pressure gradients, but it seems unlikely that suffi-cient exclusion can occur to alter the basic conclusion: the $\Delta T/T$ constraints from SZ anisotropies in hot DM models is somewhat stronger than the constraints from primordial anisotropies.

For the standard $b = 1.4$ CDM model and the $z_{nl} = 3$ hot dark matter model, parameters for the various mass structures are given in Table 3. The aver-age temperature decrement across an object is $\langle \Delta T/T \rangle_{obj}$ and the angle-averaged y-paramater giving the net Compton distortion to the CMB spectrum is $\bar{y}$; both are in units of 10^{-6}. The following physical parameters characterizing the average properties of the specific objects are also shown: mean comoving spacing d of col-lapsed objects; covering factor $\Omega_{cov}/4\pi$ (fraction of the sky covered by the objects, which equals the mean number intercepted along an average line of sight); the virial 'top hat' radius R_{vth}, related to our Gaussian virial radius R_v by $R_{vth} = 1.56 R_v$; the 3-dimensional *rms* velocity in the potential well v_{rms}; the electron pressure P_e for primordial abundances; the virial temperature T_v; the average baryon abundance n_{Bv}; the average overdensity of the objects above background δ_v; and the average redshift of collapse of the objects. Values of these parameters appropriate for bright galaxies as determined from the $b = 1.4$ CDM model are shown in square brack-ets. Our galaxy is just about one of these; they are spaced 4.6 $h^{-1} Mpc$ apart on average. Requiring $d = 4.64$ for the $R_f = 0.35\ h^{-1} Mpc$ peaks fixes the threshold value ν_t (for a given shape of the selection function) above which the peaks are selected to be bright galaxies in the simplest biased galaxy formation model. The galaxy-galaxy correlation function can then be determined, and relating this to the observed correlation function fixes the biasing parameter to be $b = 1.4$ and 1.7 for two different selection function choices. $f_c = 1.2$ is uncomfortably small for the 1.7 case.

In Figure 7.1, how the separate contributions from Poissonian statistics and clustering statistics to the SZ anisotropies enter as a function of beam smearing angle is shown for small groups ($R_f = 2\ h^{-1} Mpc$), rich clusters ($R_f = 5\ h^{-1} Mpc$), and the poor clusters ($R_f = 3.5\ h^{-1} Mpc$) which contribute most to the anisotropy level. Note that clustering plays an important role in stopping the precipitous θ_s^{-1} Poissonian decline at large θ_s, modifying it to the gentler $\theta_s^{-(\gamma-1)/2}$ fall. However, Poisson fluctuations are the most important anisotropy source over most of the range of interest of θ_s due principally to the small covering factors of these objects.

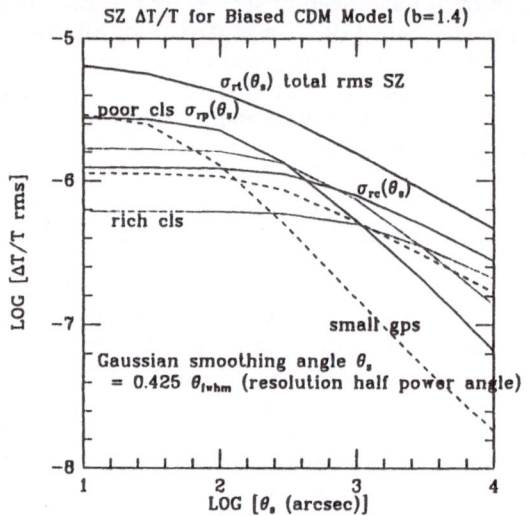

Figure 7.1

The basic lesson to be learned from these results is that the standard CDM picture gives very small Sunyaev-Zeldovich fluctuations (c.f. Ostriker and Vishniac 1986). What is also evident from Table 3 is that the basic structural properties and average spacings of objects ranging from dwarf galaxies through bright galaxies and groups to clusters are obtained with this simple spherical collapse model and the CDM fluctuation spectrum with the biasing normalization. This should undoubtedly be considered a triumph of the model. However, Birkinshaw (1987) and others claim to have observed the SZ effect in clusters with a decrement of about 1 mK, giving $\Delta T/T$ a few times 10^{-4}, much greater than the 10^{-5} value given for the average cluster in Table 3. It is not clear what the reason for the discrepency is, though it is clear that our use of average values for the clusters, especially of the baryon density, will lower the value over that expected in realistic models with finite core radii. Within the spirit of simplicity adopted for the results presented here, the SZ anisotropies are expected to be low, and it seems unlikely that the adoption of more realistic pressure profiles will alter the average rms results substantially.

7.2.8 Anisotropies from Primeval Dust: To estimate the anisotropies from dust using the CDM hierarchy, we require a model for dust production, distribution

Table 4: Dust Anisotropies from Primeval Galaxies for $\lambda = 500\mu$ ($\theta_s = 10''$) (§7.2.8)

	dwarf gals	bright gals	gals	tot
$R_f(\ h^{-1}Mpc)$	.15	.35	.35	
std biased CDM $\langle z_v \rangle \sim 5$				
$b = 1.4, f_c = 1.4, s_d = 0.5$				
$\sigma_{rpd}(10'')$	.022	.022	.034	.062
$\sigma_{rcd}(10'')$	.024	.006	.016	.045
$\sigma_{rtd}(10'')$	.032	.023	.038	.077
$\langle \Delta T/T \rangle_{angle}$	.31	.039	.15	1.1
$b = 1.4, f_c = 1.4, s_d = 1$				
σ_{rtd}	.26	.17	.28	.57
$\langle \Delta T/T \rangle_{angle}$	2.5	.31	1.2	9.1
$b = 1.4, f_c = 1.4, s_d = .25$				
$\sigma_{rtd}(10'')$	..004	.003	.005	.01
$\langle \Delta T/T \rangle_{angle}$	.039	.005	.018	.14
$b = 1.7, f_c = 1.2, s_d = 0.5$				
$\sigma_{rtd}(10'')$	.028	.020	.033	.068
$\langle \Delta T/T \rangle_{angle}$	.28	.035	.13	1.0
$b = 1, f_c = 1.69, s_d = 0.5$				
σ_{rtd}	.05	..033	.056	.12
$\langle \Delta T/T \rangle_{angle}$	.46	.056	.21	1.6
high z gal form. $\langle z_v \rangle \sim 15$				
$b = 0.5, f_c = 1.69, s_d = 0.5$				
$\sigma_{rtd}(10'')$	..31	..20	..33	.70
$\langle \Delta T/T \rangle_{angle}$	2.6	.32	1.2	9.4
high z gal form. $\langle z_v \rangle \sim 30$				
$b = 0.25, f_c = 1.69, s_d = 0.5$				
$\sigma_{rtd}(10'')$	2.0	1.2	2.0	4.4
$\langle \Delta T/T \rangle_{angle}$	15	1.8	6.9	53

and temperature and emissivity profile. The following parameters describe these uncertain numbers: (1) The dust cross section $\sim \lambda^{-\alpha}$; we adopt $\alpha = 1$. (2) A fraction ϕ_d of the local baryon density computed from the models for collapsed structures introduced in §3.7.6 are locked up in dust grains. ϕ_d is probably of order 10^{-3} for our galaxy. The computed anisotropies just scale linearly with this number $\sigma_{rd} \propto \phi_d$. We take the extremely optimistic Population I value $\phi_d = 0.1$ in the numerical results quoted here. (3) The dust is assumed to be distributed with a Gaussian profile over a fraction $s_d = R_d/R_v$ of the virial radius as determined from §7.2.6. The results are quite sensitive to varation in s_d, since it determines the total amount of dust in the universe. The cosmological density parameter in dust is $\Omega_d = \phi_d s_d^3 \Omega_{Bg}$, where Ω_{Bg} is the density of baryons in galaxies. We take $s_d = 0.5$ as our standard, but show the effects of varying s_d in Table 4. (4) The size of the grains r_d will cover some range which might extend from $\sim 10^{-3}\mu$ up to $\sim 1\mu$. The value $r_d = 0.01$ is adopted here reflecting a prejudice that at early times the grains might well be smaller than those typical now. (5) The temperature of the dust. If external radiation with a fixed fractional percentage of the CMB energy is absorbed by the galaxies, equation (3.8) shows T_d/T_c is, approximately, independent of absorption epoch. For the results in Table 4, the value $T_d = 26K(1 + z)/4$ was chosen corresponding to about 10 % of the CMB energy being liberated by primeval sources. Note that if the radiation is internally generated, absorbed by dust and reradiated in the IR by the same galaxies, then it is more appropriate to assume T_d is independent of redshift. However, since the emission occurs primarily from the collapse epoch in these models, we can take the results of Table 4 as indicative of this case with a dust temperature about $40K$ corresponding to the temperature at $\langle z_v \rangle \sim 5$.

Table 4 shows Poisson (σ_{rpd}), correlation (σ_{rcd}) and total (σ_{rtd}) anisotropies expected at $\lambda = 500\mu$ for the parameters indicated. Results for rather large dwarf galaxies, bright galaxies and for all peaks that would have collapsed on galactic scale by now are given. The total values include the sums over the dwarf and galaxy peak contributions and should not be taken too seriously. The main conclusion is that relatively plausible choices for dust and radiation parameters can lead to sizable distortions and anisotropies for the standard biased CDM model, and dwarf galaxy formation at higher redshifts, $z \sim 10 - 20$, would lead to even larger distortions. Note that the bright galaxies would not cover the sky with the biased CDM parameters, so there would be regions in which the pure pregalactic light might get through undisturbed, likely resulting in a near IR background to complement this far IR one. See Bond, Carr and Hogan (1986) for a detailed discussion.

Though these effects may be large at $\lambda \sim 500\mu$, they are small in the Rayleigh Jeans part of the CMB spectrum due to the λ^{-1} cross section law. One might expect that the power will be closer to 2 in that region, resulting in much smaller values. Nonetheless, the following values for the standard CDM model indicate that the anisotropies could be competitive with those expected from primary fluctuations even at radio wavelengths. In this table, $\Delta T/T(10'') \equiv \sigma_{rtd}/(1 + \langle \Delta T/T \rangle)$ to make the anisotropies relative to the background radiation with distortions included, a distinction which is important for temperature distortions in excess of one.

λ	3 cm	1 cm	3 mm	1 mm	500 μ	300 μ
$\Delta T/T(10'')$	2×10^{-6}	2×10^{-5}	1.4×10^{-4}	1.5×10^{-3}	0.024	0.1
$\langle \Delta T/T \rangle (distort)$	0	0	0	0.015	0.31	18

In Figure 7.2, the dependence of the single beam *rms* anisotropies as a function of smearing angle for the standard biased CDM case is shown. The correlation contribution $\sim \theta_s^{-1/2}$ for our r^{-2} choice of the correlation function takes over from the Poisson contribution at relatively small angular scales. The small coherence angle and strong wavelength dependence of primeval dust emission would unambiguously allow one to differentiate the anisotropies from primeval dust from the primary anisotropies of §5.

Lange *et al.* (1987) find tentative evidence for distortions $\langle \Delta T/T \rangle$ somewhat in excess of unity at λ about 600 μ, with no significant distortions occurring around the peak. Primeval dust emission seems a natural source to explain their data, since the parameters required are similar to those expected in standard hierarchical models of structure formation such as the biased CDM model given in this section. Ostriker and Thomson (1987) point out that Sunyaev-Zeldovich distortions from an early generation of explosions driven by electromagnetic energy liberated by superconducting cosmic strings would also lead inevitably to such distortions, and still not overproduce anisotropies in the Rayleigh Jeans part of the spectrum. By contrast, all late time energy injection models that could produce such a large distortion by Compton upscattering of CMB photons from hot gas would be expected to generate large (unobserved) Rayleigh Jeans distortions. We now turn to a discussion of anisotropies from such generic cosmic explosion models that are often invoked to explain the 'bubbly' structure of the universe.

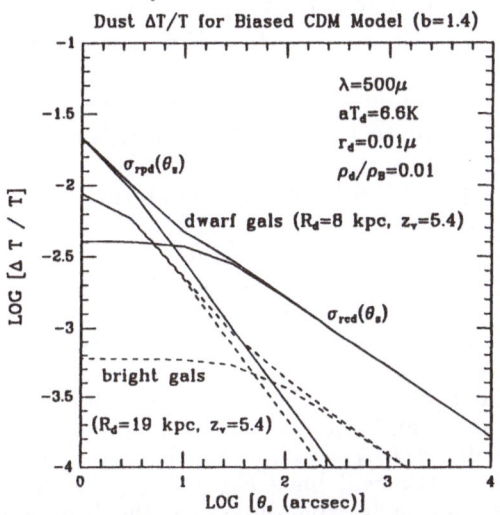

Dust ΔT/T for Biased CDM Model (b=1.4)

Figure 7.2

LOG [θ_s (arcsec)]

7.2.9 Sunyaev-Zeldovich Anisotropies from Explosions: To estimate the SZ anisotropies from explosions, we adopt a highly simplified model:

(1) The explosive structures are assumed to come into being fully developed at the explosion epoch (redshift z_x) with a shell profile of fixed *comoving* inner (R_1) and outer (R_0) radii having the form

$$V_s \mathcal{F}(r) = e^{-r^2/(2R_0^2)} - e^{-r^2/(2R_1^2)}, \qquad (7.13a)$$

$$V_s \equiv V_0 - V_1 \equiv (2\pi)^{3/2} R_0^3 - (2\pi)^{3/2} R_1^3. \qquad (7.13b)$$

V_s is the effective volume of the hot gas. Taking $R_1 = 0$ allows us to treat hot gas balls. For explosive shells, the fractional size $\Delta R/R = 2(R_0-R_1)/(R_0+R_1)$ is typically only about 3% for fully developed adiabatic cosmological blast waves (Bertschinger 1983, Ikeuchi, Tomasaka and Ostriker 1983), and could be even smaller in the cooling regime.

(2) Since the explosive structures (with comoving density of centers $n_{x*} \equiv d_x^{-3}$) are assumed not to overlap, we must assume a continuous correlation function satisfying $1 + \xi^{(c)} \approx 0$ reflecting this exclusion. A reasonable Gaussian choice which gives Poisson fluctuations in the explosion centers on very large scales and effectively excludes them from being within a distance $2\sqrt{2}s_x R_0$ of each other is $\xi^{(c)}(r) = -\exp[-r^2/(2(2s_x R_0)^2)]$. We take the parameter s_x to be one in the following. The power spectrum of fluctuations in the comoving number density of explosions is then

$$\langle |\delta n_{x*}(k)|^2 \rangle = n_{x*}(1 - \tilde{f}_x \exp[-2(ks_x R_0)^2]), \qquad (7.14)$$
$$\tilde{f}_x \equiv n_{x*}(2\pi)^{3/2}(2s_x R_0)^3 \approx (5s_x R_0/d_x)^3,$$

passing from white noise with the usual n_{x*} amplitude on short wavelengths to white noise at large wavelengths with the reduced amplitude $n_{x*}(1 - \tilde{f}_x)$ reflecting the excluded volume arising from the finite size of the structures. It is obvious from this that we can never allow $\tilde{f}_x > 1$, a constraint on the possible size of s_x. In the following we use $\tilde{f}_x$ in place of s_x to characterize the excluded region. Note that the effective filling factor of the shells is $f_x = V_s n_{x*}$, generally a much smaller number than $\tilde{f}_x$.

(3) The thermal energy content per explosive structure at redshift z is assumed to decay according to $E_{th}(z) = E_{th}(z_x)(a_x/a)^{3(\gamma-1)}$, where γ is an effective adiabatic index. In terms of this, g_c of eq. (7.5c) is $g_c = -2\sigma_T m_e^{-1} a\psi_K[q/(aT_e)]P_{ec}$, where $P_{ec} = [2Y_e E_{th}(a)/(3Y_T a^3 V_s)]$ is given in terms of the energy, volume, redshift and total and electron abundances Y_T and Y_e. The total baryon number enclosed in the region, $B = \bar{n}_{B*}V_0 = 5.2 \times 10^{69}\Omega_B h^{-1}(R_0/ h^{-1}Mpc)^3$, is conserved as the matter is swept up into the shell. We can parameterize the energy in the shell by a thermal energy per baryon $\epsilon_{th} \equiv E_{th}/B = 119$ ev $E_{th,60} \Omega_B^{-1}h[R_0/ h^{-1}Mpc]^{-3}$ and a temperature $T_s \equiv \epsilon_{th}/(1.5Y_T) = 0.4\epsilon_{th}$. This temperature decays as $(1+z)^{3(\gamma-1)}$, going as $(1+z)^2$ for $\gamma = 5/3$, the canonical case with no cooling.

The energy in a region is conserved throughout expansion provided there is no external pressure and cooling can be ignored. For a region with $\Omega = 1$, the total energy will be $E_{tot} = E_{exp}$, the energy of the explosion. A hot ball which has expanded little will have $E_{th} \approx E_{tot}$. However, if all of the matter is swept up into a thin shell expanding with the Hubble flow then the energy is $E_{tot} = E_{th} + E_{KE}/2$, where the kinetic energy of expansion of the gas in the shell per baryon is $\epsilon_{KE} \equiv E_{KE}/B \approx m_N(H(a)aR_{th0})^2/2 \approx 127 (R_0/ h^{-1}Mpc)^2(1+z_x)$ ev in terms of the top hat radius $R_{th0} = 1.56R_0$ and the nucleon mass m_N. The ratio of thermal to kinetic energy in a fully developed cosmological blast wave is generally small (Bertschinger 1983, Ikeuchi et al. 1983). However in the early phases of explosion before the blast enters the regime in which its similarity solution holds, we may expect a better description to be a ball of hot gas with the energy largely thermal. For illustration here, we treat $\epsilon_{th}/\epsilon_{KE}$ as another parameter.

For a given explosive input E_{exp}, the maximum size of the thin shell is that for which all of the explosive energy goes into kinetic energy of expansion without allowing any residual for thermal energy: $R_{0,max} \approx 1.6 h^{-1}Mpc(E_{exp,60}a_x)^{1/5}$.

Table 5: Sunyaev-Zeldovich Anisotropies from Explosions (§7.2.9)

Notes to Table 5: The single beam *rms* fluctuations $\sigma_r(\theta_s)$ for $\theta_s = 10''$, $10'$ and $60'$ and the average y-parameter are given (in units of 10^{-6}) for explosion models which are meant to reproduce the bubbly structure of the universe. (These values are appropriate to the Rayleigh-Jeans part of the spectrum, scaling with ψ_K (eq.3.7) at higher energies.)

The first line gives parameters for a standard model for explosive bubbles: redshift of bubble formation is $z_x = 7$; centers are separated on average by the spacing of rich clusters ($55\ h^{-1}Mpc$); 'top hat' radii R_{th0} are chosen to give the typical void diameter, 25 $h^{-1}Mpc$; the shell is assumed to be 1.3 $h^{-1}Mpc$ across ($\Delta R/R = 0.1$).

The thermal energy is taken to be the minimum energy per baryon required to scour out a bubble of this size at $z_x = 7$, $\epsilon_{KE}/2 = 33$ keV. (Actually, when dark matter is present, as in the cases treated here, evacuation of the gas to the shell slightly decreases the gravitational potential energy of the dark matter inside, upsetting the zero energy balance with the expansion kinetic energy of the dark matter. The required extra energy is $(\Omega_X/5)E_{KE,gas}/2$ for thin shells, a further $\sim 20\%$ drain on ϵ_{exp} over the energy $\epsilon_{KE}/2$ required by the expanding gas shell. The energy required to scour out a shell of this size is therefore really 39 keV.)

The subsequent models vary one or more of the standard parameters, those shown in bold face type. The $\tilde{f}_x = 0$ model has $s_x = 0$ to show the influence that neglecting volume exclusion has on the computed anisotropies. For $\theta \gg \theta_x$, which is a factor about 1.4 larger than the θ_0 shown, $\sigma_r \propto \theta_s^{-1}$. This certainly holds beyond $\theta_s = 60'$. The results for σ_r scale with the ratio $2\epsilon_{th}/\epsilon_{KE}$, which we have taken to be one in all but one case. The energies ϵ_{th} and ϵ_{KE} are evaluated at redshift z_x and are in kev. $R_{th0} = 1.56R_0$ and d_x are in $h^{-1}Mpc$ units.

$\sigma_r(10'')$	$\sigma_r(10')$	$\sigma_r(60')$	$\theta_0(')$	$\bar{y}$	z_x	R_{th0}	d_x	$\Delta R/R$	$\tilde{f}_x$	f_x	ϵ_{KE}	ϵ_{th}
19	12	2.5	7.1	24	**7**	**12.5**	55	**.1**	.39	.01	65	**33**
6.8	2.9	.49	4.4	14	7	**8**	30	.1	.58	.02	25	13
53	39	6.9	9.8	91	7	**17**	60	.1	.78	.03	123	61
1.9	1.2	.25	7.1	2.4	7	12.5	55	.1	.39	.01	65	**3.3**
.31	.14	.02	4.8	.68	**5**	**8**	30	.1	.58	.02	19	1
2.8	2.1	.48	9.2	4.2	**3**	12.5	55	.1	.39	.01	33	16
73	44	8.5	6.3	81	**12**	12.5	55	.1	.39	.01	106	53
18	12	2.5	7.1	24	7	12.5	55	**.01**	.39	.001	65	33
24	14	2.5	7.1	24	7	12.5	55	**1**	.39	.05	65	33
25	14	9.2	7.1	61	7	12.5	**40**	.1	.99	.03	65	33
21	14	3.2	7.1	24	7	12.5	55	.1	**0**	.01	65	33

$[5\Omega_B h^{-1}]^{-1/5}$. To scour out the typical voids seen in the CfA redshift survey (diameter $\sim 25\ h^{-1}Mpc$ hence Gaussian radii $R_0 \sim 8\ h^{-1}Mpc$) at $z_x = 5$ would require at least 2×10^{64} ergs, the explosive energy of 20 trillion supernovae, each having the typical Type II Supernova energy 10^{51} ergs.

If these are normal metal producing supernova explosions, the available explosive energy from a $10^{11}\ M_\odot$ galaxy with Population I metallicity ($Z \approx 0.02$) is $\sim 10^{60}$ ergs. Of this available energy, only a small fraction may actually escape from the galaxy on a short enough timescale to drive a coherent shock. Clean bombs which avoid heavy element production may include accretion onto black holes, super-Eddington radiation driven explosions of Very Massive Object envelopes (Bond, Arnett and Carr 1984), and perhaps even superconducting cosmic strings, which may liberate up to $\sim 10^{65}$ ergs according to Ostriker, Thompson and Witten (1986). The shocks from a group of galaxies might also sum together to make a super-shock. Therefore, we should treat the allowable energy release as a variable parameter, and see if the SZ anisotropies can constrain it in combination with the other parameters at our disposal.

With these assumptions, the radiation amplitude for $\Omega = \Omega_{nr} = 1$ universes is

$$\sigma_r = \theta_s^{-1} \frac{2Y_e \sigma_T \psi_K (H_0 d_x)^{1/2} B}{3Y_T m_e (4\pi)^{1/2} d_x^2} \epsilon_{th}(z_x)(1+z_x)^{1.75}\, W(a_x)U(\theta_s) \qquad (7.15a)$$

$$= 4.4 \times 10^{-9} [\theta_s(')]^{-1} \frac{(R_0/\ h^{-1}Mpc)^3}{(d_x/\ h^{-1}Mpc)^{3/2}} \frac{\Omega_B h \psi_K}{a_x^{1.75}} \frac{\epsilon_{th}(z_x)}{ev} WU,$$

$$W(a_x) \equiv \frac{(1 - a_x^{(6\gamma - 2.5)})^{1/2}}{(6\gamma - 2.5)^{1/2}(1 - a_x^{1/2})},$$

$$U(\theta_s) = (V_0/V_s)^2 \left[(1 + (\theta_0/\theta_s)^2)^{-1} - \tilde{f}_x(1 + (\theta_0/\theta_s)^2 + (\theta_x/\theta_s)^2)^{-1}\right]$$
$$- 2(V_0 V_1/V_s^2)\left[(1 + 0.5((\theta_0/\theta_s)^2 + (\theta_1/\theta_s)^2))^{-1}\right.$$
$$\left. - \tilde{f}_x(1 + 0.5((\theta_0/\theta_s)^2 + (\theta_1/\theta_s)^2) + (\theta_x/\theta_s)^2)^{-1}\right]$$
$$+ (V_1/V_s)^2\left[(1 + (\theta_1/\theta_s)^2)^{-1} - \tilde{f}_x(1 + (\theta_1/\theta_s)^2 + (\theta_x/\theta_s)^2)^{-1}\right],$$

$$\theta_i \equiv 0.5 H_0 R_i (1 - a_x^{1/2})^{-1},\ i = 0, 1, \quad \theta_x \equiv 0.5 H_0 \sqrt{2} s_x R_0 (1 - a_x^{1/2})^{-1},$$

$$U(\theta_s) \rightarrow 1 - \tilde{f}_x, \quad \theta_s \gg \theta_x. \qquad (7.15b)$$

Once the beam smearing angle θ_s is sufficiently greater than the angular extent of the structures (7.15b), the fluctuations follow the $\sigma_r \propto \theta_s^{-1}$ Poisson law associated with uncorrelated hot spots. The average y-distortion from these explosion models is

$$\bar{y} = \frac{2Y_e \sigma_T \psi_K B}{3Y_T m_e H_0 d_x^3} \epsilon_{th}(z_x)(1 + z_x)^{1.5} \frac{(1 - a_x^{(3\gamma - 1.5)})}{(3\gamma - 1.5)} \qquad (7.16)$$

$$\approx 7.4 \times 10^{-7}\ (R_0/d_x)^3 \Omega_B h \psi_K \frac{\epsilon_{th}(z_x)}{ev}(1 + z_x)^{1.5} \frac{(1 - a_x^{(3\gamma - 1.5)})}{(3\gamma - 1.5)}.$$

In Table 5, values of $\bar{y}$ and $\sigma_r(\theta_s)$ for the θ_s indicated are given for a variety of models characterized by the top hat radius of the shell R_{tho}, the redshift of the explosion z_x, the spacing of the explosive centers d_x, the shell width $\Delta R/R$, and the filling factors $\tilde{f}_x$ of the explosive structures and f_x of the hot gas, and the thermal energy per baryon in the shell $\epsilon_{th}(z_x)$ (which should be compared with the kinetic energy of expansion per baryon $\epsilon_{KE}(z_x)$). We assume $\gamma = 5/3$ throughout. The characteristic angular size θ_0 in arcminutes is also presented. These numbers should be regarded as indicative only. A more realistic calculation would include the detailed cosmological evolution for the radii of the blasts.

The principal result is that explosions at high redshift which generate the structure of the universe on scales $R_0 \sim 5 - 8 \ h^{-1} Mpc$ cannot be ruled out if fully developed blast wave solutions with little explosive energy in a thermal component are used. Filled balls of hot gas have somewhat higher rms fluctuations, though not necessarily high enough to be ruled out. These results should be compared with those of Hogan (1984) and Vishniac and Ostriker (1985). Hogan claims that the development phase of the blast when it is thermal energy dominated will last for a sufficiently long period of time to still rule out the formation of all of the structure in the universe from high redshift explosions. Note that the $\Delta R/R = 1$ case in Table 5 generating $25 \ h^{-1} Mpc$ gas balls spaced $55 \ h^{-1} Mpc$ apart at $z_x = 7$ does not generate uncomfortably large single beam rms anisotropies. However, when the shock energy is more concentrated, the anisotropies would be more similar to the high redshift cluster results of §7.2.7. Comparing with the *extra rich clusters* column of the $z_{nl} = 3$ HOT DM model in Table 3 demonstrates that we certainly would have very large SZ anisotropies in this $z_x = 7$ case.

Ostriker, Thompson and Witten (1986) argue that there may not be a large SZ effect from the cavities scoured out by the $\sim 10^{65}$ ergs of electromagnetic energy liberated by superconducting cosmic strings at $z_x \sim 7$ if the temperature is above 511 keV. Given that, one would still expect a SZ effect from the hot shells of matter surrounding the cavities. The results presented here, however, demonstrate that the anisotropy from the shells alone is probably not sufficient to rule out this picture of structure formation. A more detailed calculation of the anisotropies expected in specific experiments such as that of Uson and Wilkinson (1984a,b) is required to assess whether the non-Gaussian nature of the fluctuations will result in anisotropy signals in the neighbourhood of the shells greatly in excess of the single beam rms values given in Table 5.

CONCLUSIONS It should be clear from these lecture notes that the experimentalists, who are making great strides towards measuring anisotropies at the 10^{-5} level, will soon be able to topple a large number of models for the formation of structure in the universe. It is also remarkable that all angular scales are interesting, from arcseconds, suited for primeval dust generated anisotropies, through arcminutes, for anisotropies from Compton upscattering off inhomogeneous hot gas, through tens of arcminutes, probing velocity flows at the epoch of recombination, through degrees, probing entropy and gravitational potential fluctuations which are causally disconnected at the time of photon decoupling. Our field has certainly achieved the level of quantitative confrontation of observations with detailed theoretical computations of the sort described in these lectures to constrain theories. We might hope in the near future to see positive detections, providing cosmology with a tremendously powerful probe of how structure appeared in the early universe and evolved to the present.

ACKNOWLEDGMENTS No attempt has made to systematically review or refer to all of the work on CMB theory and observation produced during the three decades since the discovery of the cosmic background radiation. Hopefully, these errors of omission will be eradicated in the forthcoming review by George Efstathiou and myself. This work was supported by a Canadian Institute for Advanced Research Fellowship, a Sloan Foundation Fellowship and the Canadian NSERC.

8. REFERENCES

Bardeen, J.M. 1980, *Phys. Rev.* **D22**, 1882.
Bardeen, J.M., Steinhardt, P. and Turner, M.S. 1983, *Phys. Rev.* **D28**, 679.
Bardeen, J.M., Bond, J.R., Kaiser, N. and Szalay, A.S. 1986, *Ap. J.*, **304**, 15.
Bardeen, J.M., Bond, J.R. and Efstathiou, G. 1987, *Ap. J.*, in press.
Baym, G. and Kadanoff, L. 1962, *Quantum Statistical Mechanics*, (Benjamin).
Bertschinger, E.W. 1983, *Ap. J.* **268**, 17.
Birkinshaw, M. 1987, *Proc. XIIIth Texas Symposium on Relativistic Astrophysics*, ed. M. Ulmer (Singapore: World Scientific).
Bond, J.R., Arnett, W.D. and Carr, B.J. 1984, *Ap. J.* **280**, 825.
Bond, J.R., Carr, B.J. and Hogan, C.J. 1986, *Ap. J.* **306**, 428.
Bond, J.R., Centrella, J., Szalay, A.S., and Wilson, J., 1984, *M.N.R.A.S.* **210**, 515.
Bond, J.R. and Efstathiou, G. 1984, *Ap. J. Lett.* **285**, L45.
Bond, J.R. and Efstathiou, G. 1987, *M.N.R.A.S.* in press.
Bond, J.R. and Szalay, A.S., 1983, *Ap. J.* **277**, 443.
Bond, J.R. 1987, in preparation.
Bonometto, S.A., Caldara, A. and Lucchin, F. 1983, *Astron. Ap.* **126**, 377.
Carr, B.J., Bond, J.R., and Arnett, W.D., 1984, *Ap. J.* **277**, 445.
Chandrasekhar, S., 1960, *Radiative Transfer* (New York: Dover Publishing Co.).
Davis, M. and Peebles, P.J.E., 1983, *Ap. J.* **267**, 465.
Davis, M., Efstathiou, G., Frenk, C.S., and White, S.D.M. 1985, *Ap. J.* **292**, 371.
de Bernardis, P., Masi, M., Malagoli, A. and Melchiorri, F. 1985, *Ap. J.* **288**, 29.
Doroshkevich, A.G., and Shandarin, S.F., 1978, *M.N.R.A.S.* **182**, 27.
Doroshkevich, A.G., Zeldovich, Ya.B. and Sunyaev, R.A. 1978, *Sov. Astron.* **22**, 523.
Dressler, A., Faber, S.M., Burstein, D., Davies, R.L., Lynden-Bell, D., Terlevich, R.J. and Wegner, G. 1987, *Ap. J. Lett.* **313**, L37.
Dube, R.R., Wickes, W.C. and Wilkinson, D.T. 1977, *Ap. J. Lett.* **215**, L51.
Efstathiou, G., 1979, *M.N.R.A.S.* **187**, 117.
Efstathiou, G. and Bond, J.R. 1986, *M.N.R.A.S.* **218**,103.
Efstathiou, G. and Bond, J.R. 1987, *M.N.R.A.S.*, submitted.
Ehlers, J. 1971, in *General Relativity and Cosmology*, ed. R. Sachs (New York: Academic).
Fano, U. 1954, *Phys. Rev.* **93**, 121.
Feldman, D., Brune, W.H. and Henry, R.C. 1981, *Ap. J. Lett.* **249**, L51.
Fixsen, D.J., Cheng, E.S., and Wilkinson, D.T., 1983, *Phys. Rev. Lett.* **50**, 620.
Gould, R.J. 1984, *Ap. J.* **285**, 275.
Gush, H.P. 1981, *Phys. Rev. Lett.* **47**, 745.
Hauser, M. *et al.* 1984, *Ap. J. Lett.* **278**, L15.
Hogan, C.J. 1984, *Ap. J. Lett.* **284**, L1.
Ikeuchi, S., Tomisaki, K. and Ostriker, J.P. 1983, *Ap. J.* **265**, 538.
Kaiser, N., 1983a, *Ap. J. Lett.* **273**, L17.
Kaiser, N., 1983b, *M.N.R.A.S.* **202**, 1169.
Kompaneets, A. 1957, *Sov. Phys. JETP* **4**, 730.
Lifshitz, E.M. 1946, *JETP Letters* **16**, 587.
Lubin, P.M., Epstein, G.L., and Smoot, G.F., 1983, *Phys. Rev. Lett.* **50**, 616.
Lyubarskii, Yu.E. and Sunyaev, R.A. 1983, *Astron. Ap.* **123**, 171.
Matsumoto, T. Akiba, M. and Murakami, H. 1984, in *Advances in Space Research*, ed. G.F. Bignami and R.A. Sunyaev (Oxford: Pergamon).

Melchiorri, F., Melchiorri, B.O., Ceccarelli, C. and Pietranera, L. 1981, *Ap. J. Lett.*, **250**, L1.

Misner, C.W., Thorne, K.S. and Wheeler, J.A. 1973, *Gravitation*, (San Francisco: Freeman).

Negroponte, J. 1986, *M.N.R.A.S.* **222**, 19.

Osborn, R.K. and Yip, S. 1966, *The Foundations of Neutron Transport Theory* (New York: Gordon and Breach).

Ostriker, J.P. and Heisler, J. 1984, *Ap. J.* **278**, 1.

Ostriker, J.P. and Vishniac, E.T. 1986, *Ap. J. Lett.* **306**, L51.

Ostriker, J.P., Thompson, C. and Witten, E. 1986, preprint.

Paresce, F. and Jacobsen, P. 1980, *Nature* **288**, 119.

Peebles, P.J.E., 1968, *Ap. J.* **153**, 1.

Peebles, P.J.E., 1982, *Ap. J.* **258**, 415.

Peebles, P.J.E., 1984, *Ap. J.* **277**, 470.

Peebles, P.J.E. and Yu, J.T., 1970, *Ap. J.* **162**, 815.

Peebles, P.J.E. and Groth, E.J., 1976, *Astron. Ap.* **53**, 131.

Peebles, P.J.E., 1987, preprint.

Press, W.H and Vishniac, E.T. 1980, *Ap. J.* **236**, 323.

Rowan-Robinson, M. 1986, *M.N.R.A.S.* **219**, 737.

Shafer, R.A. and Fabian, A.C. 1983, IAU Symposium No. 104, *Early Evolution of the Universe and its Present Structure* (Dordrecht: Reidel), p.333.

Smoot, G.F., De Amici, G., Friedman, S.D., Witebsky, C., Sironi, G., Bonelli, G., Mandolesi, N., Cortliglioni, S., Morigi, G., Partridge, R.B., Danese, L., and De Zotti, G. 1985, *Ap. J. Lett.* **281**, L23.

Silk, J. and Stebbins, A. 1983, *Ap. J.* **269**, 1.

Starobinskii, A.A. 1985, *Sov. Astron. Lett.* **11**, 133.

Stewart, J.M. 1971, *Non-Equilibrium Relativistic Kinetic Theory*, lecture notes in Physics **10**, (Berlin: Springer-Verlag).

Szalay, A.S., Bond, J.R. and Silk, J. 1984, *Formation and Evolution of Galaxies and Large Scale Structure; The Third Moriond Astrophysics Meeting* (Dordrecht: Reidel), p. 499.

Toller, G.N. 1983, *Ap. J. Lett.* **266**, L79.

Uson, J.M. and Wilkinson, D.T. 1984a, *Ap. J. Lett.* **277**, L1.

Uson, J.M. and Wilkinson, D.T. 1984b, *Nature* **312**, 427.

Veryaskin, A.V., Rubakov, V.A. and Sazhin, M.V. 1983, *Sov. Astron.* **27**, 16.

Vishniac, E.T. and Ostriker, J.P. 1985, *Societa Italiana di Fisica Conference Proc.* **1**, 137.

Vishniac, E.T. 1987, preprint.

Vittorio, N. and Silk, J. 1984, *Ap. J. Lett.* **285**, L39.

Weller, C.S. 1983, *Ap. J.* **268**, 899.

White, S.D.M., Frenk, C.S., and Davis, M., 1983, *Ap. J. Lett.* **274**, L1.

White, S.D.M., Frenk, C.S., and Davis, M., 1984, *M.N.R.A.S.* **209**, 27P.

Wilson, M.L. and Silk, J. 1981, *Ap. J.* **243**, 14.

Wilson, M.L., 1983, *Ap. J.* **273**, 2.

Zeldovich Ya.B., Kurt, V.G. and Sunyaev, R.A. 1969, *Sov. Phys. JETP* **28**, 146.

Zeldovich, Ya.B. and Sunyaev, R.A. 1969, *Ap. Space Sci.* **4**, 301.

THE EARLY UNIVERSE - AN OBSERVER'S VIEW

J V Wall
Royal Greenwich Observatory
Herstmonceux Castle, Hailsham
East Sussex BN27 1RP, UK

ABSTRACT. Three aspects of an observer's early universe are discussed: primordial helium abundance as determined from observations of blue compact galaxies; the small-scale structure of the microwave background radiation as measured with the VLA; and the distribution of objects along lines of sight to distant QSOs determined from observations of Lyman-alpha and metal-line absorption systems.

1.   INTRODUCTION

As an observer, my early universe is the earliest I can see. I know that it is the late universe of particle physicists and cosmologists, and given failure at the outset, I claim a free choice of topics - within constraints. I feel obliged to discuss the very earliest things which astronomers can observe, namely primordial helium and the

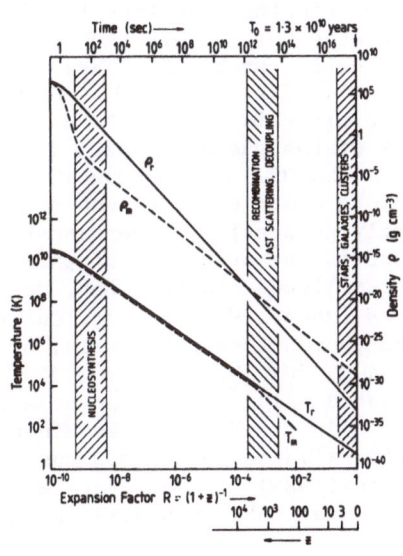

Figure 1. The temperature and density history of matter and radiation in a Friedmann universe with $\Omega = 1$, $H_0 = 50$ km s^{-1} Mpc^{-1}. The three epochs relevant to the present discussion are indicated.

W. G. Unruh and G. W. Semenoff (eds.), The Early Universe, 335–350.

microwave background radiation. In neither case do I intend a review, but rather I shall lay before you some new results from observations which I hope you find of interest. For my free choice, I shall describe recent work on QSO absorption lines – observations which have impact on formation of universal structure, which serve to connect my early universe to the present day, and which illustrate the power modern optical technology puts at the disposal of observers.

The lateness of my early universe is illustrated by my simple picture of the history of the universe in Fig 1. The three epochs on which the topics bear are indicated. <u>Primordial nucleosynthesis</u> took place during $10^2 < t < 10^3$ sec, beginning when the temperature dropped to a point below which the radiation field did not instantly shatter deuterons and ending when density dropped beyond the limit maintaining strong interactions. The <u>microwave background radiation</u> (MWB) we see from its last surface of scattering. At the epoch of recombination – when the radiation field became too weak to keep hydrogen ionized – all free electrons, the main scattering agents, were taken out of action. It is thus the epoch from which the universe became transparent and from which photons have streamed freely ever since – unless some energy-releasing process reionized the intergalactic medium between z = 1000 and now. The third epoch, z < 4, is the epoch of <u>stars and galaxies</u>, the epoch by which gravitationally bound structures have developed to be observable with present technology.

2. PRIMORDIAL HELIUM

Everywhere observers look in this third epoch they find that about 92 percent of atoms are ^{1}H, 8 percent are ^{4}He, and the rest hardly count. The mass fraction of helium, Y, thus appears to be uniformly 0.2 to 0.3. Some objects do show enhanced helium abundance; but no object has been shown convincingly to have Y < 0.2. A back-of-envelope calculation illustrates why, before the days of Hot Big-Bang (HBB) cosmology, there was a "helium problem". Take the luminosity of the Galaxy as 10^{44} ergs s^{-1} over a lifetime of 10^{10} years, providing a total of 3 x 10^{61} ergs. Suppose all this is from burning ^{1}H to ^{4}He, for which the binding energy of the nucleus is 2.5 x 10^{-5} ergs. This implies the delivery of 10^{43}g. But the mass of the Galaxy is at least 4 x 10^{44}g, and thus Y < 0.025, an order of magnitude short. Of course there are ways out, e.g. an early generation of superstars (Population III). But the simplest way out is the synthesis of primordial helium, inevitable if the universe lived through a hot, dense phase (Hoyle and Tayler 1964, Peebles 1966). What is particularly gratifying is that for a wide range of the photon-to-baryon ratio, or (equivalently) entropy per baryon, or $\Omega_b h^2$ where h = H_o/100 km s^{-1} Mpc^{-1}, some six orders of magnitude, the calculated value of Y_p lies on a plateau of 0.2 to 0.3. The action happened in the First Three Minutes (Weinberg 1977), and a recent comprehensive review of HBB nucleosynthesis is provided by Boesgaard and Steigman (1985). An accurate measure of Y_p is a very powerful test of conventional HBB, recognized as such for many years (e.g. Wagoner 1973).

Helium is difficult to destroy, and the back-of-the-envelope shows that helium is not created generously by most stars. There is therefore the possibility of determining Y_p. One of the best established methods is to observe primitive galaxies, blue compact galaxies which are dominated by HII regions. The procedure (Lequeux et al. 1979, Kunth and Sargent 1983) is to determine Y and Z (heavy-element mass-fraction) from line strengths, to plot Y vs Z, and to extrapolate the relation to Z = 0 as representative of pristine, unprocessed material. The difficulties (Davidson and Kinman 1985) should not be underestimated, but they are not insurmountable. Finding a large and clean sample is the first problem. There are ´mechanical´ problems such as requiring exceptional linearity from the detector. (Fig 2 shows a sample spectrum; look at the enormous line-ratios involved.) Obviously reddening corrections are vital. Moreover, all abundances are determined of necessity for ionized helium, and accurate ´ionization corrections´ for the amount of neutral helium are important. Finally there are problems of interpretation with a few highly individualistic objects.

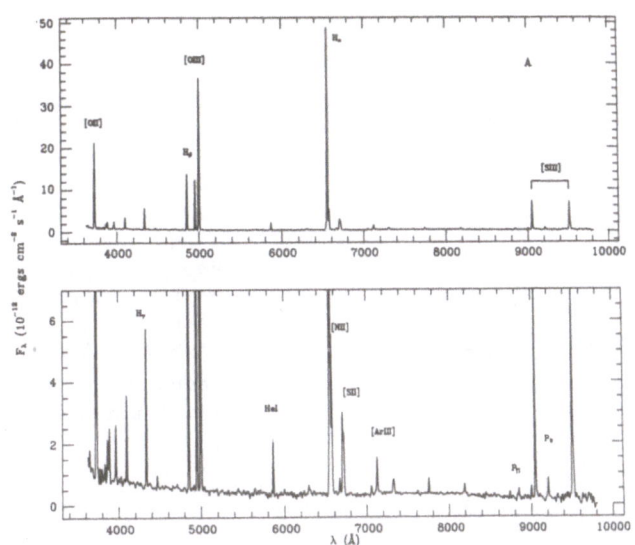

Figure 2. *The spectrum of the giant HII region NGC 604, from observations with the 2.5 m Isaac Newton Telescope, La Palma (Diaz et al. 1987). The spectrum, typical of those for which helium abundances are determined, closely resembles the spectra of blue compact galaxies, which show slightly higher excitation.*

A recent study by Pagel, Terlevich and Melnick (1986; PTM) addressed all these matters. Terlevich and Melnick conducted a spectroscopic survey of 600 emission-line galaxies found in objective-prism surveys with the UK Schmidt Telescope. From this sample, 11 new blue compacts were selected to augment the previous samples which contain a comparable number. The selection criteria included high s/n

spectra, Hβ equivalent width in excess of 100 Å, and a continuum
effective temperature > 40000 K. The latter criterion is particularly
important: PTM noted that the ionization correction is indeterminate (in
which case helium abundance is indeterminate), or the correction is zero
if the continuum is hot enough. A T_{eff} of 40000 K is hot enough.
Linearity of the photon-counting detector system was checked via the
well-known ratio of the [OIII] λλ4959,5007 doublet. Reddening
corrections were applied as usual. Lastly the results were combined with
those of other observers, while rejecting a few objects which for well-
established reasons gave less accurate results.

The PTM results appear in Fig 3. In Fig 3a it is evident that the
scatter is large, and that blue compact galaxies have a very steep
dY/dZ, steeper than the nearby irregular galaxies and HII regions. If
indeed the blue compacts are considered separately, a dY/dZ of 5.7 ±
2.7 is found, together with a value Y_p = 0.236 ± 0.005. Fig 3b
demonstrates a main conclusion of PTM - there is somewhat less scatter
in the Y vs N/H plot, and there appears to be a universal dY/d(N/H).
This plot gives dY/d(N/H) = 2.9 ± 1.5 x 10^3 and an extrapolated Y_p =
0.238 ± 0.005.

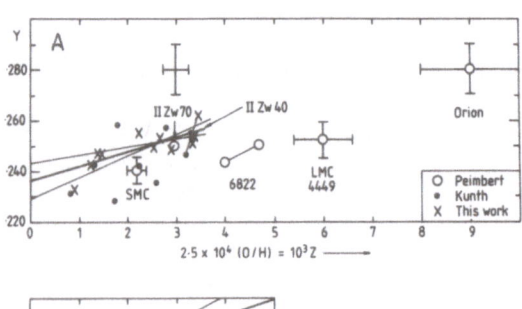

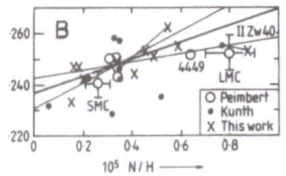

*Figure 3. Helium abundance
Y as a function of O/H (Fig
3a), and N/H (Fig 3b), for
blue compact galaxies, HII
regions in irregular
galaxies, and Orion, from
Pagel et al. 1986 ("this
work"). Note the different
slopes for different
objects in (a), and the
reduced scatter and
universal law in (b).*

There are several implications for astrophysics and cosmology.
Firstly, the value of dY/dZ for blue compacts is far too big for stellar
evolution to explain; PTM suggested that it is the result of
contamination by winds from W-R stars. Secondly, the value of Y_p has
crept back below 0.24, and is beginning to squeeze conventional HBB.
For instance, the number implies < 3 neutrino types. Three are known
now; the discovery of another would be serious. Thirdly, Ω_b < $0.05h^{-2}$.
If H_0 is 100 km s^{-1} Mpc^{-1}, there must be non-baryonic dark matter to
bind galaxy groupings, which require Ω > 0.2.

This investigation continues with a larger sample to which PTM are
now applying multivariate analysis. The new estimate is not public yet,
but an "informed source" says that it is "not higher" than the previous
estimate ....

3. THE MICROWAVE BACKGROUND

The MWB is our oldest fossil. It was a serendipitous discovery by
Penzias and Wilson (1965), immediately interpreted as relic radiation
from a hot dense phase of the universe by Dicke et al. (1965). The
prediction of residual warmth from the primaeval fireball was made
substantially earlier (Alpher & Herman 1948); it was 'lost', re-
predicted (Hoyle and Tayler 1964, Zeldovich 1965), and throughout some
of this checkered theoretical history, astronomers had the technology to
detect it. This interesting story of lost opportunities is sketched by
Weinberg (1977).

But the subsequent dedication of observers to our precious remnant
of the early universe has more than made up for the late start. The
present state is summarized in Figs 4 and 5. These demonstrate the
astonishing blandness which the wealth of observations has failed to
dislodge; the MWB remains supremely simple to describe. The radiation
appears to follow the emissivity curve for a black body at 2.74 ± .02
K, both below and above the Wien peak (~0.2 cm). The radiation is
spatially homogeneous, with two exceptions, one on the largest scale
possible, and the other on a scale of arcminutes. The large-scale
inhomogeneity is the simple dipole anisotropy, consistently defined in
magnitude and direction by balloon, high-altitude-flight, and satellite
measurements (Lubin and Villela 1986). The dipole measures the peculiar
velocity of the earth. The minor anisotropies detected are due to
spectral distortion by the hot gas in a few very rich clusters (Sunyaev
and Zeldovich 1970). It is possible in principle to determine H_0 from
such observations (Birkinshaw 1986). On other scales, observers have
set upper limits only, those of Uson and Wilkinson (1985) being the most
stringent.

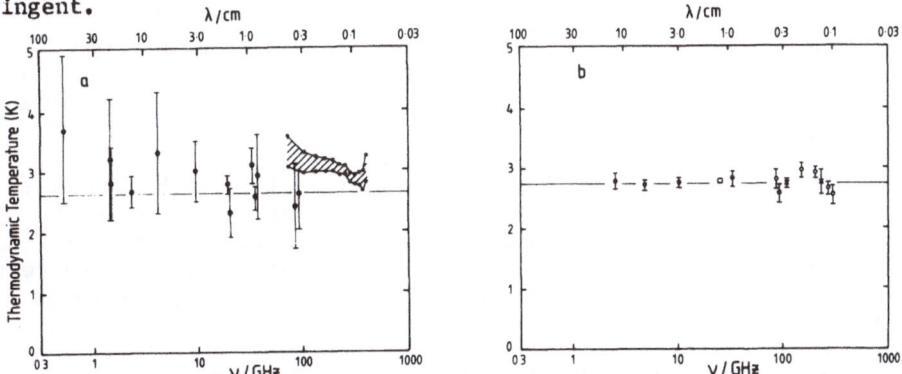

*Figure 4. Temperature measurements of the microwave background: (a)
1979 and earlier - filled circles, horn-antenna radiometers; hatched
area, balloon observations (Woody and Richards, 1979) (b) After 1979 -
filled circles, scaled-horn experiment (Smoot et al. 1985); open
circles, balloon observations (Peterson, Richards and Timusk 1985);
square, balloon observation (Johnson and Wilkinson 1987); crosses,
measurement of CN equivalent widths (Meyer and Jura 1985, Crane et al.
1986). Data are from the collections by Partridge 1986 and Johnson and
Wilkinson 1987. The solid line represents 2.74 K.*

340

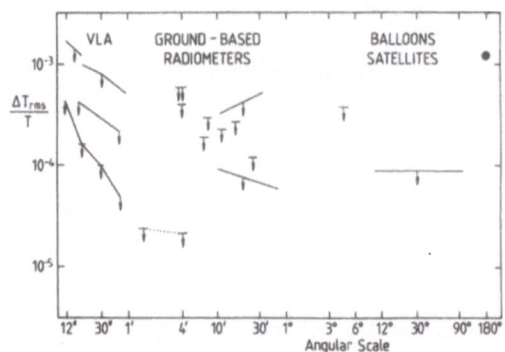

Figure 5. Measurements or upper limits for anisotropy in the microwave background, based on the collection by Partridge (1986). The only detection is that of the dipole (angular scale 180°) seen in both balloon and satellite experiments (Lubin and Villela 1986). The new small-scale VLA observations described here are shown as the connected set of four limits at the lower left. The lowest limits shown are from Uson and Wilkinson (1985)

Observers should not be discouraged – the wealth of conclusions which follow from the present observational data have been discussed by many authors, e.g. Partridge (1986), Kaiser and Silk (1986). In brief:

● T(0) = 2.74 K tells us (with $\Omega_b h^2$ and some particle physics) the primordial light-element abundances. It also permits us to calculate temperature at <u>any</u> epoch via T(z) = (z+1)T(0)

● The thermal nature of the spectrum – particularly at cm-wavelengths – severely constrains energy-releasing mechanisms (such as Population III stars) during epochs $10 < z < 10^6$ (e.g. Illarionov and Sunyaev 1975).

● The dipole vector tells us the motion of the earth, and if we can do vector arithmetic, the motion of the Galaxy relative to the cosmic backdrop. The datum is particularly crucial in considering bulk flows in the universe (Burstein <u>et al</u>. 1986), and it can be used (with other data and assumptions) to constrain Ω (e.g. Davis and Peebles 1983).

● The purity of the dipole provides a measure of universal expansion isotropy, and stringent limits to shear and rotation. It provides the horizon problem of classical HBB cosmology, one of the three major issues (together with non-observed monopoles and flatness, i.e. why is Ω anywhere near 1) resolved by second-generation inflation models.

● Limits on angular scales > 10 deg are of particular interest in that these scales involve no assumptions about early evolution (Kaiser & Silk 1986); even if the universe were reheated as late as z = 30, the event horizon is such that structure on such scales could not be causally erased. The angular scale of 0.5 to 3 degrees is of interest as the horizon limit at z~1000, and it is smoothness on greater scales which constitutes the 'horizon problem'. If no re-heating process has damaged our view of the last scattering surface at z~1000, the epoch of recombination, then the intrinsic thickness of this layer wipes out structure < 8 $\Omega^{-1/2}$ arcmin, i.e. 8 to say 20 arcmin. At all other scales the prediction of $\Delta T/T$ is model-dependent; how it is model-dependent, and the nature of the constraints which the observations provide, Kaiser and Silk (1986) set out with great clarity. Current limits rule out baryon-dominated models with adiabatic fluctuations, and usefully constrain models involving weakly-interacting dark matter. In particular, CDM models with $\Omega = 0.2$, close to that implied if galaxies trace mass, are not admissible. Biased galaxy formation (Davis et al. 1985, Bardeen 1986) and cosmic strings (Silk and Vilenkin 1984) both offer ways of forming large-scale structure in the universe with substantially lower values of $\Delta T/T$; but even these are under pressure from current limits. Kaiser and Silk offer two further general comments: (i) "The detection of anisotropy on any angular scale could provide a unique and direct measure of the primordial fluctuations from which large-scale structure evolved", and (ii) "eventual detection of $\Delta T/T$ on some angular scale is inevitable". The message to observers is very clear.

I want to tell you of recent observations with the NRAO Very Large Array (VLA) to look for fluctuations on the angular scales of an arcmin or less. The first series of such observations (Fomalont et al. 1984) was undertaken in the spirit of the foregoing Kaiser-Silk general comments. But more recently the development of the explosive scenario for galaxy formation has sharpened the goal (Ostriker and Cowie 1981, Ikeuchi 1981, Hogan 1984, Ostriker and Vishniac 1986) - $\Delta T/T \sim 10^{-5}$ on sub-arcmin scales is predicted.

A synthesis system such as the VLA provides a unique set of advantages - and of problems. Among the advantages, the ability to map stands out. Single or double-beam experiments provide scans or difference measurements, rather than actual emission contours from which fluctuation statistics can be derived. A second advantage is that the collection of time-resolved spatial samples which constitutes aperture-synthesis integration enables many numerical experiments to determine experiment characteristics and self-consistency. The sampling also permits the 'sky' to be constructed at different spatial resolutions. A third advantage is simplicity in the sense that the observations consist solely of tracking and integrating. The main disadvantage is that sky sampling is incomplete with any interferometer system, and that artifacts which this introduces, grating responses, missing spatial frequencies, must be reduced by numerical procedures such as the CLEAN algorithm. How well such algorithms work in the limits of noise and discrete-source confusion is not known (see below). There is finally the problem of unknown instrumental effects; but this is no better or no worse than other experiments in which limits are pushed beyond previous ones.

In the recent experiment (preliminary results in Kellermann

et al. 1986), the VLA was used at 5 GHz (λ = 6 cm) in D configuration, for which the longest spacing is 1.0 km, resulting in a synthesized beam of 12 arcsec FWHM. The primary beam is 9 arcmin FWHM, and in practice an area ~20 arcmin in diameter can be synthesized. The present experiment consisted of 76 hours integration spread over 8 days, with the field centre α = 00^{h}15^{m}24^s, δ = +15°33′00″ as for the earlier observation of Fomalont et al. (1984). The field centre was chosen to avoid known radio sources, and the observations were made at night.

Maps resulting from the observations are shown in Fig 6 in two different presentations. The new observations achieved the rms noise fluctuations of 4.5 μJy predicted from antenna and receiver characteristics. There is no doubt that the observations detect excess signal; the rms value of this signal closely follows the primary beam pattern (Fig 2 of Kellermann et al. 1986). The signal can be seen in Fig 6 as extra graininess near the map centre. What is it? There are three possibilities, in no particular order: 1, confusion signal due to the myriad of faint background radio sources integrated by the synthesized beam; 2, true fluctuations in the microwave background; and 3, instrumental effects.

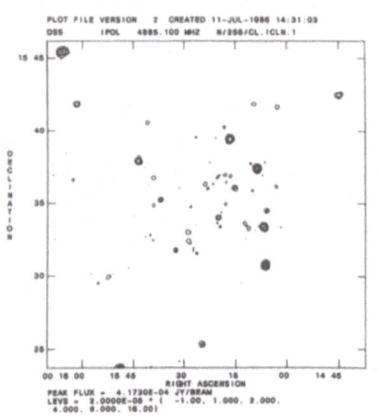

Figure 6. The map from the 76 hour integration with the VLA in D-configuration at 5 GHz, shown in contour and in grey-scale form. The peak flux density is 417 μJy. Discrete sources above 25 μJy (5 σ) have been used from this survey to extend the surface density counts of extragalactic radio sources (Kellermann et al. 1986).

With regard to the latter, the major instrumental effect limiting the previous experiment, spurious responses due to correlator offsets and cross talk between close antenna pairs, was alleviated by offsetting the phase-tracking centre outside the primary beam. Other instrumental effects may be present, but achieving the predicted noise characteristic provides some indication that these are not serious. With regard to confusion, a series of numerical experiments is in progress to determine the contribution to signal from very faint discrete sources. These experiments are carried out by (1) assuming that the sources are point-like, (2) adopting a range of extrapolated sky surface-density (log N –

log S) laws, and (3) selecting random samples according to such laws to scatter randomly onto the sky over a region covered by the primary beam, and determining the level of fluctuations resulting after the CLEANing process. The results to date show that most if not all of the excess signal can be due to confusion noise – if one needs to account for it in this way. It remains difficult to remove the larger-scale fluctuations, unless the very faint sources are significantly more extended than those which are detectable at the 25-μJy discrete-source limit of the present survey. At present the conservative approach is to quote upper limits (substantially more stringent than previous ones of Fomalont et al. 1984 and Knoke et al. 1984) resulting from the observation (plus numerical experiments) as follows (Fig 5):

Angular scale (arcsec) =	12	17	30	50
rms $\Delta T/T$ (x 10^4) =	4.5	1.6	1.0	0.5
rms ΔS (μJy) =	3.5	2.5	5	7

There may be signal present on the larger scales, > 1 arcmin, but it is premature to claim detection. Conversely there is signal present on the smaller scales, most if not all of which can be attributed to confusion. What the numerical experiments still need to delineate is how much power could be in the signal (a) if a minimum extrapolated (lowest possible) source count is assumed, and (b) if extended sources are involved. The new upper limits at small scales are pushing towards the tightest limits obtained on other scales, and are close to the predictions of explosive galaxy formation.

4. QSOs AS PROBES OF THE EARLY UNIVERSE

My final topic also bears on the development of structure in the universe. It concerns the narrow absorption lines in QSO spectra, and it has the additional appeal in the present context that the lines-of-sight along which the absorptions occur connect the early universe (my definition) to the present. In this instance, I go back to the redshift of 3.78 belonging to the QSO PKS2000-330, one-time holder of the redshift record (Peterson et al. 1982); $\Omega = 1$ implies a birthday < $0.096T_o$.

In the optical spectrum of a QSO, it is the broad emission lines which command immediate attention, and which generally define the redshift. The spectrum of a typical high-redshift QSO shows many narrow absorption features as well. Fig 7 shows a notable example. Most absorption features at wavelengths longer than Ly α emission (rest wavelength 1216 Å) are due to common ionization stages of the most abundant astrophysical elements; these so-called 'metal lines' occur at redshifts close to that of the QSO down to much lower redshifts. Most of the features shortward of emission Ly α are due to Ly α absorption in hydrogen clouds; collectively these features are known as the 'Ly α forest'. In the past the astronomical community has divided over the interpretation of these lines. One view was that the absorption lines are formed in material ejected from the QSOs (the 'intrinsic'

344

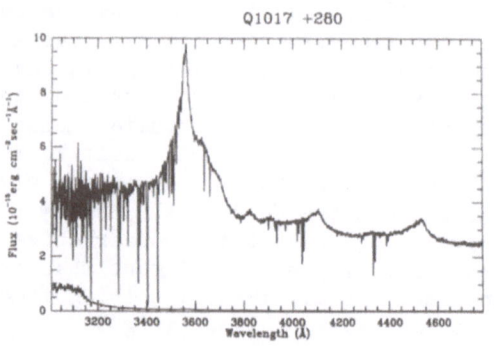

Figure 7. The optical spectrum of the QSO 1017+280, obtained on the 5 m Palomar reflector (A Boksenberg and W Sargent, unpublished data). Most of the absorptions longward of Ly α (3560 Å, rest λ 1216 Å) are 'metal lines' due to common ionization stages of the more abundant elements; the lines shortward of Ly α constitute the 'Ly α forest', and represent individual Ly α absorptions.

hypothesis), the differing redshifts implying velocities of ejection of up to 0.7c. The second view (the ´intervening´ hypothesis) was that the absorptions arise in intergalactic gas (for the Ly α forest) and the interstellar gas of intervening galaxies (for the metal-line systems). Implicit in the latter hypothesis is of course the acceptance of the cosmological interpretation of the redshift. It also requires for the metal-line systems that the gaseous haloes of galaxies have cross-sections of 4 to 10 Holmberg radii.

The controversy has been resolved in the minds of the many, resolved in favour of the ´intervening´ hypothesis, and resolved with long collaborative observation programmes involving photon-counting detectors on the world´s largest optical telescopes. The ´intervening´ hypothesis is favoured by statistical studies of the absorption-line redshift systems, which show no evidence of clumping close to the emission-line redshifts of the QSOs (Sargent et al. 1980; Young, Sargent and Boksenberg 1982a, 1982b). Direct evidence for this hypothesis comes from the comparison of the spectra of close pairs of QSOs at different redshifts (Shaver and Robertson 1983); in the spectrum of the QSO of larger emission-line redshift, absorption-line systems are present at the redshift of the other QSO. (There are QSOs, however, in which ejection is almost certainly taking place - the broad-absorption-line (BAL) QSOs, constituting perhaps 5 percent of the population. Requisite ejection velocities range up to 0.1c, and an example, 1700+518, a BAL QSO with a low z_{em}, is shown in Fig 8).

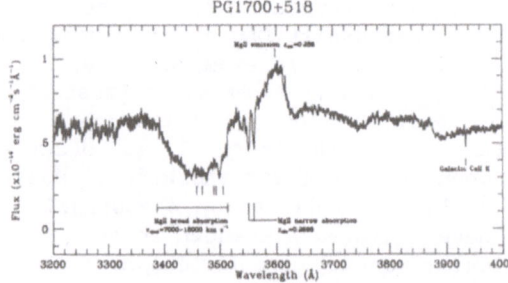

Figure 8. The optical spectrum of the broad-absorption-line QSO PG1700+518, as obtained with the 2.5 m Isaac Newton Telescope (Pettini and Boksenberg 1985).

Consider first the <u>metal-line systems</u>. The general properties to emerge from studies of these are as follows (Pettini 1985 and references cited):

- They are generally the resonance lines of the most prominent ions in the ISM; the CIV λλ1548,1550 doublet is the most common feature of all.

- They can be grouped into redshift systems.

- The redshift systems show clustering on velocity scales < 2000 km s^{-1}. The clustering is consistent with the galaxy-galaxy correlation function extrapolated back to z = 2.

- The comoving density of the systems is approximately uniform over the measurable range 0.4 < z < 3.5

The observations I want to discuss concern PKS 0215+015, an object dear to my heart ever since John Bolton and I identified it by the rashest of techniques. The stellar image at the position of the radio source is well above the Palomar Sky Survey limit. John Bolton took a second-epoch plate, and the object had vanished. Obviously this confirmed it as a variable, high-redshift QSO, and we claimed it as such (Bolton and Wall 1969). In fact it turned out to be a BL Lac object (Gaskell 1982) rich in absorption lines; an emission-line system has now been detected at z_{em} = 1.72. A superb series of observations of this object with the AAT and with IUE (Pettini <u>et al</u>. 1983, Blades <u>et al</u>. 1985) has elucidated many of the features of metal-line systems. The success of these observations stems from a programme to monitor the brightness of the object with a 26-in telescope at the RGO, and to use the epochs of peak brightness to obtain spectra of very high resolution with the AAT and with IUE. (The object does vary by about 6 magnitudes, from B = 15 to B = 21 mag, luckily for Bolton and Wall.)

Fig 9 demonstrates the importance of resolution. When observed at a resolution of 1.5 Å or 100 km s^{-1}, normally considered to be high for QSO observations, a broad doublet of CIV is seen at z_{abs} = 1.549. At peak brightness it was possible to observe the object with a resolution of 0.27 Å, or 20 km s^{-1}. The broad doublet resolved into some 7

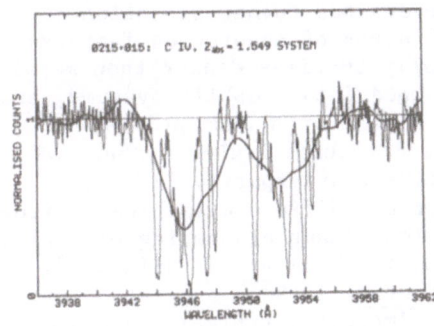

Figure 9. The complex absorption profile of the CIV doublet lines in the z = 1.549 absorption system of the BL Lac object PKS 0215+015 (Pettini <u>et al</u>. 1983). The heavy solid line shows the earlier observation with the Anglo-Australian Telescope at a resolution of 1.5 Å FWHM; the later observations are at 0.27 Å FWHM.

different components. Indeed a second redshift system at z_{abs} = 1.649 splits into 9 components. The velocity range for the former system is 300 km s^{-1}, while for the latter it exceeds 900 km s^{-1}. These two complex systems suggest that the line of sight to 0215+015 intersects metal-rich haloes of the galaxies in two rich clusters. But clusters of the requisite richness are very rare at the present epoch, and it is interesting to speculate that the observation has detected two young clusters of gas-rich galaxies or protogalaxies undergoing gravitational collapse (Pettini et al. 1983).

A particularly rich absorption-line system in 0215+015 is present at z = 1.345 (Blades et al. 1985), in which 33 lines due to H, C, N, O, Mg, Al, Si and Fe can be identified. The abundances for low-ionization material are typical of a lightly-reddened interstellar cloud in the Galactic disk. For the high-ionization species (e.g. CIV, SiIV), the abundances are similar to those for halo clouds in the Galaxy. Thus there are contributions to overall line-strengths for gas typical of both disk and halo, adding further support to the hypothesis that this system is due to an intervening galaxy. The conclusion carries the further implication that (at least some) galaxies which existed at z = 1.345, when t = $0.25T_0$, were not very dissimilar from our own Galaxy. These observations, together with similar observations of the QSO PKS 2000-330, z_{em} = 3.78, show that very significant metal enrichment took place as early as t = $0.1T_0$. It has prompted a determined effort to search for absorption systems with genuine underabundances, as indicators of truly primordial galaxies.

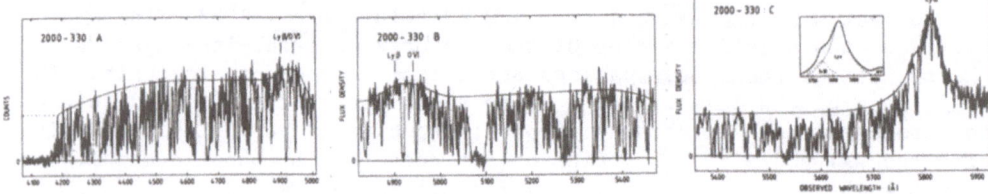

Figure 10. The Ly α absorption forest for the z = 3.78 QSO PKS 2000-330, recorded at the AAT with a spectral resolution of 1.3 Å FWHM (Hunstead et al. 1986).

The same group of meticulous observers has turned attention recently to the Lyman-alpha forest. This set of absorption features blueward of Ly α (Figs 7, 10) is typically 50 times denser than metal lines. Most of the lines cannot be grouped into redshift systems, and it is generally accepted that they are single Ly α absorptions, accompanied on rare occasions by Ly β and perhaps higher members of the series, and even more rarely by metal-line counterparts. The distribution of these individual absorptions differs dramatically from that of the metal lines: they show no significant clustering on scales down to 150 km s^{-1} (Sargent et al. 1980, Carswell et al. 1984). On the basis of these characteristics, Sargent et al. (1980) suggested that the Ly α lines are due to a population of primordial intergalactic clouds which do not cluster like galaxies. If so, the objects responsible must be fundamentally associated with formation of bound structures in the universe, and the interest of observers and theorists is assured.

An immediate question for observers concerns abundances; is the material pristine? Many attempts to measure metal abundances associated with individual Ly α systems have been made, without marked success (see Pettini 1985 for a review). Of even greater interest is the determination of the ratio D/H, which would provide a sensitive test of HBB nucleosynthesis. This task is yet more difficult for observers, but not impossible.

One task perhaps well within the grasp of observers is the question of cosmic evolution of the putative clouds. Is the co-moving density uniform, or does it show the sort of evolution which has long been known for QSOs (e.g. Schmidt and Green 1983) and extragalactic radio sources (e.g. Wall 1983)? It is astonishing that the literature contains claims of strong evolution in the sense of increased co-moving density with redshift, of no evolution, and of counter-evolution, some of these claims on the basis of the same data. The history of the controversy is sketched by Murdoch et al. (1986), and it is bewildering.

The new observations of PKS 0215+015 (Blades et al. 1985) and PKS 2000-330 (Hunstead et al. 1986) have now been combined with other published data of comparable quality to re-examine the question. In doing so, Murdoch et al. (1986) brought a thorough statistical analysis to bear on the problem. In the first instance these investigators defined a "reasonably homogeneous" sample of Ly α systems from 11 QSOs, all observed with photon-counting systems at resolutions 0.8 to 1.5 Å. The few Ly α systems with which metal lines were associated were removed, on the premise that such systems have a different spatial distribution. The best-fit (maximum-likelihood unbinned) exponent for evolution of the form $(1+z)^\gamma$ is 2.17 ± 0.36, and the expression provides a statistically satisfactory description of the data (Fig 11). Evolution is confirmed at the many-sigma level.

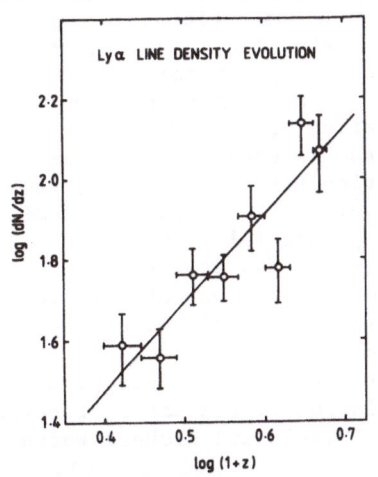

Figure 11. Log dN/dz as a function of log(1+z), where N is the number of Ly α absorptions (Murdoch et al. 1986). The data are (necessarily) binned for this presentation; the best fit to the unbinned data gives an exponent of 2.17 ± 0.36.

However, when the same process is carried out for individual QSOs (which show up to 70 Ly α systems), evolution is less evident, and indeed counter-evolution may be present. This result, too, is shown to

be statistically watertight, and it can only be understood if the trend
in individual QSOs runs counter to the overall trend. With further
statistical testing, Murdoch et al. were able to show that there is a
deficiency in Ly α systems local to each QSO, and the most plausible
physical explanation for this is that the Lyman continuum flux from each
QSO is ionizing the adjacent clouds out to a radius of 4 Mpc. This
interpretation is consistent with the occurrence of high-ionization
(e.g. NV) absorption systems within such a sphere (Weymann, Carswell and
Smith 1981). It also clarifies the conflicting results. In the
evolution investigations in which the parameters of the epoch dependence
were derived from individual QSOs, an overall best estimate obtained
from a weighted sum is bound to produce little (or even counter-)
evolution. It is salutary to note that if Murdoch et al. are correct,
the apparent disagreement among observers hid a real physical
phenomenon.

Theoretical attempts to assemble primordial hydrogen cloudlets have
developed along two lines. In the earlier of these (Ostriker and
Ikeuchi 1983, Ikeuchi and Ostriker 1986), the IGM is shock-heated to
more than 10^6 K at epochs corresponding to $10 < z < 5$. The propagation
of shocks forms dense shells which fragment; the largest ($> 10^{8.5}$ $M_\odot$)
collapse to galaxies, while the smallest ($< 10^6$ $M_\odot$) evaporate. The
intermediates expand isothermally in pressure equilibrium, and are
ionized by the integrated background from QSOs. The small neutral
fraction (10^{-5}) is responsible for the Ly α forest. Calculation of z-
dependence of cloudlet parameters (Atwood, Baldwin and Carswell 1985,
Ikeuchi and Ostriker 1986) predicts a power-law dependence of evolution
in qualitative agreement with the new results of Murdoch et al.

An alternative derives from the currently conventional Cold-Dark-
Matter-dominated Universe. Rees (1986) envisages gravitational wells of
intermediate depth in which gas, stably confined by the gravitational
field of Dark Matter, is neither able to escape nor to settle and become
self-gravitating. They inevitably produce Ly α absorptions, and indeed
such minihaloes are more likely to survive if they lie in voids, thus
displaying less clustering than normal galaxies. The model has the
advantage of dispensing with a tuned IGM, reducing constraints on it to
broad inequalities. There are several competing epoch-dependent
processes which would produce evolution in such a model. A further
attraction is that there need be no demarcation between H-only systems
and metal-line systems; once the internal velocity dispersion becomes
too high, star formation plus processing of primordial material is
inevitable. Rees argues that dark minihaloes are "an inevitable
concomitant" of CDM models; they are a prediction. How unfortunate for
theorists that the forests have been known for 15 years! Surely you
would expect me to end with the observers in front...

I am grateful to Bernard Pagel, Bruce Partridge, Max Pettini and
Roberto Terlevich for guidance and for permission to use material before
publication. I thank Beryl Andrews, Monica Everest and Richard Worth
for help preparing the manuscript.

REFERENCES

Alpher, R.A. and Herman, R. 1948, Nature, **162**, 774.

Atwood, B., Baldwin, J.A. and Carswell, R.F. 1985, Astrophys. J., **292**, 58.

Bardeen, J.M. 1986, in Inner space/outer space, eds Kolb, E.W., Turner, M.S., Lindley, D., Olive, K. and Seckel, D., University of Chicago Press, 212.

Birkinshaw, M. 1986, in 'Observational Cosmology', Proc IAU Symp 124, eds Burbidge, G. and Hewitt, A., Reidel: Dordrecht, in press.

Blades, J.C., Hunstead, R.W., Murdoch, H.S. and Pettini, M. 1985, Astrophys. J., **288**, 580.

Boesgaard, A.M. and Steigman, G. 1985, Ann. Rev. Astron. Astrophys., **23**, 319.

Bolton, J.G. and Wall, J.V. 1969, Astrophys. Lett., **3**, 177.

Burstein, D., Davies, R.L., Dressler, A., Faber, S.M., Lynden-Bell, D., Terlevich, R. and Wegner, G. 1986, in Galaxy distances and deviations from universal expansion, eds Madore, B.F. and Tully, R.B., Reidel: Dordrecht, 123.

Carswell, R.F., Morton, D.C., Smith, M.G., Stockton, A.N., Turnshek, D.A. and Weymann, R.J. 1984, Astrophys. J., **278**, 486.

Crane, P., Hegyi, D.J., Mandolesi, N. and Danks, A.C. 1986, Astrophys. J., **309**, 822.

Davidson, K. and Kinman, T.D. 1985, Astrophys. J. Suppl. Ser., **58**, 321.

Davis, M. and Peebles, P.J.E. 1983, Ann. Rev. Astron. Astrophys., **21**, 109.

Davis, M., Efstathiou, G., Frenck, C.S. and White, S.D.M. 1985, Astrophys. J., **292**, 371.

Diaz, A.I., Terlevich, E., Pagel, B.E.J., Vilchez, J.M. and Edmunds, M.G. 1987, Mon. Not. R. astr. Soc., in press.

Dicke, R.H., Peebles, P.J.E., Roll, P.G. and Wilkinson, D.T. 1965, Astrophys. J., **142**, 414.

Fomalont, E.B., Kellermann, K.I. and Wall, J.V. 1984, Astrophys. J., **277**, L23.

Gaskell, C.M. 1982, Astrophys. J., **252**, 447.

Hogan, C.J. 1984, Astrophys. J., **284**, L1.

Hoyle, F. and Tayler, R.J. 1964, Nature, **203**, 1108.

Hunstead, R.W., Murdoch, H.S., Peterson, B.A., Blades, J.C., Jauncey, D.L., Wright, A.E., Pettini, M. and Savage, A. 1986, Astrophys. J., **305**, 496.

Ikeuchi, S. 1981, Publ. astr. Soc. Japan, **33**, 211.

Ikeuchi, S. and Ostriker, J.P. 1986, Astrophys. J., **301**, 522,

Illarionov, A.F. and Sunyaev, R.A. 1975, Sov. Astron. A.J., **18**, 691.

Johnson, D.G. and Wilkinson, D.T. 1987, Astrophys. J., **313**, L1.

Kaiser, N. and Silk, J. 1986, Nature, **324**, 529.

Kellermann, K.I., Fomalont, E.B., Weistrop, D. and Wall, J.V. 1986, in Highlights of Astronomy, ed Swings J.-P., Reidel: Dordrecht, **367**.

Knoke, J.E., Partridge, R.B., Ratner, M.I. and Shapiro, I.I. 1984, Astrophys. J., **284**, 479.

Kunth, D. and Sargent, W.L.W. 1983, Astrophys. J., **273**, 81.

Lequeux, J., Peimbert, M., Rayo, J.F., Serrano, A. and Torres-Peimbert, S. 1979, Astron. Astrophys., **80**, 155.

Lubin, P. and Villela, T. 1986, in Galaxy distances and deviations from universal expansion, eds Madore, B.F. and Tully, R.B., Reidel: Dordrecht, **169**.

Meyer, D.M. and Jura, M. 1985, Astrophys. J., **297**, 119.

Murdoch, H.S., Hunstead, R.W., Pettini, M. and Blades, J.C. 1986, Astrophys. J. **309**, 19.

Ostriker, J.P. and Cowie, L.L. 1981, Astrophys. J., **243**, L127.

Ostriker, J.P. and Ikeuchi, S. 1983, Astrophys. J., **268**, L63.

Ostriker, J.P. and Vishniac, E.T. 1986, Astrophys. J., **306**, L51.

Pagel, B.E.J., Terlevich, R.J. and Melnick, J. 1986, Publ. astr. Soc. Pacif., **98**, 1005.

Partridge, R.B. 1986, in Vatican Observatory Workshop on Theory and Observational Limits in Cosmology, July 1985.

Peebles, P.J.E. 1966, Phys. Rev. Letters, **16**, 410.

Penzias, A.A. and Wilson, R.W. 1965, Astrophys. J., **142**, 419.

Peterson, B.A., Savage, A., Jauncey, D.L. and Wright, A.E. 1982, Astrophys. J., **260**, L27.

Peterson, J.B., Richards, P.L. and Timusk, T. 1985, Phys. Rev. Letters, **55**, 332.

Pettini, M. 1985, in Proc. ESO Workshop on ´Production and Distribution of C, N, O Elements´, Garching, eds Danziger, I.J., Matteucci, F. and Kjar, K., 355.

Pettini, M., Hunstead, R.W., Murdoch, H.S. and Blades, J.C. 1983, Astrophys. J., **273**, 436.

Pettini, M. and Boksenberg, A. 1985, Astrophys. J., **294**, L73.

Rees, M.J. 1986, Mon. Not. R. astr. Soc., **218**, 25P.

Sargent, W.L.W., Young, P.J., Boksenberg, A. and Tytler, D. 1980, Astrophys. J. Suppl. Ser., **42**, 41.

Schmidt, M. and Green, R.F. 1983, Astrophys. J., **269**, 352.

Shaver, P.A. and Robertson, J.G. 1983, Astrophys. J., **268**, L57.

Silk, J. and Vilenkin, A. 1984, Phys. Rev. Letters, **53**, 1700.

Smoot, G.F. et al. 1985, Astrophys. J., **291**, L23.

Sunyaev, R.A. and Zeldovich, Y.B. 1970, Astrophys. Space Sci., **7**, 1.

Uson, J.M. and Wilkinson, D.T. 1985, Nature, **312**, 427.

Wagoner, R.V. 1973, Astrophys J., **179**, 343.

Wall, J.V. 1983, in The Origin and Evolution of Galaxies, eds Jones, B.J.T. and Jones, J.E., Reidel: Dordrecht, 295.

Weinberg, S. 1977, The First Three Minutes, Andre Deutsch: London.

Weymann, R.J., Carswell, R.F. and Smith, M.G. 1981, Ann. Rev. Astron. Astrophys., **19**, 41.

Woody, D.P. and Richards, P.L. 1979, Phys. Rev. Letters, **42**, 925.

Young, P., Sargent, W.L.W. and Boksenberg, A. 1982a. Astrophys. J., **252**, 10.

Young, P., Sargent, W.L.W. and Boksenberg, A. 1982b, Astrophys. J. Suppl. Ser., **48**, 455.

Zeldovich, Y.B. 1965, Advances in Astron. Astrophys., **3**, 241.

CAN THE SOLAR NEUTRINO PROBLEM BE THE FIRST DETECTED SIGNATURE OF DARK MATTER FROM THE HALO OF OUR GALAXY? (CANDIDATES FOR SOLAR COSMIONS)

Graciela Gelmini[*]
Lyman Laboratory of Physics
Harvard University
Cambridge, MA 02138

ABSTRACT. The solar neutrino problem can be solved if the sun has accreted a sufficient number of weakly interacting particles from the dark matter of the galaxy. Dark matter particles with mass and cross-sections in a small range are required. It is an interesting problem of model building to give concrete particle candidates for these particles. We present it as an example of the deep interconnection of astrophysics and particle physics necessary to deal with dark matter candidates.

1. INTRODUCTION

The solar neutrino problem seems to indicate an inconsistency in our standard models, either the standard model of neutrinos or the standard model of the sun. There are two classes of ideas to solve this problem. The first involves changing the properties of standard neutrinos, which are massless, to allow for oscillations or decays. Masses of order $10^{-6} eV$ give oscillation lengths ("in vacuum") or order the earth-sun distance. Masses of order 10^{-2} allow for "matter amplified" resonances within the sun. Masses of order eV in models with Nambu-Goldstone bosons, Majorons, allow for decays while neutrinos arrive at the earth. The second class of ideas involves changing the solar properties. A decrease by 10% of the central temperature of the sun would decrease the ^8B flux. Weakly interacting particles, we call them "solar cosmions", with particular mass and cross section, trapped in the sun may produce such a decrease. Thus, the solar neutrino problem may be the first detected signature of dark matter from the halo of our galaxy. Solar cosmions cannot be readily assigned to conventional dark matter candidates whose properties lie outside the small range in mass and cross-section required. However realistic models can be built satisfying all existing constraints. Here we present three examples.

We start with an introduction to the solar neutrino problem itself, review the main ideas about neutrino properties which could solve this problem and pass to the "solar cosmion" model. We present the "cosmic coincidences" (which are just coincidences if the solar cosmion model is not true) which make this idea viable. We mention, then, that the solar cosmion scheme is the first astrophysical solution of the solar neutrino problem which is consistent with helioseismology data. We explain then he problems encountered in building a particle model for the solar cosmion and present some solutions. Finally we talk about experimental signatures of solar cosmions. This lecture is based on work done with Larry Hall and Jane Lin.

What is the Solar Neutrino Problem?

The solar neutrino problem is the discrepancy between the predicted and the observed rate in the only experiment performed up to now to detect neutrinos emitted from the sun. R. Davis et al., experiment has been running for more than fifteen years. It is based on the capture reaction

$$^{37}CL + \nu_e (E_\nu > 810 \; keV) \rightarrow \, ^{37}Ar + e^-$$

351

W. G. Unruh and G. W. Semenoff (eds.), The Early Universe, 351–360.
© 1988 by D. Reidel Publishing Company.

which can occur when the incident electron neutrino ν_e has an energy E_ν larger than the 810 keV threshold. Most of the solar neutrinos emitted with such energies are produced in the reaction $^8B \rightarrow {}^8Be^* + e^+ + \nu$ (with E_ν up to 14 MeV). This reaction occurs late in the proton-proton initiated chain, it is strongly temperature dependent and occurs in the inner 0.05 of the mass of the sun. The second largest contribution to the solar neutrinos detectable in Davis' experiment come from $^7Be + e^- \rightarrow {}^7Li + \nu$ ($E_\nu = 861$ keV).[1] The even rate predicted by J. Bahcall et al., based in the standard solar model, 5.8 ± 2.2 SNU [2], see Footnote 1, (from which 4.3 SNU from ^8Boron and 1.02 SNU from $^7Berilium$, three standard deviations are quoted), has been consistently above the event rate in Davis experiment 2.0 ± 0.3 SNU [4] (one standard deviation is quoted). SNU means Solar Neutrino Units, 1 $SNU = 1 \times 10^{-36}$ events/(target atom $\times$ sec).

Bahcall and collaborators claim the factor of 1/3 necessary to have an agreement between theory and experiment can not be obtained form the standard theory of how the sun shines. This being true, there are three logical possibilities to explain the discrepancy, i.e., something else happens in the 1) neutrino detection, or in the 2) neutrino propagation or in the 3) production. The first possibility is the most unlikely given the reliability of the detection technique used. We will briefly overview the second and concentrate on the third possibility.

2. NON STANDARD ν_e PROPAGATION

One explanation for the discrepancy is the existence of vacuum oscillations of the ν_e into others neutrinos.[5] The ν_e, is defined as the particle emitted from an e together with a virtual W boson, while ν_μ is emitted from a μ and ν_τ from a τ. These ν's are the "gauge" or "flavor" eigenstates. If neutrinos are massive then the flavor and mass eigenstates are not necessarily identical. States with different mass propagate differently. Thus a ν state which is a ν_e at the moment of emission,

$$\nu_e = \cos \theta \, \nu_1 + \sin \theta \, \nu_2 \qquad (1)$$

(let us take just a two neutrino system, ν_1 with mass m_1 and ν_2 with mass m_2, for simplicity) may have a nonzero component on the perpendicular flavor state $\nu_i (i = \mu$ or $\tau)$

$$\nu_i = - \sin \theta \, \nu_1 + \cos \theta \, \nu_2 \qquad (2)$$

at a later moment. Assuming that ν oscillations occur during the propagation of ν's in vacuum, between the sun and the earth, they could be important for neutrino mass-squared differences $\Delta m^2 \equiv m_1^2 - m_2^2$ as small as $\Delta m^2 \sim 10^{-11} - 10^{-10}$ eV^2, but only if the mixing angles are large. Then up to a factor 1/N can be obtained between the ν_e flux at the earth and the ν_e flux at the sun, where N is the number of neutrinos. The fact that 1/3 is needed and 3 neutrino generations are known, contributes to the beauty of this explanation.[5] Recently Mikheyev and Smirnov[6] proposed a new type of neutrino oscillations to solve the solar neutrino problem: resonance amplification of oscillations while neutrinos propagate inside the sun. This idea is based on the observation by Wolfenstein[7] that neutrino oscillations occurring in matter are modified due to the presence of electrons (and absence of muons and taos). The forward scattering amplitude of $\nu_e + e^- \rightarrow \nu_e + e^-$, which has both neutral (Z^0 mediated) and charged (W boson mediated) current contributions, is different from that for $\nu_\mu e^-$ or $\nu_\tau e^-$ scattering (which has only neutral current contributions). A resonance, in which the mixing angle becomes $\pi/4$, i.e., the mixing becomes maximal, can occur at a certain value of the electron number density n_e in matter, which depends on the vacuum oscillations length $L^{vac} = 4\pi E_\nu / |\Delta m^2|$.

$$(n_e)_{res} = \frac{\sqrt{2}\pi}{L^{vac} G_F} \cos 2\theta \qquad (3)$$

where G_F is the Fermi coupling constant and θ is defined in Eqs. 1 and 2. If the resonance value $(n_e)_{res}$ is encountered by the ν_e's in a region of the sun on their way out of it and if the variation of n_e is slow enouch (adiabatic approximation) there will be a total conversion of the ν_e into another flavor.[6,9] One can understand this process considering that the additional interaction of ν_e in matter contributes an additional term V to its energy which can be written as an additional mass-square term A.

If $k_{\nu_e}^2 + m_{\nu_e}^2 = (E_\nu - V)^2 \simeq E^2 - 2EV$ then $m_{\nu_e}^2 = 2EV \equiv A$ where $A \sim G_f N_e E_\nu$. Assume for simplicity the in-vacuum diagonal mass term for the ν_e is negligible. The mass-square matrix in the ν_e, ν_i basis in vacuum is of the form

$$m_{vac}^2 \simeq \begin{bmatrix} 0 & \delta^2 \\ \delta^2 & m^2 \end{bmatrix} \tag{4}$$

where $\delta \ll m$, if the mixing in-vacuum is small. Then the heaviest mass eigenstate in vacuum is $\nu_2 \simeq \nu_i$ with mass $m_2 = m$ and the lightest $\nu_1 \simeq \nu_e$ with mass $m_1 \simeq \delta^4/m^3$. If at the center of the sun, where n_e is maximum the condition $(A)_{max} > m^2$ is fulfilled then the mass square matrix inside the sun

$$m_{SUN}^2 = \begin{bmatrix} A & \delta^2 \\ \delta^2 & m^2 \end{bmatrix} \tag{5}$$

implies that, in a central core in the sun the mass eigenstates are $\nu_1 \simeq \nu_i$ with $m_1 \simeq m$, the lightest, and $\nu_2 \simeq \nu_e$ with $m_2 \simeq A$. The resonance occurs when $A \simeq m$. In this case ν_1 and ν_2 are maximally mixed $\nu_1 = \frac{1}{\sqrt{2}} (\nu_e \pm \nu_i)$, with masses $m_1 \simeq m \pm (\delta^2/2m)$ If the passage through the resonance region is adiabatic the vectors ν_1 and ν_2 move slowly in the ν_e, ν_μ space retaining their identity. Thus, in the interior of the sun almost entirely ν_1's (which coincide with ν_e's) are produced but when these ν_1's emerge off the resonance region, they have rotated in the ν_e, ν_μ space and are finally almost entirely ν_μ. The correct sign of V to get a resonance only occurs if $\Delta m^2 \equiv m_1^2 - m_2^2 < 0$.[8] (For small vacuum mixing θ this means that the dominant component of ν_e is the lightest). If the condition $n_e)_{res} \le (n_e)_{max}$ in the center of the sun is satisfied, the resonance occurs. This happens for every $E_\nu \ge E_{critical}$, for a given Δm^2. In order to obtain a factor of 1/3 in Davis' experiment an $E_{critical} \simeq 6\ MeV$ is needed. The expectrum of ν_e at the earth is cut practically to zero for $E_\nu > 6\ MeV$ and is left unchanged for $E_\nu < 6\ MeV$ if $-\Delta m^2 \simeq .5 \times 10^{-4}\ eV^2$ and $\sin^2 2\theta > 4 \times 10^{-4}$.[9] There is also a second set of solutions[9] satisfying $-\Delta m^2 \times \sin^2 2\theta \simeq 10^{-7.5}\ eV^2$, for which the adiabatic approximation is not valid because the variation of n_e is too rapid. The modulation of the emission expectrum is different in this case: the ratio (flux on earth/flux at emission) increases slowly with E_ν.[9]

Another idea discussed recently[10] applies if the ν_e is a Dirac fermion and has a large magnetic moment (close to the experimental bound). Then, the solar magnetic field may transform some of the left-handed ν_e (produced by weak interaction) into right-handed ν_e not detectable in the ^{37}Cl. The problem with this idea is that the necessary neutrino magnetic moment is huge ($10^{-10} e \hbar/2 m_e c$) with respect to what is predicted in elementary particle models. The same consequences would be caused by an electric dipole moment of ν_e, also huge with respect to theoretically evaluated values, close to the experimental bounds.[11]

Finally, let us mention that the neutrinos emitted by the sun may be just decaying while arriving at the earth.[12,13] Since, due to time dilation, the lifetime is proportional to the enegy, those ν_e with energy larger than some value E_0 would reach the earth, those with $E < E_0$ would have decayed. The correct factor of 1/3 can be obtained if $E_0 \simeq 10\ MeV$.[13] Models which allow for fast enough decays involve a Nambu-Goldstone boson ϕ, called Majoron, quite strongly coupled to neutrinos.[13,14] The decay $\nu_e \to \nu_i \phi$ is allowed in some of these models if there is a ν_i lighter than ν_e and the mass of ν_e is of order $1\ eV$. This possibility requires an amazing tuning between the properties of neutrinos and the sun-earth distance.

3. NON-STANDARD ^8B NEUTRINO PRODUCTION

As we mentioned above the rate of production of ^8B neutrinos depend strongly on the temperature T_{core} of the inner core containing 0.05 of the mass of the sun. A decrease by 10% of the central temperature with respect to that expected in the standard model of the sun would account for a reduction factor of 1/3 in the produced flux. Faulkner and Gilliland (in a paper unpublished in 1977, published in 1985[15]) and independently Press and Spergel[16] proposed that one way T_c could be reduced is by improving the energy transport from the neutrino producing region to the outer regions of the sun with a population of weakly interating particles. This population of particles, which we call "solar cosmions", if concentrated enough in the interior of the sun would not alter the outer photon producing region, which we see. Faulkner and Gilliland did not have a good way of getting the needed particles inside the sun. Press and Spergel[17] proposed that these particles have been trapped by the sun from the halo of our galaxy. Thus these particles would constitute the dark matter in our galaxy and simultaneously solve the solar neutrino problem. This scheme works only for particles with mass and cross-section in a small range. If nothing else, it is an extraordinary coincidence that this range exists at all. Helioseismology provides another "coincidence" if not more, the fact that solar models with solar cosmions provide a good prediction (slightly better than the standard model itself) of the separation in frequency of p-oscillation modes, well known experimentally. Let us start exploring the "cosmic coincidences"[18], that is, how nontrivial it is that this idea works at all. Let us see how many of these particles are capturable by the sun in its lifetime $\tau_\odot \simeq 5 \times 10^9 y$ and how many are needed to alter significantly the energy transport in the sun; two completely independent questions. We are dealing with particles with mass in the GeV range, sightly heavier than protons since the orbits should be concentrated in a volume smaller than the sun. Even before computing the capturable number we should worry about the capture being possible. The escape energy at a certain radius is the kinetic energy that a particle initially at rest at infinity attains at the given point if allowed to free fall. Particles in the halo of the galaxy have an initial non-zero kinetic energy E_∞. Their typical velocity is (v_{halo}) $\simeq 10^{-3}c$. These particles may be trapped in one collision if the energy they lose is larger than E_∞. The maximum energy deposited on a proton by a particle x, heavier than a proton, in an elastic collision is $\simeq (m_p/m_x)E_x$, where E_x is the total energy of the particle $E_\infty + E_{escape}$. This can be a large fraction of E_x (since m_x is not much larger than m_p) and the escape velocity in the sun goes from $\simeq 2 \times 10^{-3}c$ at the surface to $10^{-2}c$ at the center. Thus, the capture is possible because the characteristic velocities of particles in our galatic halo and the escape velocities from the sun, are similar. Considering that the local halo density is 0.4 GeV/cm^3, for a cosmion mass of 5 GeV, the flux incident on the sun, under the simplifying assumption that all cosmions move with $10^{-3}c$, is $2 \times 10^6 cm^{-2} sec^{-1}$. Let us just use the sun geometrical cross-section (even if corrections due to the sun gravitational attraction are not negligible[17]). Then, over the age of the sun 10^{46} cosmions are incident on the sun. As the sun contains 10^{57} baryons, the maximum cosmion concentration by capture from the halo is thus 10^{-11}. What is the minimum concentration of cosmions necessary to change the central solar termperature? For a cosmion to have an optimal effect on the solar energy transport it must interact once per orbit (if its mean free path λ where much smaller than the orbit size cosmions would not affect the large volume wanted, if λ were much larger cosmions would interact once every many orbits). A cosmion that interacts inside, recieves $E \sim kT_c$ from a hot proton and outside gives this energy to a cooler proton. Thus a cosmion must transport $\sim kT_c$ in its orbit period of 10^4 seconds. This should be compared with the thermal diffusion time of 10^{15} seconds, for photons and protons (which are approximately in equal numbers in the sun) to perform a similar energy transport. A concentration of cosmions of 10^{-11} per baryon number is a rough estimate of the minimum solar cosmion density which could solve the solar neutrino problem. The sun is just able to capture sufficient cosmions for this purpose, if the capture efficiency is close to one.

There is a second coincidence: the cross-section of cosmions on protons is almost weak (slightly larger, though, this is one of the problems in model building). A weak cross-section ensures that GeV mass particles, whose cosmological density derives from annihilation in the early universe, are in numbers large enough to be the dark matter in halos of galaxies, i.e., $\Omega = \rho/\rho_c \geq 0.1$ (where ρ_c is the critical density $\rho_c = 10.5 \ keV/cm^3 h^2$, and h goes from .5 to 1). The optical cross-section is such that the cosmion within the sun has a mean free path of about the solar radius. This ensures high trapping probabilities and high energy transport. Such a mean free path corresponds to a non-relativistic cosmion elastic cross-sections per baryon number in the sun of about $\sigma_{critical} \sim 10^{-36} cm^2$. A weak (through Z^0 exchange) cross-section, for example of standard Dirac neutrinos

$$\sigma_{SCATT.}^{weak} \simeq \frac{G_F^2}{\pi} \frac{m_x^2 m_B^2}{(m_x + m_B)^2} T_{3_{L_x}}^2 [(1-4\sin^2\theta_W)Z-N]^2 \qquad (6)$$

is $5\times10^{-41} cm^2$ on H and $5\times10^{-38} cm^2$ on He (here m_x is the mass of the cosmion, m_B the mass of the target $T_{3_{L_x}}$ the third component of the weak isospin of the cosmion $\sin^2\theta_W \simeq 0.23$, Z and N are the number of protons and neutrons in the target). Since σ_c is two orders of magnitude stronger than weak, we are guaranteed that the physics of solar cosmions is physics at or below the weak scale. It is physics which is bound to be highly constrained and easily testable.

Helioseismology is the study of the solar interior through oscillations in the solar atmosphere.[19] Acoustic waves (or p-waves, pressure restored modes) of order minutes have been observed as patterns of Doppler shifts on the solar surface. Also g-waves, gravity waves where buoyancy dominates, should exist but their detection is still a controversial issue. Differences in frequencies between certain modes, very well known by now for p-waves, are sensitive to the radial variation of the speed of sound in the central regions of the sun, $c(r) \sim \sqrt{T(r)/\mu(r)}$ where μ is the mean molecular weight. In the standard model the drop off of $T(r)$ with r is compensated for by a drop off in $\mu(r)$ due to nuclear burning. The resulting predictions for acoustic wave frequencies have a rough agreement with data.[19] In previous models which lowered the central temperature, the pressure was maintained by homogenizing nuclear species thereby increasing the proportion of the lighter hydrogen at the center. This flattening of μ gave incorrect predictions for acoustic wave frequencies.[20] In the cosmion case the gradient in μ due to nuclear burning is preserved, so there is no disagreement with data.[21] In fact, the cosmion contribution to energy transport has the effect of flattening the T distribution. This gives frequency predictions which are in better agreement with the helioseismology data than those of the standard solar model, although this may not be statistically significant.

The prediction of solar cosmions for g-waves is also unique since it predicts shorter periods than the standard model while all other previous modification of the sun produced larger periods. It will be amazing if these shorter periods are confirmed experimentally. But already data on acoustic waves leave solar cosmions as the only viable modification of the sun to solve the solar neutrino problem.[20,21]

4. SOLAR COSMION CANDIDATES

As we have seen, it is a highly nontrivial result that captured dark matter particles can solve the solar neutrino problem.[15,16,17,22] It is worth looking for elementary particle candidates fulfilling all the requirements for solar cosmions, given in refs. 15, 16, 17, 22 and 23. The cosmion mass must be in the range 5 to 10 GeV. If lighter than 5 GeV, the cosmions will evaporate from the sun (i.e., the rate of those cosmions populating the tail of the velocity distribution with velocities larger than the escape velocity is larger than the trapping rate). If heavier than 10 GeV, their orbits are too small to produce efficient thermal transport. The cosmion must have a nonrelativistic, elastic cross-section on protons of order $10^{-36} cm^2$. Solar cosmions must be stable and be in the universe in a quantity enough to account for the haloes of galaxies $\Omega \gtrsim 0.1$. Finally, a severe constraint on the theory is that cosmion-cosmion annihilation in the sun must be highly suppressed.[23] If cosmion annihilation can occur in the sun, then the cosmion concentration in the sun may be less than 10^{-11} even if all cosmions capturable are effectively captured. As the number of cosmions increases by trapping, the annihilation rate increases and the equilibrium concentration is achieved when the annihilation rate becomes equal to the capture rate. This leads to the constraint[23]

$$\beta\sigma_{ANNIH.} \leq 10^{-4} \sigma_{SCATT.} \simeq 10^{-40} cm^2 \qquad (7)$$

where β is the mean thermal speed of the cosmion in the sun, $\beta \sim 10^{-3}$. This requirement is nontrivial since scattering and annihilation are crossed channels, any diagram contributing to one contributes to the other, and the cross-sections tend to be of the same order. For example, for standard neutrinos $\sigma_{ANNIH.}^{weak}\beta \simeq G_F^2 m_x^2 N_A/4\pi$, where N_A is the number of annihilation channels. Using the previously given scattering cross-section, we have

$$\frac{\beta \sigma_{ANNIH.}}{\sigma_{SCATT.}} \simeq (\frac{m_x}{m_B})^2 N_A \tag{8}$$

which is larger than 1.

The first problem to overcome is the origin of the large scattering cross-section. One immediately finds that the cosmion cannot be a conventional neutrino (Majorana or Dirac), sneutrino or photino, since the scattering cross-section in all these cases is a factor of at least 50 times too small.[23] Models in which the cosmion-nucleus scattering is mediated by a new neutral gauge boson do not seem to work. For example, in the $SU(2)_L \times SU(2)_R \times U(1)_{B-L}$ theory the cosmion could be the ν_R, which has a large $U(1)_{B-L}$ charge. However, it is not surprising that the scattering cross-sections mediated by such gauge bosons, which are heavier than the Z^0, are at least two orders of magnitude too small. Since we are unwilling to consider gauge currents which have enormous charges for cosmions compared with quark and lepton charges we consider the cosmion-nucleus scattering to be scalar-mediated. The main disadvantage of this is the necessity to force flavor conservation on these scalars. We will take the cosmions to be fermions and the interactions to be mediated by new colored Higgs bosons. The models we will describe are probably not unique and are not elegant. However they demonstrate that it is possible to build realistic models for cosmions even if none of the "standard" candidates works.[24]

Let us see three ways of solving the annihilation problem.

I. The cosmion annihilation is suppressed, it fulfills Eq. 7. Take the cosmion to be a Majorana fermion so that cosmions and anticosmions are identical. Due to the presence of two identical particles in the initial state, annihilations can occur in a p-wave or in an s-wave with the amplitude proportional to the mass of the annihilation products.

For a Majorana fermion X which has Yukawa couplings g_f to fermions f and scalar ϕ

$$\beta \sigma_A / \sigma_S \sim \sum_f \frac{g_f^4 m_x^2}{m_\phi^4} \left[\beta^2, \frac{m_f^2}{m_x^2} \right] / \frac{g_{u,d}^4}{m_\phi^4} \frac{m_p^2 m_x^2}{(m_p + m_x)^2} \tag{9}$$

where σ_S is dominated by scattering from hydrogen, and m_p is the proton mass. The p-wave term is suppressed by $\beta^2 \sim 10^{-6}$, while the s-wave term is suppressed only if m_f^2 or g_f^4 is small. Thus, while the p-wave term automatically satisfies (7) even if the g_f are all comparable, the s-wave term is constrained by (7): $(g_f/g_{u,d})^4 (m_f/m_x)^2 \lesssim 10^{-4}$. This is satisfied for $f = u,d$, but requires the Yukawa couplings to heavier quarks to be much less than $g_{u,d}$.

We consider the following simple model

$$L_I = g\bar{X}(\frac{1-\gamma_5}{2})u \phi - \frac{m_x}{2} \bar{X}^c X + h.c. \tag{10}$$
$$- m_\phi^2 \phi^* \phi - \lambda(\phi^* \phi)^2,$$

where X is the cosmion field, ϕ a color triplet scalar and $(1-\gamma_5)/2$ is the right-handed projector. The quark mixing has to occur in the down sector of the theory. It is not strictly consistent to simply omit Yukawa couplings of X to c and t quarks, since they are not forbidden by a symmetry. However, it is possible that they are accidentally absent due to a family symmetry which is broken at a higher scale. Alternatively, these terms can be included with small couplings. In this model

$$\frac{\beta \sigma_A}{\sigma_S} \simeq \frac{1}{2} [\frac{m_x}{m_p}]^2 [\frac{2}{3} \beta^2 + \frac{m_u^2}{m_x^2}] \tag{11}$$

II. There is a "cosmic cosmion asymmetry", similar to the asymmetry we know already, that of baryons. Moreover, if the cosmion number C is equal to the baryon number B today, due to the difference in mass between cosmions and protons or neutrons, the density in cosmions would be $\Omega_x/\Omega_B = m_x/m_B \simeq 5$–$10$ just right to account for the galactic missing mass.

In this case the halo is composed only of cosmions and not of anticosmions (or vice versa). The interactions which contribute to $\sigma_{SCATT.}$ preserve cosmion number and do not contribute to the annihilation of two cosmions. A non grand unified model in which the cosmion X is a Dirac fermion and $B-C$ is conserved is

$$L_{II} = QQ\phi, U^c D^c \phi^*, X^c D^c \phi, XX^c + h.c., \tag{12}$$

where we have used all left-handed fields: X, X^c for cosmion and anticosmion, Q for quarks and U^c, D^c for antiquarks. ϕ is a color triplet scalar field, with a superheavy mass so that it can decay to states of differing B, differing C, but the same $B-C$ $\phi \to qX, \overline{qq}$. These out of thermal equilibrium decays therefore generate a cosmological cosmion asymmetry $\Delta C = \Delta B$. The cosmion is unstable in this model $\Gamma(X \to \overline{qqq}) \approx m_x^5/m_\phi^4$. The lifetime is longer than the age of the universe for $m_\phi > 10^{12}$ GeV. Eventually, decays of X in the halo will make the universe baryon symmetric once again.

III. We already have a cosmic asymmetry: the sun is made out of protons and neutrons (not antiprotons or antineutrons) Let us use it! Take a cosmion which is not its own antiparticle, and the halo composed of equal numbers of cosmions and anticosmions. The sun may trap a different number of cosmions than anticosmions because of differing nonrelativistic elastic scattering cross-sections off nuclei $\sigma_c \neq \sigma_{\bar{c}}$. The less abundant component in the sun may have a concentration which is drastically reduced by annihilations while the change in the concentration of the dominant component may be unimportant.

Take a Dirac fermion X. It is possible[24] to get an effective coupling with a colored scalar ϕ

$$L_{III}(\phi) = \overline{x}(g + g_5\gamma_5)u\phi + h.c. \tag{13}$$

via mixing of two colored scalars, a weak singlet and a weak doublet. The exchange of ϕ dominates σ_c and $\sigma_{\bar{c}}$ via the diagrams of fig. (1a) and (1b) respectively. The result is $\sigma_c/\sigma_{\bar{c}} = (g/g_5)^4$, which may be different from unity. For our purposes the ratio need not be a very large or small number, a factor of two is fine.

In each of the previous models the scattering cross-section for the cosmions from nuclei arises from the exchange of a colored scalar ϕ as shown in Figure (1). This cross-section is given by

$$\sigma_S = \frac{\lambda}{\pi} \frac{g^4}{m_\phi^4} \frac{m_N^2 m_x^2}{(m_N + m_x)^2}. \tag{14}$$

In this first model σ_S is dominated by hydrogen and λ is of order one. In the second and third models $\lambda = (A+Z)^2$, where A and Z are the atomic number and charge of the nucleus N, and the scattering is dominated by helium For $m_x = 5$ GeV, in the first model $m_\phi = 75$ GeV in the second and third $m_\phi = 100$ GeV, yield sufficiently large cross sections. In all our models ϕ couples only to u not to c or t. This is to avoid the $\Delta C = 2$ box diagram, with internal X and ϕ, which would give excessive D^0–$\overline{D}^0$ mixing.   What about abundances? We have no parameters to adjust to force $\Omega_x \gtrsim 0.1$. the models could have failed this cosmological consistency check, but each passed. This is perhaps surprising. A 5 GeV dirac neutrino has $\sigma_S \sim 10^{-38} cm^2$ (a factor 100 too small) and $\beta\sigma_Z \sim 10^{-36}$ cm^2 One would imagine that a model which increases σ_S by two orders of magnitude would increase $\beta\sigma_A$ by two orders of magnitude as well, giving $\Omega_x \sim 50$ which is unacceptable. This naive scaling argument is not applicable. Our models differ from the neutrino case in several important ways: the A, Z dependence of $\sigma_{S,N}$, the number of annihilation final states, and the exchange being scalar instead of vector. All these differences have the effect of lowering Ω_x compared to the above scaling argument and each of the models is cosmologically acceptable.

5. EXPERIMENTAL SIGNATURES OF SOLAR COSMIONS

How would we know if the solar cosmion idea is true? There are signatures corresponding to the particular models we presented. The $\simeq 100$ GeV colored scalar may be detected in new hadron collider machines such as SLC or the Tevatron. In models I and III cosmions and anticosmions would annihilate in the sun into two u-initiated minijets. Some neutrinos would be produced which may be detected in underground experiments.[24] Solar cosmions may be directly detected through the energy they would deposit on nuclei inside detectors working at low temperatures. Since solar cosmions have cross sections with protons larger than weak they would be easier to detect than other candidates such as photinos. If the interactino of cosmions with nuclei is coherent, such as in the examples II and III above, the coherence factor, which in those examples is $(2Z + N)^2$ (that is, the square of the number of u quarks in the nucleus), enhances considerably the cross-section with heavy nuclei. Then just prototypes of the various cryogenic experiments under study to detect dark matter should test solar cosmion models. (See the contribution of A. Drukier in these proceedings). The same cryogenic techniques may be used in the future to detect solar neutrinos through the energy they deposit in nucle in Z^0 mediated interactions. Those interactions do not distinguish flavors. Then the total number of neutrinos of any flavor would be detected. It will be possible to know, therefore, whether fewer ν_e were emitted or if they turned into other flavors. The next proposed solar neutrino experiment is based on Gallium, $^{71}Ga \rightarrow {}^{71}Ge$. Most of the expected ^{71}Ga event rate is from the low energy pp neutrinos, the flux of which can be inferred from the overall solar luminosity and is relatively insensitive to the temperature of the solar core. Gilliland, Faulkner, Press and Spergel[22] have computed the expected rate in the ^{71}Ga if solar cosmions are the solution to the ^{37}Cl neutrinos. Almost only the ^8B neutrinos are affected in this case, which give a small contribution thus the ^{71}Ga experiment should see 92 SNU instead of the 105 SNU predicted in the standard solar model. If this happens to be true, could be conclude that the solar cosmion model is true? The answer is no. Also adiabatic resonant oscillations of ν_e in the sun will cut out only the ^8B neutrinos (all ν_e with energy larger that 6 MeV will not go out of the sun as ν_e; those are only the ^8B neutrinos). Thus the ^{37}Cl and ^{71}Ga experiments, if a ^{71}Ga rate close to 100 SNU is measured, will not be able to distinguish between a small decrease of temperature at the solar core and the resonant neutrino mechanism. A measurement of the ^8B solar neutrino spectrum will allow one to distinguish between both mechanisms[26]. If solar cosmions are the solution to the solar neutrino problem, the ^8B spectrum will be diminished just by an overall normalization. In the case of resonant oscillations the shape of the ^8B is modified, missing either its high (resonant amplification) or low energy component (non-resonant amplification). The ^{71}Ga rate should be almost zero, instead, if decaying neutrinos were the solution (since only the ν_e with energy larger than 10 MeV, so only a few of the ^8B neutrinos, would arrive).

As a concluding remark let me say again that the solar cosmion model is a beautiful idea associated with amazing "coincidences", cosmic and helioseismological. If true, it has implications on elementary particles at the weak scale, thus testable. It has plenty of signatures which will be tested in the near future. It is a good example of a subject of investigation in which different areas of elementary particle physics and astrophysics and cosmology join.

* Address from October 1986: The Enrico Fermi Institute, University of Chicago, Chicago, IL 60637. On leave of absence from Department of Physics, University of Rome II, via Orazio Raimondo, Rome, Italy 00173. Work partly supported by the U.S. Department of Energy, Grant No. DE AC02 82ER-40073 and NSF Grant PHY-82-15249.

REFERENCES AND FOOTNOTES

Footnote 1. New theoretical values are 8.2 ± 2.5 SNU (3σ quoted) from which 6.4 SNU are from ^8B and 1.2 from ^7Be. [3] The experimental values are unchanged.

1. See, for example, J. Bahcall, W. Huebner, S. Lubow, P. Parker and R. Ulrich, Reviews of Modern Physics **54**, 767 (1982).

2. J. N. Bahcall AIP Conference Proc. **126**, 60 (1985).

3. J. N. Bahcall and also R. Davis contributions to the 13th Texas Symposium on Relativistic Astrophysics, Chicago, 14-19, December 1986.

4. J. K. Rowley, B. T. Cleveland and R. Davis, AIP Conf. Proc. **126**, 1 (1985).

5. B. Pontecorvo Zh. Eksper. Teor. Fiz. **34**, 247 (1958). See S. M. Bilenky and B. Pontecorvo, Phys. Rep. **41C**, 225 (1978), and references therein. for a recent discussion see L. Krauss and F. Wilczek, Phys. Rev. Lett. **55**, 122 (1985).

6. S. P. Mikheyev and A. Yu. Smirnov, 10th Int. Workshop on Weak Interactions and Neutrinos, Savonlinnea, Finland (1985); Nuovo Cimento **C9**, 17 (1986).

7. L. Wolfenstein, Phys. Rev. **D17**, 2369 (1978); Phys. Rev. **D20**, 2634 (1979).

8. The origin of the correct sign can be found in P. Langacker, J. P. Leveillé and J. Sheiman, Phys. Rev. **D27**, 1228 (1983).

9. H. A. Bethe, Phys. Rev. Lett. **567**, 1305 (1986). S. P. Rosen and J. M. Gelb; E. W. Kolb, M. S. Turner and T. P. Walker, Phys. Lett. **175B**, 478 (1986); S. Parke, Fermilab-PUB-86/67-T, May 1986 contributed to 23rd Int. Conf. on HEP, Berkeley, CA 16-23, July, 1986. S. Parke and T. Walker, Fermilab-PUB-86/107-TA, contributed to 23rd Int. Conf. on HEP, Berkeley, CA 16-23, July, 1986.
   P. Langacker, S. T. Petcov, G. Steigman and S. Toshev, CERN-TH-4421, 1986.

10. M. B. Voloshin and M. I. Vysotsky, Preprint ITEP 1, Moscow (1986).

11. L. B. Okun preprint ITEP-14, Moscow (1986); L. B. Okun, M. B. Voloshin and M. I. Vysotsky ITEP-20 and 82 (1986).

12. J. N. Bahcall, N. Cabibbo and Y. Yahil, Phys. Rev. Lett. **28**, 316 (1972).

13. J. N. Bahcall, S. T. Petcov, S. Toshev and J. W. F. Valle, Inst. for Advanced Studies, Princeton preprint-86-1185, 1986 submitted to Phys. Lett.

14. G. B. Gelmini and M. Roncadelli, Phys. Lett. **99B**, 411 (1981); J. W. F. Valle, Phys. Lett. **131B**, 87 (1983); G. B. Gelmini and J. W. F. Valle, Phys. Lett. **142b**, 181 (1984).

15. J. Faulkner and R. Gilliland, Ap. J. **B299**, 663 (1985); see also, G. Stcigman, C. Sarazin, H. Quintana and J. Faulkner Ap. J. **83**, 1050 (1978).

16. D. N. Spergel and W. H. Press, Ap. J. **294**, 663 (1985).

17. W. H. Press and D. N. Spergel, Ap. J. **296**, 679 (1985).

18. W. H. Press, Talk at the Symposium on Cosmological Processes, Univ. of Colorado, March, 1985.

19. For a review see J. Christensen-Dalsgaard, D. Gough and J. Toomre, Science **229**, 923 (1985).

20. See, J. Primack, invited talk at the 2nd ESO/CERN Symposium on Cosmology, Atronomy and Fundamental Physics, March, 1986, Santa Cruz prepritn SCIPP 86/65, June 1986, and the figure by Faulkner there contained (fig. 5).

21. J. Faulkner, D. O. Gough and M. N. Vahia, Nature **312**, 226 (1986); W Däppen, R. L. Gilliland and J. Christensen-Dalsgaard, Nature, **321**, 229 (1986).

22. R. L. Gilliland, J. Faulkner, W. H. Press and D. N. Spergel, Ap. J. **306**, 703 (1986).

23. L. M. Krauss, K. Freese, D. N. Spergel and W. H. Press, Ap. J. **299**, 1001 (1985).

24. G. Gelmini, L. Hall and M. J. Lin, Harvard University preprint HUTP-86/AO42.

25. See, for example, P. F. Smith, invited review presented at 2nd ESO/CERN Symposium on "Cosmology, Astronomy and Fundamental Physics", ESO, Garching March 1986.

26. See, S. Parke and T. Walker, in Ref. 9.

DETECTING COLD DARK MATTER CANDIDATES

Andrzej K. DRUKIER
Harvard-Smithsonian Center for Astrophysics
60 Garden Street, Cambridge, MA 02138/USA
and
Applied Research Corporation, 8201 Corporate Drive, Suite 920
Landover, MD 20785

Based on work done in collaboration with:

S. P. Ahlen[1], F. T. Avignone[2], R. H. Brodzinski[3], K. Freese[4],
G. Gelmini[5], S. Dimopoulos[6], A. Kotlicki[7], M. Legros[8],
B. W. Lynn[6], D. N. Spergel[9], G. D. Starkman[6], B. Turrell[8]

(1) Dept. of Physics, Boston Univ., Boston, MA 02215, USA
(2) Dept. of Physics, Univ. of South Carolina, Columbia, SC 29208, USA
(3) Pacific Northwest Laboratory, Richland, WA 99352, USA
(4) ITP, UC, S. Barbara, CA 93106, USA
(5) The Enrico Fermi Institute, Chicago, IL 60637, USA
(6) Dept. of Physics, Stanford Univ., Stanford, CA 94305, USA
(7) Dept. of Physics, Warsaw Univ., Warsaw, Poland
(8) Dept. of Physics, U. of Br. Columbia, Vancouver, V6T 2A6 Canada
(9) Inst. of Advanced Studies, Princeton, NJ 08544, USA

ABSTRACT. The growing synergy between astrophysics, particle physics,
and low background experiments strengthens the possibility of detecting
astrophysical non-baryonic matter. The idea of direct detection is that
an incident, massive weakly interacting particle could collide with a
nucleus and transfer an energy that could be measured. The present low
levels of background achieved by the PNL/USC Ge detector represent a new
technology which yields interesting bounds on galactic cold dark matter
and on light bosons emitted from the Sun. Further improvements require
the development of cryogenic detectors, e.g. superheated superconducting
colloid (SSCD) and crystal bolometers. We report two tests designed to
study the practicality of a superheated superconducting colloid detector
using a SQUID readout system. Furthermore, we show that in case of
particles with spin interactions, one should consider detectors based on
compounds of boron, lithium and fluorine.

1. INTRODUCTION:

One of the outstanding questions in astrophysics is understanding
what makes up 90% of the mass of the universe. We observe the
gravitational effects of this "missing mass", but see neither stars, nor

W. G. Unruh and G. W. Semenoff (eds.), The Early Universe, 361–391.

gas, nor dust. A variety of theoretical arguments suggest that this "missing mass" may not be composed of ordinary protons and neutrons, but rather exotic non-baryonic particles.

The early universe was a tremendous particle accelerator. Its high temperatures and densities were the perfect environment for producing forms of matter inaccessable at laboratory energies. If any of the particles produced in the big bang are stable, then they could survive to the present epoch and account for the "missing mass". The candidates for this "missing mass" range in size from 10^{39} gram black holes through very low mass stars down to 10^{-39} gram axions. In the middle of this range lie the GeV mass "sparticles" offered by supersymmetric theories. The effect of all these "cold dark matter" candidates on galaxy formation is very similar, so we must turn to the laboratory to elucidate the mysteries of the heavens.

When our galaxy was formed, it would have accreted not only the baryons that formed stars, but also the non-baryonic matter. Since the non-baryonic matter is not dissipative, it would not have cooled to form a disc, but rather be distributed in a spheroidal halo. We do observe the gravitational effects of this dark matter halo. If weakly interacting massive particles (WIMPs) comprise the halo dark matter, then millions of these particles are streaming through a square centimeter every second. Ultralow background experiments may soon be capable of detecting this incident flux. Actually, some limits on non-baryonic matter in our Galaxy have been already obtained using data from an underground, ultralow background Ge-semiconducting spectrometer. Several cryogenic detectors are being developed for WIMPs searches.

Stars may also serve as sources of non-baryonic matter. Axions, majorons and familons can be produced in the cores of stars. These particles can be observed (or ruled out) through their effects on stellar evolution. Several of us have been able to obtain limits on axion theories through the non-detection of axions from the Sun.

If WIMPs comprise the galactic halo, they will scatter off of the baryons in the stars in the galaxy and some of the WIMPs will become bound to stars. In stars, WIMPs can alter energy transport and produce observable effects. Several of us have explored the possibility that WIMPs could solve the solar neutrino problem. WIMPs, once captured by the Sun and the Earth, may annihilate and be a source of high energy neutrinos. These neutrinos may be detectable in underground experiments. Furthermore, annihilation of WIMP's in our galaxy can lead to a detectable flux of high energy photons.

The primary focus of this contribution is to study experimental questions associated with direct detection of cold dark matter. The second section reviews the motivation for searching for astrophysical non-baryonic matter. The third section discusses some of the candidates for cold-dark matter. The fourth section focuses on the possibilities of direct searches for WIMPs, and the use of ultralow background Ge-detectors and cryogenic detectors are briefly described. Finally, we describe what is known about three groups of WIMPs and comment on the detectability of these particles.

These five sections are conceived as a tutorial text, i.e., as providing basic information and an extensive list of references describing direct searches for dark matter candidates. The appendices contain more detailed descriptions of:

1) WIMP searches using an ultralow background Ge spectrometer;

2) experimental tests of one of the cryogenic techniques (the superheated superconducting colloid detector); and

3) optimization of cryogenic detectors of photinos.

2. ASTROPHYSICAL EVIDENCE FOR NON-BARYONIC MATTER:

Over the past few years, a consensus has developed in the astrophysical community that the mass in luminous objects (stars and gas) can account for only 10% of the mass of the galaxy[1]. Over 40 years ago, Zwicky first realized that the velocities of galaxies in clusters implied that the mass of the cluster exceeded the mass of stars in the galaxies that make up the cluster. Through the intervening years, more evidence has been accumulated that confirm this discrepancy between virial and luminous mass.

Radio observations of cold gas in spiral galaxies measure the rotation velocity in the galactic potential. This allows us to infer the radial distribution of mass. Most of the mass in the galaxy is distributed in a spheroidal component, while most of the light in the galaxy is concentrated in a disc. This spheroidal "halo component" extends to large radii where there is no stellar component seen. The ratio of dynamically detected mass-to-light at 20 kpc is 1000 times larger than the ratio of dynamically detected mass-to-light in the stellar disc. This is perhaps the strongest evidence that 90% of the galaxy's mass is in non-luminous form. Observations of hot X-ray emitting gas[2] in elliptical galaxies suggest that the dark matter is ubiquitous.

The nature of this missing mass component has been a hotly debated topic in astrophysics. Some suggest that very low mass stars of low luminosity compose the galactic halo, while others suggest that the halo mass is in the form of more exotic matter. There are two main arguments that favor the later hypothesis. Theoretical prejudice suggests that the universe lie in the balance between being open and closed. Arguments from primordial nucleosynthesis, however, suggest that baryonic matter can account for only 10% of the theoretically-expected density. This suggests that the remaining mass be in non-baryonic form[3]. Non-baryonic matter can better help to explain galaxy formation and the large scale structure of the universe[4].

The solar neutrino problem[5] ranks with the missing mass problem as one of the more important questions at the interface between particle physics and astrophysics. One of the suggested solutions of the solar neutrino problem is that weakly interacting particles from the galactic halo have been captured by the Sun and have altered energy transport in the center of the Sun. This results in lowering the ^8B neutrino flux[6]. This hypothetical particle, which we have named the "cosmion", must have its annihilation in the Sun suppressed. Explicit particle physics have been developed for the cosmion[7] and there are active experimental programs to search for these particles.

3. PARTICLE PHYSICS CANDIDATES FOR NON-BARYONIC MATTER:

Assuming that a nonbaryonic component does exist, we review some of the candidates from particle physics. The standard model of particle physics has as its fundamental constituents the following particles:

i) Three families of quarks and their associated leptons; and

ii) Gauge bosons which mediate the fundamental interactions (γ mediates electromagnetism, $W^{\pm}$, Z^0 mediate weak interactions, gluon mediates strong interactions).

But the standard model does not appear to be complete since it leaves many questions unanswered: why are there three families? how does one explain quark and lepton masses? what with the hierarchy problem? how can one incorporate gravity, etc? Several attempts at more fundamental theories have been proposed, each of them predicting the existence of new particles which serve as dark matter candidates.

The three most likely candidates are light massive neutrinos, axions, and GeV mass particles (also known as WIMP's i.e., weakly interacting massive particles). A very exciting development is the fact that all three of these candidates stand to be either discovered or ruled out as the dark matter within the next decade.

Tritium β-decay experiments as well as searches for neutrino oscillations have already been running to probe the existence of a light neutrino species, perhaps the most likely candidate since neutrinos are actually known to exist. Claims of the discovery of a nonzero neutrino mass have been made but are extremely controversial[8]. Proposed detection schemes for axions[9], the very light pseudoscalar particles necessary for the solution of the strong CP problem in theories of the strong interaction, exploit the coupling of axions to two photons to convert halo axions into photons. Several groups have designed detectors using similar ideas. If the axion coupling is stronger than that required to solve the missing mass problem, then the axion can also be detected through the axioelectric effect. The existence of other massless bosons, majorons[10] and familions[11] have been advocated by various models of theoretical physics. One of them, the majoron, could be responsible for a tantalizing result of 2β decay experiment performed by PNL/USC collaboration[12].

We will focus on recent ideas for detecting particles in the GeV mass range. To remind you why this mass range is interesting for dark matter candidates, consider the energy density of a non-baryonic species (in this case neutrinos) as a function of the particle's mass. Requiring that $\Omega < 2$ in order not to overclose the universe, one sees that two interesting mass ranges remain: $m < 200$ eV or $m > 1.5$ GeV[13]. The light neutrinos would provide hot dark matter, and the heavy ones cold dark matter. Various supersymmetric particle candidates also exhibit similar behavior, and again could have masses O(GeV). Supersymmetry predicts higher mass partners for all known particles (for every fermion a boson and vice versa). Conservation of a new quantum number in many models, R-parity, implies that the lowest mass supersymmetric particle must be stable against decay and may therefore provide the dark matter. Candidates include the photino (the supersymmetric partner of the photon), the scalar neutrino or sneutrino (the partner of the neutrino), and the Higgsino (the partner of the Higgs).

In the following section we discuss the experimental methods appropriate to searches for these cold dark matter candidates. The bounds to their abundance are quoted in Section 5.

4. METHODOLOGY OF EXPERIMENTAL SEARCHES FOR WIMPs

It is clearly important to cover the principal theoretically-motivated particles, but at the same time design the experiments to cover as wide a range as possible of alternative types of particles. This can be achieved by means of two rather general classes of experiments (see Ref. 14 for excellent review article):

(a) Searches for light bosons based on the principle that many such particles can be converted to photons by interaction with electrons or electromagnetic fields (see e.g. Ref 15); and

(b) Searches for weakly interacting particles based on the detection of low energy nuclear recoils[16].

The second method is clearly of considerable generality: although some particle types would not interact directly with electrons (e.g. the photino), all conjectured heavy particles would interact with nucleons and produce a nuclear recoil, the energy of which is governed solely by kinematics. Furthermore, atomic nuclei happen to be in the ideal mass range for optimum sensitivity to incident particles of mass 1-1000 GeV. At least some of the target material must consist of nuclei with non-zero spin, in order to be sensitive to particles with purely axial couplings (such as Majorana neutrinos or photinos).

If WIMPs comprise the "missing mass" of the galactic halo, then millions of particles are streaming through a square centimeter every second. These particles scatter elastically from nuclei producing nuclear recoils of a few keV. Since the typical WIMP has a cross-section of order 10^{-37} cm^2, a few thousands such recoils are produced in a kilogram of detector per day[17-19]. The experimental challenge is to detect these rare weak recoils amidst the background from internal radioactivity. The task is complicated by the need of using very sensitive detectors able to detect recoiling nuclei of a few keV. Unfortunately, the most popular ionization detectors, (scintillators and gas detectors) are unable to detect recoiling nuclei with $E_{recoil} \leq$ 10 keV.

One method of detecting the WIMP scattering is to measure the electron-hole pairs produced in a semiconductor by the recoiling nucleus. Ahlen et al. have obtained the first laboratory limits on galactic dark matter using the USC/PNL Ge spectrometer[20]. The low level of background in this detector and its low energy threshold make it well-suited for dark matter detection. The current limits from this detector rule out WIMPs with mass between 20 GeV and 1 TeV, and germanium scattering cross-sections of greater than 10^{-35} cm^2. To extend the limits to, say 10 GeV, we have to know better the velocity spectrum of WIMPs in our Galaxy which requires more reliable numerical calculations. It will be however difficult to extend these limits below 10 GeV, since most of the energy of nuclear recoils is transferred not to electron-hole pairs but rather to phonons in the crystal. Yet another challenge of the Ge experiment is that if background can be further rejected by a factor of 10 - 20, it may be possible to say something about photinos with $m \geq 20$ GeV. (See Appendix 1 for detailed discussion of limits obtained with semiconducting detectors).

Turning now to the detection of WIMP's with $M \leq 10$ GeV, the optimum approach here is to use the ultralow temperature detectors, and has already attracted the interest of many experimental groups. The basic idea[16,17] is that any interaction between an incident particle of mass $1 - 10^3$ GeV and a target nucleus will result in a nuclear recoil, with energy typically in the range $1 - 10^2$ keV. The recoiling nucleus moves only a short distance, giving up the majority of its energy E_r to the lattice, and the resulting temperature increase can be observed in appropriate detectors operated at very low temperature. For any of the detection schemes, a crucial problem is, of course, the reduction of background to below the signal levels estimated in the above quoted references.

We will focus on the particular detection scheme of superheated superconducting colloids, first proposed by Drukier and Stodolsky[16] as a detector for neutrino astronomy. Goodman and Witten[17] then noticed that it could be used as a dark matter detector. In a paper by Drukier, Freese and Spergel[19] we worked out the actual count rates one could expect for various particles using realistic models of the halo. The total mass of the detector is about 1 kg, with micron-size grains in a dielectric filling material. The superconducting grains are placed in a magnetic field and maintained in a superheated state. When a halo particle passes through the detector, it interacts with a nucleus inside a grain: the nucleus recoils, energy is deposited in the grain, the temperature of the grain goes up, and the grain goes normal. This change of state can be detected by the Meissner effect. Penetration of the magnetic flux into the grains produces an observable signal in the readout electronics. The flipping of a grain a few microns in size produces a flux change of typically 0.1 ϕ_0 (flux quanta) in a few centimeter superconducting loop encircling the detector. SQUIDs can detect this flux change. Demanding a signal-to-noise ratio of at least 10, we calculated the count rates in the detector for various particles. We expect 10^4 counts/day for neutrinos and Dirac neutrinos, and O(1) count/day for photinos more massive than a few GeV.

Advantages over conventional particle detectors for GeV mass particles are the sensitivity to heavy nucleus recoils with energy < 1 keV and good background rejection. The detector is small (1 kg) so one can use anticoincidence or shielding effectively, e.g., to eliminate cosmic rays. Also the signal from such small energy deposits will be the flip of one and only one grain. Most backgrounds such as charged particles will flip many grains: for example a 0.5 MeV electron with range 0.5 gm/cm^2 must pass through 100 grains to lose its energy. The worst background for high energy thresholds is natural radioactivity (^{40}K, U, Th); for low energy thresholds it is solar neutrinos. The recent experimental tests of SSCD are described in Appendix 2.

Possibly the simplest method of searching for WIMPs is to detect the temperature increase produced in a crystal by the recoiling nucleus. The existing microbolometers, (X-ray detectors using a silicon crystal coupled to an appropriate thermistor) can detect keV energy deposition[21]. Their mass, however, is only a few milligram. Two groups (Stanford Univ. and Berkeley Univ.) are developing much larger bolometers based on silicon and sapphire crystals, respectively. The technique of crystal bolometry should be further developed, but we believe that its use for WIMP detection will require the use of crystals whose thermal properties are not well known. For example, compounds of boron, lithium and fluorine would make good detectors of photinos. The problem of optimization of bolometric detectors for detection of photinos is described in Appendix 3.

One of the challenges for all of these detectors is signal recognition. It is necessary to find a characteristic of this signal that will allow its differentiation from the background. Drukier, Freese, and Spergel[19] realized that the earth's motion around the Sun produces an annual modulation in this count rate, which serves as a signature for galactic dark matter particles. The earth moves around the Sun with a velocity of 30 km/s, in an orbit inclined at 62 degree relative to the plane in which Sun revolves at 225 km/s around the galactic center. Thus in July, the detector will be moving relative to the galactic halo of dark matter with velocity 240 km/s, while in January the relative velocity will be only 210 km/s. If the detector has a moderately sharp energy threshold (dE/E < 0.5), then this "aether drift" will produce a 10% annual modulation in the signal.

An important aspect of any detector design is the ability to incorporate several materials of differing atomic number (A) and nuclear spin to allow positive identification of any apparent signal. First, the recoil energy spectrum will have a cut-off which is a function both of the incident kinetic energy and the ratio M/A, giving two methods of estimating the mass of the incident particle. Second, the pattern of variation of the spectrum with nuclear mass and spin would indicate whether the interaction is coherent or spin dependent.

5. HORS D'OUVRE, OR WHAT WE MAY HOPE TO KNOW EXPERIMENTALLY
   ABOUT WIMPs A.D. 1990.

In this section we discuss the detectability of some popular particle candidates for cold dark matter. All are weakly interacting massive particles (WIMPs):
   1) "solar cosmions";
   2) massive Dirac neutrinos and scalar neutrinos;
   3) Majorana neutrinos and photinos.
Projected developments are shown in Table 1. At least a dozen groups or collaborations in Europe and the USA are now actively studying the problems and possibilities. With the current rapid growth of interest and enthusiasm in this topic, the outlook is now optimistic that several major dark matter experimental searches will be funded within the next few years. Most probably, development will take part in a few stages: the first detectors will be relatively small, isensitive and will have a considerable radioactive background. They should, however, be able to detect cosmions and shadow matter. The following improvements will permit searches for massive scalar/Dirac neutrinos, and afterwards photinos/Majorana neutrinos.

5.1. Cosmions:

The solar neutrino mystery ranks with the missing mass problem as one of the more important issues at the interface btween particle physics and astrophysics. One recent attempt at an explanation for both the deficit in the ^{8}B neutrino flux from the sun and the missing mass in the galaxy invoked a solar population of weakly interacting particles (that have been captured from a galactic reservoir of such particles) which can improve energy transport from the center of the sun[6]. This would decrease the central temperature of the sun and would therefore reduce the ^{8}B neutrino flux. The hypothetical particle has been dubbed the cosmion. Its mass must be greater than 5 GeV to prevent its orbits from being too small to produce efficient thermal transport. The elastic cross section for the cosmion with hydrogen must be about 10^{-36} cm^2 to insure trapping from the galactic reservoir and good energy transport within the sun. Explicit particle physics models for cosmions have been given[7] which are acceptable in terms of avoiding excessive annihilations of cosmions within the sun. In one of the models[7], the cosmion is a Majorana fermion which would have a scattering rate in an appropriate detector of several thousand per kg per day. In two other models, the cosmion is a Majorana fermion which would have a scattering rate in an appropriate detector of several thousand per kg per day; in yet two other models, the cosmion is a Dirac fermion having a scattering rate of about 10^9 per kg per day in a suitable detector! These are enormous rates, which will enable us to probe for the cosmion existence. No substantial improvements in existing technology are required for the cosmion search, and this will form the first milestone in a program of dark matter detection.

5.2. Massive Dirac Neutrinos and Scalar Neutrinos:

Neutrinos with large masses occur naturally in a variety of grand unified models[22]. Lee and Weinberg[13] calculated the annihilation cross section for heavy Dirac leptons due to V - A interactions, and derived a lower bound on mass of 2 - 4 GeV in order that the mass density of neutrinos today does not exceed twice the critical density. Another dark matter candidate is the supersymmetric partner of the neutrino, which is viable in this regard only if it is the lightest supersymmetric particle (which would render it stable). An upper bound of about 10 GeV has been placed on the mass of scalar neutrinos and massive Dirac neutrinos based on the absence of signals from high energy neutrinos in proton decay experiments[23, 24]. Such high energy neutrinos could be produced from the annihilation of Dirac neutrinos or scalar neutrinos which have been captured by the Earth or Sun. However, as is the case for many indirect limits, there are many potential loopholes with this limit, and it is probably fair to say that it is not firmly established. Massive Dirac neutrinos and scalar neutrinos each interact with nuclei via the exchange of Z bosons. Goodman and Witten[17] have calculated the scattering cross sections for such particles. These can be used to calculate scattering rates in detectors for realistic models of our Galaxy[19]. For now it is sufficient to note that the rate is quite large: in a detector containing nuclei which have mass of 100 GeV (A(Ge)=76 and A(Sn)=136), the scattering rate for neutrinos having a mass of 10 GeV is about 1,000 per kg x day, if the mass density of the neutrinos is of the order of the local density of the non-luminous matter in our galaxy. For a particle velocity of the order of a few hundred km/s, the energy deposition is a few keV. The use of an ultralow background Ge detector permits us to look for particles with $M_x \geq 20$ GeV (see appendix 1). Further improvements of this semiconducting detector would permit a detection/rejection of neutrinos with $M_x \geq 10$ GeV. To search for particles with lower mass, the use of cryogenic detectors seems to be necessary.

5.3. Photinos and Majorana Neutrinos:

The photino is the supersymmetric partner of the photon. If it is the lightest supersymmetric particle, it could form the dark matter. A claim[25] for the observation of several low energy antiprotons in cosmic rays has sparked considerable interest in the possibility that these may have come from photino-photino annihilations in the galactic halo[26]. Cosmological considerations[13] limit the photino mass from below to 1.8 GeV for scalar quark masses in excess of 40 GeV. The ASP experiment[27] limits the photino mass from below to 5 GeV for a scalar electron mass m_{se} = 40 - 50 GeV, to 10 GeV for m_{se} = 25 - 35 GeV, and to 13 GeV for m_{se} = 0. Photinos interact via axial-vector couplings that are spin dependent, which implies a factor $J(J + 1)$ in the photino-nucleus cross section, where J is the nuclear spin. The photino is not believed to interact coherently with the constituent nucleons of a nucleus, and this reduces the scattering rate from that expected for scalar neutrinos or massive Dirac neutrinos. Unfortunately, the Ge detectors have only a small fraction of odd number isotopes (7.6% of ^{73}Ge) and cannot be used to obtain stringent limits on the existence of the photino. Thus the cryogenic detectors must be used. It is expected that the rate in an aluminum SSCD device would be about one per kg per day. Somewhat higher rates would occur in the lithium, boron, and fluorine bolometers under development. The problems of the choice of the optimal detector material are studied in Appendix 3.

6. CONCLUSIONS

The growing synergy between astrophysics, particle physics, and low background experiments strengthens the possibility of detecting astro-physical non-baryonic matter. It is apparent that direct experimental searches for low energy galactic neutral particle fluxes, covering all of the currently favoured "cold dark matter" candidates, should now be feasible with a few year development period. This offers an exciting opportunity for astrophysicists and particle physicists to collaborate on some challenging problems of common interest, with the possibility of creating a new area of observational physics. Advancing this goal requires collaboration between workers in these fields and low temperature physicists.

7. REFERENCES

1) For a recent review see:
"Dark Matter in the Universe," eds. J. Knapp and J. Kormendy (Reidel, 1986), proceedings of the International Astronomical Union Symposium No. 117, Princeton, June 24-28, 1985; S. M. Fall, "Dark Matter in Early-Type

Galaxies", preprint STSI 149, Dec. 1986; K. A. Olive, "Dark Matter as a Probe of the Early Universe", preprint UMN-TH-561/86, May 1986; V. C. Rubi et al., Ap. J. 255 (1978) L107; D. Burstein, V. C. Rubin, Ap. J. 297 (1985 423; J. N. Bahcall, S. Casertano, "Some Possible Regularities of the Missing Mass Problem", preprint IAS, Princeton 1985.
See also:
V. C. Rubin et al., Ap. J. 255 (1978) L107; S. M. Faber, T. S. Gallagher, Ann. Rev. Astr. Ap. 17, 135 (1979).

2) D. Fabricant, P. Gorenstein, Ap. J. 267, 535 (1983) W. Forman, C. Jones, W. Tucker, Ap. J. 293, 102 (1985).

3) For theoretical arguments see:
D. J. Hegyi, K. A. Olive, Phys. Lett. 126B (1983) L8 ibid. Ap. J. 303 (1986) 56; also contribution of M. Turner in this conference.
Concerning recent searches for "brown dwarfs":
T. J. Chester et al. "The Reddest Unidentified High Galactic Latitude IRAS 12 sources: A search for brown dwarfs and new types of sources" and M. F. Skrutskie, "Infrared Constraints on the Nature of Dark Matter: The Solar Neighborhood", both Proc. of 169th AAS Meeting, Pasadena, 1986.

4) J. Primack, SLAC preprint 3387 (1984); C. J. Hogan et al., preprint Fermilab-Conf-85/57-A, April 1985; N. Kaiser, J. Silk, Nature 324 p. 539-537, 11 Dec. 1986.

5) J. N. Bahcall et al. Rev. Mod. Phys. 54 (1982) 767; J. N. Bahcall et al. Ap. J. 292 (1985) L79; J. N. Bahcall, B. R. Holstein, Phys. Rev. C 33 (1986) 2121-2127; M. Casse et al. in "Neutrinos and the Present Day Universe", p. 49-67, Eds. Th. Montmerle, M. Spiro, CEN, Paris 1986; R. Davis, Report to the 7th Workshop on Grand Unification, ICOBAN 86, Toyoma, Japan; R. Davis, Proc. of 13th Texas Symposium, Chicago, Dec. 1986.

6) D. N. Spergel, W. H. Press, Ap. J. 294 (1985) 663; W. H. Press, D. N. Spergel, Ap. J. 296 (1985) 679; J. Faulkner, R. Gilliland, Ap. J. 299 (1985) 663; K. Freese et al., Ap. J. 299 (1985) 1001; R. L. Gilliland, J. Faulkner, W. M. Press, D. N. Spergel, Ap. J. in print.

7) G. B. Gelmini, L. J. Hall, M. J. Lin, "What is Cosmion", preprint Harvard HUTP-86/A042.

8) V. A. Lubimov et al. Phys. Lett. 94B (1980) 266; see also papers about direct mass measurements in Proc. of VIth Moriond Workshop, "Massive Neutrinos..." Eds. D. Fackler, J. Tran Thanh Van, pages 441-579, Editions Frontieres, Paris 1986.

9) R. D. Peccei and H. R. Quinn, Phys. Rev. Lett. 3B (1977) 1440; Phys. Rev. D16 (1977) 1971; S. Weinberg, Phys. Rev. Lett. 40 (1978) 223; F. Wilczek, Phys. Rev. Lett. 40 (1978) 279; J. E. Kim, Phys. Rev. Lett. 43 (1979) 103; M. A. Shifman, A. I. Vainshtein, and V. I. Zakharov, Nucl. Phys. B166 (1980) 493; M. Dine, W. Fischler, and M. Srednicki, Phys. Lett. 104B (1981) 199; J. E. Moody and F. Wilczek, Phys. Rev. D30 (1984) 130.

10) Y. Chicashige, R. N. Mohapatra and R. D. Peccei, Phys. Lett. 98B (1981) 265, Phys. Rev. Lett. 45 (1980) 1926.

11) G. B. Gelmini and M. Roncadelli, Phys. Lett. 99B (1981) 411. H. Georgi, S. L. Glashow and S. Nussinov, Nucl. Phys. B193 (1981) 297.

12) F. T. Avignone, R. L.. Brodzinsky, H. S. Miley, J. H. Reeves, "Possible evidence for neutrinoless β-decay with emission of Goldstone Bosons", preprint USC, Dec. 1986.

See also for theory. H. Georgi, S. L. Glashow, S. Nussinov, Nucl. Phys. B193 (1981) 297; J. D. Vergados, Nucl. Phys. B218 (1983) 109.

13) B. W. Lee, S. Weinberg, Phys. Rev. Lett 39 (1977) 1965.

14) P. F. Smith, "Possible Experiments for Direct Detection of Particle Candidates for the Galactic Dark Matter", preprint RAL-86-029.

15) P. Sikivie, Phys. Rev. D32 (1985) 2988, B. Moskowitz, et al., Proc. of 13th Texas Symposium on Relativistic Astrophysics, Chicago, Dec. 1986, to be published.

16) A. K. Drukier, L. Stodolsky, Phys. Rev. D30 (1984), 2295.

17) M. Goodman, E. Witten, Phys. Rev. D31 (1985), 3059; I. Wasserman, Cornell preprint, DE 3065 (1985).

18) A. K. Drukier, Acta. Physica. Polonica, B17 (1986), 229.

19) A. K. Drukier, K. Freese, D. N. Spergel, Phys. Rev., D33 (1986), 3495.

20) S. P. Ahlen, et al., preprint CFA No. 2292, March 1986 (submitted to Phys. Rev. Lett.).

21) E. Fiorini, and T. O. Niinikoski, Nucl. Instr. and Meth., 224, 83-88, 1984; S. H. Moseley and J. C. Mather, J. App. Phys., 56, 1257-1262, 1984; D. McCammon, et al. J. App. Phys., 56, 1263-1266, 1984; B. Cabrera, L. M. Krauss, and F. Wilczek, Phys. Rev. Lett., 55, 25-28, 1985.

22) See, e.g., J. Bagger, et al., Phys. Rev. Lett., 54 (1985), 2199.

23) J. Silk, K. Olive, M Srednicki, Phys. Rev. Lett. 55, (1985) 257; K. Freese, Phys. Rev. Lett., 56 (1986), 685; L. Krauss, et al., Phys. Rev., D31 (1986), 2076; B. A. Campbell et al., "Superstring Dark Matter", preprints CERN-TH.4385/86; J.S. Hagelin, K.W.Ng, K.A. Olive, Physics Lett., 180 (1986) 375; T. K. Gaisser, G. Steigman, S. TilV, "Limits on Cold Dark Matter...", preprint BA-86-42.

24) J. M. LoSecco, "Limits on the Flux of Energetic Neutrinos from the Sun," preprint UND-PDK 86-9.

25) A. Buffington, et al., Ap. J., 248 (1981), 1179.

26) J. Silk, M. Srednicki, Phys. Rev. Lett., 53 (1984), 624.

27) G. Bartha, et al., Phys. Rev. Lett., 56 (1986), 685.

TABLE 1: DIRECT SEARCHES FOR COLD DARK MATTER CANDIDATES

Generation 1: Ultralow Background Ge-Semiconductors
(1986-88)

Detector Mass:	0.7 kg
$E_{threshold}$:	10 keV
Background: (at 1- keV)	10 counts/kg*day*keV

Particles:

Dirac and scalar neutrino	M $\geq$ 15 GeV
Majorana neutrino	M $\geq$ 20 GeV, 1988?
Photino	M $\geq$ 20 GeV, 1989?
Shadow Matter	

Generation 2: Superheated Superconducting Colloid Detector (SSCD)
(1988-1990)   Crystal Bolometers (Si, Ge, GaAs)

Detector Mass:	10-100 g
$E_{threshold}$:	1 keV
Background: (At 1 keV)	1-10 counts/kg*day*keV (SSCD, Ge) 100 counts/kg*day*keV (Si)

Particles:

Solar Cosmions	M $\geq$ 2 GeV
Dirac and Scalar Neutrinos	M $\geq$ 5 GeV

Generation 3: Superheated Superconducting Colloid Detector
(1990-1992)   Crystal Bolometers (B, LiH, LiF)
           He-3 Detectors

Detector Mass:	1-10 kg
$E_{threshold}$:	0.2 keV
Background: (at 1 keV)	0.1-1 counts/kg*day*keV 0.01 for He-3 detectors

Particles:

Majorana neutrinos	M $\geq$ 2 GeV
Photinos	M $\geq$ 2 GeV

Appendix 1

BOUNDS ON GALACTIC COLD DARK MATTER PARTICLE CANDIDATES AND SOLAR AXIONS
FROM A GE-SPECTROMETER

1. THE METHOD:

The idea of direct detection is that an incident, weakly
interacting, particle could collide with a nucleus and transfer
an energy that could be measured [1,2]. Several very low temperature
detectors designed specially for this purpose are under study [3,4].
This appendix refers to the first experiment based on direct detection,
performed with already existing semiconducting detector. The ultralow
background Ge spectrometer developed by the USC/PNL[5] group for
double-beta decay is used as a detector of cold dark matter candidates
from the halo of our galaxy and of solar axions (and other light
bosons), yielding interesting bounds [6,7]. The maximum energy transfer
in a collision of a projectile particle, with velocity v, with a target
of mass M at rest is

$$T_{max} = 2\,\beta^2 \times Mm^2/(m + M)^2 \qquad (\beta = v/c) \qquad (1)$$

Typical velocities of dark matter particles in the halo should be
$v > 10^{-3}c$. Since the threshold energy expected is around 1 keV, in any
detection method considered up to now, direct searches can test for
particles of mass $m \geq 1$ GeV.

The non-relativistic cross-section for this process is

$$\sigma = (1/\pi) \times (mM|A|)^2/(m + M) \qquad (2)$$

where $|A|$ is a reduced amplitude which depends on the dynamics of the
collision. Due to the value of m = 0 (GeV), both T^{max} and σ point
towards nuclei as good targets, as opposed to electrons for which both
cross-section and energy transfer are largly suppressed.

Let us first discuss the galactic cold dark matter candidates. A
halo consisting of "cold" dark matter cannot be rotating with the disc
of the galaxy, or it would be flattened like the disc. (An upper bound
is provided by the rotation of the bulge, which is slower than the disc
even if it is probable that the halo moves much more slowly or not at
all). The sun moves with the disc. While the disc is supported against
radial collapse by its angular momentum, the dark halo is supported by
random velocity. When the matter that makes the halo collapsed to form
our galaxy, the velocities of its constituents were randomized in a
process called "violent relaxation". Since the dark matter particles
have already performed several hundred oscillations in the galactic
potential starting from slightly different initial conditions, the local

total velocity dispersion $v_{r.m.s.}$ can be inferred from the observed local circular speed: it is of order $10^{-3}c$, similar to the velocity of the sun with respect to the halo v_{o-h}. Thus, most particles have velocities between zero and $2v_{r.m.s.}$ with respect to the earth. Due to the existence of an energy threshold that can be translated into a velocity threshold, only particles with velocities $v > v_{th}$ can be detected.[3]

One of effects of a recoiling nucleus in a crystal is the production of ionization within the material. Part of the nuclear recoil energy goes into the excitation of electrons. This mechanism allows the use of an existing ultra low background germanium diode detector to search for the galactic dark matter. This type of detector was developed to look for the double-beta decay of ^{76}Ge, wherein excellent background rejection and high sensitivity are needed. These properties, together with a low energy threshold, are essential in the search for dark matter. This detector counts the number of electrons which jump into the conduction band, i.e., the number of "shallow" electron-hole pairs as a result of energy deposited either upon an electron or upon a nucleus of the Ge crystal. Unfortunately, only a small part of energy of recoiling nuclei goes into ionization.

A recoiling nucleus in a crystal gives a majority of its kinetic energy to the lattice, producing a phonon wave (ballistic phonons). This wave thermalizes after many scatterings within the restricted volume of the target, generating an increase in temperature. However, recoils with $T \geq 15$ keV are detectable using semiconducting detectors with a small gap. The lower band gap (0.69 eV at $77^{o}K$) and higher efficiency for converting electronic energy loss to electron-hole pairs (2.96 eV per pair at $77^{o}K$), make Ge detectors more appropriate than Si detectors.

2. LIMITS ON HALO COLD D.M. CANDIDATES FROM THE PNL/USC GE SPECTROMETER:

The PNL/USC group has developed a 135 cm^3 prototype intrinsic Ge detector having a background reduced by about three orders of magnitude over conventional low background gamma ray spectrometers. This is the lowest background ever achieved[5]. The detector is at a water-equivalent depth of 4000 meters in the Homestake goldmine, in order to eliminate the cosmic ray-induced background. The Ge crystals themselves are free from primordial or man-made radioactivity, and the materials used in the construction of the detector and shields were carefully studied and selected. The data on which the present analysis is based were taken with the detector surrounded by high-purity copper and 11 tons of lead to eliminate the radioactive background from the rock. Recently the shield has been upgraded by the use of 448 year-old lead (from a sunken Spanish galleon) replacing the copper inner shield which had some cosmogenic radioactivity.

When the search for dark matter was initiated in late 1985 the energy threshold was reduced to an incident electron energy of 4 keV. This corresponds to an initial nuclear recoil energy of about 15 keV. Below this threshold there was a strong increase of microphonic noise engendered by mining operation. Hardware and software have recently

been developed to reduce this noise and reduce the 4 keV threshold to
about 1 keV. The number of (e-h) pairs produced when some energy T is
deposited on an electron is a well known property of the Ge
(approximately T/3 eV). This number is larger than the one produced
when the same energy is deposited on a nucleus. Let us call R.E.F.,
(relative efficiency factor) the ratio of these two numbers. This is an
energy-dependent quantity; it is 1 only for T >> 1 MeV: for T=15-50
keV, it is about 3.5-4. Unfortunately, it is expected to become very
large at T≤ 5 keV, i.e., at an energy close to the present threshold.
If so, the energy at which this happens will be the final threshold
achievable to detect nuclear recoils (since the number of e-h pairs
produced becomes negligible). The R.E.F. for Ge detectors, shown in
Fig. 1 as function of the nuclear recoil, was calculated by evaluating
the fraction of primary Ge recoil energy lost in electronic collisions.
The calculation included the electronic loss of the second and higher
order generations of recoiling nuclei (the ones perturbed by the primary
nucleus). The theoretical values are in good agreement with the
experimental data of Chasman et al.,[8] also shown in Fig. 1, who
measured e-h yields of neutron-induced recoil nuclei within a Ge
detector. Two different assumptions regarding the behavior of the
electronic energy loss as function of the velocity of the recoiling Ge
nucleus give different results at low energies: the solid curve assumes
a linear dependence, the dashed one a kinematic threshold at 0.27 keV of
nuclear recoil energy (at which a Ge nucleus transfers a maximum energy
of 0.7 keV to an electron in a direct collision).

The results for a few weeks of low energy data from the Ge
spectrometer are shown in Fig. 2, as functions of energy deposited on
electrons (which should be multiplied by the appropriate R.E.F. to
obtain the corresponding nuclear recoil energy). To know what these
data mean for hypothetical dark matter particles, the observed count
rate must be compared with the rate predicted if the halo consisted of
those dark matter particles. This last rate, R_p, depends on the
scattering cross-section ([5]) of the candidate dark matter particles on
a Ge nucleus, on their local number density (n_x), and on the velocity
distribution f(v) of the particles:

$$R_p(T) = n_x dT \int \frac{d\sigma}{dT} (v, T) f(v) v d^3v. \tag{3}$$

Here T is the recoil energy and dT is the interval, centered on T, of
recoil energies detected in a given channel.

The model of the halo used assumes a local halo density of
$0.01 M_o/pc^3 = 0.38$ GeV/cm^3, and an isotropic gaussian distribution
f(v) with an r.m.s. of 250 km/sec and a maximum of 550 km/sec. Both
velocity values are conservative. The halo is assumed to rotate slowly
like the galactic spheroid, with a local velocity of 80 km/sec. This is
again a conservative assumption which reduces relative velocities, since
the sun moves in the same sense around the galaxy at 250 km/sec. The
integral was evaluated over all velocity phase space and a bound on n_x
was obtained for every T. The most restrictive T-dependent bound was
taken as the final bound on n_x. The best bounds come from values of T
near threshold because of the rapid decrease of the predicted rate with

increasing T. To fix ideas, stable standard Dirac neutrinos were chosen to present the bounds, since their interactions are coherent. The elastic cross sections of Dirac neutrinos of mass m and incident energy $E=m(c^2+v^2/2)$ on a nucleus of mass M with Z protons and N nucleus, at rest, through Z^0 exchange is:

$$\frac{d\sigma}{dT} = \frac{G_F^2 MC^2}{8\pi v^2} [Z(1-4sin^2 O_w)-N]^2 \times f(T,E) \tag{4}$$

with

$$f(T,E) = (1 +(1 - T/E)^2 - (MT + m^2)/E^2) \times exp(-2MTR^2/3) \tag{5}$$

where G_F is the weak coupling constant and O_w is the weak mixing angle. The exponential factor is a nuclear form factor derived from the assumption of a gaussian density of nucleons with a nucleus of radius $R < 1.2 \ A^{1/3}$ fermi, where $A = Z + N$.

Figure 3 shows the maximum halo density of heavy standard Dirac neutrinos that is consistent with the observed count rate at the 68 percent (solid line) and 95 percent (dashed line) confidence level. Standard Dirac neutrinos of mass 20 GeV < m < 1 TeV are excluded as main components of the halo. This figure can be used for other vectorial spin-independent interactions by scaling the vertical axis by the ratio of cross sections. A particle with a cross section 2 times larger should have a density 2 times lower to obtain the same interaction rate, for example. This is what happens in the case of scalar neutrinos.

The bounds on spin-dependent interacting particles are not as good as for spin independent ones because most of the Ge isotopes have zero spin. Only ^{73}Ge, with a natural abundance of 7.8 percent, has a nonzero spin J = 9/2.

Even if the bounds found in a terrestrial experiment only test the local density of dark matter, this one is considered representative of the global dark matter density in the halo. Even at small scales the velocities of the dark matter particles should be randomized, if not by violent relaxation then by phase mixing after many oscillations in the galactic potential.

Bounds on galactic abundances can be translated into bounds on cosmological abundances if it is assumed that the ratio of the total mass in non-baryonic matter to the total mass in baryonic matter in our Galaxy ($F_{Gal} = M/M_{baryon}$) is the same as in the whole universe. This may be a good assumption for cold dark matter and the interesting bounds on cosmological abundances have been obtained in ref. 6. The same paper discusses the future improvements; it seems possible to diminish the radioactive background by at least one order of magnitude. Reduction of the microphonic noise engendered by mining operations will make possible the lowering of the threshold energy to deposition of 1 keV by electrons (or T>5keV for recoiling nuclei). This would allow the detection of Dirac neutrinos with m>8GeV.

3. LIMITS ON SOLAR AXIONS (AND OTHER LIGHT BOSONS) FROM THE USC/PNL GE-SPECTROMETER:

Another use of this ultralow Ge detector is to test the couplings to electrons of light bosons emitted from the sun[7]. Interesting laboratory bounds on the coupling to electrons of pseudoscalars such as axions[9], familons[10], majorons[11,12] were obtained.

The expected rate of interaction was obtained[7] by calculating the solar axion flux, using a solar model[13] with a temperature of 1 keV, and multiplying it by the "axioelectric" cross section of Ge atoms. Only the bremsstrahlung emission process in the sun was used, even in the case of axions (the Primakoff process is suppressed due to Debye-Huckel screening in the solar plasma)[14]. The "axioelectric" effect[15] is an atomic enhancement of the interaction of scalar bosons with electrons, similar to the photoelectric effect (where the photon is replaced by a spin zero boson whose mass is zero or negligible).

In Fig. 4 the expected number of events per kg per day for germanium are plotted against the incoming axion energy (i.e., energy transferred to an electron) for $g^{-1} = 0.5 \times 10^7$ GeV (solid line) and $g^{-1} = 1 \times 10^7$ GeV (dashed line). The major contribution comes from 1 keV $< E <$ 10 keV since both the solar axion flux and the axiolectric cross section peak in this region. The first experimental points near threshold, ($E \geq 4$ keV), shown in Fig. 2, are plotted as crosses in Fig. 4. The statistical error on these data is estimated to be $\pm$ 25 percent. From this the experimental bound $g^{-1} = (F/2x_{e'}) \geq 0.5 \times 10^7$ GeV was deduced. This means $M_A \leq 6.9$ MeV.

As mentioned before, the energy threshold of the Ge detector will be lowered soon to 1 keV, embracing the region in which the expected rate is one order of magnitude larger (see Fig. 4). Furthermore the background is projected to be reduced by a factor of 10 or more. This will mean a total improvement of two orders of magnitude, at least, in the bound on the rate which depends on $(g)^4$. Thus the bound $g^{-1} > 1.8 \times 10^7$ GeV is expected soon. Further improvements are possible in the second generation of ultralow Ge detectors, with a multidetector structure and considerably higher mass, which is expected for 1987-88. One could eventually set limits $g^{-1} > 10^8$ GeV or, a more exciting possibility, detect solar axions or other light bosons. These bounds are less severe than some more speculative astrophysical bounds which rely, however, on a detailed understanding of the dynamics and evolution of red giants, white dwarfs, or other stars very different from the Sun. It should be emphasized that the bounds obtained with the Ge detector are laboratory bounds which depend on the knowledge of the best known star, the Sun.

Light bosons appear usually only as Nambu-Goldstone bosons (or pseudo N-G bosons). Let us mention however for completeness, bounds on light scalars ϕ with coupling $L = \tilde{g} \; \bar{e}e \; \phi$ to electrons. The cross section for this "scalar-electric" effect (similar to the previously treated "axioelectric" one) does not have the suppression factor $(\omega/2m_e)^2$, so that bounds on g_{scalar} are 10^3 times better than bounds on g_{axion} for E > 1 keV.

4. REFERENCES

1. A. K. Drukier, L. Stodolsky, Phys. Rev. D.
2. M. W. Goodman and E. Witten, Phys. Rev. D31, 3059 (1985).
3. A. K. Drukier, Acta Physica Polonica; and B17 (1968) 229; A. K. Drukier, D. Freese, D. N. Spergel, Phys. Rev. D. 33, 3495 (1986).

4. For a review see P. F. Smith "Possible Experiments for Direct Detection of Particle Candidates for the Galactic Dark Matter," Rutherford Lab. preprint RAL-86-029.

5. R. L. Brodzinski, D. P. Brown, J. C. Evans Jr., W. K. Hensley, J. H. Reeves, N. A. Wogman, F. T. Avignone III and H. S. Miley, Nucl. Instr. and Meth. A239, 207 (1985).

6. S. P. Ahlen, F. T. Avignone III, R. L. Brodzinski, A. K. Drukier, G. Gelmini and D. N. Spergel, "Limits on Cold Dark Matter Candidates from the Ultralow Germanium Spectrometer," Harvard Center for Astrophysics Preprint No. 2292, 1986. Submitted to Phys. Rev. Lett.

7. F. T. Avignone III, R. L. Brodzinski, A. K. Drukier, G. Gelmini, B. W. Lynn, D. N. Spergel, G. D. Starkman, "Laboratory Limits on Solar Axions from an Ultralow Background Germanium Spectrometer," SLAC preprint PUB-3872, 1986; to appear in Phys. Rev. D.

8. C. Chasman, K. W. Jones and R. A. Ristinen, Phys. Rev. Lett. 15, 245 (1965).

9. R. D. Peccei and H. R. Quinn, Phys. Rev. Lett. 3B (1977) 1440; Phys. Rev. D16 (1977) 1971. S. Weinberg, Phys. Rev. Lett. 40 (1978) 223. F. Wilczek, Phys. Rev. Lett. 40 (1978) 279. J. E. Kim, Phys. Rev. Lett. 43 (1979) 103. M. A. Shifman, A. I. Vainshtein, and V. I. Zakharov, Nucl. Phys. B166 (1980) 493. M. Dine, W. Fischler, and M. Srednicki, Phys. Lett. 104B (1981) 199. J. E. Moody and F. Wilczek, Phys. Rev. D30 (1984) 130.

10. D. B. Reiss, Phys. Lett. 115B (1982) 217; F. Wilczek, Phys. Rev. Lett. 49 (1982) 1549; B. Gelmini, S. Nussinov and T. Yanagida, Nucl. Phys. B219 (1983) 31.

11. Y. Chicashige, R. N. Mohapatra and R. D. Peccei, Phys. Lett. 98B (1981) 265, Phys. Rev. Lett. 45 (1980) 1926.

12. G. B. Gelmini and M. Roncadelli, Phys. Lett. 99B (1981) 411. H. Georgi, S. L. Glashow and S. Nussinov, Nucl. Phys. B193 (1981) 297.

13. D. S. Dicus, E. W. Kolb, V. L. Teplitz, and R. V. Wagoner, Phys. Rev. D18 (1978) 1829 and Phys. Rev. D22 (1980) 829. M. Fukujia, S. Watamura, and M. Yoshimura, Phys. Rev. 48 (1982) 1522.

14. L. M. Krauss, J. E. Moody, and F. Wilczek, Phys. Lett. B144 (1984) 391.

15. G. G. Raffelt, Phys. Rev. D33 (1986) 97.

16. S. Dimopoulos, B. W. Lynn and G. D. Starkman, Phys. Lett. 168B, 145 (1986).

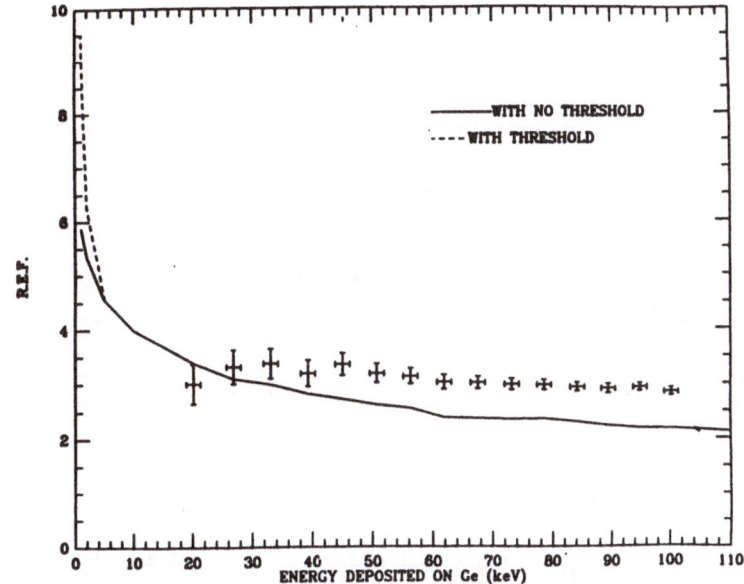

Fig. 1.   Energy dependence of relative efficiency factor (REF).
          The data points are from Ref. 11.

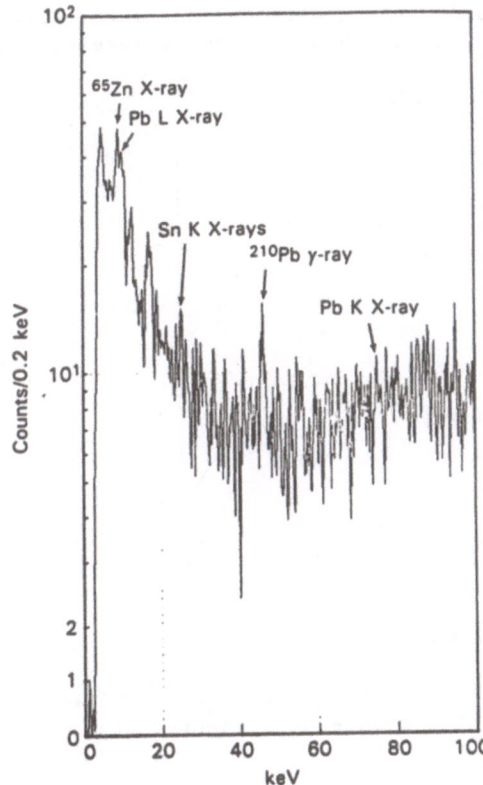

Fig. 2. Ten weeks of data
from the Ge spectrometer
(0.2 kev per channel). The
identified peaks resulted
from the radioactive decay
products in the solder point
which was recently removed.

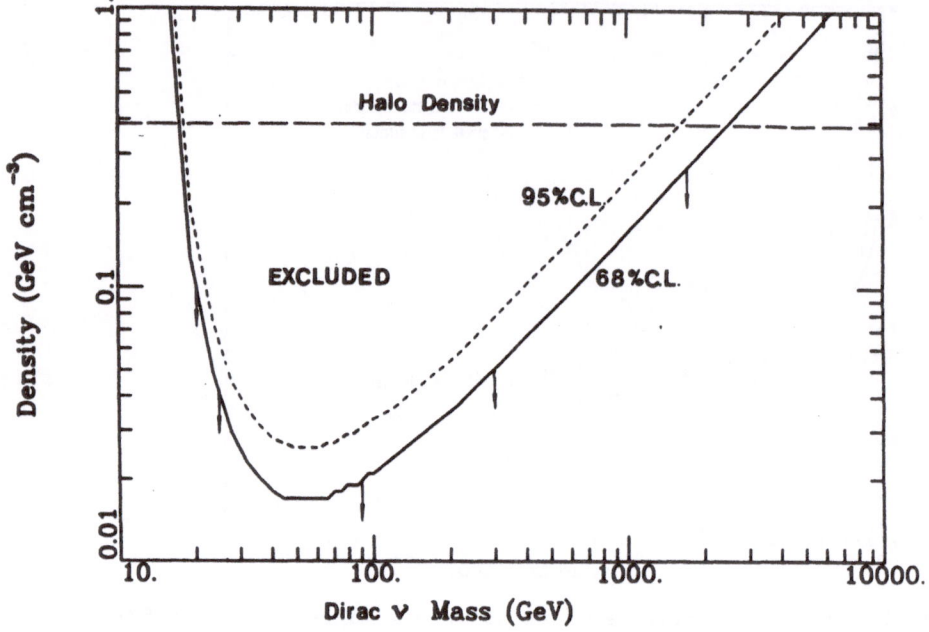

Fig. 3. Maximum halo density of standard Dirac neutrinos consistent with
the observed count rate at the 68% and 95% confidence levels.

Axion Rate for Germanium

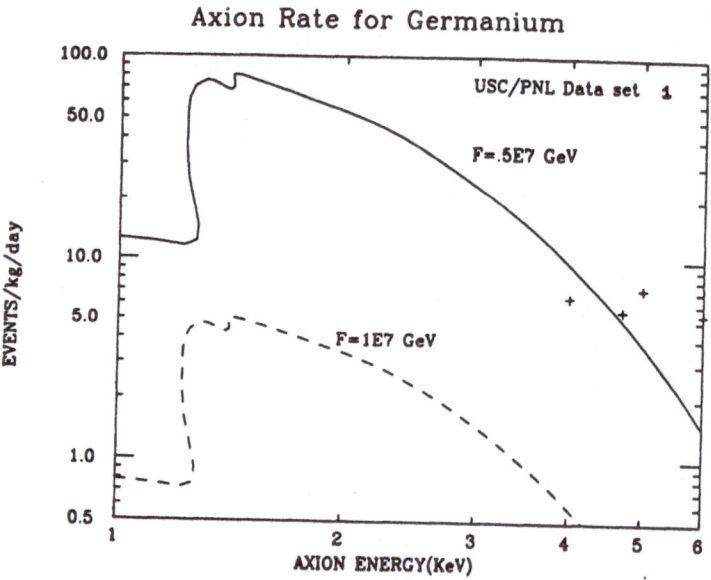

ig. 4. Solar axion events per kg per day for Ge for $F/2x'_e= 5*10^6$ GeV
(solid) and 10^7 GeV (dashes). The crosses are experimental
points from PNL/USC data.

Appendix 2

TESTS OF SUPERHEATED SUPERCONDUCTING COLLOIDS USING AN RF SQUID

ABSTRACT. We report two tests designed to study the practicality of a
superheated superconducting colloid detector using a SQUID read-out
system. In the first test, the individual "flips" of ten 15 micron
radius tin grains were observed as the temperature was swept through the
superheated superconducting normal phase transition. In the second
test, we were able to observe transitions induced by 90 keV X-rays in a
colloid of 5 micron radius grains in epoxy.

The last few years have seen an increasing interest in the
development of low temperature particle detectors, which include the
superheated superconducting colloid detector (SSCD)[1], the superconducting tunnel junction[2], the crystal bolometer,[3] and the hybrid
detector[4]. Our own interest is in the SSCD, and in this paper we will
discuss recent investigations performed at the University of British
Columbia. Although this detector was developed for applications in high
energy physics and photon detection, it is also a promising candidate
for detecting neutrinos[5] and weakly interacting massive particles
(WIMPs)[6]. The ideal SSCD consists of a large number of identical
micron-sized grains immersed in an appropriate dielectric and prepared
in a superheated superconducting state. The deposition of a few keV of
energy in a given grain will then "flip" it into the normal state
thereby producing a change of magnetic flux which can be detected in a
pickup coil. The energy needed to flip the grain depends on the grain
material, size and operating temperature.

Two tests were performed. In Test 1, the flipping of individual 15
micron radius grains was detected as the temperature was varied; in Test
2, the effect of X-radiation on a colloid sample of 5 micron radius
grains was investigated. The experiments were performed in a standard
1K pumped [4]He cryostat. A superconducting solenoid supplied an external
field (B) on the sample, and both this field and the temperature of the
sample were computer controlled to allow the sample to be taken through
various paths in the B-T phase diagram. The changes in flux produced by
grain flips were detected by a SQUID connected to the pickup coil which,
together with the 1K cryostat and the solenoid, were shielded in a
superconducting lead jacket. The signal is proportional to the magnetic
field and the grain volume, and depends on the geometry of the pickup
coil and the various inductances in the detecting system.

In this section, we describe the two tests. The samples used in
these tests were selected from batches prepared by ultrasound
disintegration of 99.99 percent pure tin. Size selection was performed
by filtering the grains through wire meshes.

Test 1 - Single Grain Signal

In this experiment, a line of ten 15 micron radius grains was
mounted on a copper strip, one end of which was attached to a regulated

heat source; the other end was heat-sunk to the 1°K pot via a large thermal resistance. With this arrangement, the temperature could be swept with each of the ten grains being at a slightly different temperature. The ten grains were selected using a microscope so that good spheres of the same size were used. Each grain was mounted on a piece of teflon tape which was then placed on a copper cold finger so that there was no electrical contact between the grain and the copper. Good thermal contact between grain, tape and copper was effected by using high thermal conductivity grease, which also encapsulated the grain.

Signals were detected using a commercial SHE System 330 RF SQUID mounted in the 4.2 K bath. The output voltage of the SQUID was digitized using a multichannel analyser (Tracor 1710) with 12-bit digitizer, and controlled by a computer. A homemade flux counter recorded the SQUID resets as well as the output voltage.

Fig. 1 shows the signals obtained from the ten 15 micron radius grains as the temperature was swept in the cycle ABCDA in the phase diagram, where A is below the supercooling transition point D, and C is above the superheating transition point B. The small temperature dependence of the SQUID signal in the A and C regions was due to paramagnetic impurities in the copper cold finger and the noise came from vibrations transmitted through the cryostat mounts. These were subsequently reduced to a level at which 5 micron radius grain signals could be observed with a signal/noise ratio of 10.

Test 2 - Radiation Test

For this experiment, a sample consisting of a few thousand 5 micron radius grains imbedded in an epoxy resin was prepared. This sample was loaded into a copper capillary connected directly to a temperature regulated copper block. The source of radiation was a ^{67}Ga placed inside the cryostat, and a 1 cm thick copper block could be moved by magnetic remote control to absorb the x-rays. When the block was in place the 90 keV x-rays were strongly attenuated (to about 2% of the unblocked radiation), while the higher energy photons suffered only small attenuation. However, the x-rays have by far the highest probability of being stopped in a grain. The experimental set-up is shown in Fig. 2.

It should be noted that no grain selection was made other than the initial filtering by the meshes. Thus there was significant variation in both the size and shape of the particles. These variations, coupled with inhomogeneity of the grain distribution within the epoxy which cause a spread of dipolar fields, produce a relatively large spread in the individual grain transition temperatures. Thus, this sample of grains is of considerably lower quality than the grains used in Test 1.

The results of radiation tests at B = 0.031 Tesla and T = 2.2 K are shown in Fig. 3 and Fig. 4, which are slow and fast scans respectively. In Fig. 3b, the aperture is closed and the signal (change in voltage) is due to low-probability transitions induced by high energy x-rays which pass through the shutter. In Fig. 3a, the aperture is open. For the fast scan with open aperture shown in Fig. 4a, the individual grain flips can be seen.

While observing grain flips, the temperature was reduced until the flipping ceased. By this procedure we estimated the maximum temperature gain caused by a x-ray hitting a grain to be approximately 50 mK. Thus the only grains that can be driven normal are those with their transition temperatures within 50 mK of the quiescent temperature. Taking this into account, together with the fraction of X-rays absorbed and the solid angle presented by the colloid sample, and defining a quantum detection efficiency (QDE) as the ratio of the grains that can and do flip when hit by a x-ray to the number that can flip, we then estimate the QDE $\geq$ 85%.

This is very encouraging to the prospect of building a SSCD to detect neutrinos and dark matter. The greatest problem to be overcome is the production of a homogeneous colloid of grains that are defectless and have uniform size and shape.

ACKNOWLEDGMENTS

This work was financially supported by the Smithsonian Astrophysical Observatory. We wish to thank Dr. M. Crooks for valuable discussions and the loan of a 1 K cryostat, Dr. M. Hasinoff and Dr. A. Olin for helpful discussions and providing the ^{67}Ga source, and the following U.B.C. summer and engineering students who, at various times, assisted on the project: D. Macquistan, E. Szarmes, N. Wright, B. Currell and J. Forrest.

REFERENCES

1. H. Bernas, et al., Phys. Lett., 24A, 721-722, 1967. A. K. Drukier, and C. Valette, Nucl. Instr. and Meth., 105, 285-287, 1972. A. K. Drukier, C. Valette, and G. Waysand, Lett. Nuovo Cimento, 14, 300-304, 1975. A. K. Drukier, and L. C. L. Yuan, Nucl. Instr. and Meth., 138, 213-226, 1976. D. Huber, C. Valette, and G. Waysand, Nucl. Instr. and Meth., 165, 201-224, 1979. A. K. Drukier, Nucl. Instr. and Meth., 173, 259-260, 1980. A. K. Drukier, Nucl. Instr. and Meth., 201, 77-84, 1982.

2. N. Coron, et. al., Nature, 314, 75-76, 1985.

3. E. Fiorini, and T. O. Niinikoski, Nucl. Instr. and Meth., 224, 83-88, 1984. S. H. Moseley, and J. C. Mather, J. App. Phys. 56, 1257-1262, 1984. D. McCammon, S. H. Moseley, J. C. Mather, and R. F. Mushotzky, J. App. Phys. 56, 1263-1266, 1984.

4. B. Cabrera, L. M. Krauss, and F. Wilczek, Phys. Rev. Lett., 55, 25-28, 1985.

5. J. Feder, and D. S. McLachlan, Phys. Rev. 177, 763-776, 1969.

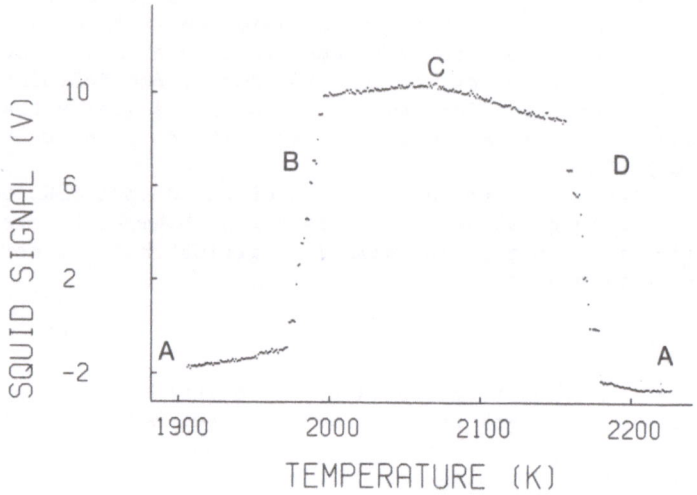

Figure 1. The sequential phase transitions of ten 15 micron radius grains. The points ABCD are shown on the B-T phase diagram in Ref. 5.

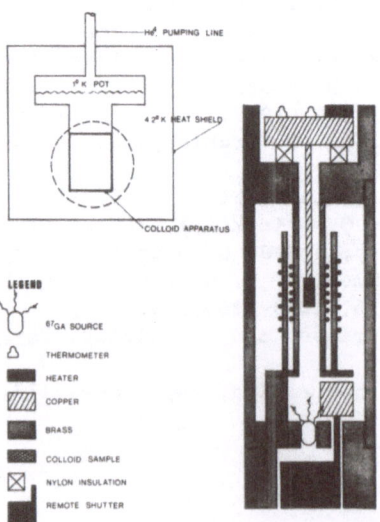

Figure 2. The design of the apparatus for Test 2. The coils around the colloid sample are the pick up coil and the superconducting solenoid.

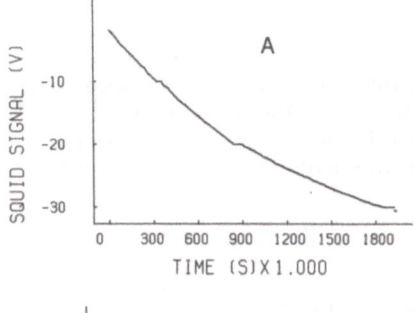

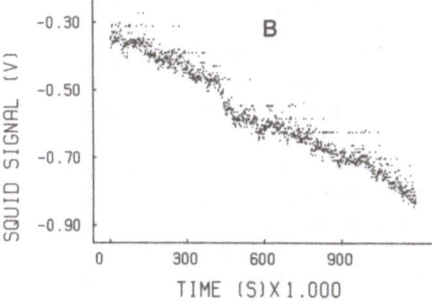

Figure 3. The SQUID signal from the superheated colloid with the x-ray shutter open (a) and closed (b). The field B = 0.031 Tesla and temperature T = 2.2 K.

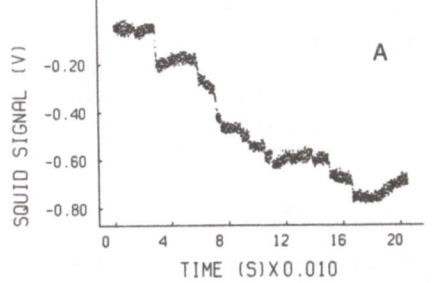

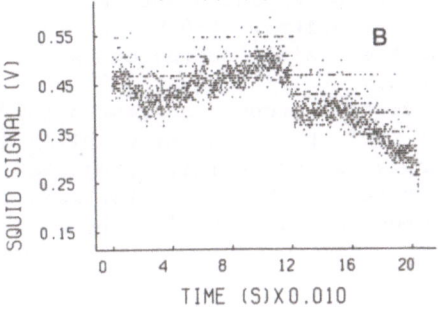

Figure 4. A faster scan showing the SQUID signal. In (a) the shutter is open; in (b) it is closed. Note the single grain flips in (a).

Appendix 3

CRYOGENIC DETECTORS OF COLD DARK MATTER CANDIDATES
Some Material Considerations

ABSTRACT. We consider the use of low temperature detectors in search of
the weakly interacting galactic halo candidate particles. We show that
in the case of particles with spin interactions, the best detectors will
be based on compounds of boron, lithium and fluorine. When compared
with silicon or germanium crystals, an improvement of 2-3 orders of
magnitude is expected.

1. INTRODUCTION

Supersymmetric theories predict high mass partners for the known
particles, the lowest mass supersymmetric particle (LSP) of which should
be stable against decay. In many theories the LSP is the photino.
Cosmological constraints establish a lower limit on its mass to be 1.8
GeV. Thus, it is quite important to provide a reliable detection
mechanism for these particles. In a previous work we found that, if a
particle with a spin-dependent interactions forms much of the galactic
halo, we can anticipate a countrate of a few events per kg x day, i.e.,
comparable with radioactive background. To improve the signal/
background ratio the photino detector should be built of light nuclei
and the details of nuclei structure are of utmost importance. The
previously proposed cryogenic detectors includes superheated
superconducting colloid and Si/Ge bolometers. Unfortunately photinos
couple very weakly with silicon or germanium.

In the next section we discuss the signal-to-background ratio in
various photino detectors. Taking into account the radioactive
background, we show that ^7Li, ^{11}B and ^{19}F have the best cross sections.
In section III we discuss the use of compounds of lithium, boron and
fluorine as cryogenic bolometers, and show that considerations of
specific heat favors boron as the best photino detector.

2. SIGNAL-TO-BACKGROUND RATIO IN DIVERSE PHOTINO DETECTORS:

The recoil energy transferred in an elastic collision between an
incident particle and nuclei as well as the realistic model of the local
distribution of velocities of dark matter candidates was discussed
elsewhere (ref. 2, Section II); the average velocity is 250 km/sec and
escape velocity is about 650 km/sec. Thus the maximal energy deposited
by a particle with mass m = 5 GeV is a few keV.

Photinos couple through axial couplings to leptons via slepton and
to quarks via squark exchange. Because of their purely axial coupling
to quarks, photinos do not scatter coherently off of nuclei. From the
formulas presented by Goodman and Witten it can be shown that for spin
dependent photino-nucleus scattering the cross section is given by

$$\sigma_{SD} = \sigma_o * \frac{m^2 M^2}{(m+M)^2} * g_f^4 \lambda^2 J(J+1),$$

(1)

where f ranges over up and down quarks, g_f is the charge of the quark partner, and J is the magnitude of the nuclear spin. λ is nucleus-dependent factor, which in the nuclear shell model is given by

$$\lambda = 0.5 * [1 + \frac{s(s+1) - 1(1 + 1)}{j(j+1)}] \qquad (2)$$

where s, 1, and j are spin, orbital, and total angular momenta. Actually, the value of σ_0 depends on the type of particle undergoing a spin-dependent interactions. For the photino and massive Majorana neutrino, the cross sections are of the same order of magnitude. Thus in the following, we discuss the detection of photinos as an example of particles with spin-dependent interactions. However, in some cases, e.g. for solar cosmions, the cross section is much higher, σ_0 (cosmions) $= 4*10^{-36} cm^2$.

Photinos interact with nuclei only through an axial vector coupling, so only the interaction with odd nuclei is important. Since the cross section scales as the fourth power of quark charge, it is much higher for elements which couple through an up-quark ($g_{up}=2/3$) than when the down-quark is important ($g_{down}=1/3$). Furthermore, the cross section is dependent upon the target mass; in an optimal target, the target mass should be the same as the photino mass. However, the photino mass is unknown so we will evaluate the detector for a range of plausible photino masses. The predicted count rate per nuclei scales as

$$FM = f_{odd} * (3g/e)^4 \, ^2J(J+1) * f(M,m) \qquad (3)$$

where $f(M,m) = Mx[m/(m+M)]^2$ and f_{odd} is the natural fractional abundance of nuclei with odd atomic weights for a given element. Calculation of $\lambda^2 J(J+1)$ involves knowledge of the shell model of nuclei; table 1 lists nuclear data for several elements which can be considered for use in photino detectors.

Table 1 suggests that it may be difficult to develop semiconducting photino detectors. Both silicon and germanium have very low cross section for plausible photino masses. Furthermore, silicon has a relatively long-lived, radioactive isotope Si^{32} ($t_{1/2}=100y, E_{max}=230keV$) which will cause background problems. Gallium has a reasonable cross section, but GaAs crystals of detector quality are available with masses up to only a few gram mass. Furthermore, the expected energy deposition in A $\geq$ 10 materials is very small; using semiconductors only photinos with m $\geq$20 GeV can be detected (see Appendix 1 for more detailed discussion).

Two types of low temperature detectors have been suggested for searches of weakly interacting particles:

- superconducting bolometers, e.g., superheated superconducting colloid (Appendix 2); and
- crystal bolometers with $\Theta_D \geq 500^\circ K$.

Both methods rely on use of the extremely low specific heat of materials in very low temperature. The recoiling nuclei deposit a few keV energy which leads to a measurable temperature increase. This change of temperature may be measured by a change in magnetic properties using extremely sensitive Superconducting Quantum Interface Devices (SQUID's). There are three superconductors Al, V and Ga with reasonable figures of merit. Unfortunately, aluminum has a long-lived radioactive isotope Al^{26}, and the expected radioactive background is about 6.8×10^5 counts/(kg * day). Vanadium and gallium are Type I superconductors and both are good candidates for photino detectors.

In this appendix we focus our attention on crystal bolometers, and more specifically on an optimal choice of target material. Two of them, Si and Ge, were discussed above; some otherwise interesting bolometer materials (e.g., berylia, diamond, sapphire, and zirconie) are radioactive and thus unsuited for photino detection. Furthermore, with the exception of sapphire, the above mentioned compounds also have fairly poor cross sections.

For crystal bolometers the figure of merit differs slightly from equation (4). First, it is more relevent to quote the specific heat per unit mass rather than per nucleus. Taking into account that energy transfer (see formula 1) is smaller for heavy nuclei, an appropriate figure of merit is given by: $FM1 = FM*[m/(m+M)]^2$. Figures of merit are evaluated in Table 2 for m =5GeV. Three elements (lithium, boron, and fluorine) are the most promising candidates for photino detectors. In the next section, we show that consideration of specific heat favors boron, which is why Table 2 includes a column with figures of merit for m =5GeV normalized to ^{11}B.

3. LITHIUM, BORON, AND FLUORINE COMPOUNDS AS PHOTINO DETECTORS:

We have argued that the best photino detectors should be based on ^7Li, ^{11}B, and ^{19}F. For example, lithium, boron, and fluorine crystalline compounds can be used as low temperature crystal bolometers. In crystals the specific heat is dominated by the lattice contribution, given by

$$C_v(T) = (2.13*10^{19}/A)*(T/\theta_D)^3 [keV/g*^\circ K]. \qquad (5)$$

The relevant thermodynamical data are provided in Table 3 for selected boron, lithium, and fluorine compounds.

It should be pointed out that the thermodynamical properties of diverse compounds are based on low, but not very low, temperature measurements. At very low temperature, say T < 0.01°K, the Debye temperature may be up to 50% different (usually higher) than quoted in Table 3. Furthermore, at very low temperatures, processes other than lattice vibrations can contribute significantly to the specific heat. Additional experimental studies of the very low temperature specific heat for selected compounds, (e.g. B, BN, LiH, LiF, and MgF_2) are necessary prerequisite for the future development of photino detectors.

Boron crystals of very high purity (a few parts in one billion) have been grown, and their Debye temperature is very high, $\Theta_D(B)=1480^\circ K$. Thus a crystal of boron with mass of one gram has specific heat $C_v=344$ KeV/$^\circ K$ at 10mK. For photinos with m =5GeV, the typical energy transfers ($E_t=0.5*E_{max}$) are 0.485 KeV, 1.94 keV and 4.4keV for v=200, 400, and 600 km/sec, respectively. Thus a few millikelvin heating is expected, which can easily be detected. Alternatively, boron nitride might be used, $\Theta_D(BN)=780^\circ K$.

Lithium is a non-superconducting metal, and has a large electronic specific heat even at very low temperatures. However, LiF will have almost as large cross section as pure lithium. Big crystals of LiF have been grown which possess an adequately high Debye temperature, $\Theta_D(LiF)=604^\circ K$. The Debye temperature of lithium hydride is even more favorable, ($\Theta_D(LiH)=739^\circ K$). Unfortunately, lithium has chemical properties similar to potassium, so that lithium based detectors may be contaminated with K-40. This may favor boron based detectors over those using lithium compounds.

Detectors based on boron and lithium may have a radioactive background due to absorption of neutrons. Naturally-occuring lithium and boron contain isotopes which strongly absorb neutrons: Li^6(7.4%, σ=950 barn) and B^{10}(18.7%, σ=3836 barn). Fortunately, the neutron induced reactions leave much more energy than the scattering of weakly interacting particles, so that some rejection based on heat pulse height analysis is possible. The use of isotopically enriched materials may, however, be necessary: these exist commercially, but are quite expensive. The problem of radioactive background induced by neutron absorption will be discussed in a future paper.

Fluorine compounds can be used as photino detectors. Some of the fluor salts are available in crystalline form. The case of LiF was already discussed: the other crystals with reasonably high Debye temperature are MgF_2($\Theta_D=364^\circ K$), NaF($\Theta_D=346^\circ K$), AlF_3($\Theta_D=343^\circ K$), CaF_2($\Theta_D=328^\circ K$). Two of these have long life isotopes producing backgrounds of 2.18×10^5 counts/kg * day) for AlF_3, and 70 counts/kg * day) for CaF_2. Due to chemical affinity, NaF may be contaminated with K-40. This leaves MgF_2 along with LiF as a good candidate for photino detection.

It should be pointed out that detector optimization is similar for all particles which have spin-dependent interactions with nuclei. For example, equation (1) can be used for photinos, massive majorana neutrinos, and some candidates for "solar cosmions". The result is the same: boron is the most efficient detector. However, the expected count rates are much larger for "Majorana type solar cosmions". A single 100g boron crystal operating in T=10 mk will produce hundreds of counts/day when used as "solar Majorana axions" detector.

Table 1: Different Elements as Photino Detectors

Element	f odd	$\lambda^2 J(J+1)$	(U/D)	FM
He^3	1.3×10^{-6}	0.91	d	0.30
Li^7	92.58%	0.50	u	1.15
B^{11}	80.22%	0.50	u	1.53
F^{19}	100.%	0.91	u	0.77
Al^{27}	100.%	0.42	u	0.25
Si^{29}	4.7%	0.91	d	0.001
$Cl^{35,37}$	100.%	0.18	u	0.08
V^{51}	99.%	0.40	u	0.13
$Ga^{67,69}$	100.%	0.50	u	0.11
Ge^{73}	7.8%	0.37	d	0.0004

Table 2: Figure of Merit for Diverse Bolometers

	FM	FM1	FM1(B^{11})	$C_v(B^{11})$	$\Delta T(B^{11})$
He^3	9.1×10^{-3}	3.7×10^{-3}	3.8×10^{-3}	--	--
Li^7	0.74	0.139	1.43	5.8	.25
B^{11}	0.906	0.097	1.0	1.0	1.0
F^{19}	2.76	0.132	1.36	6.2	0.22
Al^{27}	1.46	0.039	0.29	17.8	0.016
Si^{29}	2.3×10^{-3}	5.6×10^{-5}	5.8×10^{-4}	4.73	1.2×10^{-4}
$Cl^{35,37}$	0.76	0.013	0.13	--	--
V^{51}	1.62	0.015	0.15	12.5	0.012
$Ga^{67,69}$	2.14	0.012	0.12	15.3	7.8×10^{-3}
Ge^{73}	1.95×10^{-3}	9.1×10^{-6}	9.3×10^{-5}	9.43	9.9×10^{-6}

All figures of merit evaluated for m = 5 GeV.

Table 3: Boron, Lithium, and Flourine Compounds as Photino Detectors*

	A	B/all	$\theta D(^\circ K)$	$T_{measurement}$ $(^\circ K)$
B	5	1.0	1480.	4.2
BN	19	0.26	780.	4.2

	A	Li/all	$\theta D(^\circ K)$	$T_{measurement}$ $(^\circ K)$
Li H	8	0.88	739.	3.72
Li F	22	0.27	604.	2.31
Li_2 O	30	0.47	488.	17.06
Li Cl	42	0.06	306.	13.77
Li HF_2	42	0.06	265.	6.57

	A	F/all	$\theta D(^\circ K)$	$T_{measurement}$ $(^\circ K)$
Li F	22	.73	604.	2.31
M_g F_2	62	.61	364.	54.22
Na F	42	.45	346.	54.01
AlF_3	84	.68	343.	53.65
Ca F_2	78	.66	328.	53.51

THE EARLY UNIVERSE: HISTORICAL & PHILOSOPHICAL PERSPECTIVES

Stephen Toulmin (University of Chicago)

Cosmology is more closely linked with philosophy (e.g. the theory of knowledge) than more typical physical sciences, like solid state physics and electrodynamics. The reasons why is to throw light on the special intellectual problems facing cosmologists, just because their ambition is to discover about the physical world in its entirety. In other branches of physics, we can separate particular "hypotheses" (postsuppositions), which provide explanations of specific phenomena, from those general presuppositions that underlie the concepts governing the field or set of phenomena involved. This cannot easily be done in cosmology: the very comprehensiveness of the subject is the source of its special problems. Other fields are limited to specific kinds of phenomena and subject matters: cosmology is the science of everything at once.[1]

In appraising hypotheses, we take for granted (i.e., borrow without questioning) concepts already established in other fields (as computer science, say, borrows from solid state physics and electronics), and use these concepts to frame our current questions. So typical scientific inquiries do not call everything in doubt at once: instead, they build up theoretical constructions on foundations others which have laid in ways that can be trusted for the time being. By contrast, when we analyse presuppositions, we bring to the surface assumptions that this "trusting" and "taking for granted" concealed: the things that are presumed, in choosing one particular set of concepts, not another. This analysis may show that our assumptions are false, that they are approximations, or that they are arbitrary. (Einstein's analysis of the assumptions involved in Newtonian "simultaneity" contrasts in this respect with - e.g. - the traditional accounts of "magnetism" as a subtle fluid or emanation, as weightless corpuscles or radiation field.)

W. G. Unruh and G. W. Semenoff (eds.), The Early Universe, 393–411.
© 1988 by D. Reidel Publishing Company.

Cosmology is especially dependent on presuppositions. There, we assume that laws and theories established by studying limited systems or sets of physical objects can be extrapolated and reapplied generally, without qualification, to yield a comprehensive account of "the entirety of the universe". About any such account, four questions arise:

(1) Can it claim longterm viability, or is it the historical product of a particular phase in the history of physics?

(2) Can we accept such extrapolations naïvely, or do they in some cases take us beyond "the bounds of sense"?

(3) Can their results be accepted as fully exact, or do they provide at best approximations to the truth?

(4) Specifically, do the basic concepts of physical theory hold good right back to "the beginning of all things", or may they bend or break, as we approach that singular event?

When we discuss the early universe, all these problems come into focus the same time. With the best will and sympathy in the world, therefore, a scientifically minded philosopher or intellectual historian is forced to re-examine the assumptions embodied in the question-begging phrase, "the" "early" "universe". This essay will focus on these words, in turn. The first part looks at the assumptions underlying the idea of a universe; the second concentrates on the term, early; and the last part asks about the claims to uniqueness built into the little word, the.

§ I

The Early "Universe"

Modern cosmologists inherit from earlier "natural philosophers" an approach that treats the universe, not just as an unordered aggregate, but as a well ordered system displaying intelligible patterns. The classical Greeks had two distinct words for "the totality of things": one (ouranos) applied to it without regard to questions of order or disorder, the other (cosmos) treated it as being an essentially orderly, or "cosmetic" world. This anticipation of our modern word for facial make-up is no accident: in both senses, the adjective meant "well decked out."

Initially, the Greeks saw the universe as "orderly" in two respects, which may nowadays seem unrelated. The cosmos was so ordered as to be generally livable - a place where human beings are "at home"; but it was also so ordered as to be, specifically, intelligible - a world of which human beings can "make sense". Right up to Isaac Newton's time, these two conceptions were closely linked; and, in the sub-branch of theology still called "cosmology", they remain linked. For instance, John Calvin saw natural philosophy and theology as united in a single cosmological fabric, and came close to identifying Nature with God.[2] In the ecology movement today, too, people's view of Nature is an updated version of the Stoic cosmology: for them, "living in harmony with Nature" means recognizing, and conforming to, the ecosystems to which our species is preadapted, while ecological damage is "an offense against Nature".[3] So, scientific cosmology is not uniquely linked to astronomy: from the start, it had biological and psychological faces, as well. In a lecture written for Copernicus's 500th birthday, indeed, John Wheeler himself hailed the cosmos as "a Home for Man".[4]

In Antiquity, this link between the astronomical world, seen as the totality of things, and the natural world, adapted to human occupation, encouraged a further equation. The differences between the supposedly unchanging (superlunary) celestial world, and the changeable (sublunary) terrestrial world, dovetailed with a very different division, marking off the "immortal" from the "mortal". The fiery matter of stars was equated with the immortal substance of the soul, the base material of earthly things with the transient, corruptible flesh.[5]

Aside from that, why was astronomy so prominent in cosmology? The orderedness of the cosmos, as the natural habitat for humans, is on the face of it a matter for ecology rather than astronomy. For the Greeks as for Newton, however, the clearest example of Order in Nature was the movements of the stars. The Heavens thus offered the human spirit an "eternal" home. Like dolphin scientists looking up from the ocean into the air, and reflecting on the "supermarine" laws governing seagull flight - which is free from the "submarine" resistance of water - human beings looked up from the atmosphere into the heavens, and asked why the stars and planets behave so unlike everyday material objects in the terrestrial world. (The distinction between "superlunary" and "sublunary" was, thus, not a perverse obstacle to a rational physics: it was an intelligible first step on the road toward a comprehensive body of physical theory.)

- - - - -

From the 1680s on, the New Physics of Newton and his successors relied on intellectual methods framed forty years earlier by Galileo and Descartes, and involved substantial presuppositions. First and foremost, it presupposed the possibility that human scientists (qua "observers") can separate themselves as fully as their inquiries demanded from the natural world (qua "object observed"). The rational mental operations involved in studying natural phenomena were not themselves parts of the causal physical world in which the natural phenomena occur. Descartes' "dualism" (which Newton shared) was not an arbitrary doctrine aimed at keeping Mind out of physics: rather, it was a pragmatic response to the contrast between events, which just "happen", and thoughts, which are "correct" or "incorrect", "right" or "wrong".

Separating the Observer from the Observed was a pragmatic way of establishing procedures of scientific study and avoid the traps which earlier students of Nature (e.g. alchemists) fell into. The alchemists were never sure that the results of their observations, on (say) changing metals by heating, were not affected by their current "states of mind". The new methods of studying phenomena put scientists in a "hide", from which to record the processes of natural things without disturbing them. So, "scientific observation" came to mean a one-way coupling between Observer and Observed, in which a scientist (qua detached spectator) was influenced by the natural world, without influencing it in return; while "scientific objectivity" came to mean <u>rational detachment</u>.

On one level, the ideal of detachment was not new. The Greek word <u>theoria</u> had always had overtones of "intellectual spectatordom"; but 17th century science now separated "rational human beings" from "causal natural phenomena" in a new hardline way. For cosmology, this move had an unwelcome consequence. From the Newtonian standpoint, cosmology no longer explained "everything whatever". The rational activities of human thinkers were left out; so its scope was intrinsically limited, in a way it had not been before. Laplace made this change explicit, in his argument about the Omniscient Calculator who, given the position and velocity of every particle in the universe at the Creation, used Newton's dynamics to compute the subsequent history of the physical world. The results of his calculation only applied to the real world, supposing the computation itself did not form part of that "real world": otherwise, the physical universe would no longer operate strictly deterministically. The Omniscient Calculator did his calculations, not as an agent <u>within</u> the physical world, but <u>from outside</u> it.[6]

The other monument of 18th century cosmological theory (Immanuel Kant's <u>Universal Natural History and Theory of the Heavens</u>, 1755) made a similar point. Kant pioneered some astrophysical hypotheses that came into their own in the 20th century: nebular structure, galaxy formation, and gravitational collapse. He was responding, in speculative terms, to the observations of Thomas Wright and Maupertuis on "nebular" stars: "Given Newton's Laws, how may all these things be supposed to come into being?" His <u>a priori</u> hypothesis was that the original cosmic matter had, not a uniform, but a random distribution. Initial smoothness does not explain self-initiated star formation, since gravitation then has nothing to work on: but initial randomness creates local variations of density, which generate the spontaneous aggregation of matter around randomly distributed centers of attraction. (Still, his cosmological model did not abolish, but reinterpreted, the contrast between the "rational" world of thoughts and the "causal" world of phenomena.[7])

-   -   -   -   -

By our time, Descartes and Newton's separation of Humanity from Nature (a brand-new 17th century doctrine) is acceptable only as a first approximation: we find it as unsatisfactory as the earlier separation of the superlunary and sublunary worlds. These two divorces were in fact closely related. Plato's Timaeus saw the rational, immortal essence of human spirit as a portion of "celestial" substance trapped in mortal and transitory "terrestrial" matter. (The image recurs in later English and European literature.) Those few writers who believed that organized matter can display mental activity as "naturally" as unorganized matter displays inertia, so rejecting this part of the Cartesian or Newtonian orthodoxy, e.g. Julien de la Mettrie and Joseph Priestley, met hostility. For natural philosophers in the 17th and 18th centuries, it was evidently important to separate Rational Mind from Causal Matter for reasons of several different kinds.

Today, our task is to demolish the division between Humanity and Nature, in both theory and practice, without losing what the Newtonian era gained. This does not mean (as for Hobbes) seeing everything human (even,"ideas") as being as mechanical as everything physical. It means allowing that natural processes and phenomena can have causal aspects, without being purely and simply causal. By 17th century standards, for instance, electronic computers hardly count as "machines" at all, since their outputs are interpreted in rational terms. Yet physical processes inside computers obey normal causality: as Kant himself hinted, their workings are, at the same time, both "causal" and "rational".

The unacceptability of dividing Humanity from Nature became clear to physicists, in fact, only after Einstein, even more after Heisenberg. Yet the Newtonian cosmology might have been recognized as incomplete earlier, if scientists had reflected more carefully on the implications of Laplace's Omniscient Calculator. This image reinforced the belief that both Kant and Descartes accepted: that, by their standards, psychology could never be a "science". If we adopt a posture of rational detachment toward other human beings, and study them from a "hide", we shall learn strictly limited things about their minds. We understand fellow humans, not by avoiding interacting with them, but from those interactions; and distancing ourselves robs our understanding of all "meaning".

§ II

The "Early" Universe

"How far are cosmological world models <u>historical</u>?": the starting point for our second philosophical question is, once again, Kant. In his early years as a philosopher-physicist, Kant was entranced by Newton's success in building a system of planetary astronomy and astrophysics, and he hoped to cap this achievement by using gravitation as the key to a strictly "Newtonian" cosmology. In the <u>Universal Natural History</u>, he set aside Newton's theological account of the Creation in favor of the more naturalistic theory mentioned above.

Newton had assumed that God's Creation <u>first</u> gave stars and planets their locations and velocities, and that gravitational force was switched on only <u>afterwards</u>, to maintain a stable providential Order. This notion lent strength to the early 18th century "deists", who argued that God had created the World of Nature initially, but then turned his back and left it to run according to its own Laws; so that we can discover those Laws by studying them directly, without regard to their Divine origin. Following Descartes' <u>Principes de la Philosophie</u>, Kant set out to show that those same laws reveal natural processes by which the heavenly bodies reached their current forms and motions. Given the initially random distribution of cosmic matter, the Order imposed on the Heavens at the Creation was an historical product of the agglomeration of that cosmic matter, caused by Newton's "universal gravitation".

This daring, even hubristic speculation carried Newton's theory far beyond its previous scope; and, when Kant read David Hume's skeptical arguments, he acknowledged that he might have overstepped the mark, and reappraised his own presuppositions. The crowning product of Kant's later career, the three <u>Critiques</u>, were an outcome of this self-criticism: the need to undertake it was forced on him partly by natural philosophy, partly by ethics - "the Starry Heavens above and the Moral Law within". In the first <u>Critique</u>, he asked what discoveries Newtonian arguments could "in principle" support.

The program of the 17th century "mathematical and experimental" philosophy, formulated by Galileo and Descartes and executed by Newton, embodies a fundamental methodological problem. As "mathematical", its theoretical aim is to generate formal systems of <u>necessary</u> connections: as "experimental", it sets out to provide truly <u>empirical</u> descriptions of the World of Nature. Hume questioned whether one can have it both ways: an <u>empirical</u> body of scientific propositions could not, at the same time, be an axiomatic system of <u>necessary</u> connections, as well. If that were so, Kant's naturalistic account of the Creation in the <u>Universal Natural History</u> seemed to have been undercut, and become a piece of self deceit.

This same problem had already arisen about Plato's Theory of Ideas: "Can we recognize directly the <u>ideas</u> needed to form 'necessary' systems that provide 'true descriptions' of the world?" Descartes argued that God has implanted in us a power to recognize such "clear and distinct" ideas; and His Benevolence ensures that the resulting descriptions are <u>true</u>. ("How could an unmalicious God deceive us?") Like Descartes, Newton cast his cosmology in mathematical form; but he was not as sure about the epistemological basis of his ideas. In the "scholium and definitions" that open his <u>Principia,</u> he does not rely on simple appeals to "clarity" or "distinctness": instead, he gives complex operational definitions of the terms used in the theory that was to follow.

The success of Newton's work was such that his initial successors over-estimated it. Given the exactness of his planetary dynamics, the Laws of Motion and Gravitation (it seemed) <u>directly described</u> the events they explained; and this impression bred myths about "prediction" and "scientific method" that still bedevil late 20th century Science. Yet the solar system was far from being typical of the natural bodies available for study: rather, it was the <u>only</u> known system in which two bodies move under a single force, whose effects one can check unambiguously. In this respect, it was an <u>untypical</u>, misleadingly simple, instance of what "experience" generally allows us to observe.

In line with the times, 18th century mathematicians like Euler tried to prove that Newton's Laws of Motion and Gravitation give, not <u>one</u> true description of physical Nature, but <u>the only possible</u> one. As the Abbé Saccheri hoped to prove that Euclid's account of spatial relations alone is mathematically coherent - only to discover that non-Euclidean systems are formally as consistent as Euclid's - so, too, Euler and his rationalist colleagues hoped to prove that Newton's dynamics is the only coherent formalism capable of accounting for mechanical processes.

Early on, Kant set out to mediate the dispute between empiricism and rationalism in physics: between a naïve interpretation of Newton's theory, as a factual description of the real world, and the idealism of Leibniz, who could attach no meaning to spatial or temporal magnitudes, except as defined relative to actual material objects. For Kant, Leibniz's rationalism was arbitrary, and he leaned toward Newton's more empirical view - though even Newton explained how to identify "inertial frames" in abstract, mathematical terms, not concrete, material ones. But Hume turned Kant in a new direction. His Critique of Pure Reason pioneered an new account of the relations between "thought" and "reality", in which abstract theories and concepts gave, not a direct empirical "description" of the world, but a constructed "representation" of it.

This interpretation embodies three important insights:

(1) When we observe the natural world, the truths we discover are answers to questions that we ourselves frame: so the question is why we frame our questions in this set of terms rather than that.

(2) In interrogating Nature, the art is to find the questions that Nature "answers to": though it is we who pose them, the questions must be asked in terms to which Nature can respond.

(3) Euclid's and Newton's mathematical systems are not unique in theoretical form, but their concepts (spatial or mechanical) may yet have a preferential status in practice.

Euclidean geometry and Newtonian dynamics might not be "analytically" necessary; but they were still practically advantageous, to the point of being indispensable at least for terrestrial purposes, and so necessary in an alternative, "synthetic" sense.

This account of the relation between Matter and Mind (or facts and judgments) was initially misunderstood, and this was partly Kant's own fault. He spoke about the notorious Ding an sich (or "thing-in-itself") in terms that hinted at an invisible realm "behind" our sense impressions, implying that we know nothing about that realm, only because we cannot reach it. Yet, in contrasting "Nature as an object of rational thought and judgment" with "Nature in itself", he only needed to remind us that any representation of the facts (a map, say) includes some elements that are contributed by us (e.g. a map is drawn to a human projection system): so distinguishing "reality-as-represented" from "reality-as-it-presumably-exists, however we represent it." It is not the <u>impenetrability</u> of sense-data that is crucial, but the <u>constructedness</u> of representations.[8]

The scientific implications of Kant's innovation are best explained in Heinrich Hertz's <u>Principles of Mechanics</u>, which treats issues about our understanding of physical theory apart from those about our sensory knowledge of material objects. Where Kant wrote about the relations of a sensory <u>Vorstellung</u> ("presentation") to a <u>Gegenstand</u> ("object"), Hertz asked how a <u>Darstellung</u> ("representation") is related to a <u>Phenomän</u>, or "thing observed". Physical theories (he argued) must meet not two, but three requirements. True, they must be internally consistent, and must accommodate the relevant physical phenomena; but they also need to be "convenient" - i.e. match appropriately our intuitive understanding of the concepts accepted in the given field, and in neighboring fields.

No fixed recipe or algorithm exists to meet this last requirement: "appropriateness" is a matter for judgment. Wittgenstein spelled out the lessons of Hertz's argument in his <u>Tractatus Logico-Philosophicus</u>, and insisted that the fact that Newton's mechanics <u>can be</u> used to represent bodies in motion and other physical phenomena by itself tells us <u>nothing</u> about the world: what <u>does</u> tell us about actual phenomena is the fact that we use it to represent phenomena <u>in just that way</u> - with just those interpretations of its theoretical terms - that we <u>in fact</u> do.[9]

- - - - -

Two further examples will illustrate Hertz's point in a way that bears on the historical status of cosmology. Cartographers who map the Earth have a choice of a projection. All projections give us a means of "representing" the surface of the Earth, in ways that are "appropriate" to different purposes; and the advantages of a projection for one purpose carry with them disadvantages for others. Mercator's familiar projection system is an acceptable way (e.g.) to represent the inhabited parts of the Earth; but it has one serious deficiency, when the entire globe is shown. Mercator's projection always sends two points on the Earth to infinity - usually, the North and South Poles - so these points are <u>unrepresentable</u>: as we approach the Poles, a larger and larger amount of paper is needed to map each new mile, and an infinite amount would be needed to put the two Poles onto the resulting map.

Physicists studying extremes of cold face a similar choice in their scale of temperature. If they stay with Kelvin's definition of "absolute temperature", based on the properties of a supposed "ideal gas", they find it harder and harder to reduce the temperature of any physical system, the nearer it approaches the "absolute zero" (0°K). To cool such a system through the $9/100^{\circ}$K between $1/10^{\circ}$K and $1/100^{\circ}$K (say) takes as much effort and ingenuity as was needed earlier to cool it through (say) the $9/10^{\circ}$K between 1°K and $1/10^{\circ}$K; etc., etc. Understandably, some see the continual increase in the effort needed to cool bodies by "equal" amounts, as we approach 0°K, as an artefact of Kelvin's theoretical definition, and write it off as fictitious, like the "expansion" of the Earth's surface near the Poles, as represented on Mercator's projection. Suppose, instead, we represent temperatures on an alternative scale, $^{\circ}$t, logarithmically tied to the "absolute" scale, $^{\circ}$K - so that

$$(^{\circ}T) = \log_e(^{\circ}K).$$

On that alternative scale, the intuitive relationship between the effort needed to cool systems through "equal" intervals will be preserved as we work toward the "lowest possible" temperature; and that extreme limit will now be represented, not as "0°K", but as "$-\infty^{\circ}$T". This logarithmic definition has one further corollary. The temperature called "-300° K" is not out of reach: rather, it is <u>inconceivable</u>. The definition of the Kelvin scale robs it of meaning. If we cannot cool things down to "-300°K", that is not just too hard: thermally speaking, "there is no 'there' there".

This compression in the absolute scale of temperature close to 0° K suggestively recalls the compression in the time scale as we approach the cosmological Big Bang. Can the "passage of time", then, be similarly redefined so as to avoid this paradox? That notion was first hinted at in the 1930s by the English cosmologist, E.A. Milne, and recently revived by C.W. Misner.[10] Milne asked whether physical events on the macroscopic level might not follow different laws from those on the submicroscopic (quantum) level; and he suggested that the difference may show up in the time scales relevant to the two levels. Specifically, events on a quantum level might be measured on a time-scale ("T") logarithmically linked to the macroscopic time scale ("t"). As "t-time" goes back to Absolute Zero, "T-time" will go back to "$-\infty$": what shows up in macroscopic time as the Big Bang is an infinite time away on the quantum level.

Milne's theory is intriguing, but its complexity told against it; and few people today suppose that macroscopic and microscopic phenomena follow different physical laws. But the bare proposal to subject "cosmic time" to the same logarithmic transformation as "absolute temperature" cannot be so easily dismissed. Intuitively, that proposal is again more intelligible than the standard Big Bang view. Instead of the early stages of cosmic evolution being drastically short, of the order of 10^{-43} secs, they will stretch out so as to become, in logarithmic terms, comparable both to each other, and to those phases that take place long after the presumed Beginning of Time. That Beginning, too, is now projected back to "$-\infty$", as logarithmic temperature projects Kelvin's Absolute Zero ("$0^{\circ}K$") back without limit, into "$-\infty^{\circ} T$".

That transformation, of course, has physical implications. Kelvin's Zero is an effective limit to the meaningful range of all "temperatures": "$-300^{\circ}K$" does not designate a real temperature, different from "$+300^{\circ}K$" only in being humanly inaccessible. The same is not true of the Big Bang, regarded as "the Beginning of Time". Nobody can be sure that the Big Bang is - let alone, is <u>known to be</u> - the first moment of "all" time, rather than that particular phase of cosmic history in which we now happen to live. Nobody (that is to say) is sure that "the times before the Big Bang" are <u>ruled out by definition</u>, and so <u>inconceivable</u>, as temperatures below $0^{\circ}K$ surely are. This alternative time scale will be worth keeping in mind, as physical cosmology acquires greater sophistication and complexity. In the long run, a world-model in which observable astronomical experience comprises the <u>entirety</u> of cosmic history may carry more conviction than one in which "times before the Big Bang" form an unlimited succession of "precosmic" events forever removed from human intelligibility.

§ III

"The" Early Universe

This last point, about the possibility of enclosing all cosmic history within the period accessible to human observation, leads to cosmology's final presuppositions: those implied in discussing "the" early universe. Here as elsewhere, using the definite article ("the") suggests that our "early universe" is the one-and-only "early universe"; and we shall do well to look at these uniqueness claims, in conclusion,.

The belief that the universe is a unitary cosmos has a long history: R.G. Collingwood traced it back to the rise of monotheism. It is first visible in Antiquity, in the move from the Olympian polytheism, through the new naturalism of pre-Socratic natural philosophy, to the "rational" world pictures of the Stoics and Epicureans. On the early view, every "natural power" (the Ocean, Storms, Love or War) was an independent agency, and the actual course of events was an outcome of diplomatic compromises between Zeus and Poseidon, Aphrodite and Ares; but the Stoics saw the cosmos, instead, as a balance between different modes of Reason, embodied in inert, in living and in thinking beings. All kinds of matter were parts of the cosmic system, activated by a corresponding pneuma, or "breath". (Cosmology still had theological overtones in the 1600s. This belief was entrenched in, and rationalized by, appeals to God as the Architect and Creator of cosmic order; and it helped create the head of steam needed for the rapid evolution of physics after 1650.)

The Greeks' "rationalistic" approach to natural philosophy was a major step forward. By contrast, consider the Babylonians. At Babylon, the civil servants who calculated eclipse times and ephemerides, from the 8th century BC - what Isaiah calls the "monthly prognosticators" - used computational procedures ahead of anything the Ancient Greeks had invented for themselves: so, when Alexander the Great set off to conquer Babylonia, Aristotle sent his nephew Callisthenes with the expeditionary force, to collect information about, and records of, these techniques. Down to Alexander's conquest, however, the Babylonians regarded each of the planets as a Deity, which went its willful way: they had no project for "astrophysics", as the Greeks had, from Anaxagoras on.

From Eudoxus and Hipparchus, through Osiander and Copernicus, to the Galileo controversy, and to the arguments of the late 19th century French physicist, Pierre Duhem, astronomy shows repeated swings of the pendulum: between those who claim that the theories of Science give a "realistic" grasp of the structure of Nature, and those who see them as mere representations, that only "save appearances". But, again and again, robust realism has carried the day. Osiander might add a foreword to Copernicus' De revolutionibus, disclaiming a "realistic" intent for its arguments, but few readers found this watering down of Copernicus' claims convicing. And, when Duhem revived the position 400 years later, it was was clearly a minority view: as he admitted, this position was "the physics of a religious believer" (physique d'un croyant).

Did not Kant's analysis of Vorstellungen give Duhem a justification in advance? Not so: Kant marked "phenomena" from "the thing in itself", not (like Duhem) to reserve Reality for theologians to talk about, still less talk about "with authority". The whole lesson of his analysis was that "unrepresented reality" evades all description, theological or other. A Church can no more speak about the Ding an sich than anyone else: in Wittgenstein's phrase, "about that one must keep silent." For Kant, then, "reality" comes to mean "the World of Nature represented in a way that commands intersubjective agreement", and the ways of achieving that agreement are open to negotiation among people who have enough experience to pass judgment in the relevant field.

Even after the "critical" reformulation, some cosmological problems still generated quandaries. Kant recognized two kinds. On the one hand, questions about the Boundaries of Space and Time were tantalizing, but seemed meaningless - one may argue with equal cogency that Space does have a boundary, and that it cannot have - and likewise for the Origin of Time. Still, Kant never finally renounced his ambition to prove that our actual physical world system is the only coherent one. For instance, he cited with approval standard 18th century proofs that the distance law for gravitation can only be an inverse square law, proofs that reapply to "gravitational influence" the geometrical arguments used in high schools to this day, to "prove" that the intensity of illumination varies with the inverse square of the distance from a point source of light.

Since the late 19th century, the tide has flowed away from attempts to claim uniqueness for current theories, or even particular standpoints of observation. Einstein started down a road that is fruitful in a dozen ways, yet still has some way to go. The standpoint of general relativity differs from Kant's interpretation of physics in one crucial respect: the "invariance" of physical relations found in the experience of physicists who study the cosmos from different frames of reference is less of an empirical discovery than a posit, or intellectual demand. The world can, of course, be observed from many standpoints: Einstein tried to build a theory of nature having the same form for all frames, and this account was an Invariantentheorie, or "theory of invariants"; but it is not yet clear whether a final corroboration of general relativity will come from astronomy and cosmology, or from other parts of physics. The experience of history suggests that the latter is just as likely. The move away from classical to 20th century physics reflected many different anomalies: not just the divergence in the spectrum of the "black body" radiation, but also the puzzling properties of "cathode rays", and the infinite elasticity of molecular collisions.

- - - - -

To take the papers for this meeting as a guide, physical cosmology has away to go before it breaks with the "God's Eye View" of Laplace's Omniscient Calculator, and integrates human observers into its picture of the universe. Recent debates about an "anthropic principle" are only partial. The idea of "cosmos" always implied "a universe intelligible to human thought"; and the anthropic principle takes seriously the limits that the demand of intelligibility places on any admissible cosmology. But the further step remains, of finding ways to reinsert human beings (who theorize about, and develop representations of the cosmos) into the cosmos of which they are, at the same time, both causal constituents and rational onlookers.

Meanwhile, physical cosmology remains historical. Other historical branches of natural science (e.g. geology and organic evolution) tackled the same problems in earlier times; and cosmologists have something to learn from their experience. For example, in the late 18th century there was an inconclusive debate between two schools of historical geologists, committed to fire and water, respectively, as prime agents of geological change. Then, theory outran the available observations, and no reliable test yet existed for dating any particular rock formation. Real progress began only after 1815, when the Geological Society of London decided to set aside theoretical issues and concentrate on the empirical base of the science, using the "stratigraphic" methods devised by William Smith, the canal maker. The next twenty or thirty years saw the creation of the basic system ordering rock strata by age, and established the basis for all of the subsequent geological work by Charles Lyell and others.

When theoretical speculation revived, there was a renewed debate between "catastrophists" and "uniformitarians". Nobody doubted that the one-and-only-one Earth had one-and-only-one history, but they disagreed whether the physico-chemical agencies formerly shaping its surface had been exactly alike, in form or strength, throughout this whole history. Uniformitarians would not appeal to an agency that was not visibly at work in our own era. Catastrophists, for their part, saw in paleontology evidence of cataclysmic events separating periods when the Earth was subject to forces different in size and kind from those at work today.

Charles Lyell was a consistent uniformitarian: seeing no other way of reconstructing a convincing history of the Earth. He understood what a self-denying ordinance his method imposed on anyone who assumed (e.g.) that the Earth was much hotter in earlier times than now; but, for him, uniformitarianism was the only secure line of reconstruction available to historical geology. So, when his young admirer, Charles Darwin, sent him reports of an earthquake at Valdivia in Chile, whose effects he had witnessed just after their occurrence, this was a Godsend. The shock (Darwin said) shifted neighboring rock formations as much as 20 feet, and this greatly reduced the impact of Lyell's previous assumptions.

Nowadays, geologists are committed to neither view. Everyone now supports a weak uniformitarianism, since the processes geology appeals to must, at least, be physically and chemically intelligible. But they are also free to speculate that unique historical events may have left traces on the Earth: e.g. collisions of the Earth with wandering asteroids, such as have been cited in astronomy since Newton's pupil, William Whiston, used them to explain the Biblical Flood; and which, in the nuclear winter debate, reappeared to explain the extinction of early species.

As an outsider, I exhort cosmologists to reflect on the experience of 19th century geologists. This implies no disrespect. But the evidence of the actual character of earlier cosmological times is no fuller than that in geology around 1790-1810, when the followers of Werner and Hutton disputed the "igneous" or "aqueous" origin of the

Earth's form. By now, for instance, the implications of the "red shift" have been interpreted to the limits of plausibility. Without challenging the belief in recession, one may argue that parallel, but independent lines of evidence for dating cosmic history will give cosmology a shot in the arm more than anything else. The "epochs" of cosmic history may yet, in fact, prove as different from one another as the Époques de la Nature in Buffon's account of the Days of Creation. But, for now, it will do physical cosmology no harm to take a leaf from geology's book, suspend theoretical speculations about the Big Bang, and concentrate on giving itself a better stratigraphy.

When cosmologists puzzle out the physics and chemistry of cosmic change, Cuvier and Lyell can again teach them something. Uniformitarian today recalls the maxim known as "Ockham's Razor". One need not argue that unique events, such as planetary collisions, are never to be invoked to explain unusual changes, either in the Earth, or in the cosmos; but we should postpone these dramatic appeals until we have made sure that all more everyday and pedestrian explanations have been exhausted. We can use catastrophist theories, if and only if they are entirely indispensable: as William of Ockham might say, "cataclysms" non sunt multiplicanda praeter necessitatem.

To conclude, a philosopher or historian of science carries away from this colloquium a whole series of intellectual puzzles. Those who think of metaphysics as the most unconstrained or speculative of disciplines are misinformed: compared with cosmology, metaphysics is pedestrian and unimaginative. No mere philosopher (e.g.) dreamed up from scratch the hypothesis of the expansion of the universe generating separate, non-intercommunicating cosmic regions that move apart so fast that no light can travel between them. What does this notion imply for our modes of cosmological explanation, our demand that the cosmos be intelligible to humanity, or our basic aspiration that the ouranos be a cosmos at all? Again, does physical cosmology require revisions in general relativity? Can one integrate "observers" into the "observed" cosmos without some theoretical revisions? There were only hints in this colloquium of how this might be done: specifically, if relativistic cosmology overcomes the "distance" between the observer and the world observed, how much will be left of our everyday notion of "time"? Returning to the early phases of cosmic development (" $< 10^{-43}$ seconds after the Big Bang"), when the properties of matter were less and less like those that we can study directly, will "time" itself become a conceptual casualty, too?

Einstein focussed attention on different frames of reference and observation; Heisenberg acknowledged inevitable interactions between Observer and Observed; but these are only a first step toward escaping from the presuppositions of the classical cosmological system. Other scholars and scientists nowadays (e.g. those known as "ethnographers of science") are developing an analysis of scientific activity in which the parts physicists play in framing concepts and modes of representation are made as explicit as the "intrinsic properties" of the physical objects and systems. But it is not yet clear just how physical cosmology might follow this track. Despite all the refinements of relativity and quantum mechanics, the methods of cosmological theory today are just about as "classical" as those of Newton's world picture. (Accounts of the Big Bang hold the material cosmos at arm's length with quite as much detachment as Laplace's Omniscient Calculator.) At this point, the ideas and methods of cosmology have quite a way to go before coming fully into the 1980s.

These puzzles are special cases of a more general problem. Most cosmological arguments take physical relations that have been studied in limited parts of time or space, and reapply them uncritically to all there is. I am not challenging this extrapolation head on: I only want to keep on the agenda of cosmology a reminder that all arguments which extrapolate from particular places and times to "the entire universe" are - and will always remain - highly problematic.

---

[1] See, e.g., the introductory essay to my book, The Return to Cosmology (Berkeley, CA: 1983).

[2] See (e.g.) James Gustafson, Ethics from a Theocentric Point of View, Vol. I (Chicago, 1983), p. 258.

[3] See, e.g., The Return to Cosmology, final essay.

[4] The Greek etymology of the word "ecology" means, literally, "the science of home."

[5] See (e.g.) Plato's Timaeus.

[6] This point is effectively made by Karl Popper in several places.

[7] Kant, <u>Allgemeine Naturgeschichte und Theorie des Himmels</u> (Königsberg, 1755), tr. W. Hastie, as <u>Universal Natural History and Theory of the Heavens</u>.

[8] An <u>un</u>represented World was for Kant not yet ready to be "thought about", still less "spoken about", so it should be no surprise if we cannot describe it or discuss it!

[9] To put this point in a different way, no formalism is <u>self-applying</u>, or carries on its face the need to give it <u>one-and-only-one</u> interpretation.

[10] See C.W. Misner, "The absolute zero of time", <u>The Physical Review</u>, Vol. 186 (1969), p. 1328, and my book, <u>The Philosophy of Science</u> (London, 1953), ch. 3. I thank Dr Werner Israel of the University of Alberta for the Misner reference.

414